Statistical Mechanics of Disordered Systems

Our mathematical understanding of the statistical mechanics of disordered systems is going through a period of stunning progress. This self-contained book is a graduate-level introduction for mathematicians and for physicists interested in the mathematical foundations of the field, and can be used as a textbook for a two-semester course on mathematical statistical mechanics. It assumes only some basic knowledge of classical physics; on the mathematics side, the reader should have a good working knowledge of graduate-level probability theory.

Part I gives a concise introduction to thermodynamics and statistical mechanics, which provides the tools and concepts needed later. The main topics treated here are the classical ensembles of statistical mechanics, lattice gases and spin systems, the rigorous setting of the theory of infinite-volume Gibbs states (DLR theory), and cluster expansions for high- and low-temperature phases. Part II proceeds to disordered lattice models. It presents the general theory of random Gibbs states and metastates in the spirit of Newman–Stein and the random-field Ising model. Part III is devoted to disordered mean-field models. It begins with the random energy model as a toy example and then explains in depth the geometric structures arising in the description of the infinite-volume limit of the Gibbs states in the generalized random energy models. Finally, it presents the latest developments in the mathematical understanding of mean-field spin-glass models. In particular, recent progress towards a rigorous understanding of the replica symmetry breaking solutions of the Sherrington-Kirkpatrick spin-glass models, due to Guerra, Aizenman-Sims-Starr, and Talagrand, is reviewed in some detail. The last two chapters treat applications to non-physical systems: the Hopfield neural network model and the number partitioning problem.

ANTON BOVIER is Professor of Mathematics at the Technische Universität Berlin and the Weierstraß-Institut für Angewandte Analysis und Stochastik.

CAMBRIDGE SERIES IN STATISTICAL AND PROBABILISTIC MATHEMATICS

Editorial Board

R. Gill (Department of Mathematics, Utrecht University)
B. D. Ripley (Department of Statistics, University of Oxford)
S. Ross (Department of Industrial and Systems Engineering, University of Southern California)
B. W. Silverman (St. Peter's College, Oxford)
M. Stein (Department of Statistics, University of Chicago)

This series of high-quality upper-division textbooks and expository monographs covers all aspects of stochastic applicable mathematics. The topics range from pure and applied statistics to probability theory, operations research, optimization, and mathematical programming. The books contain clear presentations of new developments in the field and also of the state of the art in classical methods. While emphasizing rigorous treatment of theoretical methods, the books also contain applications and discussions of new techniques made possible by advances in computational practice.

Already published
1. *Bootstrap Methods and Their Application*, by A. C. Davison and D. V. Hinkley
2. *Markov Chains*, by J. Norris
3. *Asymptotic Statistics*, by A. W. van der Vaart
4. *Wavelet Methods for Time Series Analysis*, by Donald B. Percival and Andrew T. Walden
5. *Bayesian Methods*, by Thomas Leonard and John S. J. Hsu
6. *Empirical Processes in M-Estimation*, by Sara van de Geer
7. *Numerical Methods of Statistics*, by John F. Monahan
8. *A User's Guide to Measure Theoretic Probability*, by David Pollard
9. *The Estimation and Tracking of Frequency*, by B. G. Quinn and E. J. Hannan
10. *Data Analysis and Graphics using R*, by John Maindonald and John Braun
11. *Statistical Models*, by A. C. Davison
12. *Semiparametric Regression*, by D. Ruppert, M. P. Wand, R. J. Carroll
13. *Exercises in Probability*, by Loic Chaumont and Marc Yor
14. *Statistical Analysis of Stochastic Processes in Time*, by J. K. Lindsey
15. *Measure Theory and Filtering*, by Lakhdar Aggoun and Robert Elliott
16. *Essentials of Statistical Inference*, by G. A. Young and R. L. Smith
17. *Elements of Distribution Theory*, by Thomas A. Severini

Statistical Mechanics of Disordered Systems
A Mathematical Perspective

Anton Bovier

*Weierstraß-Institut für Angewandte
Analysis und Stochastik, Berlin*
and
*Institut für Mathematik,
Technische Universität Berlin*

CAMBRIDGE UNIVERSITY PRESS
Cambridge, New York, Melbourne, Madrid, Cape Town, Singapore, São Paulo

Cambridge University Press
The Edinburgh Building, Cambridge CB2 2RU, UK

Published in the United States of America by Cambridge University Press, New York

www.cambridge.org
Information on this title: www.cambridge.org/9780521849913

© A. Bovier 2006

This publication is in copyright. Subject to statutory exception
and to the provisions of relevant collective licensing agreements,
no reproduction of any part may take place without
the written permission of Cambridge University Press.

First published 2006

Printed in the United Kingdom at the University Press, Cambridge

A catalogue record for this publication is available from the British Library

ISBN-13 978-0-521-84991-3 hardback
ISBN-10 0-521-84991-8 hardback

Cambridge University Press has no responsibility for the persistence or accuracy of URLs for
external or third-party internet websites referred to in this publication, and does not guarantee that
any content on such websites is, or will remain, accurate or appropriate.

To the memory of my mother
Elisabeth Pee

Contents

Preface			*page* ix
Nomenclature			xiii
Part I	**Statistical mechanics**		**1**
1	**Introduction**		3
	1.1	Thermodynamics	4
2	**Principles of statistical mechanics**		9
	2.1	The ideal gas in one dimension	9
	2.2	The micro-canonical ensemble	13
	2.3	The canonical ensemble and the Gibbs measure	19
	2.4	Non-ideal gases in the canonical ensemble	22
	2.5	Existence of the thermodynamic limit	24
	2.6	The liquid–vapour transition and the van der Waals gas	28
	2.7	The grand canonical ensemble	31
3	**Lattice gases and spin systems**		33
	3.1	Lattice gases	33
	3.2	Spin systems	34
	3.3	Subadditivity and the existence of the free energy	36
	3.4	The one-dimensional Ising model	37
	3.5	The Curie–Weiss model	39
4	**Gibbsian formalism for lattice spin systems**		49
	4.1	Spin systems and Gibbs measures	49
	4.2	Regular interactions	52
	4.3	Structure of Gibbs measures: phase transitions	59
5	**Cluster expansions**		73
	5.1	High-temperature expansions	73
	5.2	Polymer models: the Dobrushin–Kotecký–Preiss criterion	76
	5.3	Convergence of the high-temperature expansion	82
	5.4	Low-temperature expansions	88

Part II	**Disordered systems: lattice models**	**95**
6	**Gibbsian formalism and metastates**	**97**
	6.1 Introduction	97
	6.2 Random Gibbs measures and metastates	99
	6.3 Remarks on uniqueness conditions	106
	6.4 Phase transitions	107
	6.5 The Edwards–Anderson model	109
7	**The random-field Ising model**	**111**
	7.1 The Imry–Ma argument	111
	7.2 Absence of phase transitions: the Aizenman–Wehr method	118
	7.3 The Bricmont–Kupiainen renormalization group	125
Part III	**Disordered systems: mean-field models**	**159**
8	**Disordered mean-field models**	**161**
9	**The random energy model**	**165**
	9.1 Ground-state energy and free energy	165
	9.2 Fluctuations and limit theorems	169
	9.3 The Gibbs measure	175
	9.4 The replica overlap	180
	9.5 Multi-overlaps and Ghirlanda–Guerra relations	182
10	**Derrida's generalized random energy models**	**186**
	10.1 The standard GREM and Poisson cascades	186
	10.2 Models with continuous hierarchies: the CREM	195
	10.3 Continuous state branching and coalescent processes	207
11	**The SK models and the Parisi solution**	**218**
	11.1 The existence of the free energy	218
	11.2 2nd moment methods in the SK model	220
	11.3 The Parisi solution and Guerra's bounds	227
	11.4 The Ghirlanda–Guerra relations in the SK models	238
	11.5 Applications in the p-spin SK model	240
12	**Hopfield models**	**247**
	12.1 Origins of the model	247
	12.2 Basic ideas: finite M	250
	12.3 Growing M	257
	12.4 The replica symmetric solution	265
13	**The number partitioning problem**	**285**
	13.1 Number partitioning as a spin-glass problem	285
	13.2 An extreme value theorem	287
	13.3 Application to number partitioning	289
	References	**297**
	Index	**309**

Preface

Statistical mechanics is the branch of physics that attempts to understand the laws of the behaviour of systems that are composed of very many individual components, such as gases, liquids, or crystalline solids. The statistical mechanics of disordered systems is a particularly difficult, but also particularly exciting, branch of the general subject, that is devoted to the same problem in situations when the interactions between these components are very irregular and inhomogeneous, and can only be described in terms of their statistical properties. From the mathematical point of view, statistical mechanics is, in the spirit of Dobrushin, a 'branch of probability theory', and the present book adopts this point of view, while trying not to neglect the fact that it is, after all, also a branch of physics.

This book grew out of lecture notes I compiled in 2001 for a Concentrated Advanced Course at the University of Copenhagen in the framework of the MaPhySto programme and that appeared in the MaPhySto Lecture Notes series [39] in the same year. In 2004 I taught a two-semester course on Statistical Mechanics at the Technical University of Berlin within the curriculum of mathematical physics for advanced undergraduate students, both from the physics and the mathematics departments. It occurred to me that the material I was going to cover in this course could indeed provide a suitable scope for a book, in particular as the mathematical understanding of the field was going through a period of stunning progress, and that an introductory textbook, written from a mathematical perspective, was maybe more sought after than ever. I decided to include a considerable amount of basic material on statistical mechanics, in order to make it reasonably self-contained.

Thus, Part I gives a brief introduction to statistical mechanics, starting from the basic notions of thermodynamics and the fundamental concepts of statistical mechanics. It then introduces the theory of lattice spin systems, the Gibbsian formalism in the spirit of Dobrushin, Lanford, and Ruelle, as well as some of the main tools for the analysis of the Gibbs measures, including cluster expansions.

Part II of the book deals with disordered spin systems on the lattice. It starts out with a comprehensive introduction to the formalism of *random Gibbs measures* and *metastates*. Then I discuss the extensions and limitations of the methods introduced in the first part. The bulk of this part is devoted to the random field Ising model and the question of how uniqueness and non-uniqueness can be analyzed in this case. I present the proof of uniqueness in $d = 2$ due to Aizenman and Wehr. A large section is also devoted to the renormalization group approach of Bricmont and Kupiainen, used to prove non-uniqueness in $d \geq 3$. I only comment briefly on the issue of spin-glasses with short-range interactions.

Part III is essentially devoted to mean-field models of spin-glasses. I basically treat two classes of models: Gaussian processes on the hypercube, and models of the Hopfield type. I will go to great lengths to explain in all detail the case of the *random energy model* (REM) and the *generalized random energy model* (GREM), which will give us an idea what a complete solution of such models could look like. Then I will briefly expose the deep ideas of F. Guerra and their reformulation by Aizenman, Sims, and Starr, that show how GREM-like structures can be used to provide bounds, at least for free energies, of general mean-field models in terms of hierarchical structures, essentially explaining the nature of the Parisi solution. Talagrand's proof that this bound is exact will not be given here. Finally, I discuss some of the simpler aspects of the nature of the Gibbs measures in the p-spin SK models. Much more on the SK models can be found in Talagrand's recent book [239].

The last two chapters deal with models that demonstrate the relevance of statistical mechanics beyond the classical 'physics' applications. The Hopfield model of neural networks has played a rather important rôle in the history of the subject, and I try to give a more elementary and to some extent complementary presentation to the one that can be found in Talagrand's book. The final chapter is devoted to the number partitioning problem, where a rather charming connection between a problem from combinatorial optimisation and the REM arises.

My original intention for this book had been to give full proofs of all the main results that are presented, but in the end I found that this was impracticable and that at some places the reader had to be referred to the original literature. So the practice in the book is that full proofs are given when they are reasonably easy, or where I feel they are essential for understanding. In other cases, they are omitted or only outlined.

References are given primarily with the intention to help the reader find a way into the original literature, and not with the ambition of completeness. Overall, the selection of references is due largely to my limited knowledge and memory. More generally, although I do make some comments on the history of the subject, these are by no means to be taken too seriously. This book is intended neither as an encyclopedia nor as an account of the history of the subject.

I have tried to keep the prerequisites for reading as low as reasonable. I assume very little physics background, apart from a rudimentary knowledge of classical mechanics. On the mathematics side, the reader should, however, have a good working knowledge of probability theory, roughly on the level of a graduate course.

There are a great number of very interesting topics that have been left out, either because they are treated elsewhere, or because I know too little about them, or simply for no good reason. One of these are short-range spin-glasses. A good source for this remains Newman's book [185], and at this moment I have nothing serious to add to this. Another topic I skip are Kac models. Kac models make a charming link between mean-field models and finite-range models. There has been a great deal of work concerning them in the context of disordered systems, both in the context of the random field Ising model [73, 74], the Hopfield model [46, 47], and very recently in spin-glasses [38, 104], and it would be nice to cover this. There is a beautiful new book by Errico Presutti dealing with Kac models [209], which does not, however, treat disordered systems.

I want to thank Ole E. Barndorff-Nielsen and Martin Jacobsen for the invitation to teach a course in Copenhagen in the MaPhySto programme, which ultimately triggered the writing of this book. I am deeply indebted to my collaborators, past and present, on subjects related to this book; notably J. Fröhlich (who initiated me to the subject of disordered systems more than 20 years ago), V. Gayrard, U. Glaus, Ch. Külske, I. Kurkova, M. Löwe, D. Mason, I. Merola, B. Niederhauser, P. Picco, E. Presutti, A. C. D. van Enter, and M. Zahradník, all of whom have contributed greatly to my understanding of the subject. Special thanks are due to Irina Kurkova who kindly provided most of the figures and much of the material of Chapter 10.

I am very grateful to all those who read preliminary versions and pointed out errors, misprints, and omissions, or made other comments, notably J. Černý, V. Gayrard, B. Gentz, M. Jacobsen, I. Kurkova, M. Löwe, E. Orlandi, and T. Schreiber, but in particular Aernout van Enter, who has relentlessly read various versions of the manuscript, and has provided a wealth of suggestions and corrections. I also want to thank my Editor at Cambridge University Press, Diana Gillooly, for suggesting this book project, and for constant help during the writing and production. Finally, I thank Christina van de Sand for her help in the production of the LaTeX version of the typescript.

Nomenclature

$(\Omega, \mathcal{B}, \mathbb{P})$	probability space of disorder
β	inverse temperature
$\mathcal{C}_\Gamma^*$	set of connected clusters
$\delta_x(f)$	local variation of f
$\Delta(f)$	total variation of f
$\mathbb{E}$	expectation
$\mathcal{F}$	sigma-algebra (spin variables)
$\psi_{\beta,N}, \psi_\beta$	overlap distribution
γ	connected contour
$\Gamma(\sigma)$	contour of σ
$\mathcal{G}_\Lambda$	set of polymers
$\mathcal{G}_{\beta,V,N}$	canonical distribution
$\mathcal{K}_T$	genealogical functional
$\mathcal{K}_\beta^\eta$	Aizenman–Wehr metastate
$\mathcal{K}_{\beta,\Lambda}^\eta, \mathcal{K}_\beta^\eta$	metastates
$\mathcal{K}_{\beta,N}, \mathcal{K}_\beta$	empirical distance distribution
$\mathbb{M}_Z$	median of Z
$\mathcal{M}_1(\cdot)$	space of probability measures
$\mathcal{M}_N$	set of values for magnetization
$\mathcal{M}_{\alpha,N}$	point process
μ	chemical potential
μ_β	infinite volume Gibbs measure
$\mu_{\beta,h,N}$	Gibbs measure
$\mu_{\beta,\Lambda}^{(\eta)}$	local Gibbs specification
$\mathbb{P}$	probability
Φ, Φ_A	interaction
$\phi_\beta(m)$	Curie–Weiss function
$\Phi_{\beta,N}$	log of partition function per volume
$\Phi_{\beta,N}(z)$	HS transform rate function
$\mathcal{P}$	Poisson point process
$\mathcal{P}^{(k)}$	Poisson cascade
$\mathbb{Q}_{\beta,N}$	induced measure
$\mathcal{Q}_{\beta,N}$	Hubbard–Stratonovich measure
ρ_Λ	product measure

Nomenclature

$\rho_{E,V,N}$	micro-canonical distribution
$\mathcal{R}$	Poisson point process
σ	spin configuration
σ_x	value of spin at x
$\mathcal{S}$	configuration space
$\mathcal{S}_0$	single spin space
$\mathcal{S}_\Lambda$	spin configurations in Λ
$\mathcal{S}_N$	hypercube in dimension N
$\tilde{\mathcal{G}}_{\beta,V,\mu}$	grand canonical distribution
$\tilde{\mathcal{Z}}_{\beta,V,\mu}$	grand canonical partition function
$\mathcal{W}_\alpha$	point process of masses
ξ^μ	pattern (Hopfield model)
$m_\mu(\sigma)$	overlap parameter
$B(\mathcal{S}, \mathcal{F})$	bounded measurable functions
B_{loc}	local functions
B_{smql}	quasi-local functions
$C(\mathcal{S})$	continuous functions
C_{loc}	continuous local functions
C_{ql}	continuous quasi-local functions
$d_N(\cdot,\cdot)$	hierarchical overlap
E	energy
F	Helmholtz free energy
$f(\beta, v)$	specific free energy
$F_{\beta,h,V}$	free energy in spin system
G	Gibbs free energy
H	enthalpy
$h(\cdot,\cdot)$	relative entropy
H_Λ	finite volume Hamiltonian
H_N	Hamiltonian function
$I(m)$	Cramèr entropy
$J_N(m)$	correction to Cramèr entropy
$K^\eta_{\beta,\Lambda}, K^\eta_\beta$	joint measures
m^*_β	equilibrium magnetization
$m_N(\sigma)$	empirical magnetization
N_x	control fields
p	pressure
$R_N(\cdot,\cdot)$	overlap
S	entropy
S_x, S_C	small fields
T	temperature
$u_N(x)$	scaling function for Gaussian random variable on $\mathcal{S}_N$
V	volume
$w_\Lambda(g)$	activities
X_σ	standardized Gaussian process
$Z_{\beta,V,N}$	canonical partition function
$z_{E,V,N}$	micro-canonical partition function

Part I
Statistical mechanics

1

Introduction

> L'analyse mathématique, n'est elle donc qu'un vain jeu d'esprit? Elle ne peut donner au physicien qu'un langage commode; n'est-ce pas là un médiocre service, dont on aurait pu se passer à la rigueur; et même n'est il pas à craindre que ce langage artificiel ne soit un voile interposé entre la réalité et l'œil du physicien? Loin de là, sans ce langage, la pluspart des analogies intimes des choses nous seraient demeurées à jamais inconnues; et nous aurions toujours ignoré l'harmonie interne du monde, qui est, nous le verrons, la seule véritable réalité objective.[1]
>
> Henri Poincaré, La valeur de la science.

Starting with the Newtonian revolution, the eighteenth and nineteenth centuries saw with the development of analytical mechanics an unprecedented tool for the analysis and prediction of natural phenomena. The power and precision of Hamiltonian perturbation theory allowed even the details of the motion observed in the solar system to be explained quantitatively. In practical terms, analytical mechanics made the construction of highly effective machines possible. Unsurprisingly, these successes led to the widespread belief that, ultimately, mechanics could explain the functioning of the entire universe. On the basis of this confidence, new areas of physics, outside the realm of the immediate applicability of Newtonian mechanics, became the target of the new science of theoretical (analytical) physics. One of the most important of these new fields was the theory of heat, or *thermodynamics*. One of the main principles of Newtonian mechanics was that of the conservation of energy. Now, such a principle could not hold entirely, due to the ubiquitous loss of energy through friction. Thus, all machines on earth require some source of energy. One convenient source of energy is heat, obtainable, e.g., from the burning of wood, coal, or petrol. A central objective of the theory of thermodynamics was to understand how the two types of energy, mechanical and thermal, could be converted into each other. This was originally a completely pragmatic theory, that introduced new concepts related to the phenomenon of heat, *temperature* and *entropy*, and coupled these to mechanical concepts of energy and force. Only towards the end of the nineteenth century, when the success of mechanics reached a peak, was Boltzmann, following earlier work by Bernoulli, Herapath, Joule, Krönig, Claudius,

---

[1] Approximately: So is mathematical analysis then not just a vain game of the mind? To the physicist it can only give a convenient language; but isn't that a mediocre service, which after all we could have done without; and, is it not even to be feared that this artificial language be a veil, interposed between reality and the physicist's eye? Far from that, without this language most of the intimate analogies of things would forever have remained unknown to us; and we would never have had knowledge of the internal harmony of the world, which is, as we shall see, the only true objective reality.

and Maxwell, able to give a mechanical interpretation of the thermodynamic effects on the basis of the atomistic theory. This *kinetic theory of gases* turned into what we now know as *statistical mechanics* through the work of Gibbs in the early twentieth century. It should be mentioned that this theory, that is now perfectly accepted, met considerable hostility in its early days. The first part of this book will give a short introduction to the theory of statistical mechanics.

It is not a coincidence that at the same time when statistical mechanics was created, another new discipline of physics emerged, that of *quantum mechanics*. Quantum mechanics was concerned with the inadequacies of classical mechanics on the level of microscopic physics, in particular the theory of atoms, and thus concerned the opposite side of what statistical mechanics is about. Interestingly, quantum mechanical effects could explain some deviations of the predictions of statistical mechanics from experimental observation (e.g. the problem of black body radiation that was resolved by Planck's quantum hypothesis). The basic principles of statistical mechanics can be well reconciled with quantum mechanics and give rise the the theory of *quantum statistical mechanics*. However, in many cases, a full quantum mechanical treatment of statistical mechanics turns out to be unnecessary, and much of classical mechanics applies with just some minor changes. In any case, we will here consider only the classical theory. Before approaching our main subject, let us have a very brief look at thermodynamics.

1.1 Thermodynamics

A mechanical system is characterized by essentially geometric quantities, the positions and velocities of its components (which are points of mass). If solid objects are described, the assumption of rigidity allows us to reduce their description to essentially the same kind of coordinates. Such a description does not, however, do complete justice to all the objects we can observe. Even solids are not really rigid, and may change their shape. Moreover, there are liquids, and gases, for which such a description breaks down completely. Finally, there are properties of real objects beyond their positions or velocities that may interfere with their mechanical properties, in particular their *temperature*. In fact, in a dissipative system one may observe that the temperature of a decelerating body often increases. Thermodynamics introduces a description of such new *internal* variables of the system and devises a theory allowing us to control the associated flow of energy.

The standard classical setting of thermodynamics is geared to the behaviour of a gas. A gas is thought to be enclosed in a container of a given (but possibly variable) volume, $V > 0$. This container provides the means of coupling the system to an external mechanical system. Namely, if one can make the gas change the volume of the container, the resulting motion can be used to drive a machine. Conversely, we may change the volume of the container and thus change the properties of the gas inside. Thus, we need a parameter to describe the state of the gas that reacts to the change of volume. This parameter is called the *pressure*, p. The definition of the pressure is given through the amount of mechanical energy needed to change the volume:[2]

$$dE_{\text{mech}} = -p dV \qquad (1.1)$$

---

[2] The minus sign may appear strange (as do many of the signs in thermodynamics). The point, however, is that if the volume increases, work is done by the system (transfered somewhere), so the energy of the system decreases.

Pressure is the first *intensive* variable of thermodynamics that we meet. Clearly, the relation (1.1) is not universal, but depends on further parameters. An obvious one is the total amount of gas in the container, N. Originally, N was measured in *moles*, which could be defined in terms of chemical properties of the gases. Nowadays, we know that a mole corresponds to a certain number of molecules ($\sim 6 \times 10^{23}$), and we think of N as the number of molecules in the gas. It is natural to assume that, if $V(N) = Nv$, then $p = p(v)$ should not depend on N. Hence the term *intensive*. By contrast, V is called *extensive*. It follows that E is also an extensive quantity. Like V, N can be a variable, and its change may involve a change of energy. This may not seem natural, but we should think of chemical reactions (and the possibility of having several types of molecules). By such reactions, the number of molecules will change and such a change will create or diminish a reservoir of external chemical energy (e.g. energy stored in the form of carbon). Again, we need a parameter to relate this energy change to the change in mass. We call this the *chemical potential*, μ. Then

$$dE_{\text{chem}} = \mu dN \qquad (1.2)$$

Now comes heat. Contrary to the two previous variables, volume and mass, heat is a less tangible concept. In fact, in this case the intensive variable, the *temperature*, T, is the more intuitive one. This is something we can at least feel, and to some extent also measure, e.g., using a mercury thermometer. However, we could abstract from this sensual notion and simply observe that, in order to have energy conservation, we must take into account a further internal variable property of the gas. This quantity is called *entropy*, S, and the temperature is the coefficient that relates its change to the change in energy. An important assumption is that this quantity is always non-negative. Traditionally, this *thermal energy* is called *heat* and denoted by Q, so that we have

$$dQ = TdS \qquad (1.3)$$

The principle of conservation of energy then states that any change of the parameters of the system respect the **first law of thermodynamics**:

$$dE_{\text{mech}} + dE_{\text{chem}} + dQ = dE \qquad (1.4)$$

respectively

$$dE = -pdV + \mu dN + TdS \qquad (1.5)$$

Moreover, for closed systems, i.e. for any processes that do not involve exchange of energy with some additional external system, $dE = 0$.

The main task of thermodynamics is to understand how the total energy of the system can be transformed from one type to the other in order to transform, e.g., heat into mechanical energy.

We will postulate that the state of a thermodynamic system (in equilibrium!) is described by giving the value of the three extensive variables V, N, S. Therefore we can assume that the thermodynamic state space is a three-dimensional manifold. In particular, the total energy,

$$E = E_{\text{mech}} + E_{\text{chem}} + Q \qquad (1.6)$$

will be given as a function, $E(V, N, S)$. Such a function defines the particular thermodynamic system. It then follows that the intensive variables (in equilibrium!) can be expressed as functions of the extensive variables via

$$-p(V, N, S) = \frac{\partial E(V, N, S)}{\partial V}$$

$$\mu(V, N, S) = \frac{\partial E(V, N, S)}{\partial N}$$

$$T(V, N, S) = \frac{\partial E(V, N, S)}{\partial S} \qquad (1.7)$$

These equations are called equations of state.

Remark 1.1.1 The statements above can be interpreted as follows: Suppose we fix the intensive variables p, T, μ by some mechanism to certain values, and set the extensive variables V, S, N to some initial values V_0, S_0, N_0. Then the time evolution of the system will drive these parameters to equilibrium, i.e. to the values for which equations (1.7) hold. Such processes are called *irreversible*. In contrast, *reversible* processes vary intensive and extensive parameters in such a way that the equations of state (1.7) hold both in the initial and in the final state of the process, i.e. the process passes along equilibrium states of the system. Note that this statement contains the formulation of the second law of thermodynamics.

One of the main pleasures of thermodynamics is to re-express the equations of state in terms of different sets of variables, e.g. to express V, N, S as a function of p, N, T, etc. To ensure that this is possible, one always assumes that E is a *convex* function. The function $E(V, S, N)$ is usually called the *internal energy*. Then, the desired change of variables can be achieved with the help of *Legendre transformations*.

In the example mentioned, we would like to express the energy as a function of p, T, N and to introduce a new function G with the property that $\partial G / \partial p = V$. That is, we must have that

$$\begin{aligned} dG(p, T, N) &= +V dp - S dT + \mu dN \\ &= +d(Vp) - d(ST) - p dV + T dS + \mu dN \\ &= d(Vp - ST + E) \end{aligned} \qquad (1.8)$$

Thus, we get

$$G(p, T, N) = pV(p, T, N) - TS(p, T, N) + E(V(p, T, N), N, S(p, T, N)) \qquad (1.9)$$

where the functions V and S are obtained from inverting (1.7). However, this inversion often need not be done, since an expression of the energy in the new variables is readily available. The important observation is that the fundamental function, whose derivatives provide the equations of state, is not always the energy, but its various Legendre transforms. All these functions carry interesting names, such as internal energy, free energy, enthalpy, free enthalpy, etc., which are difficult to remember. The importance of these different forms of these *thermodynamic potentials* lies in the fact that one is interested in processes where some parameters of the system are changed, while others are fixed. Computing the resulting changes is most easily done with the help of the corresponding natural potential, which typically corresponds to the conserved energy when its variables are kept fixed while the others are varied.

The function G is called the *Gibbs free energy*. Other potentials whose name it is useful to remember are

(i) the Helmholtz *free energy*,
$$F(T, V, N) = E - TS \qquad (1.10)$$

(ii) the *enthalpy*,
$$H(p, S, N) = E + pV \qquad (1.11)$$

Let us note that thermodynamics, contrary to what its name suggests, is not a theory of dynamics, but rather one of statics, or *equilibrium*. For example, the values that the intensive parameters take on when the extensive ones are fixed are equilibrium values. When performing thermodynamic calculations, one always assumes that the system takes on these equilibrium values, which is perhaps a reasonable approximation if the motion is performed very slowly. In reality, things are much more difficult.

At one point we said that the assumption of convexity allows us to invert the equations of state and to express, e.g., V as a function of p, T, N. But this is not true. It is only true if E is a *strictly convex* function. If in some region E depends linearly on V, then $p = \partial E/\partial V = const.$ on that set, and we cannot compute V as a function of p; all we know is that, for this value of p, V must lie in the said interval. In other words, V as a function of p has a jump at this value of p. If something of this type happens, we say that the system undergoes a *first-order phase transition* at this value of the parameters. Interestingly, real systems do exhibit this phenomenon. If the pressure of, say, water vapour is increased, while the temperature is not too low, at some specific value of p the volume drops down, i.e. the vapour condenses to water. It is remarkable that the formalism of thermodynamics easily allows the incorporation of such striking phenomena. If there is a phase transition, then the equations of state represent discontinuous functions. This is an unexpected feature that we are not familiar with from mechanics. This seems to indicate that classical dynamics and thermodynamics are quite different and should not have much to do with each other, as it seems inconceivable that these discontinuities should result from motions governed by Newton's equations. Therefore, phase transitions are the most remarkable phenomena in statistical mechanics, and they will be at the centre of our attention throughout this book. Even today, they represent one of the most lively topics of research in the field.

Jumps in the equations of state are the most severe singularities that are admitted in the theory, due to the convexity assumption. There are milder forms of singularities that are very interesting, where only higher derivatives of the equations of state are discontinuous. According to the order of the discontinuous derivative, such phase transitions are called second order, third order, etc. They are associated with interesting physical phenomena.

The main problem of thermodynamics is that we do not understand what entropy and temperature are, which represents the main difficulty in understanding what the thermodynamic potentials should be as functions of their parameters. In practice, they are often obtained empirically from experimental data. A derivation from first principles is of course desirable.

The preceding discussion of thermodynamics is of course very cursory. There are numerous in-depth presentations in the literature. A recent attempt to give an axiomatic foundation of thermodynamics was made in a paper by Lieb and Yngvason [169], which also contains a wealth of references.

2

Principles of statistical mechanics

> Qu'une goutte de vin tombe dans un verre d'eau; quelle que soit la loi du mouvement interne du liquide, nous le verrons bientôt se colorer d'une teinte rose uniforme et à partir de ce moment on aura beau agiter le vase, le vin et l'eau ne paraîtront plus pouvoir se séparer. Tout cela, Maxwell et Boltzmann l'ont expliqué, mais celui qui l'a vu le plus nettement, dans un livre trop peu lu parce qu'il est difficile à lire, c'est Gibbs, dans ses principes de la Mécanique Statistique.[1]
>
> *Henri Poincaré. La valeur de la science.*

About 1870, Ludwig Boltzmann proposed that the laws of thermodynamics should be derivable from mechanical first principles on the basis of the atomistic theory of matter. In this context, N moles of a gas in a container of volume V should be represented by a certain number of *atoms*, described as point particles (or possibly as slightly more complicated entities), moving under Newton's laws. Their interaction with the walls of the container is given by elastic reflection (or more complicated, partially idealized constraint-type forces), and would give rise to the observed *pressure* of the gas. In this picture, the thermal variables, temperature and entropy, should emerge as effective parameters describing the macroscopic essentials of the microscopic dynamics of the gas that would otherwise be disregarded.

2.1 The ideal gas in one dimension

To get an understanding of these ideas, it is best to consider a very simple example which can be analyzed in full detail, even if it is unrealistic. Consider N particles, all of mass m, that move on a one-dimensional line $\mathbb{R}$ and that absolutely do not interact with each other; in particular they penetrate each other freely upon impact. We denote the position and momentum of particle i by q_i and p_i. Assume further that they are confined to an interval $[0, V]$.

When reaching the boundary of this interval, they are perfectly reflected. Now let the top boundary of the interval (the piston) be movable; assume that a constant force f is acting on

---

[1] Approximately: Let a drop of wine fall into a glass of water; whatever be the law that governs the internal movement of the liquid, we will soon see it tint itself uniformly pink and from that moment on, however we may agitate the vessel, it appears that the wine and the water can separate no more. All this, Maxwell and Boltzmann have explained, but the one who saw it in the cleanest way, in a book that is too little read because it is difficult to read, is Gibbs, in his *Principles of Statistical Mechanics*.

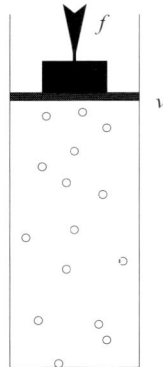

Figure 2.1 Weight on the piston exerting a force f on the piston.

this piston, as shown in Fig. 2.1. Then the container has energy $E_{\text{ext}} = fV$, if the piston's position is V. At some initial time 0 let the momenta of the particles be p_i such that

$$\frac{1}{2m} \sum_{i=1}^{N} p_i^2 = E \tag{2.1}$$

The total (conserved) energy of the system is then

$$H = fV + E \tag{2.2}$$

We will now assume that the dynamics of the system is such that (after a long time) it runs uniformly over all configurations that are compatible with the constraint that the energy of the system is constant (this is probably not the case in our system, but …). Since the kinetic energy of the particles, E, is non-negative, the position of the piston can only range over the finite interval $[0, V_{\max}]$, where $V_{\max} = H/f$. Then, the total available state space of our system is $[0, V_{\max}] \times [0, V]^N \times S^N_{\sqrt{2m(H-fV)}}$, where S^N_r denotes the $(N-1)$-dimensional sphere of radius r. Our assumptions correspond to saying that we start the process initially at random on any point of this set with equal probability, and that we will find ourselves, in the long run, uniformly distributed on this set (this distribution is called the *micro-canonical distribution* of our system). As we have explained, thermodynamics is concerned with the macroscopic observables only, and in our case this is the position of the piston, V. Finding the *equilibrium position* of this variable amounts to computing the probability distribution of the parameter V, to wit

$$\mathbb{P}[V \in dV] \equiv \frac{dV \int_{S^N_{\sqrt{2m(H-fV)}}} dp_1 \cdots dp_N \int_0^V dq_1 \cdots dq_N}{\int_0^{V_{\max}} dV \int_{S^N_{\sqrt{2m(H-fV)}}} dp_1 \cdots dp_N \int_0^V dq_1 \cdots dq_N} \tag{2.3}$$

Now $\int_0^V dq_1 \cdots dq_N = V^N$, and the surface area of the $(N-1)$-dimensional sphere being $|S^N_r| = \frac{2\pi^{N/2}}{\Gamma(N/2)} r^{N-1}$, we have that

$$\int_{S^N_{\sqrt{2m(H-fV)}}} dp_1 \cdots dp_N \int_0^V dq_1 \cdots dq_N \tag{2.4}$$

$$= V^N \frac{2\pi^{N/2}}{\Gamma(N/2)} (2m(H - fV))^{(N-1)/2}$$

Using this expression, we get that

$$\mathbb{P}[V \in dV] \equiv \frac{dV \exp\left(N \ln V + \frac{N-1}{2} \ln 2m(H - fV)\right)}{\int_0^{V_{\max}} dV \exp\left(N \ln V + \frac{N-1}{2} \ln 2m(H - fV)\right)} \quad (2.5)$$

When N is large, the integrand is sharply peaked around the value, V^*, that maximizes the exponent, $N \ln V + \frac{N-1}{2} \ln 2m(H - fV)$ (more precisely, it can be evaluated using the saddle-point method). Ignoring terms of order $1/N$, V^* is found by solving the equation

$$0 = -\frac{1}{2} \frac{f}{H - fV^*} + \frac{1}{V^*} \quad (2.6)$$

which yields

$$V^* = \frac{2}{3} \frac{H}{f} \quad (2.7)$$

To elaborate on this, let us introduce the number of particles as a parameter, and introduce the rescaled variables $v \equiv V/N$, $e \equiv E/N$, and $u \equiv H/N$. Then

$$\mathbb{P}_{u,f,N}[v \in dv] = \frac{dv \exp\left(N \ln v + \frac{N-1}{2} \ln(2m(u - fv))\right)}{\int_0^{v_{\max}} dv \exp\left(N \ln v + \frac{N-1}{2} \ln(2m(u - fv))\right)} \quad (2.8)$$

where we used that the extra terms of the form $N \ln N$ cancel between the numerator and the denominator. We now introduce the quantity

$$s(u, f; v) \equiv \ln v + \frac{1}{2} \ln 2m(u - fv) \quad (2.9)$$

so that (up to negligible terms)

$$\mathbb{P}_{u,f,N}[v \in dv] = \frac{dv \exp(Ns(u, f; v))}{\int_0^{v_{\max}} dv \exp(Ns(u, f; v))} \quad (2.10)$$

In standard probabilistic terms, (2.10) is a (strong) form of a *large deviation principle*. One says that the family of probability measures $\mathbb{P}_{u,f,N}$, indexed by N, satisfies a large deviation principle with *rate function* $-s(u, f; v)$ (where s is considered as a function of v, parametrized by u and f). We will say more about large deviations later.

For the moment we observe the appearance of a function that is related to a probability measure that has been instrumental in determining the distribution of energy between the mechanical energy and heat. This function has no purely mechanical interpretation. It is called the *entropy function*, and its *value*, computed at the equilibrium value of v, is called the entropy. In our setting the entropy appears parametrized by u and f,

$$s(u, f) = s\left(u, f; \frac{2}{3}\frac{u}{f}\right) = \frac{1}{2} \ln \frac{2}{3} um + \ln \frac{2u}{3f} \quad (2.11)$$

but since f and u determine the equilibrium value of v, and hence e, we can re-express it in the (natural) variables e and v, to get

$$s(e, v) = \ln\left(v (2em)^{1/2}\right) \quad (2.12)$$

In this form, $s(e, v)$ is the inverse of $e(s, v)$, and hence

$$\frac{\partial s(e, v)}{\partial e} = \frac{1}{\frac{\partial e(s,v)}{\partial s}} = \frac{1}{T} \tag{2.13}$$

From here we get an expression for the temperature as function of energy and volume,

$$T = 2e \tag{2.14}$$

Similarly, we can compute

$$e(s, v) = \frac{e^{2s}}{v^2} \frac{1}{2m} \tag{2.15}$$

and hence the pressure

$$p \equiv -\frac{\partial e(s, v)}{\partial v} = 2e/v \tag{2.16}$$

Comparing with (2.7), we see that everything is consistent, since, in equilibrium, $p = f$; that is, the thermodynamic pressure, p, equals the external force, f, acting on the piston.

Thus, in our simple example, we understand how the equations of thermodynamics arise, and what the meaning of the mysterious concepts of entropy and temperature is. The equilibrium state of the system is governed by the external force, and the intrinsic probability of the system to find itself in a state with a given value of the macroscopic parameter (volume). The properties of this probability distribution give rise to some effective force (the pressure) that competes with and has to be equilibrated against the external macroscopic force.

On the other hand, in our simple example, it is also easy to understand the pressure as the average force that the gas molecules exercise on the piston when they are reflected from it. Namely, each time a molecule i is reflected, its velocity changes from v_i to $-v_i$. Hence

$$t^{-1} \int_0^t f(t') dt' = t^{-1} \int_0^t dt' \sum_{i: q_i(t')=V} m \frac{d}{dt'} v_i(t') \tag{2.17}$$

$$= t^{-1} \sum_{i: q_i(t')=V, t' \in [0,t)} 2m v_i$$

It remains to compute the average number of hits of particle i at the piston. But the time between two hits is $2V/v_i$, so the number of hits is roughly $t v_i/2V$, whence

$$\lim_{t \uparrow \infty} t^{-1} \int_0^t f(t') dt' = \sum_{i=1}^N m v_i^2 / V = 2E/V \tag{2.18}$$

This yields a mechanical explanation of formula (2.16). Now we understand better why, in the constant volume ensemble, the conserved energy is the kinetic energy of the gas, whereas in the constant pressure ensemble (which we studied first), it is the sum of mechanical and kinetic energy, $E_{\text{kin}} + pV$. This also explains the appearance of the Legendre transforms of thermodynamic potentials when variables are changed.

Exercise: Repeat the computations of the example when the state space is a three-dimensional cylinder with a piston movable in the z-direction.

2.2 The micro-canonical ensemble

In our first example we have seen that we can derive thermodynamic principles from probabilistic considerations, and in particular from the assumption that the state of the system is described by a probability distribution, more precisely the uniform distribution on the submanifold of the phase space where the energy function takes a constant value. The idea that the state of a physical system with very many degrees of freedom should be described by a *probability measure* on the phase space of the underlying mechanical system is the basis of statistical mechanics. Such a probability measure will depend on a finite number of parameters, representing the thermodynamic variables of the system. Thus, each thermodynamic state, say (V, S, N), corresponds precisely to one probability measure $\rho_{(V,S,N)}$ on the state space. The rationale behind such a description is the underlying assumption that the long-time means of the dynamics (with suitable initial and boundary conditions) should converge to the ensemble averages with respect to these measures.

The micro-canonical ensemble is the most straightforward class of such thermodynamic states. First of all, we assume that the measure is concentrated on a subset of constant value for the energy. This is reasonable, since we know that for (conservative) mechanical systems the energy is conserved. Moreover, it follows from Liouville's theorem that the Hamiltonian time evolution conserves phase space volume, and thus the uniform measure is *invariant* under the time evolution. If Φ_t is the Hamiltonian flow on phase space, and A a measurable subset of the phase space P, then

$$\int_A \rho(dx) = \int_{\Phi_t(A)} \rho(dx) \equiv \int_A \Phi_t^* \rho(dx) \tag{2.19}$$

where the last equation defines Φ_t^* as the 'pull-back' of the flow Φ_t, i.e. its action on measures. Hence $\Phi_t^* \rho = \rho$, if ρ is any measure that is uniform on invariant subsets of the flow Φ_t. Of course, the fact that ρ be invariant is a necessary requirement for it to be ergodic, i.e. to ensure that for any bounded measurable function g on the support of ρ,

$$\lim_{t \uparrow \infty} \frac{1}{t} \int_0^t (\Phi_{t'}^* g)(x_0) \equiv \lim_{t \uparrow \infty} \frac{1}{t} \int_0^t g(\Phi_{t'}(x_0)) = \int g(x) \rho(dx) \tag{2.20}$$

but it is not sufficient at all. What we would need to prove in addition would be that the system is *metrically transitive*, i.e. that the energy surface does not contain further invariant subsets. There are mechanical systems for which additional conserved quantities exist (e.g. the one-dimensional ideal gas treated above), in which case we cannot expect (2.20) to hold. In the micro-canonical ensemble we take as the two other conserved quantities the volume (inasmuch as we talk about confined systems) and the number of particles. All other possible conserved quantities are ignored. There is not very much point in arguing about this. While certainly it would be nice to have a complete and rigorous derivation of ergodic theorems to justify this approach, thus providing a solid link between classical and statistical mechanics, we have to accept that this is not possible. There are very few examples where such a derivation can be given. These concern the motion of one or few hard spheres in closed boxes ('Sinai billiards') [68, 69, 225]. Worse, even if an ergodic theorem were proven, it is quite unclear why essentially instantaneous observations of a system should be related to long-term time averages. A more detailed discussion of these issues can be found in the book by Gallavotti [109]. For a philosophical discussion of the probabilistic approach taken by statistical mechanics, we refer for instance to the recent text by Guttman [135]. The important observation is that for all practical purposes, statistical mechanics seems to work marvelously well, and we will focus on the mathematical analysis of the consequences of the theory rather than on its derivation.

We can now define what we understand by a thermodynamic system.

Definition 2.2.1 A thermodynamic system involves:

(i) A parameter $N \in \mathbb{N}$ called the particle number,
(ii) a measure space P, and its product space P^N,
(iii) a Hamiltonian function $H_N : P^N \to \mathbb{R}$,
(iv) constraints depending on macroscopic parameters, such as V.

Remark 2.2.2 In the context of a gas, the space P is the phase space of a single gas molecule, H_N is the interaction, and the constraint is the indicator function that the position of all molecules should be within the container of volume V.

Definition 2.2.3 The micro-canonical ensemble of a thermodynamic system is the collection of all uniform probability distributions, $\rho_{E,V,N}$, on the sets

$$\Omega_{E,V,N} \equiv \{x \in P^N : H_N(x) = E, x \in V\} \qquad (2.21)$$

where by $x \in V$ we understand that, if $x = (p_1, q_1, \ldots, p_N, q_N)$, then $q_i \in V$, for all i. Note however that we can in general consider different types of constraints, and V can represent different subsets of the phase space. The *micro-canonical partition function* is the function[2]

$$z_{E,V,N} = \frac{1}{N!} \int dx \delta(E - H_N(x)) \mathbb{1}_{x \in V} \qquad (2.22)$$

where δ denotes the Dirac delta-function[3] on $\mathbb{R}$. This choice of the measure ensures its invariance under the Hamiltonian flow.[4] The *entropy* is defined as

$$S(E, V, N) \equiv \ln z_{E,V,N} \qquad (2.23)$$

Note that the factor $1/N!$ is introduced to take into account that particles are indistinguishable and ensures that S is proportional to N.

The micro-canonical partition function is the normalizing constant that turns the flat measure,

$$\tilde{\rho}_{E,V,N}(dx) \equiv \frac{1}{N!} dx \delta(E - H_N(x)) \mathbb{1}_{x \in V} \qquad (2.24)$$

into a *probability measure*, i.e.

$$\rho_{E,V,N}(dx) = \frac{1}{z_{E,V,N}} \tilde{\rho}_{E,V,N}(dx) \qquad (2.25)$$

---

[2] In the physics literature one introduces an additional normalizing factor (h^{dN}), where h is Planck's constant. This is done to make the classical and quantum entropies comparable. I will choose $h = 1$.
[3] The delta function is defined such that for all smooth test functions ϕ, $\int dE \phi(E) \delta(E - a) = \phi(a)$. It follows that, if Φ is a function on P^N, then $\int dE \int dx \delta(E - H_N(x)) \Phi(x) \phi(E) = \int dx \Phi(x) \phi(H_N(x))$.
[4] In many physics textbooks, one uses a soft version of this measure, namely the Lebesgue measure of the set $\{x \in P^N : |H_N(x) - E| < \epsilon\}$. For all practical purposes, this yields the same results.

It may look surprising that such a normalizing constant can have physical significance, since after all it depends on the choice of the initial unnormalized measure, that is largely arbitrary. This significance will arise from the parameter dependence of this constant, as we will see shortly.

We will assume henceforth that $S(E, V, N)$ is a differentiable function. This is certainly the case for classical systems where $H_N(x) = \sum_{i=1}^{N} \frac{p_i^2}{2m_i} + \Phi(q)$ at any energy $E > \min(\Phi)$ (see, e.g., [131]). Then the derivatives of S determine the parameters pressure, temperature, and chemical potential[5] via

$$\frac{\partial S(E, V, N)}{\partial E} = \frac{1}{T} \equiv \beta$$
$$\frac{\partial S(E, V, N)}{\partial V} = \frac{p}{T} \tag{2.26}$$
$$\frac{\partial S(E, V, N)}{\partial N} = \frac{1}{\mu}$$

Thus, computing the entropy of the system allows us to engage the full thermodynamic formalism and to compute all kinds of interesting quantities.[6]

In particular, if we want to couple the gas to a mechanical source of energy (e.g. through a movable piston at which a fixed force of strength f acts), passing to the conserved quantity $H = E + fV$, we can derive the distribution of the position of the piston as

$$\mathbb{P}(V \in dV) = \frac{dV \exp(S(V, H - fV, N))}{\int dV \exp(S(V, H - fV, N))} \tag{2.27}$$

from which the equilibrium piston position results as the value of V with *maximal entropy*, and in particular the solution of the equation

$$\frac{dS(V, H - fV, N)}{dV} = 0 \tag{2.28}$$

(with H and f fixed), as in the example treated above. We see that

$$\frac{dS(V, H - fV, N)}{dV} = \frac{\partial S}{\partial V} - \frac{f}{T} = \frac{p}{T} - \frac{f}{T} \tag{2.29}$$

so that indeed in the equilibrium position, the thermodynamic pressure p is equal to the external force f acting on the piston.

Let us look back at what precisely we have done here from the point of view of probability measures. Our underlying assumption is that the a-priori probability distribution of the system with movable piston is given by the uniform measure on the available state space,

---

[5] I set Boltzmann's constant, k, equal to 1. In physical conventions the first equation of (2.26) reads $\frac{\partial S(E,V,N)}{\partial E} = \frac{1}{kT} \equiv \beta$.

[6] It is a somewhat annoying feature of thermodynamics that the intensive variables are defined as the derivatives of the energy when the extensive quantities, in particular the entropy, are kept fixed. This leads to the fact that the inverse temperature, which appears as the derivative of the entropy, is often a more natural quantity to consider than the temperature itself. It would probably be more rational to consider the entropy as the most fundamental thermodynamic potential and to consider its derivatives as the natural intensive variables.

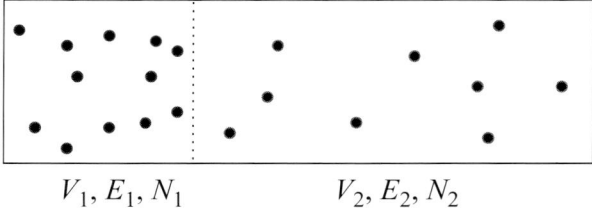

Figure 2.2 Piston containing two gases separated by a wall permitting energy transfer.

including the position of the piston, i.e. by the probability measure

$$\hat{\rho}_{H,f,N}(dx, dV) \equiv \frac{dV \tilde{\rho}_{H-fV,V,N}(dx)}{\int dV \int \tilde{\rho}_{H-fV,V,N}(dx')}$$ (2.30)

$$= \frac{dV z_{H-fV,V,N} \rho_{H-fV,V,N}(dx)}{\int dV z_{H-fV,V,N}}$$

$$= dV \rho_{H-fV,V,N}(dx) \frac{\exp(S(H-fV,V,N))}{\int dV \exp(S(H-fV,V,N))}$$

In particular, if we are only looking at the distribution of the internal degrees of freedom, we obtain the distribution

$$\check{\rho}_{H,f,N}(dx) = \int dV \rho_{H-fV,V,N}(dx) \frac{\exp(S(H-fV,V,N))}{\int dV \exp(S(H-fV,V,N))}$$

$$= \int \mathbb{P}[V \in dV] \rho_{H-fV,V,N}(dx)$$ (2.31)

which can be seen as a mixture of micro-canonical distributions. We see that the entropy governs the probability with which we see a given micro-canonical distribution $\rho_{H-fV,V,N}(dx)$ if V is allowed to vary.

In a similar way we can understand the physical significance of the temperature. To this end we consider a cylinder with a fixed piston separating the cylinder into two volumes V_1, V_2 (see Fig. 2.2). Assume that there are N_1, N_2 molecules in each partition (possibly of different types of gases). Assume that the piston allows for energy to pass from one part to the other. Then the total energy $E = E_1 + E_2$ is conserved. We would like to know the probability distribution of the value E_1. According to our assumption that the distribution of the atoms in the two containers, given the values E_i, will be the uniform distributions, ρ_{E_i,V_i,N_i}, on the sets $H_{N_i}^{(i)}(x) = E_i$, we get

$$\mathbb{P}(E_1 \in dE_1) = \frac{dE_1 \exp(S_1(V_1, E_1, N_1) + S_2(V_2, E - E_1, N_2))}{\int dE_1 \exp(S_1(V_1, E_1, N_1) + S_2(V_2, E - E_1, N_2))}$$ (2.32)

Thus, the probability distribution of E_1 has a (pronounced) maximum when

$$\frac{\partial S_1(V_1, E_1, N_1)}{\partial E_1} - \frac{\partial S_2(V_2, E - E_1, N_2)}{\partial E_1} = 0$$ (2.33)

In other words, the two systems are in equilibrium when the partial derivatives with respect to the energy of their entropies, the inverse temperatures, are the same. This is sometimes called the zeroth law of thermodynamics.

2.2 The micro-canonical ensemble

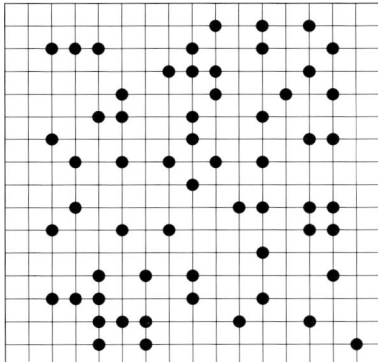

Figure 2.3 A lattice gas configuration on a square lattice.

As in the preceding discussion, we could introduce the probability distribution of the coupled systems as

$$\check{\rho}_{E,V_1,V_2,N_1,N_2}(dx_1, dx_2) \tag{2.34}$$
$$= \int_0^E dE_1 \frac{\exp(S_1(V_1, E_1, N_1) + S_2(V_2, E - E_1, N_2))}{\int dE_1 \exp(S_1(V_1, E_1, N_1) + S_2(V_2, E - E_1, N_2))}$$
$$\times \rho_{E_1,V_1,N_1}(dx_1) \rho_{E-E_1,V_2,N_2}(dx_2)$$

Note that in the previous discussion we could have replaced the second system by some artificial device with a fictitious entropy βE_2. Such a device would then force the temperature of any system that is energetically coupled to it to take the value $T = 1/\beta$. In thermodynamics this would be called a thermostat.

Example: The ideal lattice gas In the course of this book we will soon concentrate on simple systems in which the classical particles are replaced by particles with a discrete number of degrees of freedom. The simplest such system is the *ideal lattice gas*. Here we consider, instead of our usual phase space, a finite subset, Λ, of some discrete lattice $\mathbb{Z}^d$. We denote by $V \equiv |\Lambda|$ the number of vertices of this lattice. We consider a fixed number, N, of particles, whose degrees of freedom are just their positions $x_i \in \Lambda$. A configuration of particles is depicted in Fig. 2.3. We will assume that the particles can sit on top of each other, and that the energy is simply proportional to the number of particles,

$$E = \mu N \tag{2.35}$$

Then the micro-canonical partition function is simply the number of ways we can arrange N particles on the V sites of the lattice,

$$z_{E,V,N} = \frac{V^N}{N!} = \exp(N \ln V - \ln N!) \tag{2.36}$$

so that the entropy is

$$S(E, V, N) = N \ln V - \ln N! \approx N(\ln v + 1) \tag{2.37}$$

Note that, due to the strict relation between energy and particle number, there are really only two independent extensive variables in this model. We see that the pressure is

$$\frac{p}{T} = N/V = E/\mu V \tag{2.38}$$

Exercise: Consider the lattice gas with the additional constraint that no more that one particle can occupy the same site.

Interestingly, the micro-canonical entropy is equal to what one would call the entropy of the measure $\rho_{E,V,N}$. In fact, let ρ be any probability measure on the support of $\Omega_{E,V,N}$. Then the (relative) entropy of ρ (with respect to the uniform measure, $\rho_0 \equiv \tilde{\rho}_{E,V,N}$, on this set) is defined as

$$h(\rho, \rho_0) \equiv \int \ln \frac{\mathrm{d}\rho(x)}{\mathrm{d}\rho_0(x)} \mathrm{d}\rho(x) \tag{2.39}$$

Now let ρ_0 be the (unnormalized) uniform measure on the phase space of N particles in volume V and with total energy E.

It is not difficult to see that the unique minimizer of this function is the uniform *probability* measure on the same set, i.e. $\rho_{E,V,N}$, and that

$$h(\rho_{E,V,N}, \rho_0) = -S(E, V, N) \tag{2.40}$$

Thus, we can say that the micro-canonical distribution is characterized by the fact that it minimizes the relative entropy with respect to the uniform measure on the accessible state space. This statement would appear even more profound if we had not been forced to make an a priori choice of the measure ρ_0. Nonetheless, it gives an interesting interpretation of the micro-canonical distribution. Moreover, it would appear that the dynamics of a gas, started initially in any configuration (or any probability distribution concentrated on it) on the energy shell, should have a tendency to evolve towards the uniform distribution, thus *increasing entropy*. This fact appeared for quite some time an obstacle in accepting the basic premises of statistical mechanics, as it appeared in contradiction to the reversible nature of the Newtonian laws of classical mechanics.[7] Such a contradiction, however, does not exist. This may be most easily understood in the example of our moving piston. For statistical mechanics to be relevant, it must be true that, if this system is started with any position of the piston, the piston's position should evolve to its equilibrium position (rather fast), and then be seen there (almost) all the time. Indeed, few people (of any minimal level of integrity) claim to have seen huge motions of such pistons (unless someone was fiddling with the equipment). So clearly the piston movement looks rather irreversible, although everything is pure classical mechanics. Is there a contradiction? Clearly not, since our argument was based on microscopically sound reasoning: the motion of the molecules is fully reversible, and follows the laws of classical mechanics. The trajectories can, in principle, reach all points in the energetically available phase space, including those where the piston is not at its equilibrium position. However, the number of configurations where the piston is not close to this position is so ridiculously small compared to those when it is, that the occurrence of such instances is exquisitely rare if N is large. Thus, if only the motion of the piston is observed, we get the impression that there is a preferred direction in time. But

---
[7] This discussion is still not extinct today.

this is not so. If we reversed time, we would observe exactly the same phenomenon. Only by preparing very special, non-typical initial conditions, could we observe the system at a given later time in a state where the piston is not in its equilibrium position, and that is not going to be re-observed in a very very long time.

If one accepts the basic principles laid out above, we can use thermodynamics, provided we can compute the micro-canonical partition function. Unfortunately, this is not in general an easy task. In many ways, statistical mechanics is not so much a question of principles, as a question of computational techniques.

2.3 The canonical ensemble and the Gibbs measure

The difficulty of the computations in the micro-canonical ensemble comes from the fact that it always involves a constrained integral over some manifold $H_N(x) = E$ in a space of very high dimensions. It is simply a very difficult geometric problem to compute the area of a very high dimensional manifold. We have been able to do this for the sphere, and we may be able to do it for a few more examples, but in general this is hopeless. Even numerically, this is a next to impossible task. A way to get to a more accessible expression is to *change ensembles*, i.e. to consider a system where the energy is no longer fixed, but allowed to vary, while the conjugate variable, the *temperature*, is fixed. We have already seen in the previous section that this can be achieved by introducing a fictitious thermostat with which the system can exchange energy. This leads us to introduce the distribution

$$\mathcal{G}_{\beta,V,N}(dx) = \frac{\int dE e^{-\beta E} z_{E,V,N} \rho_{E,V,N}(dx)}{\int dE e^{-\beta E} \int \rho_{E,V,N}(dx)} = \frac{\frac{1}{N!} e^{-\beta H_N(x)} dx}{\frac{1}{N!} \int dx e^{-\beta H_N(x)}} \quad (2.41)$$

The denominator is called the *canonical partition function*

$$Z_{\beta,V,N} \equiv \frac{1}{N!} \int dx e^{-\beta H_N(x)} \quad (2.42)$$

where the combinatorial factor is introduced for the same reason as in the definition of the entropy in the micro-canonical ensemble. Let us investigate the thermodynamic meaning of these quantities. As we have already seen, this measure is concentrated where

$$\beta = \frac{\partial S(E, V, N)}{\partial E} \quad (2.43)$$

and then

$$Z_{\beta,V,N} = \int dE e^{-\beta E + S(E,V,N)} = \int dE e^{-\beta \tilde{F}(T,V,N;E)} \quad (2.44)$$

where $\tilde{F}$ is called the *free energy functional*

$$\tilde{F}(T, V, N, E) = E - TS(E, V, N) \quad (2.45)$$

$\tilde{F}$ is an extensive quantity, and thus the integrand in (2.44) will be sharply concentrated on the set of values of E around the equilibrium values of the energy, $E^*(V, T, N)$, at temperature β^{-1}, defined as a solution of (2.43). This suggests that

$$\ln Z_{\beta,V,N} \approx -\beta F(T, V, N) \quad (2.46)$$

where F is the thermodynamic *free energy*,

$$F(T, V, N) \equiv \tilde{F}(T, V, N, E^*(V, T, N)) \tag{2.47}$$

Equation (2.46) gives an alternative connection between thermodynamics and statistical mechanics, i.e. an alternative prescription for how to compute a thermodynamic potential from a mechanical basis. We will have to investigate this relation a little more carefully.

Theorem 2.3.1 *Assume for a statistical mechanical system that the micro-canonical entropy satisfies*

$$\lim_{N \uparrow \infty} N^{-1} S(E, V, N) = s(e, v) \tag{2.48}$$

where $\lim_{N \uparrow \infty} V/N = v$ *and* $\lim_{N \uparrow \infty} E/N = e$, *and s is a strictly concave, upper semi-continuous function such that, for all* $\beta \geq 0$,

$$\int_{s(e,v)-\beta e \leq a} \exp(N(s(e, v) - \beta e)) \, de \leq C e^{Na} \tag{2.49}$$

Define the function $f(\beta, v)$ *by*

$$\beta f(\beta, v) = \min_e (e\beta - s(e, v)) \tag{2.50}$$

Assume further that convergence in (2.48) is such that, uniformly in e, $\frac{(S(E,V,N)-E\beta)}{N(s(eN,vN)-\beta e)} \to 1$. *Then, for any β, such that $s(e, v)$ has bounded derivatives in a neighbourhood of e^*,*

$$\lim_{N \uparrow \infty} \frac{1}{\beta N} \ln Z_{\beta, V, N} = -f(\beta, v) \tag{2.51}$$

Remark 2.3.2 βf is called the Legendre transform of s. If s is differentiable and strictly concave, then

$$f(\beta, v) = e^*(v, \beta) - \beta^{-1} s(e^*(v, \beta), v) \tag{2.52}$$

where e^* is the unique solution of the equation

$$\beta = \frac{\partial s(e, v)}{\partial e} \tag{2.53}$$

Proof Basically, we have to show that the integral receives almost no contribution from values of e such that $s(e, v) - \beta e \leq (s(e^*, v) - \beta e^*) - \delta$. This is ensured by assumption (2.49). From the complement of this region, $D_\delta \equiv \{e \in \mathbb{R} : s(e, v) - \beta e > (s(e^*, v) - \beta e^*) - \delta$, it gets a contribution of the desired order, provided this set is neither too small nor too large, which our differentiability assumptions imply. Our assumption of uniform convergence ensures that, for large N, we can replace the integrand by its limit, since, for any $\epsilon > 0$, there exists $N_0 \in \mathbb{N}$ such that for all $N \geq N_0$, for all e,

$$|N^{-1}[S(E, V, N) - \beta E] - (s(e, v) - \beta e)| \leq \epsilon |s(e, v) - \beta e| \tag{2.54}$$

Therefore, for such N,

$$\int_{D_\delta^c} de \exp(-\beta eN + S(eN, V, N)) \tag{2.55}$$

$$\leq \int_{D_\delta^c} \exp(N(s(e, v) - \beta e) + [(S - \beta E) - N(s - \beta e)]) \, de$$

$$\leq \int_{D_\delta^c} \exp(N(s(e, v) - \beta e) + \epsilon |s(e, v) - \beta e|) \, de$$

$$\leq C \exp(N[s(e^*, v) - \beta e^* + \epsilon |s(e^*, v) - \beta e^*| - \delta])$$

On the other hand, for N large enough,

$$\int_{D_\delta} de \exp(-\beta eN + S(eN, V, N)) \leq C \exp(N[s(e^*, v) - \beta e^*](1 + \epsilon)) \tag{2.56}$$

and, since the function s has bounded derivatives, on a set of size N^{-1} the integrand cannot vary by more than a constant factor, for some $c > 0$,

$$\int_{D_\delta} e^{-\beta eN + S(eN, V, N)} de \geq cN^{-1} e^{N[s(e^*, v) - \beta e^*](1+\epsilon)} \tag{2.57}$$

Taking the logarithm and dividing by N, for any $\epsilon > 0$ (we chose the signs as if f were negative, otherwise they have to be reversed), we obtain that

$$-\beta f(\beta, v)(1 - \epsilon) \leq \liminf_{N \uparrow \infty} \frac{1}{N} \ln \int de \exp(-\beta eN + S(eN, V, N))$$

$$\leq \limsup_{N \uparrow \infty} \frac{1}{N} \ln \int de \exp(-\beta eN + S(eN, V, N))$$

$$\leq -\beta f(\beta, v)(1 + \epsilon) \tag{2.58}$$

which implies the assertion of the theorem. □

The measure defined by equation (2.41) is called the *Gibbs measure* or the *canonical ensemble*. Theorem 2.3.1 is a (not very strong) formulation of the *equivalence of ensembles*. As stated it justifies the use of the canonical ensemble to compute thermodynamic quantities from the canonical rather than the micro-canonical partition function, i.e. it allows us to define the free energy in terms of the logarithm of the partition function and to derive all thermodynamic quantities (including the entropy) from it via Legendre transformation. It is important to note that this equivalence holds in the limit of infinite particle number (and in consequence, infinite volume, energy, etc.). Thus, we encounter, for the first time, the notion of the *thermodynamic limit*. Then linking statistical mechanics to thermodynamics, we are really only interested in understanding what happens when the size of our systems tends to infinity. We will have to discuss this issue in far greater detail later on.

In the course of the proof of Theorem 2.3.1 we have seen (in spite of the fact that we have been rather careless) that more is true than just the fact that the free energy can be computed from the canonical partition function. Rather, we see that the Gibbs measure, even if it is a-priori supported on all possible values of the energy, really is concentrated on those states whose energy is very close to the preferred value $e^*(v, \beta)$. In fact, we should expect that

$$\mathcal{G}_{\beta, V, N} \sim \rho_{E^*(\beta, V, N), V, N} \tag{2.59}$$

in an appropriate sense when N tends to infinity. But to discuss such a question with some precision requires a more profound understanding of the meaning of the limit $N \uparrow \infty$ for measures on the phase space, a question that we will address only in Chapter 4.

The beauty of the equivalence of ensembles is that, computationally, it is much easier (even though still hard enough) to work with the Gibbs measure than with the micro-canonical measure. This should not be a surprise: working with constraints is always hard, and the canonical ensemble allows us to get rid of one annoying constraint, namely to keep the energy fixed. And the nice feature of the theorem is that it tells us that not fixing the energy is fine, because this will be taken care of effectively automatically.

Example: The classical ideal gas   Here the Hamiltonian is

$$H_N(p,q) = \sum_{i=1}^{N} \frac{p_i^2}{2m} \tag{2.60}$$

Thus, the canonical partition function is

$$\begin{aligned} Z_{V,\beta,N} &= \frac{1}{N!} \int d^N p \int_0^V d^N q \exp\left(-\beta \sum_{i=1}^{N} \frac{p_i^2}{2m}\right) \\ &= \frac{1}{N!} \left[\int dp \int_0^V dq \exp\left(-\beta \frac{p^2}{2m}\right) \right]^N \\ &= \frac{1}{N!} V^N [2\pi m/\beta]^{N/2} \end{aligned} \tag{2.61}$$

We see that this computation does not require knowledge of the formula for the surface area of the N-dimensional sphere, which we used in the micro-canonical formula. Stirling's formula states that

$$N! \sim \sqrt{2\pi N} e^{-N} N^N \tag{2.62}$$

and so

$$f(\beta, v) = \frac{-1}{\beta N} \ln Z_{V,\beta,N} \sim -\beta^{-1} \ln(e\sqrt{2\pi m/\beta} v) \tag{2.63}$$

Exercise: Compute the entropy for the one-dimensional gas from this formula. Compute the entropy directly from the micro-canonical partition function and compare. Do the same for the three-dimensional ideal gas.

2.4 Non-ideal gases in the canonical ensemble

The remarkable simplicity with which we have computed the free energy in the ideal gas could encourage us to look at non-ideal gases. Suppose we are given a Hamiltonian function

$$H_N(x) = \sum_{i=1}^{N} \frac{p_i^2}{2m} + \Phi(q_1, \ldots, q_N) \tag{2.64}$$

where, reasonably, Φ could represent a pair interaction potential of the form

$$\Phi(q_1, \ldots, q_N) = \sum_{i \neq j}^{N} \phi(q_i - q_j) \tag{2.65}$$

The pair interaction, ϕ, should incorporate at least some short-range repulsion, and possibly some weak long-range attraction. The simplest choice would be a hard-core exclusion, that just forbids the particles to penetrate each other:

$$\phi_a^{\text{h.c.}}(q) = \begin{cases} 0, & \text{if } |q| > a \\ +\infty, & \text{if } |q| \leq a \end{cases} \tag{2.66}$$

What about the partition function in this case? We have

$$Z_{\beta,V,N} = \frac{1}{N!} \int_V dq \int dp \exp\left(-\beta \sum_{i=1}^{N} \frac{p_i^2}{2m} - \beta \Phi(q_1, \ldots, q_N)\right) \tag{2.67}$$

We may feel encouraged by the fact that half of the integrals can immediately be computed:

$$\int dp \exp\left(-\sum_{i=1}^{N} \frac{p_i^2}{2m}\right) = \left(\frac{2\pi m}{\beta}\right)^{dN/2} \tag{2.68}$$

where d is the number of spatial dimensions, so that

$$Z_{\beta,V,N} = \frac{1}{N!} \left(\frac{2\pi m}{\beta}\right)^{dN/2} \int_V dq \exp(-\beta \Phi_N(q)) \tag{2.69}$$

which we could express as

$$Z_{\beta,V,N} = \frac{1}{N!} \left(\frac{2\pi m}{\beta}\right)^{dN/2} Z_{\beta,V,N}^{\text{red}} \tag{2.70}$$

where the *reduced partition function* is

$$Z_{\beta,V,N}^{\text{red}} \equiv \frac{1}{N!} \int_V dq \exp(-\beta \Phi_N(q)) \tag{2.71}$$

We see that in the non-ideal gas, we can reduce the computation of the partition function to that of a partition function involving only the positions of the particles. Of course, this gain is limited, since we cannot compute even this reduced partition function, except in very special cases.

One of these misleadingly simple cases is the one-dimensional hard-core gas. Here we have

$$Z_{\beta,V,N}^{\text{red}} = \frac{1}{N!} \int_V dq_1 \cdots dq_N e^{-\beta \sum_{i \neq j} \phi_a^{\text{hc}}(q_i - q_j)} \tag{2.72}$$

Note that the integrand takes only two values: one, if all particles are at distance at least a apart from each other, and zero else. Now in one dimension, it is easy to see how to evaluate this integral. First, there are $N!$ ways $i_1, \ldots, i_N$ to arrange the particles such that $q_{i_1} < \cdots < q_{i_N}$, each of which contributes in the same amount to the partition functions.

Then each distance between consecutive particles must be at least a. Thus

$$Z^{\text{red}}_{\beta,V,N} = \int_{(N-1)a}^{V} dq_N \int_{(N-2)a}^{q_N-a} dq_{N-1} \cdots \int_{a}^{q_3-a} dq_2 \cdots \int_{0}^{q_2-a} dq_1 \qquad (2.73)$$

Changing variables to $y_i = q_i - (i-1)a$, this can be written as

$$\begin{aligned}
Z^{\text{red}}_{\beta,V,N} &= \int_0^{V-(N-1)a} dy_N \int_0^{y_N} dy_{N-1} \cdots \int_0^{y_3} dy_2 \int_0^{y_2} dy_1 \\
&= \int_0^{V-(N-1)a} dy_N \int_0^{y_N} dy_{N-1} \cdots \int_0^{y_3} dy_2\, y_2 \qquad (2.74) \\
&= \int_0^{V-(N-1)a} dy_N \int_0^{y_N} dy_{N-1} \cdots \int_0^{y_4} \frac{y_3^2}{2} \\
&= \int_0^{V-(N-1)a} dy_N \frac{y_N^{N-1}}{(N-1)!} = \frac{(V - a(N-1))^N}{N!}
\end{aligned}$$

Thus, with $V = vN$,

$$N^{-1} \ln Z^{\text{red}}_{\beta,V,N} = \ln N(v - a(1-1/N)) - N^{-1} \ln N! \sim \ln(v-a) + 1 \qquad (2.75)$$

and so the full free energy of the hard-core gas in one dimension is (in the limit $N \uparrow \infty$)

$$f(\beta, v) = -\beta^{-1}(\ln(v-a) + 1) - \frac{1}{2\beta} \ln\left(\frac{2\pi m}{\beta}\right) \qquad (2.76)$$

We see that f has a singularity at $v = a$, which is natural, as we cannot pack too many particles into too small a volume. Moreover, we see that

$$\frac{\partial f(\beta, v)}{\partial v} = -\beta^{-1} \frac{1}{v-a} \qquad (2.77)$$

Now thermodynamically, the negative of this derivative is the pressure, i.e. the equation of state is

$$p = \beta^{-1} \frac{1}{v-a} \qquad (2.78)$$

Naturally, the pressure tends to infinity as the volume is filled up.

Not an exercise: Try to repeat the computations for dimensions larger than one.

2.5 Existence of the thermodynamic limit

When we introduced the canonical ensemble, we were assuming that the entropy per particle has a limit, as the size of the system tends to infinity. We have seen in the case of the ideal gas (and also in the one-dimensional hard-core gas) that such limits do exist. An important question for establishing the correspondence between thermodynamics and statistical mechanics is to what extent the existence of such limits is general. We will discuss the issue of thermodynamic limits at length later. At this point we want to consider the weakest version that relates only to the validity of the thermodynamic formalism.

In fact, in the preceding discussion we have been careless about the meaning of the variable N. When we introduced thermodynamics, we mentioned that this variable represents the amount of mass, measured originally in moles. At the same time we alluded to the fact that this number may also measure the number of molecules, and it was in this capacity that we treated N as a large number as soon as we started to discuss thermodynamics. It is time to make it clear that these two uses of the same variable are misleading, and that these two notions of mass are quite different.

Let us first note that in thermodynamics the variable N is (as long as we consider systems containing a single type of molecule) redundant due to the assumptions that all extensive quantities are homogeneous functions of degree one in the variable N. Thus, e.g., $E(V, S, N) = Ne(V/N, S/N) = e(v, s)$, $S(E, V, N) = Ns(V/N, E/N) = s(e, v)$, etc. The quantities $v, e, s, \ldots$ are often called specific volume, energy, entropy, etc. By this assumption, thermodynamics is really only concerned with these functions.

When introducing statistical mechanics, we had been defining entropy or free energy in terms of logarithms of partition functions with a given number of particles. This was actually imprecise. If we claim that $S(E, V, N) = \ln z_{E,V,N}$, even in the simplest example it is not strictly true that this function will be strictly a homogeneous function in N. Therefore, the true relation between statistical mechanics and thermodynamics stipulates such relations 'to leading order in N', where N is now really the number of particles. In other words, the proper relation between the thermodynamic quantities and the objects of statistical mechanics is more like

$$\lim_{N \uparrow \infty} \frac{1}{N} \ln z_{eN, vN, N} = s(e, v) \qquad (2.79)$$

respectively

$$\lim_{N \uparrow \infty} \frac{-1}{\beta N} \ln Z_{\beta, \iota N, N} = f(\beta, v) \qquad (2.80)$$

This formulation gives rise to a number of questions. The first is under which circumstances can we expect such limits to exist? The second is related to the meaning of the volume variable. When computing partition functions, we have to specify, in principle, not only the value of the 'volume' of our system, but also its precise shape (e.g. a cylinder, a cube, a sphere), as this may influence the result. On the other hand, thermodynamics does not usually concern itself too much with such shapes. For this to make sense, the limits in (2.79) and (2.80) should not be terribly dependent on the shape of the volumes of the systems along which the limit is taken. In fact, for systems with short-range interactions, it can be shown that this is true provided that the volumes are reasonable in the sense (of van Hove) that the ratio of surface to volume goes to zero.

At this point we will present one of the earliest proofs in the context of non-ideal gases. The idea goes back to van Hove [257] and the complete proof was given by Ruelle [213] and Fisher [102].

The system we will consider is a gas with Hamiltonian of the form (2.64) with a pair interaction of the form (2.65). We will consider a pair interaction with hard-core repulsion and bounded, finite-range attraction, i.e. we assume that there are real numbers, $0 < r_0 < b < \infty$, and $\epsilon > 0$, such that

$$-\epsilon < \phi(q) = \begin{cases} +\infty, & |q| < r_0 \\ \leq 0, & r_0 \leq |q| \leq b \\ = 0, & |q| > b \end{cases} \qquad (2.81)$$

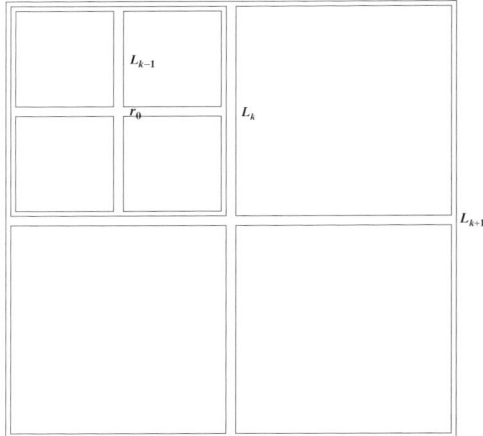

Figure 2.4 Nested sequence of boxes V_{k-1}, V_k, V_{k+1}.

Now consider a sequence of cubic boxes, V_k, of side-lengths $L_{k+1} = 2L_k + 2r_0$, as shown in Figure 2.4. This allows us to place 2^d boxes V_k into V_{k+1} in such a way that their distances from each other are r_0 and from the boundary are $r_0/2$. We choose $N_{k+1} = 2^d N_k$.

We then have that

$$\lim_{k\uparrow\infty} \frac{|V_k|}{N_k} = \lim_{k\uparrow\infty} \frac{V_0}{N_0} \prod_{l=0}^{k} \left(1 + \frac{2r_0}{L_l}\right)^d \equiv v \qquad (2.82)$$

Here the last limit exists by monotonicity and the trivial observation that $L_l > 2^l L_0$, so that (using $1 + x \leq e^x$)

$$\prod_{l=0}^{k} \left(1 + \frac{2r_0}{L_l}\right) \leq \exp\left(\sum_{l=0}^{\infty} 2r_0/L_0 2^{-l}\right) \leq \exp(4r_0/L_0) \qquad (2.83)$$

The (reduced) partition function in the $(k+1)$st step is then

$$Z_{\beta, V_{k+1}, N_{k+1}} = \frac{1}{N_{k+1}!} \int_{V_{k+1}} \exp\left(-\beta \sum_{1 \leq i \neq j \leq N_{k+1}} \phi(q_i - q_j)\right) dq_1 \cdots dq_{N_{k+1}} \qquad (2.84)$$

The key idea is now to get a lower bound by restricting the integrals over the q_i such that the particles may only be in the 2^d boxes, V_k, contained in V_{k+1} and to insist, moreover, that the number of particles in each one is equal to N_k. There is an obvious combinatorial number, $N_{k+1}!/N_k!^{2^d}$, of possible arrangements of the particles in the different boxes. Note that there is no *positive* interaction between the particles in the different sub-boxes, while the negative (attractive) interactions only increase the value of $Z_{\beta, V_{k+1}, N_{k+1}}$, compared to the situation when all interactions between these boxes are removed. These considerations show that

$$Z^{\text{red}}_{\beta, V_{k+1}, N_{k+1}} \geq \left(Z^{\text{red}}_{\beta, V_k, N_k}\right)^{2^d} \qquad (2.85)$$

2.5 Existence of the thermodynamic limit

and hence

$$a_{k+1} \equiv \frac{1}{N_{k+1}} \ln Z^{\text{red}}_{\beta, V_{k+1}, N_{k+1}} \geq \frac{1}{N_k} \ln Z^{\text{red}}_{\beta, V_k, N_k} = a_k \qquad (2.86)$$

Thus, the sequence of numbers a_k is increasing and will converge, if it is bounded from above. In fact, the only thing that might prevent this from being true is the possibility that the potential energy, for some configurations, could go to $-\infty$ faster than CN_k. Namely, the formula for the potential energy involves, in principle, N^2 terms, $\phi(q_i - q_j)$, and if all of them contributed $-\epsilon$, we would be in rather bad shape. This could happen if there were no repulsive part of the interaction, since then all particles might get very close to each other. However, due to the repulsive hard core, particles cannot get closer than a distance r_0 to each other, and thus the number of particles within the finite range b of the attractive interaction is bounded by $c(b - r_0)^d$, so that

$$\sum_{1 \leq i \neq j \leq N} \phi(q_i - q_j) \geq -c\epsilon(b - r_0)^d N \equiv -BN \qquad (2.87)$$

Thus

$$Z^{\text{red}}_{\beta, V_k, N_k} \leq \frac{1}{N_k!} \int_{V_k} dq_1 \cdots dq_{N_k} e^{\beta B N_k} \qquad (2.88)$$

$$\leq e^{N_k} N_k^{-N_k} |V_k|^{N_k} e^{\beta B N_k} \leq e^{(\beta B + 1) N_k} v^{N_k}$$

and so

$$a_k \leq (\beta B + 1) + \ln v < \infty \qquad (2.89)$$

This proves the convergence of the free energy along the special sequences N_k. It is, however, not very difficult to show that this implies convergence along arbitrary sequences, provided the shapes of the boxes are such that volume dominates surface.

Also, both the hard-core and the finite-range conditions of the potential can be relaxed. In fact it is enough to have (i) a uniform lower bound, (ii) an asymptotic lower bound $\phi(q) \geq C|q|^{-d-\epsilon}$, for some $\epsilon > 0$, as $|q| \downarrow 0$, and (iii) an asymptotic upper bound $|\phi(q)| \leq C|q|^{-d-\epsilon}$, for some $\epsilon > 0$, as $|q| \uparrow \infty$. Note that these conditions are not satisfied if the only forces present are gravity and the electrostatic forces. Fortunately, due to quantum mechanical effects, the *effective interactions* between molecules are usually less dangerous. Still, the stability condition (2.87) is quite generally a problem when working with interacting gases.

Convexity The proof of convergence outlined above yields almost as a by-product another important property of the free energy, namely convexity. Convexity of thermodynamic potentials was a postulate of thermodynamics, and is crucial for the equations of state to define single-valued functions (as long as it is strict). Certainly, convexity should be a consequence of statistical mechanics.

We will show that in our gas the free energy is convex as a function of v. To this end we use the same partition of the volume V_{k+1} as before, but this time we choose the number of particles in the different cubes to be not uniform, but instead put into half of them $N_k^1 = \rho_1 N_k$ and in the other half $N_k^2 = \rho_2 N_k$ particles. By the same argument as before, we obtain that

$$Z^{\text{red}}_{\beta, V_{k+1}, (\rho_1+\rho_2)N_{k+1}/2} \geq \left(Z^{\text{red}}_{\beta, V_k, \rho_1 N_k}\right)^{2^{d-1}} \left(Z^{\text{red}}_{\beta, V_k, \rho_2 N_k}\right)^{2^{d-1}} \quad (2.90)$$

and hence

$$\frac{1}{N_{k+1}} \ln Z^{\text{red}}_{\beta, V_{k+1}, (\rho_1+\rho_2)N_{k+1}/2} \geq \frac{1}{2}\left(\frac{1}{N_k} \ln Z^{\text{red}}_{\beta, V_k, \rho_1 N_k} + \frac{1}{N_k} \ln Z^{\text{red}}_{\beta, V_k, \rho_2 N_k}\right) \quad (2.91)$$

Since we know that $\lim_{k \uparrow \infty} \frac{1}{N_k} \ln Z^{\text{red}}_{\beta, v N_k, N_k} \equiv a(\beta, v)$ exists, it follows from (2.91) that

$$\frac{\rho_1 + \rho_2}{2} a(\beta, 2v/(\rho_1 + \rho_2)) \geq \frac{\rho_1}{2} a(\beta, v/\rho_1) + \frac{\rho_2}{2} a(\beta, v/\rho_2) \quad (2.92)$$

In other words, the function $g(\rho) \equiv \rho a(\beta, v/\rho)$ satisfies

$$g((\rho_1 + \rho_2)/2) \geq \frac{1}{2}(g(\rho_1) + g(\rho_2)) \quad (2.93)$$

Thus, g is a concave function of its argument (the inverse volume, resp. the density).

Exercise: Show that the concavity of g implies that $-a(\beta, v)$ is a convex function of v, and that thus the free energy, $f(\beta, v)$, of the class of gases considered above is a convex function of the (specific) volume. Use a different (and simpler) argument to show that the free energy is also a convex function of the temperature.

2.6 The liquid–vapour transition and the van der Waals gas

Convexity of the free energy implies that the pressure is a decreasing function of the volume. As long as f is strictly convex, the pressure is strictly increasing, and thus the function $p(v, T)$ is uniquely invertible. We have already alluded to the fact that a first-order phase transition occurs if thermodynamic potentials are not strictly convex, i.e. contain linear pieces. In our case, if for some temperature the free energy is linear on an interval $[v_1, v_2]$, this would imply that the pressure $p(v, T)$ is a constant p_0 as v is varied over this interval and, as a consequence, the inverse function is not uniquely defined: for this value of the pressure, v could be anywhere in $[v_1, v_2]$. Moreover, as p is varied over p_0, the volume (resp. the density) jumps suddenly from v_1 to v_2. This is what is actually observed in real gases: at least if the temperature is sufficiently low, there exists a critical value of the pressure at which the gas *condenses*, i.e. transforms itself into a considerably more dense *phase*, called a liquid. Such singular behaviour is called a (first-order) *phase transition*. Phase transitions are maybe the most exciting aspect of thermodynamics, since they are something quite out of the ordinary from the perspective of classical mechanics. They represent something totally new and specific to thermodynamic systems.

In the context of thermodynamics, it is easy to produce systems with phase transitions: just choose appropriate thermodynamic potentials. It is an altogether more difficult matter to reproduce phase transitions from statistical mechanics, and indeed the issue of whether this was possible continued to be debated until about the middle of the twentieth century.

So far, all the thermodynamic potentials we have computed have been strictly convex. Neither the ideal gas nor the hard-core gas in dimension one show any sign of a phase

2.6 The liquid–vapour transition and the van der Waals gas

transition. On the other hand, the van Hove gas we discussed above seems to incorporate all the main features of a real gas, and thus it should show a phase transition. Unfortunately, we cannot compute its free energy (well, we haven't really tried, and we also have no idea how we could do this. But many people have tried and there is no proof as of today that there is something like a liquid–vapour phase transition in this gas[8]). The difficulties related to classical non-ideal gases will soon lead us away to more manageable systems, but before moving there, I will at least discuss one example of a non-ideal gas that shows such a phase transition. This is the classical van der Waals gas.

The van der Waals gas was introduced in the thesis of that person in 1873 in the context of thermodynamics. That is, he proposed to modify the equation of state of the hard-core gas by adding an extra term that was to take into account the attractive part of the interaction, writing

$$\beta p = \frac{1}{v-a} - \frac{\beta}{2}\alpha v^{-2} \tag{2.94}$$

I do not know how he got his thesis accepted, because this equation violates one of the basic principles of thermodynamics, the monotonicity of pressure. But, as we will see, it can somehow produce a phase transition, and it can be obtained from statistical mechanics.

The derivation of the van der Waals equation was proposed by Ornstein in 1908, also in his thesis. He suggested introducing a potential consisting of the hard-core repulsion we have already discussed and a very long-range attraction,

$$\Phi_V^{\text{attr}}(q) \equiv -\frac{1}{V}\sum_{i\neq j}\alpha/2 = -\alpha\frac{N^2}{2V} \tag{2.95}$$

Since this term is totally independent of the configuration q, we get that

$$Z_{\beta,V,N}^{\text{red,vdw}} = e^{+\beta\alpha N/2v} Z_{\beta,V,N}^{\text{red,hc}} \tag{2.96}$$

and so

$$f_{\text{vdw}}(v,\beta) = f_{\text{hc}}(v,\beta) - \frac{\alpha}{2v} \tag{2.97}$$

At least in one dimension we can compute the free energy of the hard-core gas and thus, in dimension one,

$$f_{\text{vdw}}(v,\beta) = -\beta^{-1}(\ln(v-a) + 1) - \frac{\alpha}{2v} - \beta^{-1}\frac{1}{2}\ln\left(\frac{2\pi m}{\beta}\right) \tag{2.98}$$

from which the van der Waals equation of state (2.94) follows immediately. The function $f_{\text{vdw}}(v,\beta)$ is shown in Figure 2.5.

In dimension $d > 1$ we cannot compute the free energy of the hard-core gas, but one might accept that it will look similar to the one-dimensional case. Thus, the general conclusions should remain valid.

One can easily check that the free energy, f_{vdw}, is in general not a convex function of the volume, and that the equation of state (2.94) does not give p as a monotone function of v. Thus, we cannot invert this to obtain v as a function of p; in fact, there are values of p for

---

[8] There is, however, a proof of the existence of such a transition in a very special situation which roughly mimics such a gas, due to Lebowitz, Mazel, and Presutti [164] that dates from 1999.

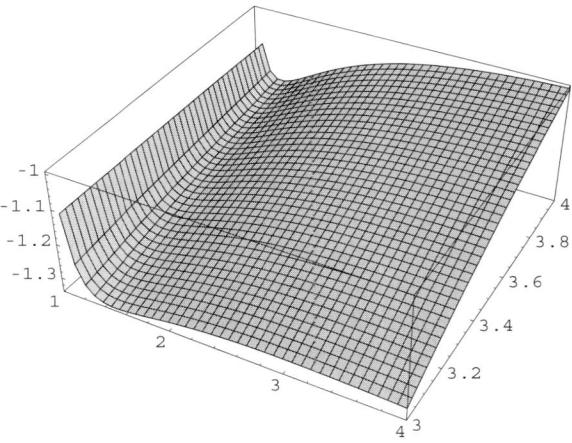

Figure 2.5 Free energy surface in the van der Waals gas.

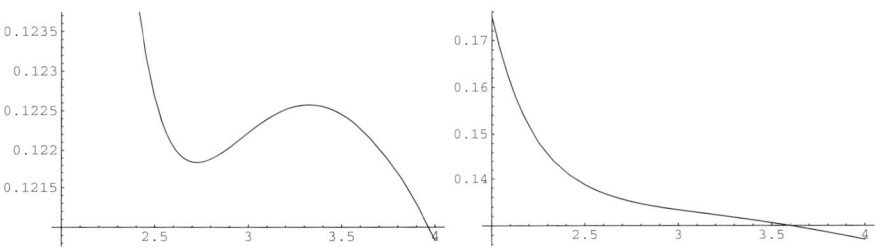

Figure 2.6 Subcritical and supercritical isotherms.

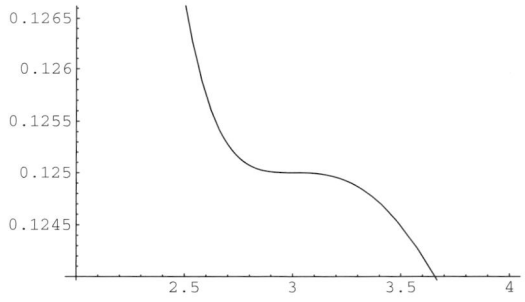

Figure 2.7 Critical isotherm.

which there are three possible values of the volume (see Figs. 2.6 and 2.7). Moreover, the pressure as a function of the volume will sometimes decrease. This is not something that anyone has ever observed in a real gas. We have a problem ...

Maxwell, in 1874, just a year after van der Waals' thesis, corrected the van der Waals theory by stating (in a few more words) that the correct free energy should be taken as the convex hull (the convex hull of a function f is the largest convex function that is less than or equal to f; it can also be obtained as the twice iterated Legendre transform of f) of the free energy f_{vdw}. It took until 1963 for it to be understood (by Kac, Uhlenbeck, and

Hemmer [147]) how this *Maxwell construction* can be derived from statistical mechanics as well. For an in-depth treatment of this theory, see the recent monograph by E. Presutti [209].

2.7 The grand canonical ensemble

Having seen the computational advantage in removing the constraint of fixed energy, it is very natural to also remove other constraints in the same way and to replace them by an exponential weighting factor. A popular ensemble that is obtained in this way is the *grand canonical ensemble*, where the particle number is no longer fixed, and instead a chemical potential is introduced in the definition of the partition function. We define the grand canonical measure on the space

$$\Omega \equiv \cup_{N=1}^{\infty} P^{\otimes N} \tag{2.99}$$

as

$$\tilde{\mathcal{G}}_{\beta,V,\mu}(dx, N) = \frac{e^{\mu\beta N} \mathcal{G}_{\beta,V,N}(dx)}{\sum_{N=1}^{\infty} e^{\mu\beta N} Z_{\beta,V,N}} \tag{2.100}$$

where the denominator is called the *grand canonical partition function*,

$$\mathcal{Z}_{\beta,V,\mu} \equiv \sum_{N=1}^{\infty} e^{\beta\mu N} Z_{\beta,V,N} \tag{2.101}$$

Clearly we will have an analogous statement to that of Theorem 2.3.1 that will affirm that the logarithm of the grand canonical partition function is related to a thermodynamic potential. However, this is a little tricky for two reasons. First, we sum over N, so it is not quite clear at first how we should pass to the thermodynamic limit. Second, we have to be careful in noting that we keep volume fixed while we sum over N. The second observation also provides the answer to the first problem: We would think of the thermodynamic limit this time as being related to letting the volume go to infinity, i.e. we will think of the volume as $V = vM$, where M is taken to infinity, and of N as $N = xM$, where $x = N/M$ runs from $1/M$ to infinity in steps of $1/M$, as N is summed over. Thus, we write

$$\mathcal{Z}_{\beta,vM,\mu} = \sum_{x=1/M}^{\infty} \exp(M\beta(\mu x - M^{-1}F(\beta, vM, Mx))) \tag{2.102}$$

The main contributions to the sum will come from values of x where the exponent has a maximum. Assume that the thermodynamic limit of the free energy exists. Then, for v and x fixed,

$$\frac{1}{M}F(\beta, vM, Mx) \to xf(\beta, v/x) \tag{2.103}$$

Thus, the leading part of the exponent has a maximum when

$$\mu = \frac{\partial}{\partial x} xf(\beta, v/x) = f(\beta, v/x) + p(\beta, v/x)v/x \tag{2.104}$$

Note that this equation fixes x, and that x/v is the mean number of particles per volume in the grand canonical ensemble. We arrive at the conclusion that

$$\lim_{V\uparrow\infty} \frac{1}{\beta V} \ln \mathcal{Z}_{\beta,V,\mu} = p \qquad (2.105)$$

i.e. the thermodynamic potential associated with the grand canonical ensemble can be thought of as pressure as a function of the chemical potential, the volume, and the temperature.

Exercise: Formulate a precise analogue of Theorem 2.3.1 that will yield the assertion (2.105).

Exercise: Compute the pressure of an ideal gas in the grand canonical ensemble and derive the ideal gas law.

3
Lattice gases and spin systems

It has long been known that iron, when raised to a certain 'critical temperature' corresponding to dull red heat, loses its susceptibility and becomes magnetically indifferent, or, more precisely, is transformed from a ferromagnetic to a paramagnetic body.

Magnetism, Encyclopedia Britannica, 11th edn.

Dealing with non-ideal gases brings a number of complications that are partly due to the continuous nature of the configuration space, combined with the need for competing interactions to account for fundamental features of realistic gases. The desire to create models where these technical aspects are simplified has led to the introduction of the idea of a *lattice gas*. The ideal lattice gas was already encountered in Chapter 2. We now turn to the study of *interacting* lattice gases.

3.1 Lattice gases

We will now proceed to introduce a lattice gas that will take into account hard-core repulsion as well as attraction. We fix a subset $V \subset \mathbb{Z}^d$ of a d-dimensional lattice. The hard-core repulsion will be taken into account by imposing the condition that each site of the lattice can be occupied by at most one particle. For an allowed configuration of particles, we then define the Hamiltonian

$$H_V(x_1, \ldots, x_N) = \sum_{i \neq j} \phi(x_i, x_j) \qquad (3.1)$$

where ϕ is some pair interaction. The canonical partition function is

$$Z_{\beta,V,N} = \frac{1}{N!} \sum_{\substack{x_1,\ldots,x_N \in V \\ x_i \neq x_j, \forall i \neq j}} e^{-\beta H_V(x_1,\ldots,x_N)} \qquad (3.2)$$

The nice feature is that the constraint $x_i \neq x_j, \forall i \neq j$, is easily taken care of by using a different parametrization of the state space. Namely, each possible configuration of particles satisfying this constraint is equivalent, up to permutation of the labels, to a function n_x, $x \in V$, that takes the value 1 if a particle is at site x, and the value 0 otherwise. Moreover, the energy of a configuration of particles corresponding to occupation numbers n can be expressed as

$$\tilde{H}_V(n) = \sum_{x \neq y \in V} \phi(x, y) n_x n_y \qquad (3.3)$$

Thus, the partition function can be written as

$$Z_{\beta,V,N} = \sum_{\substack{n_x \in \{0,1\}, x \in V \\ \sum_{x \in V} n_x = N}} e^{-\beta \tilde{H}_V(n)} \tag{3.4}$$

This formulation still involves a constraint, $\sum_{x \in V} n_x = N$, but this can be dealt with easily by passing to the grand canonical partition function

$$\mathcal{Z}_{\beta,V,\mu} = \sum_{n_n \in \{0,1\}, x \in V} e^{-\beta \tilde{H}_V(n) - \mu \sum_{x \in V} n_x} \tag{3.5}$$

This is the standard formulation of the partition function of a lattice gas. The most popular version of it is the *Ising lattice gas*, where $\phi(x, y)$ is taken as a so-called nearest-neighbour interaction:

$$\phi_{nn}(x, y) = \begin{cases} -J, & \text{if } |x - y| = 1 \\ 0, & \text{otherwise} \end{cases} \tag{3.6}$$

3.2 Spin systems

In his Ph.D. thesis in 1924, Ernst Ising[1] [144, 145] attempted to solve a model, proposed by his advisor Lenz, intended to describe the statistical mechanics of an interacting system of magnetic moments. I will not discuss the derivation of this model from quantum mechanics, but present it as a heuristic model for magnetism. The setup of the model proceeds again from a lattice, $\mathbb{Z}^d$, and a finite subset, $V \subset \mathbb{Z}^d$. This time, the lattice is more justifiable than before, since it is supposed to represent the positions of the atoms in a regular crystal. Each atom is endowed with a magnetic moment that is quantized and can take only the two values $+1$ and -1, called the *spin* of the atom. This spin variable at site $x \in V$ is denoted by σ_x. The spins are supposed to interact via an interaction potential $\tilde{\phi}(x, y)$; in addition, a magnetic field h is present. The energy of a *spin configuration* is then

$$H_V(\sigma) \equiv - \sum_{x \neq y \in V} \tilde{\phi}(x, y) \sigma_x \sigma_y - h \sum_{x \in V} \sigma_x \tag{3.7}$$

We will see that this model is mathematically equivalent to a lattice gas model. If we make the change of variables $n_x = (\sigma_x + 1)/2$, we can express $\tilde{H}_V(n)$ as a function of σ,

$$\tilde{H}_V((\sigma + 1)/2) = \sum_{x \neq y \in V} \frac{1}{4} \phi(x, y) \sigma_x \sigma_y + \sum_{x \in V} \sigma_x \sum_{y \neq x \in V} \phi(x, y) + \frac{1}{4} \sum_{x \neq y \in V} \tilde{\phi}(x, y) \tag{3.8}$$

We see that (up to the irrelevant constant term) this is exactly of the same form as the Hamiltonian of the spin system. In particular, in the case of the nearest-neighbour Ising lattice gas, we get

$$\tilde{H}_V((\sigma + 1)/2) = - \sum_{x, y \in V, |x-y|=1} \frac{J}{4} \sigma_x \sigma_y - 2dJ \sum_{x \in V} \sigma_x - J\frac{d}{2}|V| \tag{3.9}$$

---

[1] An account of the life of Ising can be found in [151] and is definitely worth reading.

where we only cheated a little bit with the terms next to the boundary of V, where the number of neighbours is not quite $2d$.

Since the lattice gas partition functions can be written as partition functions of the spin system, from our point of view the two settings are completely equivalent, and we can work in one or the other. We will mostly prefer the language of lattice spin systems, which quickly will lead to a far richer class of models.

The spin system with Hamiltonian (3.7) with the particular choice

$$\tilde{\phi}(x, y) = \begin{cases} J, & \text{if } |x - y| = 1 \\ 0, & \text{otherwise} \end{cases} \quad (3.10)$$

is known as the Ising spin system or *Ising model*. This model has played a crucial rôle in the history of statistical mechanics.

Magnetic systems have a different terminology than gases. We have already seen that the parameter that corresponds to the chemical potential in the lattice gas is here the *magnetic field*, h. The extensive variable conjugate to it is the *magnetization*, $M = \sum_{\in V} \sigma_i$, which from the lattice gas point of view corresponds to the particle number. Since magnetization times magnetic field is an electromagnetic energy, one likes to think of h and M as the magnetic analogues of p and V, whereas the size of the system, $|V|$, is the analogue of the particle number. Therefore, one usually thinks of the setting we have described, with V fixed, as a canonical partition function, rather than a grand-canonical one. The logarithm of the partition function is called the Helmholtz free energy, which is strange, since as a function of the intensive variable h it is more like a Gibbs free energy.[2] Anyway, we will call

$$F_{\beta, h, V} \equiv -\frac{1}{\beta} \ln Z_{\beta, h, V} \quad (3.11)$$

the *free energy* of the spin system.

The Ising model represent a decisive turn in the development of statistical mechanics in several ways. The most important one is that the model was invented with the clear intention of understanding a phenomenon in material science, *ferromagnetism*, that was very hard to understand from basic principles. This was quite a different ambition than the justification, or possibly derivation, of thermodynamics. The second crucial turn was the fact that the Ising model involved a considerable simplification on the level of the description of the basic microscopic degrees of freedom, reducing the state of an atom to a variable taking two values ± 1, and replacing all the complicated electromagnetic (and quantum) interactions by a simple attraction between nearest neighbours on the lattice, while still hoping to adequately reproduce the essential features of the a phenomenon that is to be understood. Today, we would justify such modelling by a *universality hypothesis*, claiming that the *collective phenomena* to be modelled are universal for wide classes of models and depend only on a few parameters, such as dimensionality, global symmetries, etc. This point of view has proven enormously successful in statistical mechanics, and

---

[2] R. Kotecký has pointed out to me that the reason for this terminology is given in the textbook by E. Stanley [229]. According to him, the terminology refers only to the classical thermodynamic variables, disregarding the magnetic ones. Then one could still think that there is a volume apart from the number of atoms in the lattice (think, e.g., of a magnetic gas or fluid), and what we now call volume remains a particle number.

without it, and the simple paradigmatic models it provoked (such as the Ising model), most of the progress of the last 80 years would not have been possible. Before we turn, in the next chapter, to the rigorous probabilistic setup of Gibbs measures for lattice spin systems, we will look at two singular situations that were studied in the early days of these models, and that gave rise to some confusion. They are the exact solution of the Ising model in one dimension, and the mean field version of the Ising model, the Curie–Weiss model of ferromagnetism.

3.3 Subadditivity and the existence of the free energy

Let us first give an instructive proof of the existence of the limit of the free energy in the Ising model. It will be useful to note that we can express the Hamiltonian in the equivalent form

$$\hat{H}_V(\sigma) = \sum_{x,y \in V} \tilde{\phi}(x,y)(\sigma_x - \sigma_y)^2 - h \sum_{x \in V} \sigma_x \tag{3.12}$$

which differs from H_V only by a constant. Now let $V = V_1 \cup V_2$, where V_i are disjoint volumes. Clearly we have that

$$Z_{\beta,V} = \sum_{\sigma_x, x \in V_1} \sum_{\tau_y, y \in V_2} \exp(-\beta \left[H_{V_1}(\sigma) + H_{V_2}(\tau) \right])$$

$$\times \exp\left(-\beta \sum_{x \in V_1} \sum_{y \in V_2} \tilde{\phi}(x,y)(\sigma_x - \tau_y)^2\right) \tag{3.13}$$

If $\tilde{\phi}(x,y) \geq 0$, this implies that

$$Z_{\beta,V} \leq Z_{\beta,V_1} Z_{\beta,V_2} \tag{3.14}$$

and therefore

$$-F_{\beta,V} \leq (-F_{\beta,V_1}) + (-F_{\beta,V_2}) \tag{3.15}$$

The property (3.14) is called *subadditivity* of the sequence $(-F_{\beta,V})$. The importance of subadditivity is that it implies convergence, through an elementary analytic fact:

Lemma 3.3.1 *Let a_n be a real-valued sequence that satisfies, for any $n, m \in \mathbb{N}$,*

$$a_{n+m} \leq a_n + a_m \tag{3.16}$$

Then $\lim_{n \uparrow \infty} n^{-1} a_n$ exists. If, moreover, $n^{-1} a_n$ is uniformly bounded from below, then the limit is finite.

By successive iteration, the lemma has an immediate extension to arrays:

Lemma 3.3.2 *Let $a_{n_1, n_2, \ldots, n_d}$, $n_i \in \mathbb{N}$, be a real-valued array that satisfies, for any n_i, $m_i \in \mathbb{N}$,*

$$a_{n_1+m_1, \ldots, n_d+m_d} \leq a_{n_1, \ldots, n_d} + a_{m_1, \ldots, m_d} \tag{3.17}$$

Then $\lim_{n\uparrow\infty}(n_1 n_2 \ldots, n_d)^{-1} a_{n_1,\ldots,n_d}$ exists.
If $a_n(n_1 n_2 \cdots, n_d)^{-1} a_{n_1,\ldots,n_d} \geq b > -\infty$, then the limit is finite.

Lemma 3.3.2 can be used straightforwardly to prove convergence of the free energy over rectangular boxes:

Proposition 3.3.3 *If the Gibbs free energy $F_{\beta,V}$ of a model satisfies the subadditivity property (3.15), and if $\sup_\sigma H_V(\sigma)/|V| \geq C > -\infty$, then, for any sequence V_n of rectangles,*

$$\lim_{n\uparrow\infty} |V_n|^{-1} F_{\beta,V_n} = f_\beta \tag{3.18}$$

exists and is finite.

Obviously this proposition gives the existence of the free energy for Ising's model, but the range of applications of Proposition 3.3.3 is far wider, and covers virtually all lattice spin systems with bounded and absolutely summable interactions. To see this, one needs to realize that strict subadditivity is not really needed, as error terms arising, e.g. from boundary conditions can easily be controlled. Further details can be found in Simon's book [224].

3.4 The one-dimensional Ising model

The thesis of E. Ising consisted in solving the one-dimensional version of the Ising model. The result was probably a disappointment for his advisor, for the solution failed to exhibit a phase transition, and so Ising (prematurely) concluded that the model was inadequate to explain ferromagnetism. It will be instructive, nonetheless, to go through this computation.

In the case $d = 1$, the Hamiltonian of the Ising model on a volume $V = \{1, \ldots, N\}$ can be written as

$$H_N(\sigma) = -J \sum_{i=1}^{N} \sigma_i \sigma_{i+1} - h \sum_{i=1}^{N} \sigma_i \tag{3.19}$$

Actually, there is a small problem that we need to discuss. It concerns the spins at the sites $i = 1$ and $i = N$. While all other spins have two neighbours, and both the terms $\sigma_{i-1}\sigma_i$ and $\sigma_i\sigma_{i+1}$ occur in the sum, for these indices one of these terms is missing. Thus, the question arises of how to deal with these boundary spins properly. We will see in the next section that this is a fundamental aspect of the problem, and we will introduce the general framework to deal with it. At the moment, we will avoid this issue in the simplest way by considering the model on a circle, i.e. we impose $\sigma_{N+1} = \sigma_1$. This is known as *periodic boundary conditions*. We will interpret (3.19) in this way. The partition function of the model then reads

$$Z_{\beta,h,N} = \sum_{\sigma_1=\pm 1,\ldots,\sigma_N=\pm 1} \exp\left(\beta J \sum_{i=1}^{N} \sigma_i \sigma_{i+1} + \beta h \sum_{i=1}^{N} \sigma_i\right)$$

$$= \sum_{\sigma_1=\pm 1,\ldots,\sigma_N=\pm 1} \prod_{i=1}^{N} \exp(\beta J \sigma_i \sigma_{i+1} + \beta h \sigma_i) \tag{3.20}$$

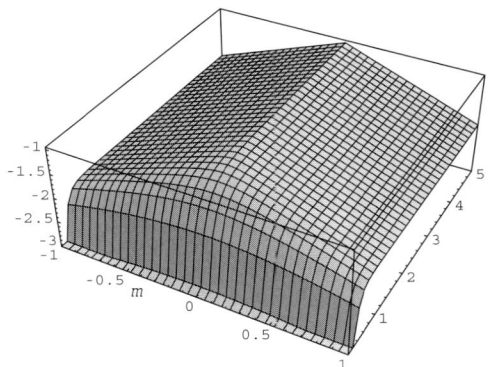

Figure 3.1 The free energy as a function of temperature and magnetic field.

Let us write, for $s, s' \in \{-1, 1\}$,

$$L(s, s') \equiv e^{\beta J s s' + \beta h s} \tag{3.21}$$

and think of it as the entries of a 2×2 matrix L (called the *transfer matrix*). Then we can write

$$Z_{\beta,h,N} = \sum_{\sigma_1 = \pm 1, \ldots, \sigma_N = \pm 1} L(\sigma_1, \sigma_2) L(\sigma_2, \sigma_3) \cdots L(\sigma_{N-1}, \sigma_N) L(\sigma_N, \sigma_1) = \mathrm{tr} L^N \tag{3.22}$$

But the trace of the matrix L^N is simply given by

$$\mathrm{tr} L^N = \lambda_1^N + \lambda_2^N \tag{3.23}$$

where λ_1, λ_2 are the two eigenvalues of the matrix L. The computation of the eigenvalues of a 2×2 matrix is a trivial exercise, and one gets

$$\lambda_1 = e^{\beta J} \cosh(\beta h) + \sqrt{e^{2\beta J} \sinh^2(\beta h) + e^{-2\beta J}} \tag{3.24}$$

$$\lambda_2 = e^{\beta J} \cosh(\beta h) - \sqrt{e^{2\beta J} \sinh^2(\beta h) + e^{-2\beta J}}$$

Since $\lambda_2/\lambda_1 < 1$, one sees easily that

$$\lim_{N \uparrow \infty} N^{-1} Z_{\beta,h,N} = \ln \lambda_1 \tag{3.25}$$

$$= \ln \left(e^{\beta J} \cosh(\beta h) + \sqrt{e^{2\beta J} \sinh^2(\beta h) + e^{-2\beta J}} \right)$$

$$= \beta J + \ln \left(\cosh(\beta h) + \sqrt{\sinh^2(\beta h) + e^{-4\beta J}} \right)$$

that is, the free energy, as depicted in Fig. 3.1, is given by the expression

$$f(\beta, h) = -J - \beta^{-1} \ln \left(\cosh(\beta h) + \sqrt{\sinh^2(\beta h) + e^{-4\beta J}} \right) \tag{3.26}$$

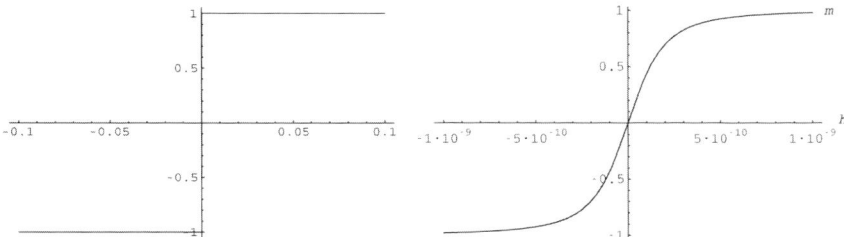

Figure 3.2 $m(h)$ at $\beta = 10$. Second plot with better resolution.

We can compute the magnetization

$$m = -\frac{\partial f}{\partial h} = \frac{\sinh(\beta h)}{\sqrt{\sinh^2(\beta h) + e^{-4\beta J}}} \quad (3.27)$$

which is a monotone and differentiable function of h, for any $0 \leq \beta < \infty$ (even if a plot with Mathematica will tend to look discontinuous if, e.g., $\beta J = 10$, as shown in Fig. 3.2).

What this result suggests is that there is no *spontaneous magnetization*. For zero external fields, the magnetization vanishes, even in the thermodynamic limit. It is not difficult to arrive at the conclusion that perhaps spontaneous magnetization is just an experimental error, and the appearance of a phase transition is misleading. The result also seems to support the following argument, which was used against the possibility of explaining phase transitions on the basis of statistical mechanics: in the Ising model, the partition function is clearly an analytic function of all parameters. Moreover, for real values of β and h, it is strictly positive, so its logarithm is also an analytic function, at least real analytic. Therefore, no jump in the derivative of the free energy can occur. The problem with this argument is that, in general, it does not survive the thermodynamic limit.

In any event, Ising drew the conclusion that something like a real phase transition, with a magnetization having a real jump-discontinuity at the values $h = 0$, cannot occur in his model.

3.5 The Curie–Weiss model

Already in 1907, Weiss [259], following the discovery of the critical temperature (Curie temperature), above which ferromagnetism disappears, by Pierre Curie in 1895, had developed a theory of ferromagnetism based on a spin system analogue of the van der Waals theory. This *Curie–Weiss model* can be cast into the language of the Ising model in a very natural way. All we need to do is to replace the nearest-neighbour pair interaction of the Ising model by another extreme choice, namely the assumption that each spin variable interacts with each other spin variable at any site of the lattice with exactly the same strength. In that case, the actual structure of the lattice becomes irrelevant, and we may simply take $V = \{1, \ldots, N\}$. The strength of the interaction should be chosen of order $1/N$, to avoid the possibility that the Hamiltonian takes on values larger than $O(N)$. Thus, the Hamiltonian of the Curie–Weiss model is

$$H_N(\sigma) = -\frac{1}{N} \sum_{1 \leq i,j \leq N} \sigma_i \sigma_j - h \sum_{i=1}^{N} \sigma_i \quad (3.28)$$

At this moment it is time to discuss the notion of *macroscopic variables* in some more detail. So far we have seen the magnetization, m, as a thermodynamic variable. It will be reasonable to define another magnetization as a *function* on the configuration space: we will call

$$m_N(\sigma) \equiv N^{-1} \sum_{i=1}^{N} \sigma_i \tag{3.29}$$

the *empirical magnetization*. Here we divided by N to have a specific magnetization. A function of this type is called a *macroscopic* function, because it depends on all spin variables, and depends on each one of them very little (we will make these notions more rigorous in the next section).

Note that the particular structure of the Curie–Weiss model entails that the Hamiltonian can be written as a function of this single macroscopic function:

$$H_N(\sigma) = -\frac{N}{2}[m_N(\sigma)]^2 - hNm_N(\sigma) \equiv N\Psi_h(m_N(\sigma)) \tag{3.30}$$

This can be considered as a defining feature of *mean field models*.

Digression Instead of considering the empirical magnetization one could study a closely related object, namely a probability distribution on the set $\{-1, 1\}$, called the *empirical spin distribution*,

$$\rho_N \equiv \frac{1}{N} \sum_{i=1}^{N} \delta_{\sigma_i} \tag{3.31}$$

If we think of the σ_i as random variables distributed according to, say, the Gibbs distribution, ρ_N is a *random probability measure*. Clearly, we have that

$$m_N(\sigma) = \int \rho_N(ds)s \equiv \rho_N(+1) - \rho_N(-1) \tag{3.32}$$

so that m_N determines ρ_N uniquely, and vice versa. This is, however, particular to the case where the spin variables take only two values. If one considers more general models, the empirical distribution contains more information than its mean value. The proper extension of the notion of mean field models to that case is then to consider Hamiltonians that are functions of the empirical distribution.

Let us now try to compute the free energy of this model. Because of the interaction term, this problem looks complicated at first. To overcome this difficulty, we do what would appear unusual from our past experience: we go from the ensemble of fixed magnetic field to that of fixed magnetization. That is, we write

$$Z_{\beta,h,N} = \sum_{m \in \mathcal{M}_N} e^{N\beta\left(\frac{m^2}{2}+mh\right)} z_{m,N} \tag{3.33}$$

where $\mathcal{M}_N$ is the set of possible values of the magnetization, i.e.

$$\mathcal{M}_N \equiv \{m \in \mathbb{R} : \exists \sigma \in \{-1, 1\}^N : m_N(\sigma) = m\} \tag{3.34}$$
$$= \{-1, -1 + 2/N, \ldots, 1 - 2/N, 1\}$$

3.5 The Curie–Weiss model

and

$$z_{m,N} \equiv \sum_{\sigma \in \{-1,1\}^N} \mathbb{I}_{m_N(\sigma)=m} \qquad (3.35)$$

is a 'micro-canonical partition function'. Fortunately, the computation of this micro-canonical partition function is easy. In fact, all possible values of m are of the form $m = 1 - 2k/N$, where k runs from 0 to N and counts the number of spins that have the value -1. Thus, the computation of $z_{m,N}$ amounts to the most elementary combinatorial problem, the counting of the number of subsets of size k in the set of the first N integers. Thus,

$$z_{m,N} = \binom{N}{N(1-m)/2} \equiv \frac{N!}{[N(1-m)/2]![N(1+m)/2]!} \qquad (3.36)$$

It is always useful to know the asymptotics of the logarithm of the binomial coefficients that I give here for future reference with more precision than we need right now. If we set, for $m \in \mathcal{M}_N$,

$$N^{-1} \ln z_{m,N} = \ln 2 - I(m) - J_N(m) \qquad (3.37)$$

where

$$I(m) = \frac{1+m}{2} \ln(1+m) + \frac{1-m}{2} \ln(1-m) \qquad (3.38)$$

Then

$$J_N(m) = \frac{1}{2N} \ln \frac{1-m^2}{4} + \frac{\ln N + \ln(2\pi)}{2N} + O\left(N^{-2}\left(\frac{1}{1-m} + \frac{1}{1+m}\right)\right) \qquad (3.39)$$

Equation (3.39) is obtained using the asymptotic expansion for the logarithm of the Gamma function. The function $I(x)$ is called *Cramèr's entropy function* and is worth memorizing. Note that by its nature it is a relative entropy. The function J_N is of lesser importance, since it is very small.

Some elementary properties of I are useful to know: first, I is symmetric, convex, and takes its unique minimum, 0, at 0. Moreover $I(1) = I(-1) = \ln 2$. Its derivative, $I'(m) = \text{arcth}(m)$, exists in $(-1, 1)$. While I is not uniformly Lipschitz continuous on $[-1, 1]$, it has the following property:

Lemma 3.5.1 *There exists $C < \infty$ such that for any interval $\Delta \subset [-1, 1]$ with $|\Delta| < 0.1$, $\max_{x,y \in \Delta} |I(x) - I(y)| \leq C|\Delta| |\ln |\Delta||$.*

We would like to say that $\lim_{N \uparrow \infty} \frac{1}{N} \ln z_{m,N} = \ln 2 + I(m)$. But there is a small problem, due to the fact that the relation (3.37) does only hold on the N-dependent set $\mathcal{M}_N$. Otherwise, $\ln z_{m,N} = -\infty$. A precise asymptotic statement could be the following:

Lemma 3.5.2 *For any $m \in [-1, 1]$,*

$$\lim_{\epsilon \downarrow 0} \lim_{N \uparrow \infty} \frac{1}{N} \ln \sum_{m \in \mathcal{M}_M : |m - \tilde{m}| < \epsilon} z_{m,N} = \ln 2 + I(\tilde{m}) \qquad (3.40)$$

Proof The proof is elementary from properties of $z_{m,N}$ and $I(m)$ mentioned above and is left to the reader. □

In probability theory, the following formulation of Lemma 3.5.2 is known as *Cramèr's theorem*. It is the simplest so-called *large deviation principle* [98]:

Lemma 3.5.3 *Let $A \in \mathcal{B}(\mathbb{R})$ be a Borel-subset of the real line. Define a probability measure p_N by $p_N(A) \equiv 2^{-N} \sum_{m \in \mathcal{M}_N \cap A} z_{m,N}$, and let $I(m)$ be defined as in (3.38). Then*

$$- \inf_{m \in A} I(m) \leq \liminf_{N \uparrow \infty} \frac{1}{N} \ln p_N(A) \tag{3.41}$$

$$\leq \limsup_{N \uparrow \infty} \frac{1}{N} \ln p_N(A) \leq - \inf_{m \in \bar{A}} I(m)$$

Moreover, I is convex, lower semi-continuous, Lipschitz continuous on $(-1, 1)$, bounded on $[-1, 1]$, and equal to $+\infty$ on $[-1, 1]^c$.

Remark 3.5.4 The classical interpretation of the preceding theorem is the following. The spin variables $\sigma_i = \pm 1$ are *independent, identically distributed* binary random variables taking the values ± 1 with equal probability. $m_N(\sigma)$ is the normalized sum of the first N of these random variables. p_N denotes the probability distribution of the random variable m_N, which is inherited from the probability distribution of the family of random variables σ_i. It is well known, by the law of large numbers, that p_N will concentrate on the value $m = 0$, as N tends to ∞. A large deviation principle states in a precise manner how small the probability will be that m_N takes on different values. In fact, the probability that m_N will be in a set A that does not contain 0 will be of the order $\exp(-Nc(A))$, and the value of $c(A)$ is precisely the smallest value that the function $I(m)$ takes on the set A.

The computation of the canonical partition function is now easy:

$$Z_{\beta,h,N} = \sum_{m \in \mathcal{M}_N} \binom{N}{N(1-m)/2} \exp\left(N\beta\left(\frac{m^2}{2} + hm\right)\right) \binom{N}{N(1-m)/2} \tag{3.42}$$

and by the same type of argument that was used in the proof of Theorem 2.3.1 we get the following:

Lemma 3.5.5 *For any temperature β^{-1} and magnetic field h,*

$$\lim_{N \uparrow \infty} \frac{-1}{\beta N} \ln Z_{\beta,h,N} = \inf_{m \in [0,1]} (-m^2/2 + hm - \beta^{-1}(\ln 2 - I(m)))$$

$$= f(\beta, h) \tag{3.43}$$

Proof We give the simplest proof, which, however, contains some valuable lessons. We first prove an upper bound for $Z_{\beta,h,N}$:

$$Z_{\beta,h,N} \leq N \max_{m \in \mathcal{M}_N} \exp\left(N\beta\left(\frac{m^2}{2} + hm\right)\right) \binom{N}{N(1-m)/2} \tag{3.44}$$

$$\leq N \max_{m \in [-1,1]} \exp\left(N\beta\left(\frac{m^2}{2} + hm\right) + N(\ln 2 - I(m) - J_N(m))\right)$$

Hence

$$N^{-1} \ln Z_{\beta,h,N} \tag{3.45}$$
$$\leq N^{-1} \ln N + \max_{m \in [-1,1]} \left(\beta \left(\frac{m^2}{2} + hm \right) + \ln 2 - I(m) - J_N(m) \right)$$
$$\leq \ln 2 + \sup_{m \in [-1,1]} \left(\beta \left(\frac{m^2}{2} + hm \right) - I(m) \right) + N^{-1} O(\ln N)$$

so that

$$\limsup_{N \uparrow \infty} N^{-1} \ln Z_{\beta,h,N} \leq \beta \sup_{m \in [-1,1]} \left(\frac{m^2}{2} + hm - \beta^{-1} I(m) \right) + \ln 2 \tag{3.46}$$

This already looks good. Now all we need is a matching lower bound. It can be found simply by using the property that the sum is bigger than its parts:

$$Z_{\beta,h,N} \geq \max_{m \in \mathcal{M}_N} \exp\left(N\beta \left(\frac{m^2}{2} + hm \right) \right) \binom{N}{N(1-m)/2} \tag{3.47}$$

We see that we will be in business, up to the small problem that we need to pass from the max over $\mathcal{M}_N$ to the max over $[-1, 1]$, after inserting the bound for the binomial coefficient in terms of $I(m)$. In fact, we get that

$$N^{-1} \ln Z_{\beta,h,N} \geq \ln 2 + \beta \max_{m \in \mathcal{M}_N} \left(\frac{m^2}{2} + hm - \beta^{-1} I(m) \right) - O(\ln N/N) \tag{3.48}$$

for any N. Now, we can easily check that

$$\max_{m \in \mathcal{M}_N} \left| \left(\frac{m^2}{2} + hm - \beta^{-1} I(m) \right) \right. \tag{3.49}$$
$$\left. - \sup_{m' \in [0,1], |m'-m| \leq 2/N} \left(\frac{m^2}{2} + hm - \beta^{-1} I(m) \right) \right| \leq C \ln N / N$$

so that

$$\liminf_{N \uparrow \infty} \frac{1}{\beta N} \ln Z_{\beta,h,N} \geq \beta^{-1} \ln 2 + \sup_{m \in [-1,1]} \left(\frac{m^2}{2} + hm - \beta^{-1} I(m) \right) \tag{3.50}$$

and the assertion of the lemma follows immediately. □

Remark 3.5.6 The function $g(\beta, m) \equiv -m^2/2 - \beta^{-1}(\ln 2 - I(m))$ should be rightfully called the *Helmholtz free energy* for zero magnetic field (see above), since by our calculations,

$$\lim_{\epsilon \downarrow 0} \lim_{N \uparrow \infty} \frac{-1}{\beta N} \ln \sum_{\tilde{m}: |\tilde{m}-m| < \epsilon} \tilde{Z}_{\beta,\tilde{m},N} = g(\beta, m) \tag{3.51}$$

where

$$\tilde{Z}_{\beta,\tilde{m},N} = \sum_{\sigma \in \{-1,1\}^N} e^{\beta H_N(\sigma)} \mathbb{I}_{m_N(\sigma)=m} \tag{3.52}$$

for $h = 0$. Thermodynamically, the function $f(\beta, h)$ is then the *Gibbs free energy*, and the assertion of the lemma would then be that the Helmholtz free energy is given by this particular function, and that the Gibbs free energy is its *Legendre transform*. The Helmholtz free energy is closely related to the rate function of a large deviation principle for the distribution of the magnetization under the Gibbs distribution. Namely, if we define the Gibbs distribution on the space of spin configurations

$$\mu_{\beta,h,N}(\sigma) \equiv \frac{e^{-\beta H_N(\sigma)}}{Z_{\beta,h,N}} \tag{3.53}$$

and denote by $\tilde{p}_{\beta,h,N}(A) \equiv \mu_{\beta,h,N}(\{m_N(\sigma) \in A\})$ the law of m_N under this distribution, then we obtain very easily:

Lemma 3.5.7 *Let $\tilde{p}_{\beta,h,N}$ be the law of $m_N(\sigma)$ under the Gibbs distribution. Then the family of probability measures $\tilde{p}_{\beta,h,N}$ satisfies a large deviation principle, i.e. for all Borel subsets of $\mathbb{R}$,*

$$-\inf_{m \in A}(g(\beta, m) - hm) + f(\beta, h) \leq \liminf_{N \uparrow \infty} \frac{1}{\beta N} \ln \tilde{p}_{\beta,h,N}(A) \tag{3.54}$$

$$\leq \limsup_{N \uparrow \infty} \frac{1}{\beta N} \ln \tilde{p}_{\beta,h,N}(A)$$

$$\leq -\inf_{m \in \bar{A}}(g(\beta, m) - hm) + f(\beta, h)$$

We see that the thermodynamic interpretation of equilibrium emerges very nicely: the equilibrium value of the magnetization, $m(\beta, h)$, for a given temperature and magnetic field, is the value of m for which the rate function in Lemma 3.5.7 vanishes, i.e. which satisfies the equation

$$g(\beta, m(\beta, h)) - hm(\beta, h) = f(\beta, h) \tag{3.55}$$

(which is the usual thermodynamic relation between the Gibbs and the Helmholtz free energy). By the definition of f (see (3.43)), this is the case whenever $m(\beta, h)$ realises the infimum in (3.43). If $g(\beta, m)$ is strictly convex, this infimum is unique, and, as long as g is convex, it is the set on which $\frac{\partial g(\beta, m)}{\partial m} = h$.

Note that, in our case, $g(\beta, m)$ is not a convex function of m if $\beta > 1$, as can be seen in Fig. 3.3.

In fact, g has two local minima, situated at the values $\pm m^*_\beta$, where m^*_β is defined as the largest solution of the equation

$$m = \tanh \beta m \tag{3.56}$$

Moreover, the function g is symmetric, and so takes the same value at both minima. As a consequence, the minimizer of the function $g(\beta, m) - mh$, the magnetization as a function of the magnetic field, is not unique at the value $h = 0$ (and only at this value). For $h > 0$, the minimizer is the positive solution of $m = \tanh(\beta(m + h))$, while for negative h it is the negative solution. Consequently, the magnetization has a jump discontinuity at $h = 0$, where it jumps by $2m^*_\beta$.

As in the van der Waals gas, the Curie–Weiss model exhibits a first-order phase transition, unlike the one-dimensional Ising model. But, also as in the van der Waals gas, the basic hypothesis of thermodynamics, namely the convexity of the free energy (here g), is violated. Ising could have argued that the fact that the interaction in this model has infinite

3.5 The Curie–Weiss model

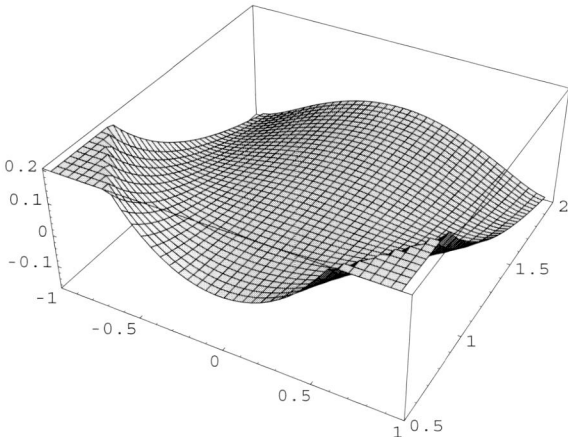

Figure 3.3 The function $g(\beta, m)$.

range (i.e. all spins interact with the same strength), which clearly is responsible for the non-convexity, is also responsible for the appearance of the phase transition.

Before we turn to some further investigations of the Curie–Weiss model, let us discuss the physical implications of the non-convexity problem. The usual argument against the unphysical nature of non-convex g goes as follows. Given g, the magnetic field (as the analogue of the pressure) should be

$$h(m, \beta) = \frac{\partial g(\beta, m)}{\partial m} = m - \beta^{-1} I'(m) \qquad (3.57)$$

This *isotherm* is not a monotone function of the magnetization, that is, there are regions of the magnetization where the magnetic field drops when the magnetization increases, which looks funny. Also, it is clear that this function is not invertible, so we could argue that we cannot compute m as a function of the magnetic field. But from our more probabilistic point of view, things are not so bad, after all. The equilibrium value of $m(\beta, h)$ as function of β and h is the minimizer of the function $g(\beta, m) - hm$, which is uniquely defined, except at $h = 0$. The values in the interval $(-m^*(\beta), m^*(\beta))$ are unphysical, i.e. for no value of the magnetic field will the system attain an equilibrium magnetization in this interval. In fact, Maxwell's cure to replace the non-convex Helmholtz free energy by its convex hull also works here. This then basically allows any value of the magnetization in that interval, if $h = 0$. If one were to look more closely into the probability distribution of m_N in a lattice model in dimension d, one would, however, discover that the intermediate values of the magnetization are considerably less probable than the extremal ones, albeit only by a factor of order $\exp(-N^{1-1/d})$. So from a thermodynamic point of view, the Curie–Weiss model is not such a bad model after all. The main drawback appears if one wants to analyze the behaviour of systems where the magnetization is forced by a constraint to lie in the forbidden interval. Real physical systems will exhibit what is called *phase separation*, i.e. the system will select a sub-volume, where the magnetization takes the value $+m^*$, while in the complement it will take the value $-m^*$ in such a way that the total magnetization has the enforced value. The precise details of phase separation have been understood from

More on the CW model Our solution of the Curie–Weiss model relied on the fact that we could solve the combinatorial problem of counting the number of spin configurations having a given magnetization m. There is a nice trick, called the Hubbard–Stratonovich transformation [140, 230], that allows us to compute the Gibbs free energy directly, without having to solve any combinatorial problem.

Recall that we want to compute

$$Z_{\beta,h,N} = \sum_{\sigma \in \{-1,1\}^N} \exp(\beta N m_N^2(\sigma)/2 + \beta N h m_N(\sigma)) \tag{3.58}$$

The difficulty in performing the sum is due to the presence of the quadratic term in the exponent. But there is a simple identity that allows us to solve this issue, namely

$$\frac{1}{\sqrt{2\pi}} \int dz\, e^{-z^2/2 + yz} = e^{y^2/2} \tag{3.59}$$

Applying this yields

$$Z_{\beta,h,N} = \sum_{\sigma \in \{-1,1\}^N} \frac{1}{\sqrt{2\pi}} \int dz\, e^{-z^2/2 + (\sqrt{N\beta} z + \beta h N) m_N(\sigma)} \tag{3.60}$$

$$= \sum_{\sigma \in \{-1,1\}^N} \sqrt{\frac{\beta N}{2\pi}} \int dz\, e^{-\beta N z^2/2 + (z+h)\beta \sum_{i=1}^N \sigma_i}$$

$$= \sqrt{\frac{N}{2\pi}} \int dz\, e^{-N\beta z^2/2 + N \ln[2\cosh(\beta(z+h))]}$$

Lemma 3.5.8 *For any β, h,*

$$\lim_{N \uparrow \infty} \frac{1}{\beta N} \ln Z_{\beta,h,N} = -\inf_{z \in \mathbb{R}}(z^2/2 - \beta^{-1} \ln \cosh \beta(z+h)) + \beta^{-1} \ln 2 \tag{3.61}$$

The proof of the lemma is very simple and will be skipped. Clearly, the variational formula (3.61) must represent the same function as (3.43). In particular, the minimizer is the solution of the equation $x = \beta \tanh \beta(x+h)$ that has the same sign as h, i.e. is precisely $m(\beta, h)$.

Exercise: Critical behaviour in the CW model We have seen that a first-order phase transition appears in the Curie–Weiss model for $\beta > \beta_c = 1$. Analyze the behaviour of the thermodynamic functions in the vicinity of this critical point (see Fig. 3.4).

(i) Compute the spontaneous magnetization $m^*(\beta)$ as a function of $\beta - \beta_c$ as $\beta \downarrow \beta_c$.
(ii) Compute the *specific heat*, $c(h, T) \equiv -T \frac{\partial^2 f(\beta,h)}{\partial T^2}$, and its asymptotic expansion for $\beta > \beta_c$ when $h = 0$.
(iii) Compute the *susceptibility*, $\chi = \frac{\partial m(\beta,h)}{\partial h}$, at $h = 0$, for $\beta < \beta_c$ and find its leading-order behaviour in powers of $\beta_c - \beta$.
(iv) For $\beta = \beta_c$, compute the leading behaviour of $m(\beta_c, h)$ as $h \downarrow 0$.

3.5 The Curie–Weiss model

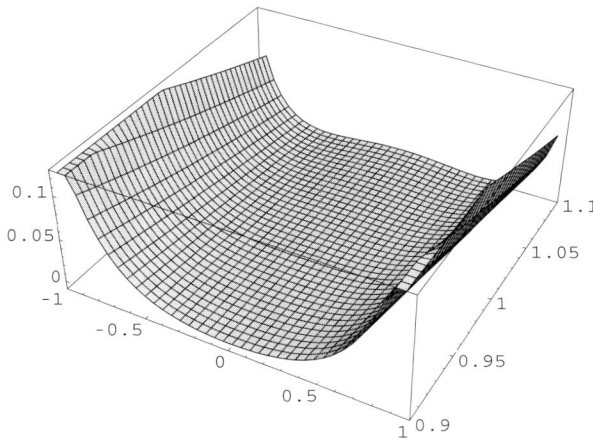

Figure 3.4 The function $g(\beta, m)$ for values of β near the critical value 1.

Exercise: Fluctuations in the CW model We have seen that thermodynamic computations amount to proving large deviation principles for thermodynamic variables. One can look at finer properties of the distribution functions of such variables. For instance, we know that the value of the magnetization $m_N(\sigma)$ will sharply concentrate on its equilibrium value $m(\beta, h)$.

Consider the family of random variables $X_N \equiv \sqrt{N}(m_N(\sigma) - m(\beta, h))$ distributed according to the Gibbs measure $\mu_{\beta, h, N}$.

(i) If $\beta < 1$, show that X_N converges in distribution to a centered Gaussian random variable and compute its variance.
(ii) Do the same for $\beta < 1$ and $h > 0$.
(iii) Compute the behaviour of the variance of X_N for $h = 0$ as $\beta \uparrow 1$, and for $\beta = 1$ as $h \downarrow 0$.
(iv) For $\beta = 1$ and $h = 0$, how should one rescale the magnetization to obtain a random variable with a non-trivial distribution? Compute the distribution of the properly rescaled variable as well as you can.
(v) If $\beta > 1$ and $h = 0$, try to formulate a limit theorem for the fluctuations of the magnetization.

The Curie–Weiss model has proven to be an easily solvable model that exhibits a first-order (and as shown in the exercise, a second-order) phase transition. However, the question of whether long-range order can appear in a short-range model remains open.

The two-dimensional Ising model In 1944, Onsager [195] produced an exact solution of the two-dimensional Ising model with zero magnetic field. From this solution, the existence of a phase transition could be concluded, and even the precise asymptotics near the critical temperature could be inferred. The two-dimensional Ising model has been of paramount importance in the theory of *critical phenomena*, resp. *second-order phase transitions*, because its exact solution provided an example that showed that, in general, *critical exponents* are different from those found in the mean field model. Later, starting with the work of Lieb on the ice-model [168] and Baxter [20] on the eight-vertex model, it was found that the Ising model is a special case of a much wider class of two-dimensional models that permit exact solutions. Exact solubility of non-mean field models is, however, a particular, and somewhat accidental property, and we will not discuss this topic in this book. Note that more recently the two-dimensional Ising model has also played an important rôle as the first model where a rigorous treatment of the phase separation problem could be given [88].

4

The Gibbsian formalism for lattice spin systems

> The word 'statistic' is derived from the Latin *status*, which, in the middle ages, had come to mean 'state' in the political sense. 'Statistics', therefore, originally denoted inquiries into the condition of a state.
>
> *Statistics, Encyclopedia Britannica, 11th edn.*

We will now turn to the investigation of the rigorous probabilistic formalism of the statistical mechanics of lattice spin systems, or lattice gases. The literature on this subject is well developed and the interested student can find in-depth material for further reading in [125, 207, 208, 224, 226], and the classical monographs by Ruelle [214, 215]. A nice short introduction with a particular aim in view is also given in the first sections of the paper [250].

4.1 Spin systems and Gibbs measures

As mentioned in the last chapter, the idea of the spin system was born in about 1920 in an attempt to understand the phenomenon of ferromagnetism. At that time it was understood that ferromagnetism should be due to the alignment of the elementary magnetic moments ('spins') of the (iron) atoms, that persists even after an external field is turned off. The phenomenon is temperature dependent: if one heats the material, the coherent alignment is lost. It was understood that the magnetic moments should exert an 'attractive' ('ferromagnetic') interaction towards each other, which, however, is of short range. The question was then, how such a short-range interaction could sustain the observed very long-range coherent behaviour of the material, and why such an effect should depend on the temperature.

Recall that the Ising model can be defined via a *Hamiltonian*, H, that assigns to each configuration, $\sigma \equiv \{\sigma_x\}_{x \in \mathbb{Z}^d}$, the energy

$$H(\sigma) \equiv - \sum_{\substack{x,y \in \mathbb{Z}^d \\ \|x-y\|_1 = 1}} \sigma_x \sigma_y - h \sum_{x \in \mathbb{Z}^d} \sigma_x \tag{4.1}$$

In the last chapter we only considered systems that were confined to some finite volume Λ, whose size would be taken to infinity when taking the *thermodynamic limit*. We will now take a different point of view. In fact, our aim will be to define systems, or more precisely Gibbs measures, directly in the *infinite volume*. This touches on an important fundamental issue of statistical mechanics, which we will have occasion to discuss repeatedly. It is tempting to formulate this as an (informal) axiom:

A system composed of a very large number of degrees of freedom can be well approximated by an infinite system.

We will have to see how to interpret this statement and what its limitations are later. I would ask you to accept this for the moment and take it as an excuse for the otherwise seemingly unreasonable struggle we will enter into to describe infinite systems.

The basic axiom of statistical mechanics is, as we have seen, that the (equilibrium) properties of a system shall be described by specifying a probability measure on the space of configurations, in our case $\{-1, +1\}^{\mathbb{Z}^d}$. From what we have learned so far, the appropriate candidate for such a measure should be the Gibbs measure, as it is parametrized only by intensive variables. We will therefore accept as another axiom that the proper measure to choose is the *Gibbs measure*, which formally is given by

$$\mu_\beta(d\sigma) = \frac{1}{Z_\beta} e^{-\beta H(\sigma)} \rho(d\sigma) \qquad (4.2)$$

where Z_β is a normalizing constant and ρ is the uniform measure on the configuration space. Again, this expression makes no sense for the infinite system, but would make perfect sense if we replaced $\mathbb{Z}^d$ by a finite set, Λ, everywhere.[1]

We will see how to obtain a sensible version of (4.2) in the infinite-volume setting. We start with the 'a-priori' measure, ρ, that is supposed to describe the non-interacting system. In finite volumes, the uniform measure on the finite space $\{-1, +1\}^\Lambda$ is the product Bernoulli measure

$$\rho_\Lambda(\sigma_\Lambda = s_\Lambda) = \prod_{x \in \Lambda} \rho_x(\sigma_x = s_x) \qquad (4.3)$$

where $\rho_x(\sigma_x = +1) = \rho_x(\sigma_x = -1) = 1/2$. There is a standard construction to extend this to infinite volume. First, we turn $\mathcal{S} \equiv \{-1, +1\}^{\mathbb{Z}^d}$ into a measure space by equipping it with the product topology of the discrete topology on $\{-1, +1\}$. The corresponding sigma-algebra, $\mathcal{F}$, is then just the product sigma-algebra. The measure ρ is then defined by specifying that, for all cylinder events $\mathcal{A}_\Lambda$ (i.e. events that for some finite set $\Lambda \subset \mathbb{Z}^d$ depend only on the values of the variables σ_x with $x \in \Lambda$),

$$\rho(\mathcal{A}_\Lambda) = \rho_\Lambda(\mathcal{A}_\Lambda) \qquad (4.4)$$

with ρ_Λ defined in (4.3). In this way we have set up an a-priori probability space, $(\mathcal{S}, \mathcal{F}, \rho)$, describing a system of non-interacting spins. It is worth noting that this set-up is not totally innocent and reflects a certain *physical* attitude towards our problem. Namely, the choice to consider the system as truly infinite and to use the product topology implies that we consider the individual degrees of freedom, or finite collections of them, as the main physical observables, which can be measured. While this appears natural, it should not be forgotten

---

[1] Here we are touching a crucial point. The problem with a finite-volume description is that it appears to be unable to reflect the very phenomenon we want to describe, namely the existence of several phases, i.e. the persistence of magnetized states after the magnetic field has been turned off. The argument was brought forward that a single formula could not possibly describe different physical states at the same time. The question is indeed quite intricate and full understanding requires consideration of the dynamical aspects of the problem. On the level of the equilibrium theory, the issue is however, as we will see, solved precisely and elegantly by the adoption of the infinite-volume axiom.

that this has important implications in the interpretation of the infinite-volume results as asymptotic results for large systems, which may not in all cases be the most desirable ones.[2]

To continue the interpretation of (4.2), one might be tempted also to specify the measure μ_β by prescribing the finite-dimensional marginals, e.g., by demanding that $\mu_{\beta,\Lambda}(d\sigma_\Lambda) = Z_{\beta,\Lambda}^{-1}\exp(-\beta H_\Lambda(\sigma_\Lambda))\rho_\Lambda(d\sigma_\Lambda)$, with $H_\Lambda(\sigma_\Lambda)$ the restriction of (4.1) to the finite volume Λ. The problem with this, however, is the compatibility conditions that are required for such a set of measures to specify a measure on $(\mathcal{S},\mathcal{F})$; Kolmogorov's theorem would require that for $\Lambda \subset \Lambda'$, $\mu_{\beta,\Lambda}(\mathcal{A}_\Lambda) = \mu_{\beta,\Lambda'}(\mathcal{A}_\Lambda)$. While in the case of the non-interacting system this is trivially checked, this will not hold in the interacting case.

Exercise: Prove this fact. Check explicitly that the compatibility conditions do not hold in the case when Λ, Λ' consist of 1 resp. 2 points!

Since there appears no other feasible way to specify marginal measures, we need a better idea. Actually, there are not too many choices: if we cannot fix marginals, we can try to fix conditional distributions. This seems quite natural today from the point of view of the theory of Markov processes, but it was only realized in 1968–9 by Roland L. Dobrushin [93, 94] (and shortly after that by O. Lanford and D. Ruelle [161]), and is now seen as one the cornerstones of the foundation of modern mathematical statistical mechanics. To understand this construction, we have to return to (4.1) and give a new interpretation to this formal expression. The Hamiltonian should measure the energy of a configuration; this makes no sense in infinite volume, but what we could ask is: what is the energy of an *infinite-volume* configuration *within a finite-volume* Λ? A natural definition of this quantity is

$$H_\Lambda(\sigma) \equiv -\sum_{\substack{x \vee y \in \Lambda \\ \|x-y\|_1 = 1}} \sigma_x \sigma_y - h \sum_{x \in \Lambda} \sigma_x \tag{4.5}$$

Note that this expresssion, in contrast to the formula (4.1), contains the energy corresponding to the interaction between spins in Λ with those outside Λ (which here involves only spins in the boundary of Λ). The notion of finite-volume restriction given by (4.5) has the nice feature that it is compatible under iteration: if $\Lambda' \supset \Lambda$, then

$$(H_{\Lambda'})_\Lambda(\sigma) = H_\Lambda(\sigma) \tag{4.6}$$

Equation (4.5) will furnish our standard interpretation of a Hamiltonian function H; we will always consider it as a function $H : (\Lambda, \sigma) \to H_\Lambda(\sigma)$ from the pairs consisting of finite subsets of $\mathbb{Z}^d$ and configurations in $\mathcal{S}$ to the real numbers. This allows us to define, for any fixed configuration of spins $\eta \in \mathcal{S}$ and finite subset $\Lambda \subset \mathbb{Z}^d$, a probability measure

$$\mu_\Lambda^\eta(d\sigma_\Lambda) = \frac{1}{Z_{\beta,\Lambda}^\eta} e^{-\beta H_\Lambda((\sigma_\Lambda, \eta_{\Lambda^c}))} \rho_\Lambda(d\sigma_\Lambda) \tag{4.7}$$

Equation (4.7) defines a much richer class of measures than just the marginals. The idea is that these should be the family of conditional probabilities of some measure, μ_β, defined on the infinite-volume space. The point is that they satisfy automatically the compatibility

---

[2] For instance, it might be that one is interested in collections of variables that are composed of enormously many local variables. It may then be that an appropriate description requires intermediate divergent ('mesoscopic') scales in between the 'macroscopic' volume and the microscopic degrees of freedom. This would require a slightly different approach to the problem.

conditions required for conditional probabilities (see below), and so have a chance to be conditional probabilities of some infinite-volume measure. Dobrushin's idea was to start from this observation to define the notion of the infinite-volume Gibbs measure, i.e. as the proper definition for the formal expression (4.2):

A probability measure μ_β on $(S, \mathcal{F})$ is a Gibbs measure for the Hamiltonian H and inverse temperature β if and only if its conditional distributions (given the configurations in the complement of any finite set Λ) are given by (4.7).

Two questions immediately pose themselves:

(i) Does such a measure exist?
(ii) If it exists, is it uniquely specified?

We will see soon that there is a large class of systems for which the existence of such a measure can be shown. That means that Dobrushin's formalism is meaningful and defines a rich theory. The second question provides all the charm of the Gibbsian formalism: there are situations when the infinite-volume measure is not uniquely specified and when *several* infinite-volume measures exist for the same Hamiltonian and the same temperature.[3] This observation will furnish the explanation for the strikingly different behaviour of a ferromagnet at high and low temperatures: if $d \geq 2$, the temperature is low, and $h = 0$, there will be measures describing a state with positive magnetization and one with negative magnetization, and the system can be in either of them; at high temperatures, however, there is always a unique Gibbs measure.

Before we continue the investigation of these two questions in the Ising model, we will provide a more general and more formal set-up of the preceding discussion.

4.2 Regular interactions

4.2.1 Some topological background

We will now describe the general framework of spin systems with so-called *regular interactions*. Our setting will always be lattice systems and our lattice will always be $\mathbb{Z}^d$. Λ will always denote a finite subset of $\mathbb{Z}^d$. Spins will take values in a set S_0 that will always be a complete separable metric space. In most of our examples it will just be the set $\{-1, 1\}$. We equip S_0 with its sigma-algebra generated by the open sets in the metric topology (resp. the discrete topology in the case when S_0 is a discrete set), $\mathcal{F}_0$, to obtain a measure space $(S_0, \mathcal{F}_0)$. To complete the description of the single-spin space, we add a (probability) measure ρ_0, the so-called a-priori distribution of the spin. This gives a single-site (probability) space $(S_0, \mathcal{F}_0, \rho_0)$.

As discussed in the previous paragraph, we first want to furnish the setting for infinitely many non-interacting spins. To do this, we consider the infinite-product space

$$S \equiv S_0^{\mathbb{Z}^d} \tag{4.8}$$

---

[3] This could be re-phrased as saying that the one (meaningless) formula (4.2) defines several (meaningful) Gibbs measures. This resolves the (serious) dispute in the first half of the twentieth century on the question of whether statistical mechanics could possibly account for phase transitions. See the very amusing citations in the prologue of Ueltschi's thesis [241].

4.2 Regular interactions

which we turn into a complete separable space by equipping it with the *product topology*. This is done by saying that the open sets are generated by the *cylinder sets* $B_{\epsilon,\Lambda}(\sigma)$, defined as

$$B_{\epsilon,\Lambda}(\sigma) \equiv \left\{ \sigma' \in \mathcal{S} \, \bigg| \max_{x \in \Lambda} |\sigma_x - \sigma'_x| < \epsilon \right\} \quad (4.9)$$

where $\sigma \in \mathcal{S}$, $\Lambda \subset \mathbb{Z}^d$, and $\epsilon \in \mathbb{R}_+$. The product topology of a metric space is metrizable, and $\mathcal{S}$ is a complete separable metric space if $\mathcal{S}_0$ is. The Borel sigma-algebra of $\mathcal{S}$, $\mathcal{F}$, is the product sigma-algebra

$$\mathcal{F} = \mathcal{F}_0^{\mathbb{Z}^d} \quad (4.10)$$

An important fact is *Tychonov's theorem* [125]:

Theorem 4.2.1 *If $\mathcal{S}_0$ is compact then the space $\mathcal{S}$ equipped with the product topology is compact.*

A particularly important consequence in the case when $\mathcal{S}_0$ is a compact, separable metric space is that the same holds true for the product space, and hence any sequence in that space has a convergent subsequence.

Exercise: Consider the space $\{-1, 1\}^{\mathbb{N}}$. Show by direct construction that any sequence $\sigma^{(n)} \in \{-1, 1\}^{\mathbb{N}}$ has a convergent subsequence. (Hint: Show that $\{-1, 1\}^{\mathbb{N}}$ can be given the structure of a partially ordered set, and use this order to construct a bounded, increasing subsequence.)

We will use the notation $\mathcal{S}_\Lambda \equiv \mathcal{S}_0^\Lambda$ and $\mathcal{F}_\Lambda \equiv \mathcal{F}_0^\Lambda$ for the finite-volume configuration space and the sigma-algebra of local events. Note that we identify $\mathcal{F}_\Lambda \subset \mathcal{F}$ with the sub-sigma-algebra of events depending only on the co-ordinates σ_x, $x \in \Lambda$. We will call an event that is measurable with respect to $\mathcal{F}_\Lambda$, for some finite Λ, a *local*, or *cylinder*, event. A sequence of volumes, $\Lambda_1 \subset \Lambda_2 \subset \cdots \subset \Lambda_n \subset \cdots \subset \mathbb{Z}^d$, with the property that, for any finite $\Lambda' \subset \mathbb{Z}^d$, there exists n such that $\Lambda' \subset \Lambda_n$ will be called an *increasing and absorbing sequence*. The corresponding family of sigma-algebras, $\mathcal{F}_{\Lambda_n}$, forms a filtration of the sigma-algebra $\mathcal{F}$. Similarly, we write $\mathcal{S}_{\Lambda^c} \equiv \mathcal{S}_0^{\mathbb{Z}^d \setminus \Lambda}$ and $\mathcal{F}_{\Lambda^c} \equiv \mathcal{F}_0^{\mathbb{Z}^d \setminus \Lambda}$. A special rôle will be played later by the so-called 'tail sigma-algebra' $\mathcal{F}^t \equiv \cap_{\Lambda \subset \mathbb{Z}^d} \mathcal{F}_{\Lambda^c}$. The events in $\mathcal{F}^t$ will be called tail-events or non-local events.

We will refer to various spaces of (real valued) functions on $\mathcal{S}$ in the sequel. In the physical terminology, such functions are sometimes referred to as *observables*. The largest space one usually considers is $B(\mathcal{S}, \mathcal{F})$, the space of bounded, measurable functions. (A function, f, from a measure space, $\mathcal{S}$, into the real numbers is called measurable if, for any Borel set $B \subset \mathcal{B}(\mathbb{R})$, the set $\mathcal{A} \equiv \{\sigma : f(\sigma) \in B\}$ is contained in $\mathcal{F}$.)

Correspondingly, we write $B(\mathcal{S}, \mathcal{F}_\Lambda)$ for bounded functions measurable with respect to $\mathcal{F}_\Lambda$, i.e. depending only on the values of the spins in Λ. Functions that are in some $B(\mathcal{S}, \mathcal{F}_\Lambda)$ are called *local* or *cylinder* functions; we denote their space by

$$B_{\text{loc}}(\mathcal{S}) \equiv \cup_{\Lambda \subset \mathbb{Z}^d} B(\mathcal{S}, \mathcal{F}_\Lambda) \quad (4.11)$$

A slight enlargement of the space of local functions are the so-called *quasi-local* functions, $B_{\text{ql}}(\mathcal{S})$; this is the closure of the set of local functions under uniform convergence.

Quasi-local functions are characterized by the property that

$$\lim_{\Lambda \uparrow \mathbb{Z}^d} \sup_{\substack{\sigma, \sigma' \in \mathcal{S} \\ \sigma_\Lambda = \sigma'_\Lambda}} |f(\sigma) - f(\sigma')| = 0 \quad (4.12)$$

We also introduce the spaces of continuous, local continuous, and quasi-local continuous functions, $C(\mathcal{S})$, $C_{\text{loc}}(\mathcal{S}, \mathcal{F}) = C(\mathcal{S}) \cap \mathcal{B}_{\text{loc}}(\mathcal{S}, \mathcal{F})$, and $C_{\text{ql}} = C(\mathcal{S}) \cap \mathcal{B}_{\text{ql}}(\mathcal{S}, \mathcal{F})$.

The reader should be warned that in general (i.e. under the hypothesis that $\mathcal{S}_0$ is just a complete separable metric space), neither are all quasi-local functions continuous, nor all continuous functions quasi-local (see, e.g., [250] for nice examples). However, under stronger hypotheses on $\mathcal{S}_0$, the different spaces acquire relations:

Lemma 4.2.2
 (i) If $\mathcal{S}_0$ is compact, then $C(\mathcal{S}) = C_{\text{ql}}(\mathcal{S}) \subset \mathcal{B}_{\text{ql}}(\mathcal{S})$.
 (ii) If $\mathcal{S}_0$ is discrete, then $\mathcal{B}_{\text{ql}}(\mathcal{S}) = C_{\text{ql}}(\mathcal{S}) \subset C(\mathcal{S})$.
 (iii) If $\mathcal{S}_0$ is finite, then $C(\mathcal{S}) = \mathcal{B}_{\text{ql}}(\mathcal{S}) = C_{\text{ql}}(\mathcal{S})$.

Proof Left as an exercise. □

Remark 4.2.3 Since we are mostly interested in finite-spin spaces, quasi-locality will be the essential aspect of continuity in the product topology.

We now turn to the space $\mathcal{M}_1(\mathcal{S}, \mathcal{F})$ of probability measures on $(\mathcal{S}, \mathcal{F})$ and its topological structure. There are several possibilities for equipping this space with a topology. The most convenient and commonly used one is that of weak convergence with respect to continuous functions. This topology is generated by the open balls

$$B_{f,\epsilon}(\mu) \equiv \left\{ \mu' \in \mathcal{M}_1(\mathcal{S}, \mathcal{F}) \,\big|\, |\mu(f) - \mu(f')| < \epsilon \right\} \quad (4.13)$$

where $f \in C(\mathcal{S})$, $\epsilon \in \mathbb{R}_+$, $\mu \in \mathcal{M}_1(\mathcal{S}, \mathcal{F})$. The main advantage of this topology is that it turns $\mathcal{M}_1(\mathcal{S}, \mathcal{F})$ into a complete separable metric space, and moreover, if $\mathcal{S}_0$ is compact, then $\mathcal{M}_1(\mathcal{S}, \mathcal{F})$ is compact.[4]

4.2.2 Local specifications and Gibbs measures

We now introduce a very large class of Hamiltonians for which the Gibbsian theory can be set up. We first define the concept of an *interaction*.

Definition 4.2.4 An interaction is a family $\Phi \equiv \{\Phi_A\}_{A \subset \mathbb{Z}^d}$ where $\Phi_A \in B(\mathcal{S}, \mathcal{F}_A)$. If all $\Phi_A \in C(\mathcal{S}, \mathcal{F}_A)$, then the interaction is called continuous.

An interaction is called regular if, for all $x \in \mathbb{Z}^d$, there exists a constant c such that

$$\sum_{A \ni x} \|\Phi_A\|_\infty \leq c < \infty \quad (4.14)$$

Remark 4.2.5 What we call 'regular' interaction is called 'absolutely summable' interaction in Georgii's book [125]. In most of the standard literature one finds the stronger

---

[4] Note that Georgii's book [125] uses a stronger topology than the weak topology on measures. There, balls are defined with quasi-local, but not necessary continuous functions. In this topology the space of probability measures over $\mathcal{S}$ is not necessarily compact. However, if $\mathcal{S}_0$ is a finite space, the two notions coincide.

condition that

$$\||\Phi\|| \equiv \sup_{x \in \mathbb{Z}^d} \sum_{A \ni x} \|\Phi_A\|_\infty < \infty \tag{4.15}$$

With this definition the set of all regular interactions equipped with the norm $\||\cdot\||$ forms a Banach space, $\mathcal{B}_0$, while the weaker condition we use makes the set of regular interactions only into a Fréchet space [125]. In the case of translation-invariant interactions, both conditions coincide. However, in the case of random systems, the stronger condition (4.15) would introduce some unnatural restrictions on the class of admissible interactions.

Remark 4.2.6 Unbounded interactions occur naturally in two settings: in the case of non-compact state space (e.g. 'Gaussian models', interface models) or as so called 'hard-core' exclusions to describe models in which certain configurations are forbidden (e.g. so-called 'subshifts of finite type'). While some of such models can be treated quite well, they require special work and we will not discuss them here.

From an interaction one constructs a Hamiltonian by setting, for all finite volumes $\Lambda \subset \mathbb{Z}^d$,

$$H_\Lambda(\sigma) \equiv - \sum_{A \cap \Lambda \neq \emptyset} \Phi_A(\sigma) \tag{4.16}$$

If Φ is in $\mathcal{B}_0$, H_Λ is guaranteed to satisfy the bound

$$\|H_\Lambda\|_\infty \leq C|\Lambda| \tag{4.17}$$

for some $C < \infty$. Moreover, it is easy to check that H_Λ is a quasi-local function, and, if Φ is continuous, a continuous quasi-local function, for any finite Λ.

The Hamiltonians defined in this way share most of the nice properties of the Ising Hamiltonian defined in Section 4.1, and we can proceed to use them to construct Gibbs measures. We begin with the definition of *local specifications*:

Definition 4.2.7 A local specification is a family of probability kernels, $\{\mu_{\Lambda,\beta}^{(\cdot)}\}_{\Lambda \subset \mathbb{Z}^d}$, such that:

(i) For all Λ and all $\mathcal{A} \in \mathcal{F}$, $\mu_{\Lambda,\beta}^{(\cdot)}(\mathcal{A})$ is a $\mathcal{F}_{\Lambda^c}$-measurable function.
(ii) For any $\eta \in \mathcal{S}$, $\mu_{\Lambda,\beta}^\eta$ is a probability measure on $(\mathcal{S}, \mathcal{F})$.
(iii) For any pair of volumes, Λ, Λ', with $\Lambda \subset \Lambda'$, and any measurable function, f,

$$\int \mu_{\Lambda',\beta}^\eta(d\sigma') \mu_{\Lambda,\beta}^{(\sigma'_{\Lambda'},\eta_{\Lambda'^c})}(d\sigma) f((\sigma_\Lambda, \sigma'_{\Lambda' \setminus \Lambda}, \eta_{\Lambda'^c})) \tag{4.18}$$
$$= \int \mu_{\Lambda',\beta}^\eta(d\sigma') f((\sigma'_{\Lambda'}, \eta_{\Lambda'^c}))$$

where we use the notation $(\sigma_\Lambda, \eta_{\Lambda^c})$ to denote the configuration that equals σ_x if $x \in \Lambda$, and η_x, if $x \in \Lambda^c$.

The most important point is that local specifications satisfy compatibility conditions analogous to conditional expectations. Given a regular interaction, we can now construct local specifications for the Gibbs measures to come.

Lemma 4.2.8 *If Φ is a regular interaction, then the formula*

$$\int \mu_{\Lambda,\beta}^{\eta}(d\sigma) f(\sigma) \equiv \int \rho_{\Lambda}(d\sigma_{\Lambda}) \frac{e^{-\beta H_{\Lambda}((\sigma_{\Lambda}, \eta_{\Lambda^c}))}}{Z_{\Lambda,\beta}^{\eta}} f((\sigma_{\Lambda}, \eta_{\Lambda^c})) \qquad (4.19)$$

defines a local specification, called the Gibbs specification for the interaction Φ at inverse temperature β.

Proof Left as an exercise. The crucial point is that we have (4.6). □

We will use a shorthand notation for relations like (4.18) and symbolize this equation by

$$\mu_{\Lambda',\beta}^{(\cdot)} \mu_{\Lambda,\beta}^{(\cdot)} = \mu_{\Lambda',\beta}^{(\cdot)} \qquad (4.20)$$

As we mentioned, the notion of local specifications is closely related to that of *conditional expectations*. Since this is fundamental in what follows, let us recall some standard definitions (see, e.g., [78]).

Definition 4.2.9 Let $(S, \mathcal{F}, \mu)$ be a probability space, f an $\mathcal{F}$-measurable function (a 'random variable'), and $\mathcal{G} \subset \mathcal{F}$ a sub-sigma-algebra. We call a function $g \equiv \mu(f|\mathcal{G})$ a conditional expectation of f, given $\mathcal{G}$, if and only if

(i) g is $\mathcal{G}$-measurable, and
(ii) for any $\mathcal{G}$-measurable function, h, it holds that

$$\mu(hg) = \mu(hf) \qquad (4.21)$$

In our setting, if $\mathcal{F}$ is a product sigma-algebra, and $\mathcal{G} = \mathcal{F}_{\Lambda}$, then this means that $\mu(f|\mathcal{F}_{\Lambda})$ is obtained from f by integrating over all variables σ_x with $x \notin \Lambda$ while keeping the variables σ_x with $x \in \Lambda$ fixed.

Conditional expectations are defined uniquely up to sets of measure zero; i.e. any $\mathcal{G}$-measurable function, g', for which $g' = \mu(f|\mathcal{G})$, μ-almost surely, is a *version* of the conditional expectation.

Conditional expectations satisfy a compatibility condition.

Lemma 4.2.10 *Let $\mathcal{F} \supset \mathcal{G}' \supset \mathcal{G}$, and f a $\mathcal{F}$-measurable function. Let $g = \mu(f|\mathcal{G})$ and $g' = \mu(f|\mathcal{G}')$ be conditional expectations of f with respect to $\mathcal{G}$ and $\mathcal{G}'$, respectively. Then*

$$\mu(g'|\mathcal{G}) = g, \quad \mu\text{-a.s.} \qquad (4.22)$$

Proof We just have to show that $\mu(g'|\mathcal{G})$ is the conditional expectation of f with respect to $\mathcal{G}$. Obviously it is $\mathcal{G}$-measurable. It remains to show that the second defining property holds. But, if h is $\mathcal{G}$-measurable,

$$\mu(h\mu(g'|\mathcal{G})) = \mu(hg') = \mu(h\mu(f|\mathcal{G}')) = \mu(hf) \qquad (4.23)$$

which was to be shown. □

It is natural to associate to a conditional expectation the notion of a regular conditional probability *distribution*.

Definition 4.2.11 Given two sigma-algebras $\mathcal{F} \supset \mathcal{G}$, a regular conditional distribution is a function $\mu_{\mathcal{G}}^{\eta}$ such that

(i) for each $\eta \in \mathcal{S}$, $\mu_{\mathcal{G}}^{\eta}$ is a probability measure on $\mathcal{F}$, and
(ii) for each $A \in \mathcal{F}$, $\mu_{\mathcal{G}}^{\eta}(A)$ is a $\mathcal{G}$-measurable function such that for almost all η, $\mu_{\mathcal{G}}^{\eta}(A) = \mu(\mathbb{I}_A|\mathcal{G})(\eta)$.

The existence of regular conditional distributions is ensured in all situations we will be concerned with, in particular whenever the underlying probability spaces are Polish spaces (see, e.g., [28, 78]).

We see that local specifications are 'conditional expectations waiting for a measure'; thus nothing is more natural than to define infinite-volume Gibbs measures as follows:

Definition 4.2.12 Let $\{\mu_{\Lambda,\beta}^{(\cdot)}\}$ be a local specification. A measure μ_β is called compatible with this local specification if and only if, for all $\Lambda \subset \mathbb{Z}^d$ and all $f \in \mathcal{B}(\mathcal{S}, \mathcal{F})$,

$$\mu_\beta(f|\mathcal{F}_{\Lambda^c}) = \mu_{\Lambda,\beta}^{(\cdot)}(f), \quad \mu_\beta\text{-a.s.} \tag{4.24}$$

A measure μ_β which is compatible with the local Gibbs specification for the regular interaction Φ and a-priori measure ρ at inverse temperature β is called a Gibbs measure corresponding to Φ and ρ at inverse temperature β.

Remark 4.2.13 We see that the local specifications of a Gibbs measure provide an explicit version of their regular conditional distributions, as they exist for all η. One might be content with a weaker notion of Gibbs states, where local specifications are defined only for almost all $\eta \in \mathcal{S}$. The associated concepts of weaker notions of Gibbs measures are currently under active debate (see, e.g., [92, 171]).

Theorem 4.2.14 *A probability measure μ_β is a Gibbs measure for Φ, ρ, β if and only if, for all $\Lambda \subset \mathbb{Z}^d$,*

$$\mu_\beta \mu_{\Lambda,\beta}^{(\cdot)} = \mu_\beta \tag{4.25}$$

Proof Obviously, (4.25) holds if $\mu_{\Lambda,\beta}^{(\cdot)}(f)$ is the conditional probability $\mu_\beta(f|\mathcal{F}_{\Lambda^c})$, by definition. We only have to show the converse. But the local specifications are by construction $\mathcal{F}_{\Lambda^c}$-measurable, so that property (i) of Definition 4.2.9 is satisfied. To show that property (ii) holds, apply (4.25) with a function $f'(\eta) = f(\eta)h(\eta_{\Lambda^c})$ where h is $\mathcal{F}_{\Lambda^c}$-measurable. This shows that $\mu_{\Lambda,\beta}^{(\cdot)}(f)$ satisfies the second requirement of a conditional expectation of f. This proves the theorem. □

The equations (4.25) are called the DLR equations after Dobrushin, Lanford and Ruelle, to whom this construction is due. We have now achieved a rigorous definition of what the symbolic expression (4.2) is supposed to mean. Of course, this should be completed by an observation saying that such Gibbs measures exist in typical situations. This will turn out to be easy.

Theorem 4.2.15 *Let Φ be a continuous regular interaction and let $\mu_{\Lambda,\beta}^{(\cdot)}$ be the corresponding Gibbs specification. Let Λ_n be an increasing and absorbing sequence of finite volumes. If, for some $\eta \in \mathcal{S}$, the sequence of measures $\mu_{\Lambda_n,\beta}^{\eta}$ converges weakly to some probability measure ν, then ν is a Gibbs measure with respect to to Φ, ρ, β.*

Proof Let f be a continuous function. By hypothesis, we have that

$$\mu_{\Lambda_n,\beta}^{\eta}(f) \to \nu(f), \quad \text{as } n \uparrow \infty \tag{4.26}$$

On the other hand, for all $\Lambda_n \supset \Lambda$,

$$\mu^\eta_{\Lambda_n,\beta} \mu^{(\cdot)}_{\Lambda,\beta}(f) = \mu^\eta_{\Lambda_n,\beta}(f) \tag{4.27}$$

We would like to assert that $\mu^\eta_{\Lambda_n,\beta} \mu^{(\cdot)}_{\Lambda,\beta}(f)$ converges to $\nu \mu^{(\cdot)}_{\Lambda,\beta}(f)$, since this would immediately imply that ν satisfies the DLR equations (4.25) and hence is a Gibbs measure. To be able to make this assertion, we would need to know that $\mu^{(\cdot)}_{\Lambda,\beta}(f)$ is a continuous function. The property of a specification to map continuous functions to continuous functions is called the *Feller property*.

Lemma 4.2.16 *The local specifications of a continuous regular interaction have the Feller property.*

Proof We must show that, if $\eta_n \to \eta$, then $\mu^{\eta_n}_{\Lambda,\beta}(f) \to \mu^\eta_{\Lambda,\beta}(f)$. A simple consideration shows that, since f is continuous, this property follows if

$$H_\Lambda(\sigma_\Lambda, \eta_{n,\Lambda^c}) \to H_\Lambda(\sigma_\Lambda, \eta_{\Lambda^c}) \tag{4.28}$$

But H_Λ is by assumption a uniformly convergent sum of continuous functions, so it is itself continuous. Then (4.28) is immediate. □

The proof of Theorem 4.2.15 is now obvious. □

Exercise: Local specifications have even nicer properties than the Feller property. In particular, they are 'quasi-local', in the sense that they map local functions into quasi-local functions. This is expanded on in [250]. Prove the quasi-locality of local specifications and fill in the details in the proof of Lemma 4.2.16.

The constructive criterion of Theorem 4.2.15 now gives us a cheap existence result:

Corollary 4.2.17 *Assume that S_0 is compact and Φ is regular and continuous. Then there exists at least one Gibbs measure for any $0 \leq \beta < \infty$.*

Proof By Tychonov's theorem S is compact. The set of probability measures on a compact space is compact with respect to the weak topology, and so any sequence $\mu^\eta_{\Lambda_n,\beta}$ must have convergent subsequences. Any one of them provides a Gibbs measure, by Theorem 4.2.15. □

Remark 4.2.18 There are models with non-compact state space for which no Gibbs measure exists.

Theorem 4.2.15 is of absolutely central importance in the theory of Gibbs measures, since it gives a way to construct infinite-volume Gibbs measures. Physicists would even view this as the definition of infinite-volume Gibbs measures (and we will have to return to this attitude later when we discuss mean field models). The procedure of taking increasing sequences of finite-volume measures is called 'passing to the thermodynamic limit'. It is instructive to compare the physical 'approximation' statement contained in the DLR equations and in the weak limit construction. The DLR equations can be interpreted in the sense that, if we consider a physical finite system, when we apply 'boundary conditions'[5] and weigh these

---

[5] In the formal discussion we fixed configurations in the entire complement of Λ. Of course for models with short-range interactions, like the Ising model, the inside of a volume Λ depends only on the configuration on a layer of width one around Λ. Thus it is physically feasible to emulate the effect of the exterior of Λ just by boundary conditions.

with the infinite-volume measure μ_β, then the finite-volume measure within Λ will look exactly like the infinite-volume measure $\mu_{\Lambda,\beta}$. On the other hand, the constructive criterion of Theorem 4.2.14 means that there are suitable configurations η and suitable volumes Λ such that, if we fix boundary conditions η, the finite-volume measure looks, for large Λ, very much like an infinite-volume Gibbs state ν. It is experimentally not very feasible to apply boundary conditions weighted according to some Gibbs measure, while the second alternative seems a bit more realistic. But here difficulties will arise if the dependence on the boundary conditions and on the volumes is too dramatic. Such situations will arise in disordered systems.

Let us note that there is a different approach that characterizes Gibbs measures in terms of a *variational principle*. Such characterizations always carry a philosophical appeal as they appear to justify the particular choice of Gibbs measures as principal objects of interest. Excellent references are again [125] and [224], but also [146], and the recent lecture notes by Ch. Pfister [203]. Although several important notions linking statistical mechanics, thermodynamics, and the theory of large deviations arise in this context, we will not pursue this theme here.

4.3 Structure of Gibbs measures: phase transitions

In the previous section we established the concept of infinite-volume Gibbs measures and established the existence of such measures for a large class of systems. The next natural question is to understand the circumstances under which for a given interaction and a given temperature there exists a unique Gibbs measure, and when this is not the case. We have already seen that the possibility that the local specifications might be compatible with several Gibbs measures precisely provides for the possibility of describing phase transitions in this framework, and therefore this will be the case that we shall be most interested in. Nonetheless, it is important to understand under what conditions one must expect uniqueness. For this reason we start our discussion with some results on uniqueness conditions.

4.3.1 Dobrushin's uniqueness criterion

In a certain sense one should expect that, as a rule, a local specification is compatible with only one Gibbs measure. But there are specific interactions (or specific values of the parameters of an interaction), where this rule is violated.[6] However, there are general conditions that preclude this degenerate situation; vaguely, these conditions say that 'βH is small'; in this case one can see the Gibbs measure as a weak perturbation of the a-priori measure ρ. There are several ways of establishing such conditions. Possibly the most elegant one is due to Dobrushin, which we will present here. Our treatment follows closely that given in Simon's book [224], where the interested reader may find more material.

---

[6] The so-called *Gibbs phase rule* states that coexistence of several Gibbs measures should occur only on submanifolds of lower dimension in the space of interactions. A precise mathematical justification, or even formulation, of this rule is still missing (see [217] for a recent detailed discussion).

Let us introduce the *total variation distance* of two measures ν, μ by

$$\|\nu - \mu\| \equiv 2 \sup_{\mathcal{A} \in \mathcal{F}} |\nu(\mathcal{A}) - \mu(\mathcal{A})| \tag{4.29}$$

Theorem 4.3.1 *Let* $\mu^{(\cdot)}_{\Lambda,\beta}$ *be a local specification satisfying the Feller property. Set, for* $x, y \in \mathbb{Z}^d$,

$$\rho_{x,y} \equiv \frac{1}{2} \sup_{\substack{\eta,\eta' \\ \forall z \neq x\, \eta_z = \eta'_z}} \|\mu^{\eta}_{y,\beta} - \mu^{\eta'}_{y,\beta}\| \tag{4.30}$$

If $\sup_{y \in \mathbb{Z}^d} \sum_{x \in \mathbb{Z}^d} \rho_{x,y} < 1$, *then the local specification is compatible with at most one Gibbs measure.*

Proof For a continuous function f, we define its *variation* at x

$$\delta_x(f) = \sup_{\substack{\eta,\eta' \\ \forall z \neq x\, \eta_z = \eta'_z}} |f(\eta) - f(\eta')| \tag{4.31}$$

and the *total variation*

$$\Delta(f) \equiv \sum_{x \in \mathbb{Z}^d} \delta_x(f) \tag{4.32}$$

We define the set of functions of finite total variation $\mathcal{T} \equiv \{f \in C(\mathcal{S}) | \Delta(f) < \infty\}$. It is easy to check that this set is a dense subset of $C(\mathcal{S})$. The idea of the proof is:

(i) Show that Δ is a semi-norm and $\Delta(f) = 0 \Rightarrow f = const.$
(ii) Construct a contraction $\mathbb{T}$ with respect to Δ such that any solution of the DLR equations is $\mathbb{T}$-invariant.

Then it holds that, for any solution of the DLR equations, $\mu(f) = \mu(\mathbb{T}f) = \mu(\mathbb{T}^n f) \to c(f)$, independent of which one we choose. But the value on continuous functions determines μ, so all solutions of the DLR equations are identical.

To simplify notation we drop the reference to β in the course of the proof. Let us first establish (ii). To construct the map $\mathbb{T}$, let $x_1, x_2, \ldots, x_n, \ldots$ be an enumeration of all points in $\mathbb{Z}^d$ (this implies that x_n must disappear to infinity as $n \uparrow \infty$). Set

$$\mathbb{T}f \equiv \lim_{n \uparrow \infty} \mu^{(\cdot)}_{x_1} \ldots \mu^{(\cdot)}_{x_n}(f) \tag{4.33}$$

For any continuous function, the limit in (4.33) exists in norm. (Exercise: Prove this fact. Hint: Check the convergence first on local functions!) This implies that $\mathbb{T}$ maps continuous functions to continuous functions, which is a crucial property we will use.

It is obvious by construction that, if μ satisfies the DLR-equation with respect to the specification $\mu^{(\cdot)}_\Lambda$, then

$$\mu(\mathbb{T}f) = \mu(f) \tag{4.34}$$

It remains to show that $\mathbb{T}$ is a contraction with respect to Δ, if

$$\sup_{y \in \mathbb{Z}^d} \sum_{x \in \mathbb{Z}^d} \rho_{x,y} \leq \alpha < 1 \tag{4.35}$$

4.3 Structure of Gibbs measures

In fact, we will show that, under this hypothesis, $\Delta(\mathbb{T}f) \leq \alpha \Delta(f)$ for any continuous function f. We first look at $\delta_x(\mu_y(f))$.

Lemma 4.3.2 *Let $f \in \mathbb{T}$. Then*

(i)
$$\delta_x(\mu_x(f)) = 0 \tag{4.36}$$

(ii) For any $y \neq x$,

$$\delta_x(\mu_y(f)) \leq \delta_x(f) + \rho_{x,y}\delta_y(f) \tag{4.37}$$

Proof Obviously, $\delta_x(\mu_x(f)) = 0$, since $\mu_x(f)$ does not depend on η_x. Now let $x \neq y$. Then

$$\delta_x(\mu_y(f)) \equiv \sup_{\substack{\eta,\eta' \\ \forall_{z \neq x} \eta_z = \eta'_z}} \left|\mu_y^\eta(f) - \mu_y^{\eta'}(f)\right| \tag{4.38}$$

$$= \sup_{\substack{\eta,\eta' \\ \forall_{z \neq x} \eta_z = \eta'_z}} \left|\int f(\sigma_y, \eta_{y^c})\mu_y^\eta(d\sigma_y) - \int f(\sigma_y, \eta'_{y^c})\mu_y^{\eta'}(d\sigma_y) \right.$$

$$\left. + \int f(\sigma_y, \eta'_{y^c})\left(\mu_y^\eta(d\sigma_y) - \mu_y^{\eta'}(d\sigma_y)\right)\right|$$

$$\leq \sup_{\substack{\eta,\eta' \\ \forall_{z \neq x} \eta_z = \eta'_z}} \int \left|f(\sigma_y, \eta_{y^c}) - f(\sigma_y, \eta'_{y^c})\right|\mu_y^\eta(d\sigma_y)$$

$$+ \sup_{\substack{\eta,\eta' \\ \forall_{z \neq x} \eta_z = \eta'_z}} \left|\int f(\sigma_y, \eta'_{y^c})\left(\mu_y^\eta(d\sigma_y) - \mu_y^{\eta'}(d\sigma_y)\right)\right|$$

Clearly,

$$\sup_{\substack{\eta,\eta' \\ \forall_{z \neq x} \eta_z = \eta'_z}} \int \left|f(\sigma_y, \eta_{y^c}) - f(\sigma_y, \eta'_{y^c})\right|\mu_y^\eta(d\sigma_y) \leq \delta_x(f) \tag{4.39}$$

To treat the second term, we take advantage of the fact that any constant integrated against the difference of the two probability measures gives zero, so that

$$\left|\int f(\sigma_y, \eta'_{y^c})\left(\mu_y^\eta(d\sigma_y) - \mu_y^{\eta'}(d\sigma_y)\right)\right| \tag{4.40}$$

$$= \int \left|\left(f(\sigma_y, \eta'_{y^c}) - \inf_{\tau_y} f(\tau_y, \eta_{y^c})\right)\left(\mu_y^\eta(d\sigma_y) - \mu_y^{\eta'}(d\sigma_y)\right)\right|$$

$$\leq \sup_{\substack{\eta,\eta' \\ \forall_{z \neq y} \eta_z = \eta'_z}} |f(\eta) - f(\eta')| \sup_{\substack{\eta,\eta' \\ \forall_{z \neq x} \eta_z = \eta'_z}} \sup_{\mathcal{A} \in \mathcal{F}} \left|\mu_y^\eta(\mathcal{A}) - \mu_y^{\eta'}(\mathcal{A})\right|$$

$$= \frac{1}{2}\|\mu_y^\eta - \mu_y^{\eta'}\|\delta_y(f)$$

Combining the two estimates gives (ii). □

Lemma 4.3.3 *Under the hypothesis $\sup_{y \in \mathbb{Z}^d} \sum_{x \in \mathbb{Z}^d} \rho_{x,y} \leq \alpha$, for all $n \in \mathbb{N}$,*

$$\Delta\left(\mu_{x_1}^{(\cdot)} \cdots \mu_{x_n}^{(\cdot)} f\right) \leq \alpha \sum_{i=1}^n \delta_{x_i}(f) + \sum_{j \geq n+1} \delta_{x_j}(f) \tag{4.41}$$

Proof By induction. For $n = 0$, (4.41) is just the definition of Δ. Assume that (4.41) holds for n. Then

$$\Delta\big(\mu^{(\cdot)}_{x_1}\cdots\mu^{(\cdot)}_{x_n}\mu^{(\cdot)}_{x_{n+1}}f\big) \leq \alpha\sum_{i=1}^{n}\delta_{x_i}\big(\mu^{(\cdot)}_{x_{n+1}}f\big) + \sum_{j\geq n+1}\delta_{x_j}\big(\mu^{(\cdot)}_{x_{n+1}}f\big) \tag{4.42}$$

$$\leq \alpha\sum_{i=1}^{n}\big[\delta_{x_i}(f) + \rho_{x_i,x_{n+1}}\delta_{x_{n+1}}(f)\big] + \sum_{j\geq n+2}\big[\delta_{x_j}(f) + \rho_{x_j,x_{n+1}}\delta_{x_{n+1}}(f)\big]$$

$$= \alpha\sum_{i=1}^{n}\delta_{x_i}(f) + \sum_{i=1}^{\infty}\rho_{x_i,x_{n+1}}\delta_{x_{n+1}}(f) + \sum_{j\geq n+2}\delta_{x_j}(f) \leq \alpha\sum_{i=1}^{n+1}\delta_{x_i}(f) + \sum_{j\geq n+2}\delta_{x_j}(f)$$

so that (4.37) holds for $n + 1$. Note that the fact that $\delta_{x_{n+1}}(\mu^{(\cdot)}_{x_{n+1}}f) = 0$ was used crucially: it allowed us to omit the term $j = n + 1$ in the second sum. This proves the lemma. $\square$

Passing to the limit $n \uparrow \infty$ yields the desired estimate

$$\Delta(\mathbb{T}f) \leq \alpha\Delta(f) \tag{4.43}$$

It remains to be proven that $\Delta(f) = 0$ implies that $f = const$. We will show that $\Delta(f) \geq \sup(f) - \inf(f)$. Now, since f is continuous, for any $\epsilon > 0$ there exists a finite Λ and configurations ω^+, ω^- with $\omega^+_{\Lambda^c} = \omega^-_{\Lambda^c}$ such that

$$\sup(f) \leq f(\omega^+) + \epsilon, \tag{4.44}$$
$$\inf(f) \geq f(\omega^-) - \epsilon$$

But, using a simple telescopic expansion,

$$f(\omega^+) - f(\omega^-) \leq \sum_{x\in\Lambda}\delta_x(f) \leq \Delta(f) \tag{4.45}$$

Thus, $\sup(f) - \inf(f) \leq \Delta(f) + 2\epsilon$, for all ϵ, which implies the claimed bound. This concludes the proof of the theorem. $\square$

For Gibbs specifications with respect to regular interactions, the uniqueness criterion in Dobrushin's theorem becomes

$$\sup_{x\in\mathbb{Z}^d}\sum_{A\ni x}(|A| - 1)\|\Phi_A(\sigma)\|_\infty < \beta^{-1} \tag{4.46}$$

Thus it applies if the temperature β^{-1} is sufficiently 'high'.

If we apply this criterion formally in the Curie–Weiss model, we get the correct conditions $\beta < 1$ for uniqueness. This can be turned in a precise argument by considering so-called Kac interactions, where $\Phi_{x,y}(\sigma) = \gamma^d\phi((\gamma(x - y))\sigma_x\sigma_y$, and $\phi(u)$ is a non-negative, rapidly decaying function, normalized such that $\int d^d u\phi(u) = 1$. In this case, Dobrushin's criterion again gives $\beta < 1$, and it can be shown that the condition is optimal in the limit $\gamma \downarrow 0$ if $d \geq 1$.

Exercise: Compute the bound on the temperature for which Dobrushin's criterion applies in the Ising model (4.1).

The techniques of the Dobrushin uniqueness theorem can be pushed further to get more information about the unique Gibbs measure; in particular it allows us to prove the decay of correlations. Since this is not of immediate concern for us, we will not go into it. The interested reader is referred to the very clear exposition in Simon's book [224].

4.3.2 The Peierls argument

Having established a condition for uniqueness, it is natural to seek situations where uniqueness does not hold. As we mentioned earlier, this possibility was disbelieved for a long time and the solid establishment of the fact that such situations occur in reasonable models like the Ising model was one of the triumphs of statistical mechanics.

In contrast to the very general uniqueness criterion, situations with coexisting Gibbs measures are much more evasive and require case-by-case study. There exist a number of tools for investigating this problem in many situations, the most powerful being what is called the *Pirogov–Sinai theory* [204, 205], but, even in its most recent developments, it is far from being able to give a reasonably complete answer for a class of interactions as large as, e.g., the regular interactions.[7] We will discuss this theory briefly in Chapter 5.

The basis of most methods for proving the existence of multiple Gibbs states is the *Peierls argument*. We will explain this in the context in which it was originally derived, the Ising model, and discuss extensions later.

The basic intuition for the large β (low temperature) behaviour of the Ising model is that the Gibbs measure should in this case strongly favour configurations with minimal H. If $h \neq 0$, one sees that there is a unique configuration, $\sigma_x = \text{sign}(h)$, minimizing H, whereas for $h = 0$ there are two degenerate minima, $\sigma_x \equiv +1$ and $\sigma_x \equiv -1$. It is a natural idea to characterize a configuration by its deviations from an optimal one. This leads to the concept of the *contour*. We denote by $<xy>$ an edge of the lattice $\mathbb{Z}^d$ and by $<xy>^*$ the corresponding *dual plaquette*, i.e. the unique $(d-1)$-dimensional facet that cuts the edge in the middle. We set

$$\Gamma(\sigma) \equiv \{<xy>^* \,|\, \sigma_x \sigma_y = -1\} \tag{4.47}$$

Note that $\Gamma(\sigma)$ forms a surface in $\mathbb{R}^d$. The following properties are immediate from the definition:

Lemma 4.3.4 *Let Γ be the surface defined above, and let $\partial \Gamma$ denote its $(d-2)$-dimensional boundary.*

(i) $\partial \Gamma(\sigma) = \emptyset$ for all $\sigma \in \mathcal{S}$. Note that $\Gamma(\sigma)$ may have unbounded connected components.
(ii) Let Γ be a surface in the dual lattice such that $\partial \Gamma = \emptyset$. Then there are exactly two configurations, σ and $-\sigma$, such that $\Gamma(\sigma) = \Gamma(-\sigma) = \Gamma$.

Any Γ can be decomposed into its connected components, γ_i, called *contours*. We write $\gamma_i \in \Gamma$ if γ_i is a connected component of Γ. Any contour γ_i satisfies $\partial \gamma_i = \emptyset$. That is, each contour is either a finite, closed or an infinite, unbounded surface. We denote by int γ the volume enclosed by γ, and we write $|\gamma|$ for the number of plaquettes in γ.

---
[7] Of course it would be unreasonable to expect such a theory in any general form to exist.

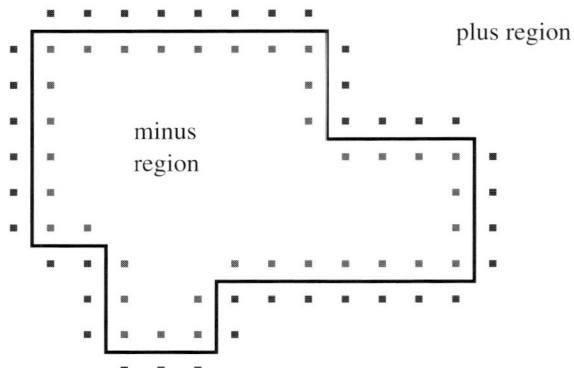

Figure 4.1 A contour (solid line) and its interior and exterior boundary.

The following theorem goes back to Peierls [201]. Its rigorous proof is due to Dobrushin [89] and Griffiths [130].

Theorem 4.3.5 *Let μ_β be a Gibbs measure for the Ising model (4.1) with $h = 0$ and ρ the symmetric product measure defined in (4.3). Assume that $d \geq 2$. Then there is $\beta_d < \infty$ such that for all $\beta > \beta_d$*

$$\mu_\beta \left[\exists_{\gamma \in \Gamma(\sigma): 0 \in \text{int } \gamma} \right] < \frac{1}{2} \tag{4.48}$$

The proof of this theorem is almost immediate from the following:

Lemma 4.3.6 *Let μ_β be a Gibbs measure for the Ising model with $h = 0$. Let γ be a finite contour. Then*

$$\mu_\beta \left[\gamma \in \Gamma(\sigma) \right] \leq 2 e^{-2\beta |\gamma|} \tag{4.49}$$

Proof We present the proof as an application of the DLR construction. Recall that γ is finite and thus closed. We will denote by γ^{in} and γ^{out} the layer of sites in $\mathbb{Z}^d$ adjacent to γ in the interior, resp. the exterior of γ, and call them the interior and exterior boundaries of the contour (see Fig. 4.1).

Clearly we have

$$\mu_\beta \left[\gamma \subset \Gamma(\sigma) \right] \equiv \mu[\sigma_{\gamma^{\text{out}}} = +1, \sigma_{\gamma^{\text{in}}} = -1] + \mu_\beta[\sigma_{\gamma^{\text{out}}} = -1, \sigma_{\gamma^{\text{in}}} = +1] \tag{4.50}$$

The DLR equations give

$$\mu_\beta[\sigma_{\gamma^{\text{out}}} = +1, \sigma_{\gamma^{\text{in}}} = -1] = \mu_\beta[\sigma_{\gamma^{\text{out}}} = +1] \mu_{\text{int } \gamma, \beta}^{+1}[\sigma_{\gamma^{\text{in}}} = -1] \tag{4.51}$$

4.3 Structure of Gibbs measures

But

$$\mu_{\mathrm{int}\,\gamma,\beta}^{+1}[\sigma_{\gamma^{\mathrm{in}}} = -1] \tag{4.52}$$

$$= \frac{\mathbb{E}_{\sigma_{\mathrm{int}(\gamma)\backslash\gamma^{\mathrm{in}}}}\rho(\sigma_{\gamma^{\mathrm{in}}} = -1)e^{-\beta H_{\mathrm{int}(\gamma)}\left(\sigma_{\mathrm{int}(\gamma)\backslash\gamma^{\mathrm{in}}}, -1_{\gamma^{\mathrm{in}}}, +1_{\gamma^{\mathrm{out}}}\right)}}{\mathbb{E}_{\sigma_{\gamma^{\mathrm{in}}}}\mathbb{E}_{\sigma_{\mathrm{int}(\gamma)\backslash\gamma^{\mathrm{in}}}}e^{-\beta H_{\mathrm{int}(\gamma)}\left(\sigma_{\mathrm{int}(\gamma)\backslash\gamma^{\mathrm{in}}}, \sigma_{\gamma^{\mathrm{in}}}, +1_{\gamma^{\mathrm{out}}}\right)}}$$

$$= \frac{e^{-\beta|\gamma|}Z_{\mathrm{int}(\gamma)\backslash\gamma^{\mathrm{in}}}^{(-1)}\rho(\sigma_{\gamma^{\mathrm{in}}} = -1)}{\mathbb{E}_{\sigma_{\gamma^{\mathrm{in}}}}e^{\beta\sum_{x\in\gamma^{\mathrm{in}}, y\in\gamma^{\mathrm{out}}}\sigma_y}Z_{\mathrm{int}(\gamma)\backslash\gamma^{\mathrm{in}}}^{\sigma_{\gamma^{\mathrm{in}}}}}$$

$$\leq e^{-2\beta|\gamma|}\frac{Z_{\mathrm{int}(\gamma)\backslash\gamma^{\mathrm{in}}}^{(-1)}}{Z_{\mathrm{int}(\gamma)\backslash\gamma^{\mathrm{in}}}^{(+1)}} = e^{-2\beta|\gamma|}$$

In the last line we used the symmetry of H_Λ under the global change $\sigma_x \to -\sigma_x$ to replace the ratio of the two partition functions with spin-flip related boundary conditions by one. If $h \neq 0$, this would not have been possible. The second term in (4.50) is treated in the same way. Thus (4.49) follows. □

Proof (of Theorem 4.3.5).   The proof of the theorem follows from the trivial estimate

$$\mu_\beta\left[\exists_{\gamma\in\Gamma(\sigma):0\in\mathrm{int}\,\gamma}\right] \leq \sum_{\gamma:0\in\mathrm{int}\,\gamma}\mu_\beta[\gamma\in\Gamma(\sigma)] \tag{4.53}$$

and (roughly) counting the number of contours of area k that enclose the origin. Let

$$\#\{\gamma : 0 \in \mathrm{int}\,\gamma, |\gamma| = k\} \equiv C(d, k) \tag{4.54}$$

It is a simple exercise to show that $C(2, k) \leq k3^k$. Obviously, any path γ of length k can be constructed as follows: choose a starting point within the square of side-length k centered at the origin. Then build up the path stepwise, noting that there are at most three possible moves at each step. Finally, note that each closed path constructed in this way is counted k times, because each of the points it visits can be considered the starting point. Not taking into account that the path has to be closed gives immediately the estimate above. This argument can be improved, and extended to any dimension; in this way, Ruelle [214] obtained that for any $d \geq 2$, $C(d, k) \leq 3^k$. In high dimension, this has been improved by Lebowitz and Mazel [165] to $C(d, k) \leq \exp(k 64 \ln d/d)$.

Thus, using Ruelle's bound,

$$\mu_\beta\left[\exists_{\gamma\in\Gamma(\sigma):0\in\mathrm{int}\gamma}\right] \leq \sum_{k=2d}^{\infty} e^{-k(2\beta-\ln 3)} \tag{4.55}$$

so choosing β a little larger than $\frac{1}{2}\ln 3$ we get the claimed estimate. □

Notice that Theorem 4.3.5 does *not* imply that there are no *infinite* contours with positive probability.

Theorem 4.3.5 brings us very close to showing the existence of at least two Gibbs states. Intuitively, it implies that, with probability greater than $1/2$, the spin at the origin has the same sign as 'the spins at infinity', which in turn could be plus one or minus one. Most importantly, the spin at the origin is correlated to those at infinity, establishing the existence of long-range correlation.

Theorem 4.3.7 *Consider the Ising model for parameters where the conclusion of Theorem 4.3.5 hold. Then there exist (at least) two extremal Gibbs measures μ_β^+ and μ_β^- satisfying $\mu^+(\sigma_0) = -\mu^-(\sigma_0) > 0$.*

Proof Let $\Lambda_n \uparrow \mathbb{Z}^d$ be a sequence of volumes such that the sequence of local specifications μ_{β,Λ_n}^+ converges to a Gibbs measure μ_β^+, where $+$ stands for the constant configuration $\eta_x \equiv +1$, $\forall x \in \mathbb{Z}^d$. Then for any n, $\mu_{\beta,\Lambda_n}^+(\sigma_0 = -1) \leq \mu_{\beta,\Lambda_n}^+ (\exists \gamma : 0 \in \text{int} \gamma) < \frac{1}{2}$, uniformly in n, as the proof of Theorem 4.3.5 applies unchanged to μ_{β,Λ_n}^+. On the other hand, $\mathbb{I}_{\sigma_0 = -1}$ is a local function, so

$$\mu_\beta^+(\sigma_0 = -1) \leq \lim_{n \uparrow \infty} \mu_{\beta,\Lambda_n}^+ \left(\exists_{\gamma \in \Gamma(\sigma)} : 0 \in \text{int } \gamma \right) < \frac{1}{2} \quad (4.56)$$

which implies the theorem. □

On a qualitative level, we have now solved Ising's problem: the Ising model in dimension two and higher has a unique Gibbs state with decaying correlations at high temperatures, while at low temperature there are at least two extremal ones, which exhibit spontaneous magnetization. Thus, the phenomenon of a phase transition in ferromagnets is reproduced by this simple system with short-range interaction.

I said earlier that the Peierls argument is the basis of most proofs of the existence of multiple Gibbs states. This is true in the sense that whenever one wants to prove such a fact, one will want to introduce some notion of contours that characterize a locally unlikely configuration; one will then want to conclude that 'typical' configurations do not contain large regions where configurations are atypical, and finally one will want to use that there are several choices for configurations not containing large undesirable regions. What is lacking then is an argument showing that these 'good' regions are equally likely; on a more technical level, this corresponds to being able to pass from the one-but-last line in (4.52) to the last one. In the Ising model we were helped by the spin-flip symmetry of the problem. This should be considered accidental, as should be the fact that the ratio of the two partition functions appearing in (4.51) is equal to one. In fact, they are equal because the parameter h was chosen equal to zero. In a situation without symmetry, one should expect that there will be some value of h (or other parameters of the model) for which the ratio of the partition function is close enough to one, for all γ. This is a subtle issue and at the heart of the *Pirogov–Sinai theory* [204, 205, 262, 263]. Most methods for analyzing such problems in detail rely on perturbative methods that in statistical mechanics go by the name of *cluster expansions*. Chapter 5 will be devoted to such methods.

Having seen that the non-uniqueness of Gibbs states does in fact occur, we are motivated to investigate the structure of the set of Gibbs states more closely.

By the characterization of Gibbs measures through the DLR equations it is obvious that, if μ_β, μ_β' are any two Gibbs measures for the same local specification, their convex combinations, $p\mu_\beta + (1-p)\mu_\beta'$, $p \in [0, 1]$, are also Gibbs measures. Thus, the set of Gibbs measures for a local specification forms a closed convex set. One calls the extremal points of this set *extremal Gibbs measures* or *pure states*.[8]

The following gives an important characterization of extremal Gibbs measures.

---

[8] The name 'pure state' is sometimes reserved for extremal translation-invariant Gibbs measures.

4.3 Structure of Gibbs measures

Proposition 4.3.8 *A Gibbs measure μ_β is extremal if and only if it is trivial on the tail sigma-field $\mathcal{F}^t$, i.e. if, for all $\mathcal{A} \in \mathcal{F}^t$, $\mu_\beta(\mathcal{A}) \in \{0, 1\}$.*

To prove this proposition, we need two important observations:
The first states that a Gibbs measure is characterized by its value on the tail sigma-field.

Proposition 4.3.9 *Let μ_β and ν_β be two Gibbs measures for the same specification. If, for all $\mathcal{A} \in \mathcal{F}^t$, $\nu_\beta(\mathcal{A}) = \mu_\beta(\mathcal{A})$, then $\nu_\beta = \mu_\beta$.*

Proof   Again we use the DLR equations. Let f be any local function. Since, for any Λ,

$$\mu_\beta(f) = \mu_\beta\bigl(\mu_{\beta,\Lambda}^{(\cdot)}(f)\bigr) \tag{4.57}$$
$$\nu_\beta(f) = \nu_\beta\bigl(\mu_{\beta,\Lambda}^{(\cdot)}(f)\bigr)$$

the lemma follows if $\lim_{\Lambda\uparrow\mathbb{Z}^d}\mu_{\beta,\Lambda}^{(\cdot)}(f)$ is measurable with respect to $\mathcal{F}^t$. But, by definition, $\mu_{\beta,\Lambda}^{(\cdot)}(f)$ is measurable with respect to $\mathcal{F}_{\Lambda^c}$, and so $\lim_{\Lambda\uparrow\mathbb{Z}^d}\mu_{\beta,\Lambda}^{(\cdot)}(f)$ is measurable with respect to $\cap_{\Lambda\uparrow\mathbb{Z}^d}\mathcal{F}_{\Lambda^c}$, i.e. $\mathcal{F}^t$. □

The second observation is:

Lemma 4.3.10 *Let μ be a Gibbs measure, and $\mathcal{A} \in \mathcal{F}^t$ with $\mu(\mathcal{A}) > 0$. The conditioned measure $\mu(\cdot|\mathcal{A})$ is also a Gibbs measure for the same specification.*

Proof   We again consider a local function f. Then

$$\mu(f|\mathcal{A}) \equiv \frac{\mu(f\mathbb{I}_\mathcal{A})}{\mu(\mathcal{A})} = \frac{\mu\mu_\Lambda^{(\cdot)}(f\mathbb{I}_\mathcal{A})}{\mu(\mathcal{A})} = \frac{\mu\mathbb{I}_\mathcal{A}\mu_\Lambda^{(\cdot)}(f)}{\mu(\mathcal{A})}$$
$$= \mu\bigl(\mu_\Lambda^{(\cdot)}(f)|\mathcal{A}\bigr) \tag{4.58}$$

for any Λ; so $\mu(\cdot|\mathcal{A})$ satisfies the DLR equations. □

Proof   (of Proposition 4.3.8). Assume that μ is trivial on the tail field and $\mu = p\mu' + (1-p)\mu''$, for $p \in (0, 1)$. Then, for any $\mathcal{A} \in \mathcal{F}^t$, by Lemma 4.3.9,

$$p\mu'(\mathcal{A}) + (1-p)\mu''(\mathcal{A}) \in \{0, 1\} \tag{4.59}$$

But this can only hold if $\mu'(\mathcal{A}) = \mu''(\mathcal{A}) \in \{0, 1\}$, and so $\mu' = \mu''$.

To prove the converse, assume that μ is not trivial on the tail field. Then there exists $\mathcal{A} \in \mathcal{F}^t$ with $\mu(\mathcal{A}) = p \in (0, 1)$. So, by Lemma 4.3.10,

$$\mu = p\mu(\cdot|\mathcal{A}) + (1-p)\mu(\cdot|\mathcal{A}^c) \tag{4.60}$$

and, by Lemma 4.3.10, $\mu(\cdot|\mathcal{A})$ and $\mu(\cdot|\mathcal{A}^c)$ are Gibbs measures, so μ is not extremal. This concludes the proof of the proposition. □

Tail field triviality is equivalent to a certain uniform decay of correlations, which is a common alternative characterization of extremal Gibbs measures:

Corollary 4.3.11 *A Gibbs measure μ is trivial on the tail sigma-field if and only if, for all $\mathcal{A} \in \mathcal{F}$,*

$$\lim_{\Lambda\uparrow\mathbb{Z}^d} \sup_{\mathcal{B}\in\mathcal{F}_{\Lambda^c}} |\mu(\mathcal{A}\cup\mathcal{B}) - \mu(\mathcal{A})\mu(\mathcal{B})| = 0 \tag{4.61}$$

4.3.3 The FKG inequalities and monotonicity

The Peierls argument gave us the possibility of proving the existence of more than one Gibbs measure in the Ising model. Still, even this argument is not constructive in the sense that it allows us to exhibit particular sequences of finite volume measures that will actually converge to different extremal Gibbs states. Of course it is a natural guess that this should be the case if we take, for instance, a sequence of increasing cubes, and choose as boundary conditions the configurations $\eta_x \equiv +1$ and $\eta_x \equiv -1$, for all $x \in \mathbb{Z}^d$, respectively. Strangely enough, this is not that easy to prove and requires the help of so-called *correlation inequalities*, which in turn rely strongly on specific properties of the model at hand. The FKG inequalities, named after Fortuin, Kasteleyn, and Ginibre [103] are amongst the most useful ones. We will briefly discuss them and some of their applications. For more material, see [84].

Definition 4.3.12 Let the single-spin space S be a linearly ordered set. We say that a probability measure μ on S_Λ for a finite $\Lambda \subset \mathbb{Z}^d$ satisfies the *FKG inequalities* or *is positively correlated* if, for all bounded, $\mathcal{F}_\Lambda$-measurable functions f, g that are non-decreasing with respect to the partial order on S_Λ induced by the order on S, it holds that

$$\mu(fg) \geq \mu(f)\mu(g) \tag{4.62}$$

Remark 4.3.13 The assertion (4.62) is trivial in the case when the underlying probability space is a *completely* ordered set, e.g. if Λ is a single point and S_0 is a subset of $\mathbb{R}$. In that case one just observes that

$$\mu(fg) - \mu(f)\mu(g) \tag{4.63}$$
$$= \frac{1}{2} \int \mu(d\sigma) \int \mu(d\tau)(f(\sigma) - f(\tau))(g(\sigma) - g(\tau)) \geq 0$$

where the last inequality follows since, if both f and g are increasing, then whenever σ and τ are comparable the two factors in the integral have the same sign. But on a completely ordered space, this is always the case.

Theorem 4.3.14 [103] *Assume that the cardinality of S_0 is 2, and consider a ferromagnetic pair interaction. Then any finite-volume Gibbs measure for this interaction satisfies the FKG inequalities.*

Proof We will give a proof following Battle and Rosen [17] as given in Ellis [98] for ferromagnetic Ising models with the Hamiltonian

$$H_\Lambda(\sigma) = -\sum_{x \vee y \in \Lambda} J_{x,y} \sigma_x \sigma_y - \sum_{x \in \Lambda} h_x \sigma_x \tag{4.64}$$

where all $J_{x,y} \geq 0$ and $\sup_x \sum_{y \in \mathbb{Z}^x} J_{x,y} < \infty$. It will be convenient to consider the local specifications, $\mu_{\beta,\Lambda}^{(\eta)}$, as functions of real-valued variables η_x, $x \in \Lambda^c$, rather than only $\{-1, +1\}$-valued variables. The proof then proceeds by induction over the size of the volume Λ. Note first that if $|\Lambda| = 1$, the assertion

$$\mu_{\beta,\{x\}}^\eta(fg) \geq \mu_{\beta,\{x\}}^\eta(f) \mu_{\beta,\{x\}}^\eta(g) \tag{4.65}$$

holds trivially, as we just remarked. Assume that the assertion holds for $\Lambda \subset \mathbb{Z}^d$. Take any $y \in \Lambda^c$ and set $\Lambda' = \Lambda \cup \{y\}$. We want to show that the assertion follows for any $\mu_{\beta,\Lambda'}^\eta$, and any two non-decreasing, bounded and $\mathcal{F}_{\Lambda'}$-measurable functions f, g. Notice first that, by the compatibility of local specifications,

$$\mu_{\beta,\Lambda'}^\eta(fg) = \sum_{\eta_y = \pm 1} \mu_{\beta,\Lambda'}^\eta(\sigma_y = \eta_y)\big(\mu_{\beta,\Lambda}^\eta(fg)\big) \quad (4.66)$$

$$\geq \sum_{\eta_y = \pm 1} \mu_{\beta,\Lambda'}^\eta(\sigma_y = \eta_y)\big(\mu_{\beta,\Lambda}^\eta(f)\big)\big(\mu_{\beta,\Lambda}^\eta(g)\big)$$

where we used the induction hypothesis. Since the sum over η_y satisfies FKG trivially, we only need to show that $\mu_{\beta,\Lambda}^\eta(f)$ is a monotone function of the variable η_y if f is monotone. Suppressing all variables except η_y in the notation, this task reduces to showing that $\mu_{\beta,\Lambda}^{+1}(f(+1)) \geq \mu_{\beta,\Lambda}^{-1}(f(-1))$. Since $f(-1) \leq f(+1)$, we may as well show the stronger

$$\mu_{\beta,\Lambda}^{+1}(f(+1)) \geq \mu_{\beta,\Lambda}^{-1}(f(+1)) \quad (4.67)$$

Recalling that η_y may be considered as a real variable, (4.67) follows in turn from

$$\frac{d}{d\eta_y}\mu_{\beta,\Lambda}^{\eta_y}(f(+1)) = \mu_{\beta,\Lambda}^{\eta_y}\left(f(+1)\sum_{z \in \Lambda} J_{zy}\sigma_y\right) - \mu_{\beta,\Lambda}^{\eta_y}\left(\sum_{z \in \Lambda} J_{zy}\sigma_y\right)\mu_{\beta,\Lambda}^{\eta_y}(f(+1)) \geq 0 \quad (4.68)$$

where the first equality follows from explicit differentiation, and the second inequality holds because $\sum_{z \in \Lambda} J_{zy}\sigma_y$ is a non-decreasing function since all J_{xy} are positive. This concludes the argument. $\square$

We will now show how the FKG inequalities can be used to prove interesting facts about the Gibbs measures.

Lemma 4.3.15 *Let $\mu_{\beta,\Lambda}^\eta$ be local specifications for a Gibbs measure that satisfies the FKG inequalities. Denote by $+$ the spin configuration $\eta_x = +1, \forall x \in \mathbb{Z}^d$. Then*

(i) For any $\Lambda \subset \mathbb{Z}^d$, any $\eta \in \mathcal{S}$, and any increasing function $f : \mathcal{S}_\Lambda \to \mathbb{R}$,

$$\mu_{\beta,\Lambda}^+(f) \geq \mu_{\beta,\Lambda}^\eta(f) \quad (4.69)$$

(ii) For any $\Lambda_2 \supset \Lambda_1$, and any increasing function $f : \mathcal{S}_{\Lambda_1} \to \mathbb{R}$,

$$\mu_{\beta,\Lambda_2}^+(f) \leq \mu_{\beta,\Lambda_1}^+(f) \quad (4.70)$$

Proof For the proof we only consider the case where $\mathcal{S}_0 = \{-1, 1\}$. We first prove (i). Let $x \in \Lambda^c$, and consider η_x as an element of $[-1, 1]$. We will show that $\mu_{\beta,\Lambda}^\eta(f)$ is increasing in η_x. If this is true, (4.69) is immediate. Now compute

$$\frac{\partial}{\partial \eta_x}\mu_{\beta,\Lambda}^\eta(f) = \sum_{y \in \Lambda} \beta J_{xy}\big(\mu_{\beta,\Lambda}^\eta(\sigma_y f) - \mu_{\beta,\Lambda}^\eta(\sigma_y)\mu_{\beta,\Lambda}^\eta(f)\big) \quad (4.71)$$

Since all J_{xy} are positive, and since σ_y is an increasing function, by the FKG inequalities the right-hand side of (4.71) is non-negative and (i) is proven.

To prove (ii), consider $\mu^+_{\beta,\Lambda_2}(\mathbb{1}_{+1_{\Lambda_2\setminus\Lambda_1}}f)$. By FKG,

$$\mu^+_{\beta,\Lambda_2}\big(\mathbb{1}_{+1_{\Lambda_2\setminus\Lambda_1}}f\big) \geq \mu^+_{\beta,\Lambda_2}\big(\mathbb{1}_{+1_{\Lambda_2\setminus\Lambda_1}}\big)\mu^+_{\beta,\Lambda_2}(f) \qquad (4.72)$$

$$= \exp\left(\beta \sum_{\substack{x,y\in\Lambda_1^c \\ x\vee y\in\Lambda_2\setminus\Lambda_1}} J_{xy} - \beta \sum_{x\in\Lambda_2\setminus\Lambda_1} h_x\right) \frac{Z^+_{\beta,\Lambda_1}}{Z^+_{\beta,\Lambda_2}} \mu^+_{\beta,\Lambda_2}(f)$$

where the equality uses the DLR equations. On the other hand, applying the DLR equations directly to the left-hand side of (4.72), we get

$$\mu^+_{\beta,\Lambda_2}\big(\mathbb{1}_{+1_{\Lambda_2\setminus\Lambda_1}}f\big) = \mu^+_{\beta,\Lambda_1}(f)\exp\left(\beta \sum_{\substack{x,y\in\Lambda_1^c \\ x\vee y\in\Lambda_2\setminus\Lambda_1}} J_{xy} - \beta \sum_{x\in\Lambda_2\setminus\Lambda_1} h_x\right) \frac{Z^+_{\beta,\Lambda_1}}{Z^+_{\beta,\Lambda_2}} \qquad (4.73)$$

and combining both observations we have (ii). $\square$

An immediate consequence of this lemma is:

Corollary 4.3.16 *Under the hypothesis of Lemma 4.3.15:*

(i) *For any increasing and absorbing sequence of volumes $\Lambda_n \subset \mathbb{Z}^d$, the limit*

$$\lim_{n\uparrow\infty} \mu^+_{\beta,\Lambda_n} \equiv \mu^+_\beta \qquad (4.74)$$

exists and is independent of the particular sequence.

(ii) *The Gibbs measure μ^+_β is extremal.*

(iii) *Similarly, the limit*

$$\lim_{n\uparrow\infty} \mu^-_{\beta,\Lambda_n} \equiv \mu^-_\beta \qquad (4.75)$$

exists, is independent of the sequence Λ_n and is an extremal Gibbs measure.

(iv) *For all Gibbs measures for the same interaction and temperature, and any increasing, bounded, continuous function f,*

$$\mu^-_\beta(f) \leq \mu_\beta(f) \leq \mu^+_\beta(f) \qquad (4.76)$$

Proof Note that compactness and monotonicity (4.70) imply that, for all increasing, bounded, continuous functions, for any sequence, Λ_n, of increasing and absorbing sequences, the limit $\mu^+_{\beta,\Lambda_n}(f)$ exists. Let Λ_n, Λ'_n be two such sequences. Since both sequences are absorbing, it follows that there exist infinite subsequences n_k and n'_k such that, for all $k \in \mathbb{N}$, $\Lambda_{n_k} \subset \Lambda'_{n'_k} \subset \Lambda_{n_{k+1}}$. But this implies that

$$\lim_{n\uparrow\infty} \mu^+_{\beta,\Lambda_n}(f) = \lim_{k\uparrow\infty} \mu^+_{\beta,\Lambda_{n_k}}(f) \qquad (4.77)$$

$$\geq \lim_{k\uparrow\infty} \mu^+_{\beta,\Lambda'_{n'_k}}(f) = \lim_{n\uparrow\infty} \mu^+_{\beta,\Lambda'_n}(f)$$

and

$$\lim_{n\uparrow\infty} \mu^+_{\beta,\Lambda_n}(f) = \lim_{k\uparrow\infty} \mu^+_{\beta,\Lambda_{n_{k+1}}}(f) \qquad (4.78)$$

$$\leq \lim_{k\uparrow\infty} \mu^+_{\beta,\Lambda'_{n'_k}}(f) = \lim_{n\uparrow\infty} \mu^+_{\beta,\Lambda'_n}(f)$$

and so
$$\lim_{n\uparrow\infty} \mu^+_{\beta,\Lambda_n}(f) = \lim_{n\uparrow\infty} \mu^+_{\beta,\Lambda'_n}(f) \tag{4.79}$$

Thus, all possible limit points of $\mu^+_{\beta,\Lambda}$ coincide on the set of increasing, bounded, continuous functions. But then, by standard approximation arguments, the limits coincide on all bounded continuous functions, which implies that the limiting measures exist and are independent of the subsequences chosen. This proves (i). To prove (ii), assume that μ^+_β is not extremal. Then there exist two distinct Gibbs measures μ and ν such that $\mu^+_\beta = \alpha\mu_\beta + (1-\alpha)\nu_\beta$, with $\alpha > 0$. In particular, for f increasing,
$$\mu^+_\beta(f) = \alpha\mu_\beta(f) + (1-\alpha)\nu_\beta(f) \tag{4.80}$$

Now, by (4.69) and the DLR equations, for any local increasing function f, for all Λ so large that f is $\mathcal{F}_\Lambda$-measurable, for any Gibbs measure ν_β,
$$\nu_\beta(f) = \nu_\beta\big(\mu_{\beta,\Lambda}(f)\big) \leq \mu^+_{\beta,\Lambda}(f) \tag{4.81}$$

Since $\mu^+_{\beta,\Lambda}$ converges to μ^+_β, this implies that
$$\nu_\beta(f) \leq \mu^+_\beta(f) \tag{4.82}$$

Thus, (4.80) can only hold if both $\mu_\beta(f)$ and $\nu_\beta(f)$ are equal to $\mu^+_\beta(f)$. But then, by the same argument as before, we conclude that $\mu_\beta = \nu_\beta = \mu^+_\beta$, contradicting the assumption that μ_β and ν_β are different. This proves (ii). (iii) is obvious by repeating all arguments with decreasing functions, which also yields the complementary version of (4.82), which implies (iv). □

As a final result we will show that in the presence of FKG inequalities, the uniqueness of the Gibbs state can be tied to a so-called order parameter. If μ is a Gibbs measure, we set
$$m^\mu \equiv \lim_{\Lambda\uparrow\infty} \frac{1}{|\Lambda|} \sum_{x\in\Lambda} \mu(\sigma_x) \tag{4.83}$$

provided the limit exists. We will also use the notation $m^\pm_\beta = m^{\mu^\pm_\beta}$.

Proposition 4.3.17 *Consider a translation-invariant system for which the FKG inequalities hold. Then the two measures μ^+_β and μ^-_β coincide if and only if $m^+_\beta = m^-_\beta$.*

This result is due to Lebowitz and Martin-Löf [163] and Ruelle [213]. We give a proof in the Ising case following Preston [207]. It is based on the following simple lemma:

Lemma 4.3.18 *Consider a model with Ising spins for which the FKG inequalities hold. Then for any finite sets $A, B \subset \Lambda$,*
$$\mu^+_\beta(\sigma_{A\cup B} = +1) - \mu^-_\beta(\sigma_{A\cup B} = +1) \tag{4.84}$$
$$\leq \mu^+_\beta(\sigma_A = +1) - \mu^-_\beta(\sigma_A = +1) + \mu^+_\beta(\sigma_B = +1) - \mu^-_\beta(\sigma_B = +1)$$

(where $\sigma_A = +1$ is shorthand for $\forall_{x\in A}\sigma_x = +1$).

Proof Notice the set-equality
$$\mathbb{I}_{\sigma_A=+1 \wedge \sigma_B=+1} = \mathbb{I}_{\sigma_A=+1} + \mathbb{I}_{\sigma_B=+1} - \mathbb{I}_{\sigma_A=+1 \vee \sigma_B=+1} \tag{4.85}$$

This implies that

$$\mu_\beta^+(\sigma_{A\cup B} = +1) - \mu_\beta^-(\sigma_{A\cup B} = +1) \tag{4.86}$$
$$= \mu_\beta^+(\sigma_A = +1) - \mu_\beta^-(\sigma_A = +1) + \mu_\beta^+(\sigma_B = +1) - \mu_\beta^-(\sigma_B = +1)$$
$$+ \mu_\beta^-(\sigma_A = +1 \vee \sigma_B = +1) - \mu_\beta^+(\sigma_A = +1 \vee \sigma_B = +1)$$

But $\{\sigma_A = +1 \vee \sigma_B = +1\}$ is an increasing event, and so, by (4.76),

$$\mu_\beta^-(\sigma_A = +1 \vee \sigma_B = +1) - \mu_\beta^+(\sigma_A = +1 \vee \sigma_B = +1) \leq 0 \tag{4.87}$$

This implies the assertion of the lemma.

In the Ising model, all local functions can expressed in terms of the indicator functions $\mathbb{1}_{\sigma_A = +1}$, for finite $A \subset \Lambda$. By repeated application of Lemma 4.3.18, we get

$$0 \leq \mu_\beta^+(\sigma_A = +1) - \mu_\beta^-(\sigma_A = +1) \leq \sum_{x \in A} \mu_\beta^+(\sigma_x = +1) - \mu_\beta^-(\sigma_x = +1) \tag{4.88}$$

Therefore, if, for all x, $\mu_\beta^+(\sigma_x = +1) = \mu_\beta^-(\sigma_x = +1)$, it follows indeed that $\mu_\beta^+ = \mu_\beta^-$. This concludes the proof of Proposition 4.3.17. □

The (macroscopic) functions m^μ are called *order parameters* because their values allow us to decide (in this model) on the uniqueness, respectively co-existence, of phases. One can generalize this notion to other models, and one may set up a general theory that is able to produce interesting abstract results (see [125]). Recall that, after all, extremal Gibbs measures are characterized by their values on the tail sigma-field, i.e. by their values on macroscopic functions. The general philosophy would thus be to identify a (hopefully) finite set of macroscopic functions whose values suffice to characterize all possible Gibbs states of the system. We will not enter this subject here, but will have occasion to return to the notion of order parameters in our discussion of disordered systems.

Remark 4.3.19 One would tend to believe that, in the Ising model, the Gibbs measures $\mu_\beta^\pm$ should be the only extremal Gibbs states. However, this turns out to be true only in dimension $d \leq 2$, as was proven by Aizenman [2] and Higushi [138] (see also [126] for a simplified proof). In dimension $d \geq 3$, it is only true that these two states exhaust the *translation-invariant* extremal Gibbs states. This was first proven for low enough temperatures by Gallavotti and Miracle-Solé [111], and only very recently by Bodineau for all $\beta \neq \beta_c$ [30]. Dobrushin [90] (see also van Beijeren [243]) showed that in $d \geq 3$, for low enough temperatures, there exist further non translation-invariant states (called *Dobrushin states*), that describe states with an interface separating two half-spaces where spins are predominantly positive, respectively negative in the corresponding regions. They can be constructed with mixed boundary conditions (e.g. $\eta_x = +1$, $x_3 \geq 0$, $\eta_x = -1$, $x_3 < 0$). The full classification of extremal states in $d \geq 3$ is not known.

5

Cluster expansions

> Derrière la série de Fourier, d'autres séries analogues sont entrées dans la domaine de l'analyse; elles y sont entrées par la même porte; elles ont été imaginées en vue des applications.[1]
>
> Henri Poincaré, *La valeur de la science.*

Most computational methods in statistical mechanics rely upon perturbation theory around situations that are well understood. The simplest one is, as always, the ideal gas. Expansions around the ideal gas are known as *high-temperature* or *weak-coupling expansions*. The other type of expansions concern the situation when the Gibbs measure concentrates near a single *ground-state* configuration. Such expansions are known as *low-temperature expansions*. Technically, in both cases, they involve a reformulation of the model in terms of what is called a *polymer model*. We begin with the high-temperature case, which is both simpler and less model-dependent than the low-temperature case, and show how a polymer model is derived.

5.1 High-temperature expansions

We place ourselves in the context of regular interactions, and we assume that β will be small. In this situation, we can expect that our Gibbs measure should behave like a product measure. To analyze such a situation, we will always study the local specifications, establishing that they depend only weakly on boundary conditions. The first, and in a sense generic, step consists in computing the partition function

$$Z^{\eta}_{\Lambda,\beta} = \int d\rho_\Lambda(\sigma_\Lambda) \exp\left(\beta \sum_{A \cap \Lambda \neq \emptyset} \Phi_A(\sigma_\Lambda, \eta_{\Lambda^c})\right) \tag{5.1}$$

The basic idea of the high-temperature expansion is to use the trivial formula

$$\exp\left(\beta \sum_{A \cap \Lambda \neq \emptyset} \Phi_A(\sigma_\Lambda, \eta_{\Lambda^c})\right) = \prod_{A \cap \Lambda \neq \emptyset} \exp\left(\beta \Phi_A(\sigma_\Lambda, \eta_{\Lambda^c})\right) \tag{5.2}$$

$$= \prod_{A \cap \Lambda \neq \emptyset} (1 + \exp(\beta \Phi_A(\sigma_\Lambda, \eta_{\Lambda^c})) - 1)$$

---

[1] Approximately: After the Fourier series, other series have entered the domain of analysis; they entered by the same door; they have been imagined in view of applications.

to think of $e^{\beta \Phi_A(\sigma_\Lambda, \eta_{\Lambda^c})} - 1$ as being small, and consequently to expand the product over the A's into a sum

$$\prod_{A \cap \Lambda \neq \emptyset} \left(1 + e^{\beta \Phi_A(\sigma_\Lambda, \eta_{\Lambda^c})} - 1\right) = \sum_{G \in \mathcal{G}_\Lambda} \prod_{A \in G} \left(e^{\beta \Phi_A(\sigma_\Lambda, \eta_{\Lambda^c})} - 1\right) \quad (5.3)$$

where $\mathcal{G}_\Lambda$ is the set of all subsets of the collection of all finite sets $A \subset \mathbb{Z}^d$ that intersect Λ (without repetition). Of course, in concrete examples, this set can be reduced to those sets A for which $\Phi_A \neq 0$. The elements of the set $\mathcal{G}_\Lambda$ will be called *polymers*.

Example: In the case of the Ising model with zero magnetic field, the only relevant sets A are pairs of nearest-neighbour bonds $<x,y>$. The sum is then over all collections of subsets built from such bonds. These are nicely interpreted as graphs on the lattice. In this case, the collection of possible polymers intersecting Λ is finite.

Definition 5.1.1 If $G = (A_1, \ldots, A_n)$ is a collection of sets, we call $\underline{G} = \cup_{i=1}^n A_i$ the support of G. We say that a polymer $G \subset \mathcal{G}_\Lambda$ is *connected* if it cannot be decomposed into two subcollections, whose supports are non-intersecting, i.e. if, for any decomposition $G = (g, g')$, $\underline{g} \cap \underline{g}' \neq \emptyset$. Two connected polymers are called non-intersecting if their supports have empty intersection.

Remark 5.1.2 Note that in this definition the constituent sets A that make polymers are considered connected.

Lemma 5.1.3 *Any polymer $G \in \Lambda$ can be uniquely decomposed into a collection of mutually non-intersecting connected polymers $g_1, \ldots, g_k$ such that $G = \cup_{i=1}^k g_i$.*

Proof Any $G \in \mathcal{G}_\Lambda$ is of the form $G = (A_1, \ldots, A_n)$, where the A_i are subsets of $\mathbb{Z}^d$ that intersect Λ. If G is connected, we are done and $G = g$. Otherwise, we may pick A_1 and look for the largest subset $(A_1 = A_{i_1}, \ldots, A_{i_l})$ that is connected. Call this connected polymer g_1. Now all other subsets $A_j \in G$ do not intersect any of these A_{i_k}. Then pick any of the remaining A_j and form the maximal connected set g_2, etc. In the end we obtain a decomposition $G = (g_1, \ldots, g_k)$ into connected polymers such that, for any g_i, g_j in the collection, the supports of g_i and g_j do not intersect. We have to verify that this decomposition is unique. Thus assume that there are two different ways to decompose G, say $G = (g_1 \ldots, g_k)$ and $G = (g'_1 \ldots, g'_{k'})$. If these decompositions are different, there must be one g', say g'_1, such that g'_1 is not equal to any of the g_i; in particular, there must be a $g_i \neq g'_1$ such that $g'_1 \cap g_i \neq \emptyset$. Still there must be B that is an element of the symmetric difference $g_i \Delta g'_1 \equiv (g_i \cup g'_1) \setminus (g_i \cap g'_1)$. Assume without loss of generality that this set $B \in g_i$. But now $B \subset g'_j$ for some $j \neq 1$, while it is not in $g'_1 \ni A$. But there is a connected cluster in G containing both A and B, namely g_i, and so it follows that g'_1 and g'_j are intersecting, contradicting the hypothesis that they are maximally connected components. $\square$

Definition 5.1.4 Let g be a connected polymer. We define the set $\underline{g} \equiv \cup_{A \in g} A$ to be the support of g. Then the activity $w_\Lambda^\eta(g)$ is defined as

$$w_\Lambda^\eta(g) = \int d\rho_{\underline{g} \cap \Lambda} \sigma_{\underline{g} \cap \Lambda} \prod_{A \in g} \left(e^{\beta \Phi_A(\sigma_{A \cap \Lambda}, \eta_{A \cap \Lambda^c})} - 1\right) \quad (5.4)$$

5.1 High-temperature expansions

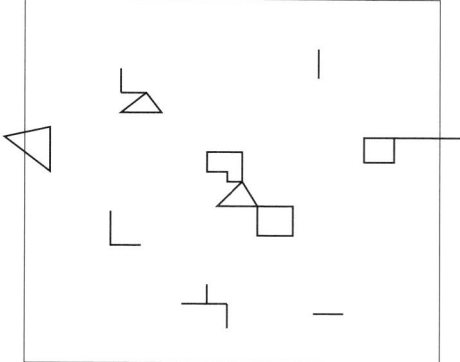

Figure 5.1 A compatible collection of polymers.

Lemma 5.1.5 *Let $G = (g_1, \ldots, g_n)$ be a polymer with connected components g_i. Then*

$$\int \mathrm{d}\rho_\Lambda(\sigma_\Lambda) \prod_{A \in G} \left(\mathrm{e}^{\beta \Phi_A(\sigma_{A \cap \Lambda}, \eta_{A \cap \Lambda^c})} - 1 \right) = \prod_{i=1}^{n} w_\Lambda^\eta(g_i) \tag{5.5}$$

Proof This formula follows from the simple observation that, by the definition of connectivity, different components g_i involve integrations only over spin variables on disjoint sets $\underline{g}$. □

Remark 5.1.6 Note that $w_\Lambda(g)$ depends on η_{Λ^c} only if g is a connected component of G that intersects the complement of Λ.

A simple consequence of Lemma 5.1.5 is:

Theorem 5.1.7 *(Polymer representation.)* *With the definition above*

$$Z_{\Lambda, \beta}^\eta = \sum_{n=0}^{\infty} \frac{1}{n!} \sum_{g_1, g_2, \ldots, g_n} \mathbb{I}_{\forall i \neq j\, g_i \not\sim g_j} \prod_{i=1}^{n} w_\Lambda^\eta(g_i) \tag{5.6}$$

where the sum is over connected polymers and $g \not\sim g'$ means g and g' are not connected.

Remark 5.1.8 The factor $1/n!$ takes into account the fact that relabeling the connected polymers gives the same polymer G. One frequently expresses the sum in (5.6) as a sum over *compatible collections of connected polymers*, where compatible just means that all elements of the collection are mutually disconnected. A compatible collection of contours is depicted in Fig. 5.1.

The formulation of the partition function given by Theorem 5.1.7 can be seen as a particular instance of a more general class of sums where connected polymers are elements of certain sets endowed with a compatibility relation, and with certain weights, called activities. The question one wants to pose then is: under what conditions can these sums be evaluated through convergent series? Thus, before continuing our investigation of high-temperature expansions, we will address this general question in an abstract context.

5.2 Polymer models: the Dobrushin–Kotecký–Preiss criterion

Abstract polymer models are constructed as follows. Assume that there is a countable set Γ endowed with the structure of a simple,[2] loop-free[3] graph $\mathcal{G}$. Of course, without loss of generality, we can take Γ to be the set of natural numbers or a subset thereof. Moreover, we will assume that $\mathbb{N}$ is endowed with the structure of an infinite graph, $\mathcal{G}_\infty$, once and for all, and any subset, $\Gamma \subset \mathbb{N}$, is naturally endowed with the induced graph. We say that $g \sim g'$, if and only if (g, g') is an edge of $\mathcal{G}$. Otherwise, we write $g \not\sim g'$. Let, furthermore, $w : \Gamma \to \mathbb{C}$ be a complex-valued function on Γ. We define a function $Z_\Gamma \equiv Z_\Gamma((w(g), \gamma \in \Gamma))$ on $\mathbb{C}^\Gamma$ through

$$Z_\Gamma \equiv \sum_{n=0}^\infty \frac{1}{n!} \sum_{g_1,\ldots,g_n \subset \Gamma} \mathbb{1}_{\forall_{i \neq j} g_i \not\sim g_j} \prod_{i=1}^n w(g_i) \tag{5.7}$$

Clearly (5.6) is a special case of such a function. It will be useful to think of the sum in (5.7) as a sum over all *completely disconnected subsets of* Γ. To make this notion precise, we will say that $G \subset \Gamma$ is completely disconnected if the subgraph induced by $\mathcal{G}$ on G has no edges. Let us denote the set of completely disconnected subsets of Γ by $\mathcal{D}_\Gamma$,

$$\mathcal{D}_\Gamma \equiv \cup_{\ell=0}^{|\Gamma|} \{(g_1, \ldots, g_\ell) \subset \Gamma : \forall_{i \neq j \leq \ell} \, g_i \not\sim g_j\} \tag{5.8}$$

Then (5.7) can be written as

$$Z_\Gamma = \sum_{G \in \mathcal{D}_\Gamma} \prod_{g \in G} w(g) \tag{5.9}$$

Our aim is to show under which conditions it is true that the logarithm of Z_Γ can be written as a convergent power series in the (complex) variables $w(g)$. Here the logarithm of a complex number $z = a + ib$ with $a > 0$ will be understood to be $\ln(a + ib) = \frac{1}{2}\ln(a^2 + b^2) + i \, \text{arcth}(b/a)$. We will make sure in the course of the proof that we need compute logarithms only on this domain of the complex plane.[4]

That is, we will seek to write

$$\ln Z_\Gamma = \sum_{C \in \mathcal{C}} K_C \prod_{g \in C} w(g) \tag{5.10}$$

where the sum should run over a suitable set $\mathcal{C}$ and the K_C are constants. Stated as such, both expressions (5.9) and (5.10) will be infinite most of the time when Γ is infinite. If Γ is finite, the expression for the partition function is necessarily finite, but it will most likely diverge as the size of Γ tends to infinity. But even if Γ is finite, we will need conditions for the logarithm to be representable as a convergent series.[5] What we really would like to obtain is a condition that allows us to write (5.10) as a (possibly infinite) sum of expressions that are under suitable conditions finite all the time, and that suffice to compute the free energy per volume as a finite expression in the limit as Λ goes to infinity.

---

[2] I.e. each edge appears only once.   [3] I.e. (g, g) is not an edge of $\mathcal{G}$.
[4] That means, in particular, that we will identify a domain in $\mathbb{C}^{|\Gamma|}$ in the variables $w(g)$, containing the origin in the variables $w(g)$, on which the real part of the partition function does not vanish.
[5] Already if $\Gamma = \{1\}$, we have $Z = 1 + w_1$, but $\ln Z = \ln(1 + w_1)$ will be an absolutely convergent series in w_1 only if $|w_1| < 1$.

A natural candidate for an expression that may remain finite is a sum over connected sets[6] containing a given element. It remains to guess what could be a candidate for the set $\mathcal{C}$. Since taking the logarithm involves power series, it is natural to guess that we will end up finding sums over terms where the elements of Γ can occur arbitrarily many times. Thus we consider the set of all collections of elements of Γ with repetition, such that the induced graph of this set is connected. Formally, we think of these sets as multi-indices $\boldsymbol{n} = (n_1, \ldots, n_{|\Gamma|})$, where n_g counts the number of occurrences of g. Frequently, one refers to multi-indices as *clusters*. Thus, we can define

$$\mathcal{C}_\Gamma^* \equiv \left\{ \boldsymbol{n} \in \mathbb{N}_0^\Gamma : \{g \in \Gamma : n_g \geq 1\} \text{ is connected}\right\} \tag{5.11}$$

Theorem 5.2.1 *Let Γ be any finite subset of $\mathbb{N}$, and let $a : \mathbb{N} \to \mathbb{R}_+$ be chosen arbitrarily. Let $P_\Gamma^a \subset \mathbb{C}^\Gamma$ be the set of complex numbers $w(g), g \in \Gamma$, such that, for any $g \in \Gamma$, $|w(g)|e^{a(g)} < 1$, and*

$$\sum_{g' \sim g} \left(-\ln\left(1 - |w(g')|e^{a(g')}\right) \right) \leq a(g) \tag{5.12}$$

Then, on P_Γ^a, $\ln Z_\Gamma$ is well defined and analytic. In particular, there are constants $K_{\boldsymbol{n}}$ such that

$$\ln Z_\Gamma = \sum_{\boldsymbol{n} \in \mathcal{C}_\Gamma^*} K_{\boldsymbol{n}} \prod_{g' \in \Gamma} w(g')^{n_{g'}} \tag{5.13}$$

and for any $g \in \Gamma$,

$$\sum_{\boldsymbol{n} \in \mathcal{C}_\Gamma^* : n_g \geq 1} K_{\boldsymbol{n}} | \prod_{g' \in \Gamma} |w(g')|^{n_{g'}} \leq -\ln\left(1 - |w(g)|e^{a(g)}\right) \tag{5.14}$$

Remark 5.2.2 Note that the sets P_Γ^a, for any choice of a, are non-empty and contain a neighbourhood of the origin, In general, P_Γ^a is a poly-disc. Moreover, if $w \in P_\mathbb{N}^a$, then any of its projections to $\mathbb{C}^\Gamma$ will be in P_Γ^a.

Remark 5.2.3 Equation (5.13) is called a *cluster expansion* or *Mayer expansion*.

Our first observation is that the constant $K_{\boldsymbol{n}}$ is independent of Γ and depends only on $\boldsymbol{n}$.

Lemma 5.2.4 *Let $\boldsymbol{n} \in \mathcal{C}_\mathbb{N}$, and let $\Gamma_{\boldsymbol{n}}$ denote the subset of $\mathbb{N}$ on which $\boldsymbol{n}$ is non-zero, i.e.*

$$\Gamma_{\boldsymbol{n}} \equiv \{g \in \Gamma : n_g \geq 1\} \tag{5.15}$$

Assume that $\Gamma_{\boldsymbol{n}}$ is finite, and that all $|w(g)|$ are so small that $\ln Z_\Gamma$ has a convergent expansion of the form (5.13). Then, for all $\Gamma \supseteq \Gamma_{\boldsymbol{n}}$,

$$K_{\boldsymbol{n}} = \frac{1}{\prod_{g \in \Gamma_{\boldsymbol{n}}} n_g!} \frac{\partial^{\sum_{g \in \Gamma_{\boldsymbol{n}}} n_g}}{\prod_{g \in \Gamma_{\boldsymbol{n}}} \partial^{n_g} w(g)} \ln Z_\Gamma \bigg|_{w(g)=0, \forall g \in \Gamma} \tag{5.16}$$

Proof Considering Z_Γ as a polynomial in the variables $w(g), g \in \Gamma$, the identity (5.16), with $Z_{\Gamma_{\boldsymbol{n}}}$ replaced by Z_Γ, follows from Taylor's formula, with a-priori Γ-dependent $K_{\boldsymbol{n}}$.

---

[6] We say that a subset $C \subset \Gamma$ is connected if the induced graph on C is connected.

But now write

$$Z_\Gamma = \sum_{G \in \mathcal{D}_{\Gamma_n}} \prod_{g \in G} w(g) + \sum_{\substack{G \in \mathcal{D}_\Gamma \\ G \cap (\Gamma \setminus \Gamma_n) \neq \emptyset}} \prod_{g \in G} w(g) = Z_{\Gamma_n} + Z_\Gamma^{\Gamma_n} \quad (5.17)$$

where

$$Z_\Gamma^{\Gamma_n} \equiv \sum_{\substack{G \in \mathcal{D}_\Gamma \\ G \cap (\Gamma \setminus \Gamma_n) \neq \emptyset}} \prod_{g \in G} w(g) \quad (5.18)$$

Thus

$$\ln Z_\Gamma = \ln Z_{\Gamma_n} + \ln\left(1 + \frac{Z_\Gamma^{\Gamma_n}}{Z_{\Gamma_n}}\right) \quad (5.19)$$

But

$$\ln\left(1 + \frac{Z_\Gamma^{\Gamma_n}}{Z_{\Gamma_n}}\right) = \sum_{\ell=1}^{\infty} \frac{(-1)^\ell}{\ell} \left(\frac{Z_\Gamma^{\Gamma_n}}{Z_{\Gamma_n}}\right)^\ell \quad (5.20)$$

and, expanding further, all terms appearing contain some factor $w(g)$ with $g \in \Gamma \setminus \Gamma_n$. None of the differentiations in (5.16) removes such a factor, and thus, setting $w(g) = 0$ in the end, all terms vanish, so that

$$\frac{\partial^{\sum_{g \in \Gamma_n} n_g}}{\prod_{g \in \Gamma_n} \partial^{n_g} w(g)} \left(1 + \frac{Z_\Gamma^{\Gamma_n}}{Z_{\Gamma_n}}\right)\bigg|_{w(g)=0, \forall g \in \Gamma} = 0 \quad (5.21)$$

so that we get (5.16). This proves the lemma. □

Remark 5.2.5 The estimate (5.14) implies that the functions

$$\sum_{n \in \mathcal{C}_\Gamma^* : n_g \geq 1} K_n \prod_{g' \in \Gamma} w(g')^{n_{g'}} \quad (5.22)$$

are convergent series for any $\Gamma \subset \mathbb{N}$ whenever the variables $w(g)$ satisfy the hypothesis (5.12). Thus, these functions are *holomorphic functions* of the $|\Gamma|$ complex variables in the respective poly-disc. Due to the observation of the preceding lemma, we can also define these functions for $\Gamma = \mathbb{N}$ and obtain, due to the uniformity of the estimates (5.14), convergent sums.

Corollary 5.2.6 *Assume that there is a function* $a : \mathbb{N} \to \mathbb{R}_+$ *such that for any* $g \in \mathbb{N}$,

$$\sum_{\mathbb{N} \ni g' \sim g} \left(-\ln\left(1 - |w(g')|e^{a(g')}\right)\right) \leq a(g) \quad (5.23)$$

Then, for any function $w : \mathbb{N} \to \mathbb{C}$ *such that* $w(g')$ *satisfies condition (5.23), for any* $g \in \mathbb{N}$, *the series*

$$F(g) \equiv \sum_{n \in \mathcal{C}_\mathbb{N}^* : n_g \geq 1} K_n \prod_{g' \in \mathbb{N}} w(g')^{n_{g'}} \quad (5.24)$$

where K_n *is defined by the right-hand side of (5.16), is absolutely convergent, and represents an analytic function of any of the variables* $w(g')$. *Moreover, if* $w(g) = w_g(z)$ *are holomorphic functions of a complex variable* z, *and if* $D \subset \mathbb{C}$ *is a domain such that, for* all $z \in D$, (5.12) is satisfied for all $g \in \mathbb{N}$, then the series $F(g)$, as a function of z, is holomorphic on D.

Proof (of Theorem 5.2.1)   The theorem will be proven by induction over the cardinality of the sets Γ, i.e. we will assume that the theorem holds for all sets of cardinality $|\Gamma| = N$ and then deduce it for all sets of cardinality $N + 1$.

It is already instructive to verify the theorem for the case $N = 1$. Here the hypothesis is void, while the assertion states that

$$\sum_{n=1}^{\infty} |K_n w(1)^n| \leq -\ln(1 - |w(1)|) \tag{5.25}$$

where $\ln(1 + w(1)) = \sum_{n=1}^{\infty} K_n w(1)^n$. Clearly, in this case $K_n = \frac{(-1)^n}{n}$, which implies (5.25) with equality.

The key identity that will allow us to carry through the induction is a formula similar to (5.17). Let Γ_N be any set of cardinality N, and let $g \notin \Gamma_N$. Set $\Gamma_{N+1} = \Gamma_N \cup \{g\}$. Then any $G \in \mathcal{D}_{\Gamma_{N+1}}$ is either a completely disconnected set of elements taken only from Γ_N, and thus is an element of $\mathcal{D}_{\Gamma_N}$, or the collection G contains g, and is completely disconnected in $\Gamma \cup g$. Thus

$$Z_{\Gamma_{N+1}} = \sum_{G \in \mathcal{D}_{\Gamma_N}} \prod_{g' \in G} w(g') + \sum_{\substack{G \in \mathcal{D}_{\Gamma_{N+1}} \\ g \in G}} \prod_{g' \in G} w(g') \tag{5.26}$$

$$= Z_{\Gamma_N} + w(g) Z_{\Gamma_N^g}$$

Here we have defined

$$\Gamma_N^g \equiv \{g' \in \Gamma_N : g' \not\sim g\} \tag{5.27}$$

the subset of elements of Γ that are not connected to g. The first equality in (5.26) is obvious. To see the second one, note that for any G in the second sum, one of its elements is g. Thus we can write $G = (g, G')$. Since g can occur only once in G, G' is made from elements of Γ_N. Moreover, since $G \in \mathcal{D}_{\Gamma_{N+1}}$, none of these elements may be connected to g, so in fact G' is made from elements of Γ_N^g. Moreover, these elements must be completely disconnected, which means that

$$\sum_{\substack{G \in \mathcal{D}_{\Gamma_{N+1}} \\ g \in G}} \prod_{g' \in G} w(g') = \sum_{G' \in \mathcal{D}_{\Gamma_N^g}} w(g) \prod_{g' \in G'} w(g') = Z_{\Gamma_N^g} w(g) \tag{5.28}$$

Now the nice thing is that both Γ_N and Γ_N^g are sets of no more than N elements, and thus the induction hypothesis can be applied to both partition functions on the right-hand side of (5.26). We want to write

$$\ln Z_{\Gamma_{N+1}} = \ln Z_{\Gamma_N} + \ln\left(1 + w(g) \frac{Z_{\Gamma_N^g}}{Z_{\Gamma_N}}\right) \tag{5.29}$$

To be able to do so, we must ensure that the term $\left|w(g)\frac{Z_{\Gamma_N^g}}{Z_{\Gamma_N}}\right|$ is strictly smaller than one. But (with the abbreviation $W_n \equiv \prod_{g' \in \Gamma_n} w(g')^{n_{g'}}$),

$$\frac{Z_{\Gamma_N^g}}{Z_{\Gamma_N}} = \exp\left(\sum_{n \in \mathcal{C}^*_{\Gamma_N^g}} K_n W_n - \sum_{n \in \mathcal{C}^*_{\Gamma_N}} K_n W_n\right) \tag{5.30}$$

$$= \exp\left(-\sum_{n \in \mathcal{C}^*_{\Gamma_N}\setminus \mathcal{C}^*_{\Gamma_N^g}} K_n W_n\right)$$

where we used the fact that, if $\Gamma' \subset \Gamma$, then $\mathcal{C}^*_{\Gamma'} \subset \mathcal{C}^*_{\Gamma}$. The set $\mathcal{C}^*_{\Gamma_N}\setminus \mathcal{C}^*_{\Gamma_N^g}$ is obviously the set of all connected multi-indices that contain at least one element that is connected to g,

$$\mathcal{C}^*_{\Gamma_N}\setminus \mathcal{C}^*_{\Gamma_N^g} = \{n \in \mathcal{C}^*_{\Gamma_N}, \exists g' \in \Gamma : g' \sim g, \wedge n_{g'} \geq 1\} \tag{5.31}$$

This allows us to bound

$$\left|\frac{Z_{\Gamma_N^g}}{Z_{\Gamma_N}}\right| \leq \exp\left(+\sum_{n \in \mathcal{C}^*_{\Gamma_N}\setminus \mathcal{C}^*_{\Gamma_N^g}} |K_n||W_n|\right) \tag{5.32}$$

Now

$$\sum_{n \in \mathcal{C}^*_{\Gamma_N}\setminus \mathcal{C}^*_{\Gamma_N^g}} |K_n||W_n| \leq \sum_{g' \in \Gamma, g' \sim g}\sum_{n \in \mathcal{C}^*_{\Gamma_N}, n_{g'} \geq 1} |K_n||W_n| \tag{5.33}$$

$$\leq \sum_{g' \in \Gamma_N, g' \sim g}\left|\ln\left(1 - |w_{g'}|e^{a(g')}\right)\right| \leq a(g)$$

where the last-but-one inequality uses the induction hypothesis, and the last inequality uses the condition (5.12). Equation (5.33) implies in particular that, on $P^a_{\Gamma_N \cup g}$,

$$|w(g)|\left|\frac{Z_{\Gamma_N^g}}{Z_{\Gamma_N}}\right| \leq |w(g)|e^{a(g)} < 1 \tag{5.34}$$

Under these conditions, we can write, using (5.29),

$$\ln Z_{\Gamma_{N+1}} = \sum_{n \in \mathcal{C}^*_{\Gamma_N}} K_n W_n \tag{5.35}$$

$$-\sum_{k=1}^{\infty} \frac{(-1)^k}{k}\left(w(g)\exp\left(-\sum_{n \in \mathcal{C}^*_{\Gamma_N}\setminus \mathcal{C}^*_{\Gamma_N^g}} K_n W_n\right)\right)^k$$

Expanding the powers in the second term, it is manifest that we will obtain an expression that is a polynomial in $w(g')$, where each monomial will involve at least one power of $w(g)$, and where the corresponding multi-index belongs to $\mathcal{C}^*_{\Gamma_{N+1}}$.

Now write the obvious formula

$$\ln Z_{\Gamma_{N+1}} = \sum_{\substack{n \in \mathcal{C}^*_{\Gamma_{N+1}} \\ n_g = 0}} K_n W_n + \sum_{\substack{n \in \mathcal{C}^*_{\Gamma_{N+1}} \\ n_g \geq 1}} K_n W_n \tag{5.36}$$

Since the set

$$\{n \in \mathcal{C}^*_{\Gamma_{N+1}}, n_g = 0\} = \mathcal{C}^*_{\Gamma_N} \tag{5.37}$$

we recognize the first term in (5.36) as the logarithm of the partition function Z_{Γ_N}, and thus deduce that

$$\sum_{\substack{n \in \mathcal{C}^*_{\Gamma_{N+1}} \\ n_g \geq 1}} K_n W_n = \ln\left(1 + w(g) \frac{Z_{\Gamma_N^g}}{Z_{\Gamma_N}}\right) \tag{5.38}$$

Inserting the expansion (5.33), we see that

$$\sum_{\substack{n \in \mathcal{C}^*_{\Gamma_{N+1}} \\ n_g \geq 1}} |K_n||W_n| \leq \sum_{k=1}^{\infty} \frac{1}{k} \left(|w(g)| \exp\left(\sum_{n \in \mathcal{C}^*_{\Gamma_N} \setminus \mathcal{C}^*_{\Gamma_N^g}} |K_n||W_n|\right)\right)^k$$

$$= -\ln\left(1 - |w(g)| \exp\left(\sum_{n \in \mathcal{C}^*_{\Gamma_N} \setminus \mathcal{C}^*_{\Gamma_N^g}} |K_n||W_n|\right)\right)$$

$$\leq -\ln\left(1 - |w(g)|e^{a(g)}\right) \tag{5.39}$$

which is indeed the assertion of the theorem for Γ_{N+1}. Since the foregoing argument holds for any $g \notin \Gamma$, the inductive step is concluded and the theorem proven. □

Remark 5.2.7 The first proof of the convergence of the high-temperature expansion in a lattice model, due to Gallavotti, Miracle-Solé and Robinson [110, 112], did not use polymer models but was based on the Kirkwood-Salsburg equations [149]. The notion of a polymer model was introduced by Gruber and Kunz [132]. The idea of the Mayer expansion was introduced in the physical literature, probably by Mayer [175], in the study of interacting gases. Convergence of the Mayer expansion in polymer models was first proven for repulsive gases by Rota [212], and later by Cammarota [70]. These proofs were based on combinatorial bounds on the coefficients K_n, which were represented as sums over connected graphs (controlling the possible connectivity structure of the multi-indices n). The key observation was that these sums could in turn be bounded by sums over trees, of which there are sufficiently few to achieve convergence. These techniques were simplified and stream-lined in works of Battle [15], Glimm-Jaffe [129], Federbush [16, 99], Brydges [65, 67], V. A. Malyshev [172], and others. A good exposition of this combinatorial approach is given in Simon's book [224]. A formulation of the convergence condition similar to that of our Theorem 5.2.1 appeared in 1984 in a paper by R. Kotecký and D. Preiss [155], and their analogue of the condition (5.12) is known as the *Kotecký–Preiss criterion*. Their proof makes use of the so-called Möbius inversion formula, which allows us to express the logarithm of a sum over disconnected sets as a sum over connected clusters. This was the first major step towards a reduction of combinatorial efforts in the convergence proofs.

The present form of Theorem 5.2.1 was first proven by R. L. Dobrushin [91], who also initiated the idea of proving convergence by induction over the set of polymers. The observation of Lemma 5.2.4 is also due to him. The main difference between his proof and the one presented here is that he used the Cauchy integral representation for the coefficients

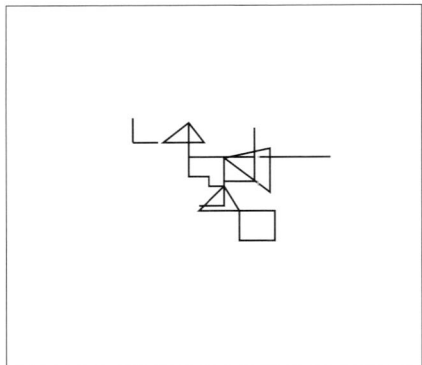

Figure 5.2 A connected cluster of polymers.

K_n to obtain bounds and to prove convergence. The idea of proving the estimates necessary for convergence directly by induction is due to M. Zahradník, and was first used in a paper by F. Nardi, E. Olivieri, and M. Zahradník [181], with a different form of the hypothesis (5.12) (that gives slightly worse estimates on the domain of analyticity). Independently, S. Miracle-Solé [180] gave a very similar proof. This was further elaborated in the paper [61], where we observed that the method of proof can also yield the conditions in Dobrushin's form, i.e. Theorem 5.2.1. The first purely inductive proof of the theorem in this form is, however, due to A. Sokal [228]. An extensive discussion, dealing with more general models, and making connections to Lovasz' Lemma in graph theory, can be found in a recent paper by Scott and A. Sokal [219]. A concise exposition that also covers the case of continuous state space is given by Ueltschi in [242].

5.3 Convergence of the high-temperature expansion

We will now use the general convergence criterion for the polymer model to obtain a convergence criterion for the high-temperature expansion. The polymers are now the connected polymers g from Section 5.1, and the graph on this set is derived from the connectivity defined in Definition 5.1.1. A connected cluster of such polymers in shown in Fig. 5.2. All we need to do is to establish criteria for the interaction under which in this context (5.12) holds.

First, we will use the bound

$$|w_\Lambda^\eta(g)| \leq \prod_{A \in g} \left(e^{\beta \|\Phi_A\|_\infty} - 1 \right) \equiv \prod_{A \in g} v(A) \tag{5.40}$$

We will choose $a(g) = \sum_{A \in g} \tilde{a}(A)$ with $\tilde{a}(A) = c|A|$, and c to be determined later. We will assume that, with this choice, v is such that there exists $K < \infty$ such that, for all $k \geq 0$,

$$\sup_{x \in \mathbb{Z}^d} \sum_{A \ni x} v(A) e^{\tilde{a}(A)} |A|^k \leq k! K \tag{5.41}$$

Lemma 5.3.1 *Assume that the temperature and activities are such that (5.41) is satisfied with $\tilde{a}(A) = \frac{4K}{1-2K}|A|$. Assume that Φ is a translation-invariant interaction. Then the polymer activities satisfy the Dobrushin–Kotecký–Preiss criterion (5.12).*

Proof To simplify the argument, we will use that, for $0 \leq x \leq 1/2$, $-\ln(1-x) \leq 2x$, and show that the stronger condition

$$\sum_{g' \sim g} 2v(g') e^{a(g')} \leq a(g) \tag{5.42}$$

holds for suitable choices of a (where $v(g) \equiv \prod_{A \in g} v(A)$). The key idea of the proof is to use the fact that, if $g' = (A_1, \ldots, A_k)$ is a connected polymer, then there exists a tree[7] on the set $\{1, \ldots, k\}$ such that, if (i, j) is an edge of the tree, then $A_i \cap A_j \neq \emptyset$. While there are several ways to assign a tree to a polymer, it is possible to choose a rule that makes this assignment univalent. Moreover, if g' is connected to g, then (at least) one of its components intersects $\underline{g}$. Without loss of generality, we can assume that this component is A_1. We will single out the vertex 1 of the tree and consider it to be its *root*. In the sequel, all trees appearing will be understood to be rooted in the origin. Next observe that, if Φ is translation invariant, then so is $v(A)$. Thus we get the first simple estimate

$$\sum_{g' \sim g} 2v(g') e^{a(g')} \leq |\underline{g}| \sum_{g' \ni 0} 2v(g') e^{a(g')} \tag{5.43}$$

From this estimate it is clear that we will need to choose $a(g) \geq |\underline{g}|$. On the other hand, we will succeed if we can show that

$$\sum_{g' \ni 0} 2v(g') e^{a(g')} \leq const. \tag{5.44}$$

Now

$$\sum_{g' \ni 0} 2v(g') e^{a(g')} = \sum_{k=1}^{\infty} \frac{1}{(k-1)!} \sum_{t} \sum_{\substack{A_1, A_2, \ldots, A_k \\ t(A_1, \ldots, A_k) = t}} 2 \prod_{i=1}^{k} v(A_i) e^{\tilde{a}(A_i)} \tag{5.45}$$

The idea is to sum over the sets A_i starting from the leaves (= vertices of coordination number one that are not the root) of the tree and to use (5.41). Note that a vertex ℓ to which $c - 1$ leaves are attached produces a factor

$$|A_\ell|^{c-1} \left(\sum_{A \ni 0} v(A) e^{\tilde{a}(A)} \right)^{c-1} \leq |A_\ell|^{c-1} K^{c-1} \tag{5.46}$$

This explains the necessity of having the conditions (5.41) with $k > 0$. It also shows that it is important to keep track of the coordination numbers of the vertices of the tree t. Therefore we will sum first over the possible assignment of coordination numbers,[8] $c_1, \ldots, c_k$ (satisfying $\sum_{i=1}^{k} c_i = 2(k-1)$), and then over all trees with these coordination numbers:

$$\sum_{t} \sum_{\substack{A_1, A_2, \ldots, A_k \\ t(A_1, \ldots, A_k) = t}} 2 \prod_{i=1}^{k} v(A_i) e^{\tilde{a}(A_i)} = \sum_{c_1, \ldots, c_k} \sum_{t: c_i(t) = c_i} \sum_{\substack{A_1, A_2, \ldots, A_k \\ t(A_1, \ldots, A_k) = t}} 2 \prod_{i=1}^{k} v(A_i) e^{\tilde{a}(A_i)} \tag{5.47}$$

---

[7] A tree is a graph that is connected and contains no loop.
[8] I learned this from a lecture given by Jürg Fröhlich at the ETH Zurich in 1982/3, of which there are handwritten lecture notes by Giovanni Felder. I am not aware of any earlier publication of this trick.

Summing successively over all A_i, starting from the leaves, then the leaves of what is left, and finally over the root, we get the estimate

$$\sum_{\substack{A_1,A_2,\ldots,A_k \\ t(A_1,\ldots,A_k)=t}} 2\prod_{i=1}^{k} v(A_i)e^{\tilde{a}(A_i)} \leq K^k c_1! \prod_{i=2}^{k} (c_i - 1)! \qquad (5.48)$$

Finally, we must count the number of trees that have given coordination numbers. This is a classical formula, due to Cayley [21], and it states that

$$\sum_{t:c_i(t)=c_i} 1 = \frac{(k-2)!}{(c_1-1)!\cdots(c_k-1)!} \qquad (5.49)$$

Remark 5.3.2 The proof of this formula is simple and uses induction over the number of vertices. It can be found in Simon [224].

It follows that

$$\sum_{t}\sum_{\substack{A_1,A_2,\ldots,A_k \\ t(A_1,\ldots,A_k)=t}} 2\prod_{i=1}^{k} v(A_i)e^{\tilde{a}(A_i)}$$

$$\leq 2 \sum_{\substack{c_1,\ldots,c_k \\ \sum c_i = 2(k-1)}} \frac{(k-2)!}{(c_1-1)!\ldots(c_k-1)!} K^k c_1! \prod_{i=2}^{k}(c_i-1)!$$

$$= 2 \sum_{\substack{c_1,\ldots,c_k \\ \sum c_i = 2(k-1)}} c_1(k-2)! K^k \qquad (5.50)$$

Finally, we use that $c_1 \leq k-1$, and that

$$\sum_{\substack{c_1,\ldots,c_k \geq 1 \\ \sum c_i = 2(k-1)}} 1 = \binom{2(k-1)}{k} \leq 2^k \qquad (5.51)$$

to see that

$$\sum_{k=1}^{\infty} \frac{1}{(k-1)!} \sum_{t}\sum_{\substack{A_1,A_2,\ldots,A_k \\ t(A_1,\ldots,A_k)=t}} 2\prod_{i=1}^{k} v(A_i)e^{\tilde{a}(A_i)} \qquad (5.52)$$

$$\leq \sum_{k=1}^{\infty} 2(2K)^k = \frac{4K}{1-2K} \qquad (5.53)$$

Thus, we can choose $a(g) = \sum_{A \in g} \frac{4K}{1-2K}|A|$ and impose as a condition on $v(A)$ that

$$\sum_{A \ni 0} |v(A)||A|^k e^{\frac{4K}{1-2K}|A|} \leq K k! \qquad (5.54)$$

$\square$

Exercise: Assume that $\sum_{A:|A|=M} \|\Phi_A\|_\infty \leq C e^{-\delta M}$, with $\delta > 0$. Show that there exists $\beta_0 > 0$ such that, for $\beta < \beta_0$, the DKP criterion is satisfied.

We will now show a few implications of Lemma 5.3.1. The first is, naturally, the existence of a convergent expansion for the free energy.

5.3 Convergence of the high-temperature expansion

Theorem 5.3.3 *Assume that we are given a regular, translation-invariant interaction Φ, and assume that the hypothesis of Lemma 5.3.1 is satisfied. Assume that Λ_n is an increasing and absorbing sequence of subsets of $\mathbb{Z}^d$ and $\lim_{n\uparrow\infty}\frac{|\partial\Lambda_n|}{|\Lambda_n|}=0$. Then, for any $\eta\in\mathcal{S}$,*

$$\lim_{n\uparrow\infty}\frac{1}{|\Lambda_n|}\ln Z_{\beta,\Lambda_n}^{\eta}=\sum_{\mathbf{n}:\underline{\mathbf{n}}\ni 0}\frac{1}{|\underline{\mathbf{n}}|}K_{\mathbf{n}}\prod_{g:n_g\geq 1}(w_\beta(g))^{n_g} \tag{5.55}$$

Here we have set $\underline{\mathbf{n}}=\bigcup_{g:n_g\geq 1}\underline{g}$.

Proof Under our assumptions, for any given Λ, it holds that

$$\ln Z_{\beta,\Lambda}^{\eta}=\sum_{\mathbf{n}\in\mathcal{C}_{\Gamma_\Lambda}^*}K_{\mathbf{n}}\prod_{g:n_g\geq 1}(w_\Lambda^\eta(g))^{n_g} \tag{5.56}$$

where Γ_Λ are all connected polymers that can be constructed from sets A intersecting Λ. It will be convenient to split this set into those polymers that are contained in Λ, and those that intersect both Λ and its complement. Note that, if g is contained in Λ, then its activity does not depend on either Λ or η, so that

$$\ln Z_{\beta,\Lambda}^{\eta}=\sum_{\substack{\mathbf{n}\in\mathcal{C}_{\Gamma_\Lambda}^*\\\underline{\mathbf{n}}\cap\Lambda^c\neq\emptyset}}K_{\mathbf{n}}\prod_{g:n_g\geq 1}(w_\Lambda^\eta(g))^{n_g}+\sum_{\substack{\mathbf{n}\in\mathcal{C}_{\Gamma_\Lambda}^*\\\underline{\mathbf{n}}\subset\Lambda}}K_{\mathbf{n}}\prod_{g:n_g\geq 1}(w(g))^{n_g} \tag{5.57}$$

The second sum can be written as

$$\sum_{\substack{\mathbf{n}\in\mathcal{C}_{\Gamma_\Lambda}^*\\\underline{\mathbf{n}}\subset\Lambda}}K_{\mathbf{n}}\prod_{g:n_g\geq 1}(w(g))^{n_g}=\sum_{x\in\Lambda}\sum_{\substack{\mathbf{n}\in\mathcal{C}_{\Gamma_\Lambda}^*\\x\in\underline{\mathbf{n}}\subset\Lambda}}\frac{K_{\mathbf{n}}}{|\underline{\mathbf{n}}|}\prod_{g:n_g\geq 1}(w(g))^{n_g} \tag{5.58}$$

$$=\sum_{x\in\Lambda}\sum_{\substack{\mathbf{n}\in\mathcal{C}_{\Gamma_{\mathbb{Z}^d}}^*\\x\in\underline{\mathbf{n}}}}\frac{K_{\mathbf{n}}}{|\underline{\mathbf{n}}|}\prod_{g:n_g\geq 1}(w(g))^{n_g}-\sum_{x\in\Lambda}\sum_{\substack{\mathbf{n}\in\mathcal{C}_{\Gamma_{\mathbb{Z}^d}}^*\\x\in\underline{\mathbf{n}}\cap\Lambda^c\neq\emptyset}}\frac{K_{\mathbf{n}}}{|\underline{\mathbf{n}}|}\prod_{g:n_g\geq 1}(w(g))^{n_g}$$

$$=|\Lambda|\sum_{\substack{\mathbf{n}\in\mathcal{C}_{\Gamma_{\mathbb{Z}^d}}^*\\0\in\underline{\mathbf{n}}}}\frac{K_{\mathbf{n}}}{|\underline{\mathbf{n}}|}\prod_{g:n_g\geq 1}(w(g))^{n_g}-\sum_{\substack{\mathbf{n}\in\mathcal{C}_{\Gamma_\Lambda}^*\\\underline{\mathbf{n}}\cap\Lambda\neq\emptyset\wedge\underline{\mathbf{n}}\cap\Lambda^c\neq\emptyset}}K_{\mathbf{n}}\prod_{g:n_g\geq 1}(w(g))^{n_g}$$

Note that the addition or subtraction of clusters that are not contained in Λ produces a term that is fully translation invariant and that yields the desired expression for the infinite-volume free energy density in terms of a convergent series, while the second sum involves only clusters that cross the boundary of Λ and thus will be seen to give a contribution that vanishes in the infinite-volume limit. We still have to show that the coefficient of $|\Lambda|$ in (5.58) is uniformly bounded. But, using the assertion of Theorem 5.2.1,

$$\left|\sum_{\substack{\mathbf{n}\in\mathcal{C}_{\Gamma_\Lambda}^*\\0\in\underline{\mathbf{n}}}}\frac{K_{\mathbf{n}}}{|\underline{\mathbf{n}}|}\prod_{g:n_g\geq 1}(w(g))^{n_g}\right|\leq\sum_{g':\underline{g'}\ni 0}\sum_{\substack{\mathbf{n}\in\mathcal{C}_{\Gamma_{\mathbb{Z}^d}}^*\\n_{g'}\geq 1}}\frac{|K_{\mathbf{n}}|}{|\underline{\mathbf{n}}|}\prod_{g:n_g\geq 1}|w(g)|^{n_g}$$

$$\leq\sum_{g':\underline{g'}\ni 0}\left(-\ln\left(1-|w(g)|e^{a(g)}\right)\right)$$

$$\leq a(0) \tag{5.59}$$

where the last inequality is obtained by identifying 0 with the support of a one-site polymer situated at the origin.

It remains to show that the first term in (5.57) and the second summand in (5.58) tend to zero when divided by $|\Lambda|$. But this follows easily, since these sums only involve clusters that intersect the boundary of Λ. Thus we get, using the same arguments as before, e.g.

$$\left| \sum_{x \in \Lambda} \sum_{\substack{\underline{n} \in \mathcal{C}^*_{\Gamma_{\mathbb{Z}^d}} \\ x \in \underline{n} \cap \Lambda^c \neq \emptyset}} \frac{K_{\underline{n}}}{|\underline{n}|} \prod_{g : n_g \geq 1} (w(g))^{n_g} \right| \leq \sum_{x \in \partial \Lambda} \left| \sum_{\substack{\underline{n} \in \mathcal{C}^*_{\Gamma_{\mathbb{Z}^d}} \\ x \in \underline{n}}} \frac{K_{\underline{n}}}{|\underline{n}|} \prod_{g : n_g \geq 1} (w(g))^{n_g} \right| \leq |\partial \Lambda| a(0) \tag{5.60}$$

which tends to zero when divided by $|\Lambda|$. $\square$

From the computation of partition functions we can easily pass to computing correlation functions, i.e. expectation values. It will be enough to compute probabilities of events of the form $\{\sigma_x = \eta_x, x \in D\}$, $D \subset \mathbb{Z}^d$, and these can be expressed as ratios of partition functions $Z^\eta_{\beta, \Lambda'_n}$ and $Z^\eta_{\beta, \Lambda_n}$, where $\Lambda'_n = \Lambda_n \setminus D$. Applying the cluster expansion in both the numerator and denominator, we see that there is a huge cancellation of terms, and only multi-indices that intersect D will survive:

$$\frac{Z^\eta_{\beta, \Lambda'_n}}{Z^\eta_{\beta, \Lambda_n}} = \exp \left(\sum_{\substack{\underline{n} \in \mathcal{C}^*_{\Gamma_{\Lambda_n}} \\ \underline{n} \cap D \neq \emptyset}} K_{\underline{n}} \left(\prod_{g : n_g \geq 1} \left(w^\eta_{\Lambda'_n}(g) \right)^{n_g} - \prod_{g : n_g \geq 1} \left(w^\eta_{\Lambda_n}(g) \right)^{n_g} \right) \right) \tag{5.61}$$

Since the respective sums converge absolutely, we can take the limit $n \uparrow \infty$ and obtain expressions that do not depend on the boundary conditions η, except for η_D,

$$\lim_{n \uparrow \infty} \frac{Z^\eta_{\beta, \Lambda'_n}}{Z^\eta_{\beta, \Lambda_n}} = \mu_\beta (\{\sigma_x = \eta_x, x \in D\}) \tag{5.62}$$

$$= \exp \left(\sum_{\substack{\underline{n} \in \mathcal{C}^*_{\Gamma_{\mathbb{Z}^d}} \\ \underline{n} \cap D \neq \emptyset}} K_{\underline{n}} \left(\prod_{g : n_g \geq 1} \left(w^{\eta_D}_{D^c}(g) \right)^{n_g} - \prod_{g : n_g \geq 1} (w_\beta(g))^{n_g} \right) \right)$$

Note that (5.62) gives an alternative proof of the uniqueness of the Gibbs measure for regular interactions at high temperatures, since the explicit expressions for marginals of the Gibbs measure in the thermodynamic limit are independent of the sequence of volumes and the boundary conditions.

The explicit expressions for these correlations may look quite horrible, but they are not as bad as they appear. After all, all sums are rapidly converging, and computing a few terms already tends to give reasonable approximations. Explicit computation to high orders (say 20) is, however, a tedious task, to which a large number of people have devoted a great deal of work. The objective of such computations has often been to try to get information beyond the regime of high temperatures where convergence is assured, and even to use analytic extrapolation ideas (e.g. Padé approximants) to guess the nature of the singularities of the partition function at second-order phase transitions.

Example: The Curie–Weiss model Although the cluster expansion is not the tool of choice for doing computations in mean field models, it may be interesting to see it in action in a model where we already know everything, namely the Curie–Weiss model. We recall the

5.3 Convergence of the high-temperature expansion

that the partition function (we set $h = 0$) is (with a slight modification from the convention of Chapter 3)

$$Z_{\beta,N} = 2^{-N} \sum_{\sigma \in \{-1,1\}^N} e^{\frac{\beta}{N} \sum_{i<j} \sigma_i \sigma_j} \tag{5.63}$$

We will now use some special features of Ising spins, in particular the fact that

$$e^{\beta \sigma_i \sigma_j / N} = \cosh(\beta/N)(1 + \sigma_i \sigma_j \tanh(\beta/N)) \tag{5.64}$$

which allows us to write

$$Z_{\beta,N} = [\cosh(\beta/N)]^{\frac{N(N-1)}{2}} \hat{Z}_{\beta,N} \tag{5.65}$$

with

$$\hat{Z}_{\beta,N} \equiv 2^{-N} \sum_{\sigma \in \{-1,1\}^N} \sum_{G \in \mathcal{G}_N} \prod_{(i,j) \in G} (\sigma_i \sigma_j \tanh(\beta/N)) \tag{5.66}$$

where $\mathcal{G}_N$ is the collection of all subsets of pairs (i, j), $i, j \in \{1, \ldots, N\}$, with $i \neq j$, i.e. the collection of all simple, loop-free graphs on N vertices. The reduced partition function, $\hat{Z}_{\beta,N}$, can be written in the form (5.6), with

$$w_N(g) \equiv [\tanh(\beta/N)]^{b(g)} 2^{-|g|} \sum_{\sigma_i, i \in g} \prod_{(i,j) \in g} \sigma_i \sigma_j \tag{5.67}$$

and where the sum runs over connected graphs g on N vertices and $b(g)$ denotes the number of edges in the graph g. It is easy to see that the sum over σ in (5.67) vanishes unless all vertices i in γ have a coordination number that is even, in which case the sum is equal to $2^{|g|}$. Thus

$$w_N(g) \equiv [\tanh(\beta/N)]^{b(g)} \mathbb{1}_{\{g \text{ has only even coordination numbers}\}} \tag{5.68}$$

We can now check the DKP criterion. We can take $a(g) = c|g|$, and so we only have to bound $\sum_{g \ni 1} |w_N(g)| e^{c|g|}$ to find c. To control this sum, we note that any g containing 1 can be obtained by performing a walk on $\{1, \ldots, N\}$ starting at 1 that does not trace back immediately, and that in the last step returns to 1. Thus, summing over all such walks certainly gives us an upper bound. We get, using that $|g| \leq b(g)$ and that the shortest closed g has three edges,

$$\sum_{g \ni 1} |w(g)| e^{c|g|} \leq \sum_{k=3}^{N(N-1)/2} (N-1)^{k-1} [\tanh(\beta/N)]^k e^{ck} \tag{5.69}$$

$$\leq \tanh(\beta/N) e^c \sum_{k=2}^{\infty} N^k \tanh^k(\beta/N) e^{ck}$$

$$\leq \tanh(\beta/N) e^{3c} \frac{N^2 \tanh^2(\beta/N)}{1 - N \tanh(\beta/N) e^c}$$

For large N, $N \tanh(\beta/N) \leq \beta$, so that

$$\sum_{g \sim g'} |w(g)| e^{c|g|} \leq |g'| N^{-1} \beta^3 e^{3c} \frac{1}{1 - \beta e^c} \tag{5.70}$$

so that the condition for c becomes

$$N^{-1}\beta^3 e^{3c} \frac{1}{1-\bar\beta e^c} \leq c \qquad (5.71)$$

which can be satisfied for any $\beta < 1$ with $c = c'/N$, $c' \sim 1/(1-\beta)$. Thus we see that the convergence of the high-temperature expansion is ensured whenever $\beta < 1$, in agreement with the exact results. Note also that all terms in the expansion will be of order $1/N$, in agreement with the fact that in the Curie–Weiss model, with our normalization, the free energy is equal to zero for $\beta \leq 1$.

Remark 5.3.4 The fact that we can control the DKP criterion right up to the critical point $\beta = 1$ is quite atypical. It is due to the fact that the activities of all high-temperature polymers are a factor of $1/N$ smaller than the inverse of their number, since only loops contribute. This allowed us to choose $a(g) \sim N^{-1}|g|$. As soon as we add a magnetic field to the Hamiltonian (i.e. break the spin-flip symmetry), this feature disappears and we get more restrictive conditions for the convergence of the high-temperature expansion. Of course, if we are careful enough with the combinatorics, we should be able to recover a domain of convergence of the form $\beta < 1 - O(h)$ if h is small.

5.4 Low-temperature expansions

The ideas developed for high-temperature expansions, together with what we observed in the Peierls argument, suggest that it should also be possible to develop systematic convergent expansions for the free energy (and hence expectation values) in the limit of low temperatures. In fact, the Peierls representation of the Ising model at low temperatures suggests that we should write the partition function as a sum over geometric entities – contours separating regions of positive and negative values of σ – that are connected, mutually disjoint, and have small activities, just as for the polymers of the high-temperature expansion.

5.4.1 The Ising model at zero field

Let us first focus on the Ising model with zero magnetic field. It is convenient to write the Hamiltonian in a slightly different form as

$$H_\Lambda(\sigma) = \sum_{<x,y>\cap\Lambda\neq\emptyset} \mathbb{1}_{\sigma_x\neq\sigma_y} \qquad (5.72)$$

We will for simplicity only consider constant $+1$ or -1 boundary conditions outside of Λ. Then the Hamiltonian is just the volume of the set

$$\Gamma(\sigma) = \{<x,y> \cap\Lambda \neq \emptyset : \sigma_x \neq \sigma_y\} \qquad (5.73)$$

The partition function can then be written as

$$Z_{\beta,\Lambda} = \sum_\Gamma \sum_{\sigma:\Gamma(\sigma)=\Gamma} e^{-\beta|\Gamma|} \qquad (5.74)$$

The set Γ can be decomposed into connected subsets $\gamma_1, \ldots, \gamma_g$ that are called *contours*. In the Ising model we can think of them (see Section 4.3.2) as *closed loops* on the dual lattice separating domains of spins with constant sign, which is the reason for the name 'contours'. Thus,

$$Z_{\beta,\Lambda} = \sum_{k=0}^{\infty} \frac{1}{k!} \sum_{\gamma_1,\ldots,\gamma_k} \sum_{\sigma : \Gamma(\sigma)=(\gamma_1,\ldots,\gamma_k)} \prod_{\ell=1}^{k} e^{-\beta|\gamma_\ell|} \qquad (5.75)$$

$$= \sum_{k=0}^{\infty} \frac{1}{k!} \sum_{\gamma_1,\ldots,\gamma_k} \prod_{i \neq j} \mathbb{1}_{\gamma_i \not\sim \gamma_j} \prod_{\ell=1}^{k} e^{-\beta|\gamma_\ell|}$$

where we used that the spin configurations are uniquely determined (for a given sign of the boundary conditions) by the contours, and that all configurations of mutually disjoint loops are compatible with some spin configuration. We see that the form of this partition function is the same as that of the high-temperature expansion in Theorem 5.1.7. Moreover, the activities are small if β tends to infinity, so that we can analyze this partition function in exactly the same way as the high-temperature expansion, using the general theory of polymer models. It is thus easy to show that, for large enough values of β, we obtain an absolutely convergent expansion for the free energy per site, and it is also easy to show the existence of two Gibbs states, as was already proven using the Peierls argument.

However, this example is misleadingly simple. In particular, the perfect symmetry of the interaction is used to remove all reference to the spin configurations. Already when we add a magnetic field term to the Hamiltonian, this symmetry is lost and it is impossible to obtain such a simple representation, since we need to keep track of the sign of the spins outside the contours. Nonetheless, low-temperature expansions using the concept of contours are the basis of the most powerful machinery for analyzing the phase diagrams of spin systems at low temperatures, the *Pirogov–Sinai theory* [204, 205]. I will not give a detailed exposition of this theory here, because an excellent pedagogical exposition is being written by Ch. Borgs and R. Kotecký [37], which the interested reader should consult. I will only explain the general setup of contour models and outline the main ideas in an informal way.

5.4.2 Ground-states and contours

In the zero-field Ising model, the contour representation can be seen intuitively as an expansion around the obvious *ground-states* of the model, namely the configurations that are constant, either $+1$ or -1. This notion of ground-states requires, however, a bit more care, since we are dealing with infinite systems. The standard definition reads [6]:

Definition 5.4.1 Let Φ be a regular interaction, and H_Λ the corresponding finite-volume Hamiltonians. A configuration $\eta \in \mathcal{S}$ is called a ground-state for Φ if and only if for all finite $\Lambda \subset \mathbb{Z}^d$

$$H_\Lambda(\eta) = \inf_{\sigma_\Lambda \in \mathcal{S}_\Lambda} H_\Lambda(\sigma_\Lambda, \eta_{\Lambda^c}) \qquad (5.76)$$

Examples: In the Ising model with zero field, the states that are constant are two obvious ground-states in the above sense. But they are not the only ones. Another example is

$$\eta_x = \begin{cases} +1, & \text{if } x_1 \geq a \\ -1, & \text{if } x_1 < a \end{cases} \qquad (5.77)$$

as the reader can easily verify. In fact, the set of ground-states is typically extremely rich. Below we will be mostly concerned with translation-invariant ground-states, which form a small subset of all ground-states.

The basic idea of low-temperature expansions is that, for large β, one should be able to construct Gibbs states that are essentially supported on perturbations of the ground-states; at least one would expect this for the translation-invariant ground-states. It may be surprising that this turns out to be not exactly true.

The idea of a *contour representation* is to assign to a configuration $\eta \in \mathcal{S}$ a partition of $\mathbb{Z}^d$ into regions where the configuration looks locally like one of the translation-invariant ground-states, and regions where it does not. The latter are called *contours* (by analogy with the Ising example). They should carry considerable excess energy, and therefore their appearance in the Gibbs measure should be suppressed. The precise implementation of this general idea is unfortunately model dependent, and can be very cumbersome. In the following we will consider only a relatively simple context, where we assume the following to be true:

(i) The state space $\mathcal{S}_0$ is finite.
(ii) The interaction is finite range, i.e. there exists $R < \infty$ such that $\Phi_A \equiv 0$ whenever diam$(A) > R$.
(iii) There exists a set $Q \subset \mathcal{S}_0$ such that the constant configurations $\sigma_x \equiv q, q \in Q$, are the only periodic ground-states of Φ.

In this setting, we can decide whether, locally at $x \in \mathbb{Z}^d$, a configuration σ 'looks like' one of its ground-states by looking only at a finite neighbourhood of radius R of x.

We define $c(x) = q$ if $\eta_z = q$ for all z such that $|x - z| \leq R$. One sometimes calls such points q-points. Otherwise, $c(x)$ is undefined. We set

$$\underline{\Gamma}(\eta) \equiv \{z \in \mathbb{Z}^d : \exists y : |z - y| \leq R : c(y) \notin Q\} \tag{5.78}$$

and

$$\Lambda_q(\eta) \equiv \{z \in \underline{\Gamma}(\eta) \wedge c(x) = q\} \tag{5.79}$$

Observe that the actual non-q-points are surrounded by large layers of q-points. This is done to avoid any interaction between different *connected components* of contours. Figure 5.3 depicts a collection of contours and the sets of q-points.

Let us assume that $H_\Lambda(q) = 0$, and introduce the Hamiltonian with additional fields,

$$\tilde{H}_\Lambda(\sigma) = H_\Lambda(\sigma) + \sum_{x \in \Lambda} e_q \mathbb{I}_{\sigma_x = q} \tag{5.80}$$

The important observation is that we can represent the energy of a configuration σ in the form

$$\tilde{H}_\Lambda(\sigma) = \tilde{H}_{\Lambda \cap \underline{\Gamma}(\sigma)}(\sigma) + \sum_{q \in Q} e_q |\Lambda_q| \tag{5.81}$$

where for lighter notation we write Λ_q for $\Lambda \cap \Lambda_q$. Moreover, by construction, given $\underline{\Gamma}(\sigma)$, $\tilde{H}_{\Lambda \cap \underline{\Gamma}(\sigma)}(\sigma)$ depends only on the configuration on $\underline{\Gamma}$. If $(\underline{\gamma}_1, \ldots, \underline{\gamma}_n)$ are the connected

Figure 5.3 A collection of contours in a model with four colours.

components of $\underline{\Gamma}$, then

$$\tilde{H}_{\Lambda \cap \Gamma(\sigma)}(\sigma) = \sum_{i=1}^{n} \tilde{H}_{\Lambda \cap \underline{\gamma}_i(\sigma)}(\sigma) \tag{5.82}$$

This allows us to represent the partition function in the form

$$\begin{aligned}
Z_{n,\Lambda}^{\eta} &= \mathbb{E}_{\sigma_\Lambda} \prod_i e^{-\beta \sum_i H_{\Lambda \cap \underline{\gamma}_i(\sigma)}(\sigma)} \prod_{q \in Q} e^{\beta c_q |\Lambda_q|} \tag{5.83}\\
&= \sum_{\underline{\Gamma}} \mathbb{E}_{\sigma_\Lambda} \mathbb{I}_{\underline{\Gamma}(\sigma_\Lambda,\eta_{\Lambda^c})=\underline{\Gamma}} \prod_i e^{-\beta \sum_i H_{\Lambda \cap \underline{\gamma}_i}(\sigma)} \prod_{q \in Q} e^{\beta c_q |\Lambda_q|}\\
&\equiv \sum_{\underline{\Gamma}} \mathbb{E}_{\sigma_\Lambda} \mathbb{I}_{\underline{\Gamma}(\sigma_\Lambda,\eta_{\Lambda^c})=\underline{\Gamma}} \prod_i w(\gamma_i) \prod_{q \in Q} e^{\beta c_q |\Lambda_q|}
\end{aligned}$$

where the $w(\gamma_i)$ can be thought of as activities of the contours. We say that contours satisfy a *Peierls condition* if

$$\mathbb{E}_\sigma \mathbb{I}_{\Gamma(\sigma)=\gamma} w(\gamma) \leq \exp(-C|\underline{\gamma}|) \tag{5.84}$$

i.e. when their activities are exponentially suppressed.

The representation (5.83) looks similar to that of a polymer model, and we might hope that the Peierls condition, with large enough C, would ensure convergence of the expansion for the logarithm. However, there is an important difference: the configurations on connected components of the boundaries of the supports of contours may have different constant values, and a configuration of connected contours can arise only if these values can be matched. For example, in the Ising model, we can think of an annulus-shaped contour that is constant $+1$ outside and constant -1 inside. In the interior of the annulus we can then only have contours that at their outer boundaries are -1. In cases where there are more ground-states, the situation is similar and more complicated.

In the Ising model with zero magnetic field, this problem did not seem to be important because we did not need to keep track of whether a contour separated $+$ from $-$ or vice versa, since there was perfect symmetry between the two. As soon as this symmetry is broken (either in the weights or by the presence of the fields c_q), the task of controlling the expansion becomes much more difficult. To get an idea of what is going on, recall the estimate (4.52) in the proof of the Peierls argument. Right before the last step we had obtained

$$\mu_{\text{int}\,\gamma,\beta}^{+1}[\sigma_{\gamma^{\text{in}}} = -1] \leq e^{-2\beta|\gamma|} \frac{Z_{\text{int}(\gamma)\setminus\gamma^{\text{in}}}^{(-1)}}{Z_{\text{int}(\gamma)\setminus\gamma^{\text{in}}}^{(+1)}} \tag{5.85}$$

Then we used the symmetry of the Hamiltonian under spin flip to deduce that the ratio of partition functions is equal to one. What if this symmetry is broken, e.g. by a magnetic field? Clearly, if the field is positive, we would expect the partition function with minus boundary conditions to be smaller than the one with plus boundary conditions, and so the estimate is only strengthened. But if the magnetic field is negative, the converse is true, and the ratio of partition functions spoils our estimate. In fact, at low temperatures, the spins have the option of following the sign of the boundary condition, in which case we would get a bound of the type

$$\frac{Z_{\text{int}(\gamma)\setminus\gamma^{\text{in}}}^{(-1)}}{Z_{\text{int}(\gamma)\setminus\gamma^{\text{in}}}^{(+1)}} \sim \exp(+2\beta h|\text{int}(\gamma)|) \tag{5.86}$$

or they flip to -1 within γ, in which case we get a bound

$$\frac{Z_{\text{int}(\gamma)\setminus\gamma^{\text{in}}}^{(-1)}}{Z_{\text{int}(\gamma)\setminus\gamma^{\text{in}}}^{(+1)}} \sim \exp(+2\beta|\gamma|) \tag{5.87}$$

which may or may not offset the exponentially small prefactor. In the case where $h|\text{int}(\gamma)| \ll |\gamma|$, the estimate (5.86) suggests that the contour γ is still unlikely. This happens when γ is small. The reason is that flipping the spins cannot produce enough energetic gain from the magnetic field to offset the cost paid for the resulting interaction energy. But it may be advantageous to create a large contour, whose interior is large compared to its surface, to take advantage of the corresponding gain in magnetic field energy. It is thus possible that, under plus boundary conditions, the system likes to create an enormous contour right near the boundary, so that in the inside it looks just the same as the system with minus boundary conditions. In this way phase coexistence is destroyed by a magnetic field term.

Example: A three-state model To get an intuitive feeling for what happens at low temperatures in the absence of symmetries, the following model serves as a standard example (see [64]). We consider spin variables $\sigma_i \in \{-1, 0, 1\}$, and a Hamiltonian with nearest-neighbour interaction

$$H_\Lambda(\sigma) = \sum_{\langle x,y \rangle \int \Lambda \neq} |\sigma_x - \sigma_y| \tag{5.88}$$

Clearly, this model has three translation-invariant ground-states, $\sigma \equiv \pm 1$ and $\sigma \equiv 0$. Thus, we have three candidates for low-temperature Gibbs states. We must ask whether all of them will exist at low enough temperatures. Let us make a formal computation of the free energies associated to these states by expanding in perturbations about the constant configurations. We will only keep track of the smallest perturbations, which consist of having the spin at one site taking a deviant value. The crucial observation is that, in the case of the ± 1 configurations, there are two such contributions with excess energy $2d$ and $4d$, respectively, whereas in the zero configuration there are two contributions both having energy $2d$. Thus, to leading order

$$|\Lambda|^{-1} \ln Z^{\pm}_{\beta,\Lambda} \sim e^{-2d\beta} + e^{-4d\beta} \qquad (5.89)$$

whereas

$$|\Lambda|^{-1} \ln Z^{0}_{\beta,\Lambda} \sim 2e^{-2d\beta} \qquad (5.90)$$

Thus, in a -1 phase, it may be advantageous to create a large contour flipping to the zero configuration, since the ratio of the partition functions inside will produce a factor $\exp(|\text{int }\gamma|(e^{-2d\beta} - e^{-4d\beta}))$ that can compensate for the price in contour energy $\exp(-\beta|\gamma|)$. In fact, in this model, at zero external field, there is a unique phase corresponding to a perturbation of the zero configuration. It may look paradoxical that this phase is *stabilized*, because of a larger number of low-energy perturbations, i.e. because it is less rigid than the other ground-states.

It should be clear that a rigorous analysis of the preceding discussion is rather complicated. The key observation of the Pirogov–Sinai theory is that even the equality of all ground-state energy densities e_q does not ensure that there will be a Gibbs state that is a perturbation of the corresponding ground-state. Rather, in the absence of symmetries, to ensure the coexistence all phases it will in general be necessary to fine-tune the values of e_q in a temperature-dependent way. In fact, what has to be done is to adjust the values of the e_q (by adding a magnetic field) in such a way that the *metastable free energies* corresponding to these phases become equal. The definition of the concept of metastable free energies is subtle. Roughly, it corresponds to computing the free energy in a low-temperature expansion around a given ground-state while suppressing the contributions from large contours (that might lead to divergences). One can understand that, if these metastable free energies are all equal, one has artificially restored a symmetry between the phases, in the sense that the ratios of partition functions as in (5.85) are all almost equal to 1 (and, in particular, not significant against the exponential preceding it). Therefore, in such a situation, contours are again unlikely, and the different phases can coexist. One way to make this rigorous is to use recursive partial summation of contours starting from small and moving up to larger scales. As this procedure, and even the precise statement of the results, is quite involved, we will not enter further into the details of this method. A good exposition can be found in [37], see also [263]. We will have occasion to revisit low-temperature expansions and iterative methods of a similar kind in the analysis of the random-field Ising model later on.

Part II

Disordered systems: lattice models

6

Gibbsian formalism and metastates

> Longtemps les objets dont s'occupent les mathématiciens étaient pour la plus-part mal définis; on croyait les connaître, parce qu'on se les représentait avec le sens ou l'imagination; mais on n'en avait qu'une image grossière et non une idée précise sur laquelle le raisonnement pût avoir prise.[1]
>
> Henri Poincaré, La valeur de la science.

We now turn to the main topic of this book, disordered systems. We split this into two parts, treating in turn *lattice models* and *mean-field models*. From the physical point of view, the former should be more *realistic* and hence more relevant, so it is natural that we present the general mathematical framework in this context. However, the number of concrete problems one can to this day solve rigorously is quite limited, so that the examples we will treat can mostly be considered as random perturbations of the Ising model. In the second part we will be able to look, in a simplified setting, at more complex, genuinely random models, that rather surprisingly will turn out to produce fascinating mathematics, but also lead to applications that are beyond the standard scope of physics.

6.1 Introduction

We have seen that statistical mechanics is a theory that treats the dynamical degrees of freedom of a large system as random variables, distributed according to the Gibbs measure. The basic rationale behind this idea is that, on the time-scales on which the system is observed, the dynamics relaxes to equilibrium and, in particular, forgets the details of the initial conditions. Such an assumption can of course not always be satisfied, as it requires the microscopic degrees of freedom to vary sufficiently fast. A typical example where this would fail are solid alloys. Assume that we have a material made of a mixture of two types of atoms, say gold and iron, that at low temperatures forms a crystalline solid. Then some lattice sites will have to be occupied by iron atoms, while the others are occupied by gold atoms. If we melt or just heat the system, the atoms become mobile and quickly change places, so that we might describe the system by some Gibbs distribution. However, at low temperatures, the motion of atoms between different sites is strongly suppressed (for reasons that we will not discuss here), and, over large time-scales, the microscopic realization of

---

[1] Approximately: For a long time the objects that mathematicians dealt with were mostly ill defined; one believed one knew them, because one represented them with the senses and imagination; but one had but a rough picture and not a precise idea on which reasoning could take hold.

the gold–iron mixture will not change. One says that the positions of the atoms are 'frozen', and the system will not be in thermal equilibrium.

However, the positions of the atoms are not the only degrees of freedom of the system. The iron atoms have magnetic moments, and we might be interested in the magnetic properties of the system. But the orientations of the magnetic moments are not 'frozen', and their behaviour could be very well described by a Gibbs measure. However, the description of this system must take into account the positions of the iron atoms, as the interaction between them depends on their mutual distances. Thus, assuming that we knew the positions, x_i, of the iron atoms, we could write a (formal) Hamiltonian for the spin degrees of freedom of the form

$$H(\sigma;x) = -\sum_{i,j} \sigma_i \sigma_j \phi(x_i, x_j) \tag{6.1}$$

Again, given the positions x_i, we would then write the Gibbs measure

$$\mu_\beta(\cdot;x) = \frac{e^{-\beta H(\sigma;x)}}{Z_\beta(x)} \tag{6.2}$$

where the partition function also depends on the positions x. We would call such a system *spatially inhomogeneous*, or *disordered*. The point is that it would be fairly impractical to study all possible systems for all possible arrangements of the x_i; thus we should hope that the microscopic details of these arrangements do not matter too much, and that only certain statistical properties are relevant. In other words, we would like to *model* the spatial inhomogeneity by a *random process*, i.e. model a disordered system as a *random model*, by introducing some probability distribution, $\mathbb{P}_x$, on the space of possible realizations of the iron positions. This new type of randomness is often called *quenched* randomness, a term derived from the idea that the solid alloy has been produced by rapid cooling through immersion in water, a process that in metallurgy is called *quenching*. One should be well aware that this new type of randomness is conceptually very different from the randomness we have encountered previously. When dealing with the dynamical variables, we hope that time averages will converge to averages with respect to the Gibbs measures, whereas for quenched randomness we cannot expect such a thing to happen.

What we may hope – and we will learn to what extent this is justified – is that certain properties of our materials depend little on the microscopic realizations, and are the same for *almost all* realizations of the disorder (essentially through a spatial effective averaging). Thus, a first reflex is to consider *averages* with respect to the disorder. Here there are two notions conventionally used in the physics literature that we need to clarify from the outset:

- *Quenched average*. This is the proper way to average: one computes for fixed disorder variables thermodynamic quantities, such as the Gibbs measure of the free energy, and then performs an average over the disorder; e.g. the *quenched free energy* is given as

$$F_{\beta,\Lambda}^{\text{quenched}} \equiv \beta^{-1} \mathbb{E}_x \ln Z_{\beta,\Lambda}(x) \tag{6.3}$$

- *Annealed average*. One computes the average of the partition function and the unnormalized averages of the dynamical variables first and normalizes later. This yields, e.g., the *annealed* free energy

$$F_{\beta,\Lambda}^{\text{annealed}} \equiv \beta^{-1} \ln \mathbb{E}_x Z_{\beta,\Lambda}(x) \tag{6.4}$$

Upon reflection, this procedure corresponds to treating the disorder variables as dynamical variables on an equal footing with the other degrees of freedom and thus disregards the fact that they do not equilibrate on the same time-scale. Thus this is inappropriate in the situations we want to describe. Of course, one can always try and see whether this will yield by accident the same results anyway.

After these preliminary discussions we will now proceed to a rigorous setup of the Gibbsian formalism for quenched random models.

6.2 Random Gibbs measures and metastates

We will now give a definition of disordered lattice spin systems. This will not be as general as possible, as we allow disorder only in the interactions, but not in the lattice structure or the spin spaces. As in Chapter 4, we consider a lattice, $\mathbb{Z}^d$, a single-site spin space, $(\mathcal{S}_0, \mathcal{F}_0, \nu_0)$, and the corresponding a-priori product space, $(\mathcal{S}, \mathcal{F}, \nu)$. As a new ingredient, we add a (rich enough) probability space, $(\Omega, \mathcal{B}, \mathbb{P})$, where Ω will always be assumed to be a Polish space. On this probability space we construct a *random interaction* as follows:

Definition 6.2.1 A random interaction, Φ, is a family, $\{\Phi_A\}_{A \subset \mathbb{Z}^d}$, of random variables on $(\Omega, \mathcal{B}, \mathbb{P})$ taking values in $B(\mathcal{S}, \mathcal{F}_A)$, i.e. measurable maps $\Phi_A : \Omega \ni \omega \to \Phi_A[\omega] \in B(\mathcal{S}, \mathcal{F}_A)$. A random interaction is called *regular* if, for $\mathbb{P}$-almost all ω, for any $x \in \mathbb{Z}^d$, there exists a finite constant $c_x[\omega]$ such that

$$\sum_{A \ni x} \|\Phi_A[\omega]\|_\infty \leq c_x[\omega] < \infty \tag{6.5}$$

A regular random interaction is called *continuous* if, for each $A \subset \Lambda$, Φ_A is jointly continuous in the variables η and ω.

In the present section we discuss only regular random interactions. Some interesting physical systems do correspond to irregular random interactions. In particular, many real spin-glasses have a non-absolutely summable interaction, called the RKKY interaction. See [107, 108, 249, 264] for some rigorous results.

Remark 6.2.2 In most examples one assumes that the random interaction has the property that Φ_A and Φ_B are *independent* if $A \neq B$, or, at least, if $A \cap B = \emptyset$. In fact, in all examples of interest, Ω is a product space of the form $\Omega = E^{\mathbb{Z}^d}$, where $E \subseteq \mathbb{R}^k$.

Given a random interaction, it is straightforward to define random finite-volume Hamiltonians, $H_\Lambda[\omega]$, as in the deterministic case. Note that, for regular random interactions, H_Λ is a random variable that takes values in the space $\mathcal{B}_{ql}(\mathcal{S})$, i.e. the mapping $\omega \to H_\Lambda[\omega]$ is measurable. If, moreover, the Φ_A are continuous functions of ω, then the local Hamiltonians are also continuous functions of ω.

Next we need to define the random analogue of local specifications. A natural definition is the following:

Definition 6.2.3 A *random local specification* is a family of probability kernels, $\{\mu_{\beta,\Lambda}^{(\cdot)}[\omega]\}_{\Lambda \subset \mathbb{Z}^d}$, depending on a random parameter, ω, such that:

(i) For all $\Lambda \subset \mathbb{Z}^d$ and $\mathcal{A} \in \mathcal{F}$, $\mu_{\beta,\Lambda}^{(\cdot)}(\mathcal{A})$ is a measurable function with respect to the product sigma-algebra $\mathcal{F}_{\Lambda^c} \times \mathcal{B}$.

(ii) For $\mathbb{P}$-almost all ω, for all $\eta \in \mathcal{S}$, $\mu^{(\eta)}_{\Lambda,\beta}[\omega](d\sigma)$ is a probability measure on $\mathcal{S}$.
(iii) For $\mathbb{P}$-almost all ω, the family $\{\mu^{(\cdot)}_{\beta,\Lambda}[\omega]\}_{\Lambda\subset\mathbb{Z}^d}$ is a Gibbs specification for the interaction $\Phi[\omega]$ and inverse temperature β.
(iv) The random local specification is called *continuous* if, for any finite Λ, $\mu^\eta_{\beta,\Lambda}[\omega]$ is jointly continuous in η and ω.

A regular random interaction should naturally give rise to a *random Gibbs specification*. In fact, we have:

Lemma 6.2.4 *Let Φ be a regular random interaction. Then the formula*

$$\mu^{(\eta)}_{\Lambda,\beta}[\omega](d\sigma) \equiv \frac{1}{Z^\eta_{\beta,\Lambda}[\omega]} e^{-\beta H_\Lambda[\omega](\sigma_\Lambda,\eta_{\Lambda^c})} \rho_\Lambda(d\sigma_\Lambda) \delta_{\eta_{\Lambda^c}}(d\sigma_{\Lambda^c}) \tag{6.6}$$

defines a random local specification, called a random Gibbs specification. Moreover, if Φ is continuous, then the Gibbs specification is continuous.

The important point is that the maps $\omega \to \mu^{(\cdot)}_{\Lambda,\beta}[\omega]$ are again measurable in all appropriate senses.

We now feel ready to define random infinite-volume Gibbs measures. The following is surely reasonable:

Definition 6.2.5 *A measurable map, $\mu_\beta : \Omega \to \mathcal{M}_1(\mathcal{S},\mathcal{F})$, is called a random Gibbs measure for the regular random interaction Φ at inverse temperature β if, for $\mathbb{P}$-almost all ω, $\mu_\beta[\omega]$ is compatible with the random local specification $\{\mu^{(\cdot)}_{\beta,\Lambda}[\omega]\}_{\Lambda\subset\mathbb{Z}^d}$ for this interaction.*

The first question one must ask concerns the existence of such random Gibbs measures. One would expect that, for compact state space, the same argument as in the deterministic situation should provide an affirmative answer. Indeed, it is clear that, for almost all ω, any sequence, $\mu^\eta_{\beta,\Lambda_n}[\omega]$, taken along an increasing and absorbing sequence of volumes, possesses limit points, and, therefore, there exist subsequences, $\Lambda_{n[\omega]}$, such that $\mu^\eta_{\beta,\Lambda_{n[\omega]}}[\omega]$ converges to a Gibbs measure for the interaction $\Phi[\omega]$. The only open question is then whether such limits can provide a *measurable* map from Ω to the Gibbs measures? This is non-trivial, due to the fact that the subsequences $\Lambda_n[\omega]$ must in general depend on the realization of the disorder!

This question may at first sound like some irrelevant mathematical sophistication, and indeed this problem was mostly disregarded in the literature. To my knowledge, it was first discussed in a paper by van Enter and Griffiths [251] and studied in more detail by Aizenman and Wehr [8], but it is the merit of Ch. Newman and D. Stein [183, 185, 186, 187, 188, 189, 190] to have brought the intrinsic *physical relevance* of this issue to light. Note that the problem is solved immediately if there are deterministic sequences, Λ_n, along which the local specifications converge. Newman and Stein pointed out that, in very strongly disordered systems such as spin-glasses, such deterministic sequences might not exist.

In more pragmatic terms, the construction of infinite-volume Gibbs measures via limits along random subsequences can be criticised by its lack of actual approximative power. An infinite-volume Gibbs measure is supposed to approximate reasonably a very large system under controlled conditions. If, however, this approximation is only valid for certain very special finite volumes that depend on the specific realization of the disorder, while for other volumes the system is described by other measures, just knowing the set of all infinite-volume measures is surely not enough.

As far as proving existence of random Gibbs measures is concerned, there is a simple way out of the random subsequence problem. This goes by extending the local specifications to probability measures, $K^{\eta}_{\beta,\Lambda}$, on $\Omega \times \mathcal{S}$, in such a way that the marginal distribution of $K^{\eta}_{\beta,\Lambda}$ on Ω is simply $\mathbb{P}$, while the conditional distribution, given $\mathcal{B}$, is $\mu^{(\eta)}_{\beta,\Lambda}[\omega]$. The measures K^{η}_{β} are sometimes called *joint measures*.

Theorem 6.2.6 *Let Φ be a continuous regular random interaction. Let $K^{(\cdot)}_{\beta,\Lambda}$ be the corresponding measure defined as above. Then*

(i) *If, for some increasing and absorbing sequence Λ_n and some $\eta \in \mathcal{S}$, the weak limit $\lim_{n \uparrow \infty} K^{\eta}_{\beta,\Lambda_n} \equiv K^{\eta}_{\beta}$ exists, then its regular conditional distribution $K^{\eta}_{\beta}(\cdot | \mathcal{B} \times \mathcal{S})$, given $\mathcal{B}$, is a random Gibbs measure for the interaction Φ.*
(ii) *If $\mathcal{S}$ is compact, and if $\mathbb{P}$ is tight in the sense that $\forall \epsilon > 0, \exists \Omega_\epsilon \subset \Omega$, which is compact, and $\mathbb{P}[\Omega_\epsilon] \geq 1 - \epsilon$, then there exist increasing and absorbing sequences Λ_n such that the hypothesis of (i) is satisfied.*

Proof The proof of this theorem is both simple and instructive. Note first that the existence of a regular conditional distribution is ensured if Ω and $\mathcal{S}$ are Polish spaces. Let $f \in C(\mathcal{S}, \mathcal{F})$ be a continuous function. We must show that, a.s.

$$K^{\eta}_{\beta}(f | \mathcal{B} \times \mathcal{S})[\omega] = K^{\eta}_{\beta}(\mu^{(\cdot)}_{\beta,\Lambda}[\omega](f) | \mathcal{B} \times \mathcal{S})[\omega] \qquad (6.7)$$

Set $g(\omega, \sigma) \equiv \mu^{(\sigma)}_{\beta,\Lambda}[\omega](f(\omega, \sigma))$. Let $\mathcal{B}_k, k \in \mathbb{N}$ be a filtration of the sigma-algebra $\mathcal{B}$ where $\mathcal{B}_k$ is generated by the interaction potentials Φ_A with $A \subset \Lambda_k$ with Λ_k some increasing and absorbing sequence of volumes. The important point is to realize that, for continuous functions h on $\Omega \times \Sigma$,

$$K^{\eta}_{\beta}(h | \mathcal{B} \times \mathcal{S})[\omega] = \lim_{k \uparrow \infty} \lim_{n \uparrow \infty} K^{\eta}_{\beta,\Lambda_n}(f | \mathcal{B}_k \times \mathcal{S})[\omega] \qquad (6.8)$$

But for any fixed Λ, and n large enough,

$$\mathbb{E}[\mu^{(\eta)}_{\beta,\Lambda_n}(f) | \mathcal{B}_k \times \Sigma][\omega] = \mathbb{E}[\mu^{(\eta)}_{\beta,\Lambda_n}(\mu^{(\cdot)}_{\beta,\Lambda}(f)) | \mathcal{B}_k \times \Sigma][\omega]$$
$$= \mathbb{E}[\mu^{(\eta)}_{\beta,\Lambda_n}(\mu^{(\cdot)}_{\beta,\Lambda}[\omega](f)) | \mathcal{B}_k \times \Sigma][\omega]$$
$$+ \mathbb{E}[\mu^{(\eta)}_{\beta,\Lambda_n}(\mu^{(\cdot)}_{\beta,\Lambda}(f) - \mu^{(\cdot)}_{\beta,\Lambda}[\omega](f)) | \mathcal{B}_k \times \Sigma][\omega] \qquad (6.9)$$

The first term converges to $K^{\eta}_{\beta}(\mu^{(\eta)}_{\beta,\Lambda}[\omega](f) | \mathcal{B} \times \mathcal{S})[\omega]$, while for the last we observe that, due to the continuity of the local specifications in ω, uniformly in n,

$$\mathbb{E}\left[\mu^{(\eta)}_{\beta,\Lambda_n}\left(\mu^{(\eta)}_{\beta,\Lambda}(f) - \mu^{(\eta)}_{\beta,\Lambda}[\omega](f)\right)\right| \qquad (6.10)$$
$$\leq \sup_{\omega' \in \mathcal{B}_k[\omega]} \sup_{\eta \in \mathcal{S}} \left| \mu^{(\eta)}_{\beta,\Lambda}[\omega'](f) - \mu^{(\eta)}_{\beta,\Lambda}[\omega](f) \right| \downarrow 0$$

as $k \uparrow \infty$. Here $\mathcal{B}_k[\omega]$ denotes the set of all $\omega' \in \Omega$ that have the same projection to $\mathcal{B}_k$ as ω, more formally

$$\mathcal{B}_k[\omega] \equiv \{\omega' \in \Omega \,|\, \forall_{A \in \mathcal{B}_k : \omega \in A} : \omega' \in A\} \qquad (6.11)$$

This proves (i). To prove (ii), fix any $\epsilon > 0$. If f is a bounded, continuous function on $\Omega \times S$, then

$$\int K_{\beta,\Lambda}(d\omega, d\sigma) f(\omega, \sigma) = \mathbb{E} \int \mu_{\beta,\Lambda}[\omega](d\sigma) f(\omega, \sigma) \quad (6.12)$$

$$= \mathbb{E} \mathbb{I}_{\Omega_\epsilon} \int \mu_{\beta,\Lambda}[\omega](d\sigma) f(\omega, \sigma) + \mathbb{E} \mathbb{I}_{\Omega_\epsilon^c} \int \mu_{\beta,\Lambda}[\omega](d\sigma) f(\omega, \sigma)$$

The second term is by hypothesis bounded by $C\epsilon$, i.e. as small as desired, while the first is (up to a constant) a sequence of probability measures on the compact space $\Omega_\epsilon \times S$, and hence there are subsequences $\Lambda_{n_k^\epsilon}$ such that $K_{\Lambda_{n_k^\epsilon}}^\epsilon(f) \equiv \mathbb{E} \mathbb{I}_{\Omega_\epsilon} \int \mu_{\beta,\Lambda_{n_k^\epsilon}}[\omega](d\sigma) f(\omega, \sigma) \to K_\beta^\epsilon(f)$. Now take a sequence $\epsilon_k \downarrow 0$. By successively thinning out subsequences, one can find a sequence Λ_n such that $K_{\Lambda_n}^{\epsilon_k}(f)$ converges, for any k. Then (6.12) implies that

$$\left| \limsup_{n \uparrow \infty} \int K_{\beta,\Lambda_n}(d\omega, d\sigma) f(\omega, \sigma) - \liminf_{n \uparrow \infty} \int K_{\beta,\Lambda}(d\omega, d\sigma) f(\omega, \sigma) \right| \leq \epsilon_k \quad (6.13)$$

for any k. Thus, $\int K_{\beta,\Lambda_n}(d\omega, d\sigma) f(\omega, \sigma)$ converges, and (ii) is proven. □

Remark 6.2.7 There has recently been some interest in the question as to whether the joint measures K_β^η can themselves be considered as Gibbs measures on the extended space $S \times \Omega$ [159, 160, 245]. The answer turns out to be that, while they are never Gibbs measures in the strict sense, in many cases they are *weakly Gibbsian*, i.e. there exists an almost surely absolutely summable interaction, for which the $K_{\beta,\Lambda}^\eta$ are local specifications.

Theorem 6.2.6 appears to solve our problems concerning the proper Gibbsian setup for random systems. We understand what a random infinite-volume Gibbs measure is and we can prove their existence in reasonable generality. Moreover, there is a constructive procedure that allows us to obtain such measures through infinite-volume limits. However, upon closer inspection, the construction is not fully satisfactory. As can be seen from the proof of Theorem 6.2.6, the measures $K_\beta^\eta(\cdot|\mathcal{B} \times S)$ effectively still contain an averaging over the realization of the disorder 'at infinity'. As a result they will often be mixed states. Such states then do not describe the result of the observations of one sample of the material at given conditions, but the average over many samples that have been prepared to look alike locally.

While we have come to understand that it may not be realistic to construct a state that predicts the outcome of observations on a single (infinite) sample, it would already be more satisfactory to obtain a probability distribution for these predictions (i.e. a random probability measure), rather than just a mean prediction (and average over probability measures). This suggests the extension of the preceding construction to a measure-valued setting. That is, rather than consider measures on the space $\Omega \times S$, we introduce measures $\mathcal{K}_{\beta,\Lambda}^\eta$ on the space $\Omega \times \mathcal{M}_1(S)$, defined in such a way that the marginal distribution of $\mathcal{K}_{\beta,\Lambda}^\eta$ on Ω is again $\mathbb{P}$, while the conditional distribution, given $\mathcal{B}$, is $\delta_{\mu_{\beta,\Lambda}^{(\eta)}[\omega]}$, the Dirac-measure concentrated on the corresponding local specification. We will introduce the symbolic notation

$$\mathcal{K}_{\beta,\Lambda}^\eta \equiv \mathbb{P} \times \delta_{\mu_{\beta,\Lambda}^{(\eta)}[\omega]} \quad (6.14)$$

One has the following analogue of Theorem 6.2.6:

Theorem 6.2.8 *Let Φ be a continuous regular random interaction. Let $\mathcal{K}^{(\cdot)}_{\beta,\Lambda}$ be the corresponding measure defined as above. Then:*

(i) *If, for some increasing and absorbing sequence Λ_n and some $\eta \in \mathcal{S}$, the weak limit $\lim_{n\uparrow\infty} \mathcal{K}^{\eta}_{\beta,\Lambda_n} \equiv \mathcal{K}^{\eta}_{\beta}$ exists, then its regular conditional distribution $\mathcal{K}^{\eta}_{\beta}(\cdot|\mathcal{B}\times\mathcal{S})$, given $\mathcal{B}$, is a probability distribution on $\mathcal{M}_1(\mathcal{S})$, that, for almost all ω, gives full measure to the set of infinite-volume Gibbs measures corresponding to the interaction $\Phi[\omega]$ at inverse temperature β. Moreover,*

$$K^{\eta}_{\beta}(\cdot|\mathcal{B}\times\mathcal{S}) = \mathcal{K}^{\eta}_{\beta}(\mu|\mathcal{B}\times\mathcal{S}) \quad (6.15)$$

(ii) *If $\mathcal{S}$ is compact and $\mathbb{P}$ is tight, then there exist increasing and absorbing sequences Λ_n such that the hypothesis of (i) is satisfied for any η.*

Remark 6.2.9 The regular conditional distribution

$$\kappa^{\eta}_{\beta} \equiv \mathcal{K}^{\eta}_{\beta}(\cdot|\mathcal{B}\times\mathcal{S}) \quad (6.16)$$

is called the Aizenman–Wehr *metastate* (following the suggestion of Newman and Stein [187]).

Proof The proof of this theorem is even simpler than that of Theorem 6.2.6. Note that assertion (i) will follow if for any bounded continuous function $f : \mathcal{S} \to \mathbb{R}$, and any finite $\Lambda \subset \mathbb{Z}^d$, we can show that

$$\mathbb{E}\int \kappa^{\eta}_{\beta}(d\mu)[\omega]|\mu(f) - \mu(\mu^{(\cdot)}_{\beta,\Lambda}[\omega](f))| = 0 \quad (6.17)$$

Let us set $h(\omega,\mu) \equiv |\mu(f) - \mu(\mu^{(\cdot)}_{\beta,\Lambda}[\omega](f))|$. We need to check that h is a continuous function on $\Omega \times \mathcal{M}_1(\mathcal{S})$. By definition of the weak topology, $\mu(g)$ is a continuous function of μ if g is continuous. By Lemma 4.2.16, $\mu^{\eta}_{\beta,\Lambda}[\omega](f)$ is jointly continuous in η and ω. Thus, both $\mu(f)$ and $\mu(\mu^{(\cdot)}_{\beta,\Lambda}[\omega](f))$ are continuous in μ, and hence h is a bounded continuous function of μ and ω. But then,

$$\mathcal{K}^{\eta}_{\beta}(h) = \lim_{n\uparrow\infty} \mathcal{K}^{\eta}_{\beta,\Lambda_n}(h) = 0 \quad (6.18)$$

by the compatibility relations of local specifications. But h, being non-negative, must be zero $\mathcal{K}^{\eta}_{\beta}$-almost surely, so (6.17) holds, proving (i). Assertion (ii) follows exactly as in the proof of Theorem 6.2.6. $\square$

Apart from the Aizenman–Wehr metastate, Newman and Stein propose another version of the metastate that they call the *empirical metastate* as follows. Define the random empirical measures $\kappa^{em}_N(\cdot)[\omega]$ on $(\mathcal{M}_1(\mathcal{S}^{\infty}))$, to be given by

$$\kappa^{em}_N(\cdot)[\omega] \equiv \frac{1}{N}\sum_{n=1}^{N} \delta_{\mu_{\Lambda_n}[\omega]} \quad (6.19)$$

In [187] it was proven that, for sufficiently sparse sequences Λ_n and subsequences N_k, it is true that almost surely

$$\lim_{i\uparrow\infty} \kappa^{em}_{N_k}(\cdot)[\omega] = \kappa(\cdot)[\omega] \quad (6.20)$$

Newman and Stein conjectured that in many situations the use of sparse subsequences would not be necessary to achieve convergence. However, Külske [157] has exhibited some simple mean-field examples where almost sure convergence only holds for very sparse (exponentially spaced) subsequences. He also showed that, for more slowly growing sequences, convergence in law can be proven in these cases.

Illustration At this stage the reader may rightly pause and ask whether all this abstract formalism is really necessary, or whether, in reasonable situations, we could avoid it completely? To answer this question, we need to look at specific results and, above all, at examples. Before turning to the difficult analysis of metastates in specific spin systems, it may be worthwhile to transplant the formalism developed above to the more familiar context of sums of i.i.d. random variables.

Let $(\Omega, \mathcal{F}, \mathbb{P})$ be a probability space, and let $\{X_i\}_{i\in\mathbb{N}}$ be a family of i.i.d. centered random variables with variance one; let $\mathcal{F}_n$ be the sigma-algebra generated by $X_1, \ldots, X_n$ and let $\mathcal{F} \equiv \lim_{n\uparrow\infty} \mathcal{F}_n$. Define the random variables $G_n \equiv \frac{1}{\sqrt{n}} \sum_{i=1}^n X_i$. We may define the joint law, K_n, of G_n and the X_i, as a probability measure on $\mathbb{R} \otimes \Omega$. Clearly, this measure converges to some measure K whose marginal on $\mathbb{R}$ will be the standard normal distribution. However, we can say more, namely:

Lemma 6.2.10 *In the example described above,*

(i) $\lim_{n\uparrow\infty} K_n = \mathbb{P} \times \mathcal{N}(0, 1)$, *where* $\mathcal{N}(0, 1)$ *denotes the normal distribution, and*
(ii) the conditional measure $\kappa(\cdot)[\omega] \equiv K(\cdot|\mathcal{F})[\omega] = \mathcal{N}(0, 1)$, *a.s.*

Proof All we have to understand is that indeed the limiting measure K is a product measure; then (ii) is immediate. Let f be a continuous function on $\Omega \times \mathbb{R}$, where we identify Ω with $\mathbb{R}^\mathbb{N}$. We must show that $K_n(f) \to \mathbb{E}_X \mathbb{E}_g f(X, g)$, where g is a standard Gaussian random variable, independent of X. Since local functions are dense in the set of continuous functions, it is enough to assume that f depends only on finitely many coordinates, say $X_1, \ldots, X_k$, and G_n. Then

$$K_n(f) = \mathbb{E} f(X_1, \ldots, X_k, G_n) = \mathbb{E} f\left(X_1, \ldots, X_k, \frac{1}{\sqrt{n}} \sum_{i=1}^n X_i\right)$$

$$= \mathbb{E} f\left(X_1, \ldots, X_k, \frac{1}{\sqrt{n-k}} \sum_{i=k+1}^n X_i\right) + \mathbb{E}\left[f\left(X_1, \ldots, X_k, \frac{1}{\sqrt{n}} \sum_{i=1}^n X_i\right)\right.$$

$$\left. - f\left(X_1, \ldots, X_k, \frac{1}{\sqrt{n-k}} \sum_{i=k+1}^n X_i\right)\right] \quad (6.21)$$

Clearly, by the central limit theorem,

$$\lim_{n\uparrow\infty} \mathbb{E} f\left(X_1, \ldots, X_k, \frac{1}{\sqrt{n-k}} \sum_{i=k+1}^n X_i\right) = \mathbb{E}\mathbb{E}_g f(X_1, \ldots, X_k, g) \quad (6.22)$$

while the remaining terms tend to zero, as $n \uparrow \infty$, by continuity of f. This proves the lemma. □

Let us now look at the empirical metastate in our example. Here the empirical metastate corresponds to

$$\kappa_N^{\text{em}}(\cdot)[\omega] \equiv \frac{1}{N}\sum_{n=1}^N \delta_{G_n[\omega]} \qquad (6.23)$$

We will prove that the following lemma holds:

Lemma 6.2.11 *Let G_n and $\kappa_N^{\text{em}}(\cdot)[\omega]$ be defined above. Let B_t, $t \in [0,1]$ denote a standard Brownian motion. Then the random measures κ_N^{em} converge in law to the measure $\kappa^{\text{em}} = \int_0^1 dt\, \delta_{t^{-1/2}B_t}$.*

Proof We will see that, quite clearly, this result relates to Lemma 6.2.10 as the Invariance Principle does to the CLT, and, indeed, its proof is essentially an immediate consequence of Donsker's theorem. Donsker's theorem (see [136] for a formulation in more generality than needed in this chapter) asserts the following: Let $\eta_n(t)$ denote the continuous function on $[0,1]$ that, for $t = k/n$, is given by

$$\eta_n(k/n) \equiv \frac{1}{\sqrt{n}}\sum_{i=1}^k X_i \qquad (6.24)$$

and that interpolates linearly between these values. Then $\eta_n(t)$ converges in distribution to standard Brownian motion, in the sense that, for any continuous functional $F : C([0,1]) \to \mathbb{R}$, it is true that $F(\eta_n)$ converges in law to $F(B)$. We have to prove that, for any bounded continuous function f,

$$\frac{1}{N}\sum_{n=1}^N \delta_{G_n[\omega]}(f) \equiv \frac{1}{N}\sum_{n=1}^N f(\eta_n(n/N)/\sqrt{n/N}) \qquad (6.25)$$

$$\to \int_0^1 dt\, f(B_t/\sqrt{t}) \equiv \int_0^1 dt\, \delta_{B_t/\sqrt{t}}(f)$$

To see this, simply define the continuous functionals F and F_N by

$$F(\eta) \equiv \int_0^1 dt\, f(\eta(t)/\sqrt{t}) \qquad (6.26)$$

and

$$F_N(\eta) \equiv \frac{1}{N}\sum_{n=1}^N f(\eta(n/N)/\sqrt{n/N}) \qquad (6.27)$$

We have to show that in distribution $F(B) - F_N(\eta_N)$ converges to zero. But

$$F(B) - F_N(\eta_N) = F(B) - F(\eta_N) + F(\eta_N) - F_N(\eta_N) \qquad (6.28)$$

By the invariance principle, $F(B) - F(\eta_N)$ converges to zero in distribution while $F(\eta_N) - F_N(\eta_N)$ converges to zero since F_N is the Riemann sum approximation to F. B_t is measurable with respect to the tail sigma-algebra of the X_i, so that the conditioning on $\mathcal{F}$ has no effect. □

Exercise: Consider the random field version of the Curie–Weiss model, i.e. the mean-field model with Hamiltonian

$$H_N(\sigma)[\omega] \equiv -\frac{1}{2N}\sum_{i,j=1}^{N}\sigma_i\sigma_j - \delta\sum_{i=1}^{N} h_i[\omega]\sigma_i \qquad (6.29)$$

where the h_i are i.i.d. random variables taking the values ± 1 with probability $1/2$.

(i) Introduce the random sets $\Lambda_+[\omega] \equiv \{i : h_i[\omega] = +1\}$ and $\Lambda_-[\omega] = \{1,\ldots,N\}\setminus \Lambda_+[\omega]$. Define the (empirical) magnetizations of these two sets, $m_\pm(\sigma) \equiv \frac{1}{|\Lambda_\pm|}\sum_{i\in\Lambda_\pm}\sigma_i$. Express $H_N(\sigma)$ in terms of these quantities.

(ii) Compute an expression for the distribution of the variables $m(\sigma) \equiv (m_+(\sigma), m_-(\sigma))$ under the canonical Gibbs measure.

(iii) Show that, if $\delta < 1$, there is a critical value $\beta_c = \beta_c(\delta)$ such that, for $\beta < \beta_c$, there exist two distinct points $m^*, \bar m^* \in [-1,1]^2$, with $(\bar m^*_+, \bar m^*_-) = (-m^*_-, -m^*_+)$, such that for any $\epsilon > 0$, $\lim_{N\uparrow\infty}\mu_{\beta,N}(\{|m(\sigma) - m^*| < \epsilon\} \vee \{|m(\sigma) - \bar m^*| < \epsilon\}) = 1$, a.s.

(iv) Show that for almost all ω there exists a random subsequence $N_k[\omega]$ such that $\lim_{k\uparrow\infty}\mu_{\beta,N_k[\omega]}[\omega](\{|m(\sigma) - m^*| < \epsilon\}) = 1$. Are there also subsequences such that the limit is $1/2$?

(v) Now consider $q_N \equiv \mu_{\beta,N}(\{|m(\sigma) - m^*| < \epsilon\})$ as a sequence of random variables. What is the limit of its distribution as $N \uparrow \infty$? Use this result to formulate a result on the convergence in distribution of the Gibbs measures $\mu_{\beta,\delta,N}$.

(vi*) Give an expression for the Aizenman–Wehr metastate.

(vii*) Consider the empirical distribution of the random variables q_N, i.e. $\frac{1}{N}\sum_{n=1}^{N}\delta_{q_n}$. What is the limit of this probability measure as $N \uparrow \infty$?

All the concepts introduced above have been worked out explicitly for two non-trivial disordered models, the random field Curie–Weiss model and the Hopfield model with finitely many patterns (see Chapter 12), by Külske [157, 158]. Explicit constructions of metastates for lattice models are largely lacking. The only example is the two-dimensional Ising model with random boundary conditions that was studied in [246, 247].

6.3 Remarks on uniqueness conditions

As in the case of deterministic interactions, having established existence of Gibbs states, the next basic question is that of uniqueness. As in the deterministic case, uniqueness conditions can be formulated in a quite general setting for 'weak' enough interactions. Indeed, Theorem 4.3.1 can be applied directly for any given realization of the disorder. However, a simple application of such a criterion will not capture the particularities of a disordered system, and will therefore give bad answers in most interesting examples. The reason for this lies in the fact that the criterion of Theorem 4.3.1 is formulated in terms of a supremum over $y \in \mathbb{Z}^d$; in a translation-invariant situation, this is appropriate, but, in a random system, we will often find that, while for most points the condition will be satisfied, there may exist rare random points where it is violated. Extensions of Dobrushin's criteria have been developed by Bassalygo and Dobrushin [14], van den Berg and Maes [244], and Gielis [128]. Uniqueness for weak interactions in the class of regular interactions can be proven with the help of cluster expansion techniques rather easily. The best results in this direction are due to Klein and Masooman [150], while the basic ideas go back to Berretti [22] and Fröhlich and Imbrie [106].

It should be pointed out that the most interesting problems in high-temperature disordered systems concern the case of non-regular interactions. For example, Fröhlich and Zegarlinski [107, 108, 264] have proven uniqueness (in a weak sense), for mean zero square integrable interactions of mean zero, in the Ising case.

6.4 Phase transitions

The most interesting questions in disordered systems concern again the case of non-uniqueness of Gibbs measures, i.e. phase transitions. Already in the case of deterministic models, we have seen that there is no general theory for the classification of the extremal Gibbs states in the low-temperature regime; in the case of disordered systems the situation is even more difficult. Basically, one should distinguish between two situations:

(1) Small random perturbations of a deterministic model (whose phase structure is known).
(2) Strongly disordered models.

Of course, this distinction is a bit vague. Nonetheless, we say that we are in situation (1) if we can represent the Hamiltonian in the form

$$H[\omega](\sigma) = H^{(0)}(\sigma) + H^{(1)}[\omega](\sigma) \tag{6.30}$$

where $H^{(0)}$ is a non-random Hamiltonian (corresponding to a regular interaction) and $H^{(1)}$ is a random Hamiltonian corresponding to a regular random interaction, that is 'small' in some sense. The main question is then whether the phase diagram of H is a continuous deformation of that of $H^{(0)}$, or not. In particular, if $H^{(0)}$ has a first-order phase transition, will the same be true for H?

There are situations when this question can be answered easily; they occur when the different extremal states of $H^{(0)}$ are related by a symmetry group and if, for any realization of the disorder, this symmetry is respected by the random perturbation $H^{(1)}[\omega]$. The classical example of this situation is the *dilute Ising model*. The Hamiltonian of this model is given (formally) by

$$H[\omega](\sigma) = -\sum_{|i-j|=1} J_{ij}[\omega]\sigma_i\sigma_j \equiv H^{\text{Ising}}(\sigma) + \sum_{|i-j|=1}(1-J_{ij})[\omega]\sigma_i\sigma_j \tag{6.31}$$

where J_{ij} are i.i.d. random variables taking the values 0 and 1 with probabilities p and $1-p$, respectively.[2] If p is small, we may consider this as a small perturbation of the standard Ising model. We will show that the Peierls argument (Theorem 4.3.5) applies with just minor modifications, as was observed in [12].

Theorem 6.4.1 *Let μ_β be a Gibbs measure for the dilute Ising model defined by (6.31) and assume that $d \geq 2$. Then there exists $p_0 > 0$ such that, for all $p \leq p_0$, there exists*

---

[2] The precise distribution of the J_{ij} plays no rôle for the arguments that follow; it is enough to have $\mathbb{E}J_{ij} = J_0 > 0$, and $\text{var}(J_{ij}) \ll J_0$.

$\beta(p) < \infty$ such that, for $\beta \geq \beta(p)$,

$$\mathbb{P}\left[\mu_\beta\left[\exists_{\gamma \in \Gamma(\sigma): 0 \in \text{int}\gamma}\right] < \tfrac{1}{2}\right] > 0 \tag{6.32}$$

Proof Define the random contour energy $E(\gamma)$ by

$$E(\gamma) \equiv \sum_{<ij>^* \in \gamma} J_{ij} \tag{6.33}$$

Repeating the proof of Lemma 4.3.6 mutatis mutandis, one gets immediately the estimate

$$\mu_\beta[\omega][\gamma \in \Gamma(\sigma)] \leq e^{-2\beta E[\omega](\gamma)} \tag{6.34}$$

By the law of large numbers, for large γ, $E(\gamma)$ will tend to be proportional to $|\gamma|$; indeed we have that

$$\mathbb{P}[E(\gamma) = x|\gamma|] = \binom{|\gamma|}{x|\gamma|}(1-p)^{x|\gamma|} p^{(1-x)|\gamma|} \tag{6.35}$$

for $x|\gamma|$ integer. Now define the event

$$\mathcal{A} \equiv \{\exists_{\gamma: 0 \in \text{int}\gamma} : E(\gamma) < |\gamma|/2\} \tag{6.36}$$

Clearly, $\mathbb{P}[\mathcal{A}] \leq \sum_{\gamma: 0 \in \text{int}\gamma} \mathbb{P}[E(\gamma) < |\gamma|/2]$, and, by the crudest estimate, using (6.35), $\mathbb{P}[E(\gamma) < |\gamma|/2] \leq 2^{|\gamma|} p^{|\gamma|/2}$. Recalling (4.54), we get that

$$\mathbb{P}[\mathcal{A}] \leq \sum_{k=2d}^{\infty} 3^k 2^k p^{-k/2} \leq \frac{(36p)^d}{1 - 6\sqrt{p}} \tag{6.37}$$

if $p < 1/36$. But, if $\omega \in \mathcal{A}^c$,

$$\mu_\beta[\omega][\exists_{\gamma \in \Gamma(\sigma): 0 \in \text{int}\gamma}] \leq \sum_{\gamma: 0 \in \text{int}\gamma} \mu_\beta[\omega][\gamma \in \Gamma(\sigma)] \leq \sum_{\gamma: 0 \in \text{int}\gamma} e^{-\beta|\gamma|} \tag{6.38}$$

which is smaller than $1/2$ if β is large enough. Thus, for such β,

$$\mathbb{P}\left[\mu_\beta[\exists_{\gamma \in \Gamma(\sigma): 0 \in \text{int}\gamma}] < \tfrac{1}{2}\right] \geq \mathbb{P}[\mathcal{A}^c] \mathbb{P}\left[\mu_\beta\left[\exists_{\gamma \in \Gamma(\sigma): 0 \in \text{int}\gamma}\right] < \tfrac{1}{2}\right] \big| \mathcal{A}^c$$

$$\geq 1 - \frac{(36p)^d}{1 - 6\sqrt{p}} \tag{6.39}$$

which can be made as close to 1 as desired if p is small enough. □

From Theorem 6.4.1 we can deduce the existence of at least two distinct random Gibbs states.

Corollary 6.4.2 *In the dilute Ising model, for any $d \geq 2$, there exists $p_0 > 0$ such that, for all $p \leq p_0$, there exists $\beta(p) > 0$ such that, for all $\beta \geq \beta(p)$, with probability one, there exist at least two extremal random Gibbs states.*

Proof Theorem 6.4.1 implies, by the arguments put forward in Section 4.3, that there exist at least two extremal Gibbs states with positive probability. However, the number of extremal Gibbs measures for a given random interaction (with sufficient ergodic properties which are trivially satisfied here) is an almost sure constant ([185], Proposition 4.4). The argument goes in two steps: first, one shows that the number of *extremal Gibbs states* for given values of the temperature is a $\mathcal{B}$-measurable function. Next, it is clear that the number of extremal Gibbs states, for a given realization of the disorder, is a translation-invariant quantity (where the translation group $\mathbb{Z}^d$ acts on Ω in such a way that, for $x \in \mathbb{Z}^d$ and $\omega \in \Omega$, $T_x \omega$ is defined such that, for all $A \subset \mathbb{Z}^d$, $\Phi_A[T_x\omega](\sigma) = \Phi_{A-x}[\omega]$). But in all cases considered, the measure $\mathbb{P}$ is stationary and ergodic under the group of translations $\mathbb{Z}^d$, and thus, by the ergodic theorem (see, e.g., Appendix A3 of [125]), any translation-invariant function is almost surely a constant [78]. □

Remark 6.4.3 In the dilute Ising model, since all couplings J_{ij} are non-negative, the FKG inequalities hold, and thus, according to Corollary 4.3.16, we can construct two random Gibbs measures, $\mu_\beta^\pm[\omega]$, as limits of local specifications with pure plus, resp. pure minus boundary conditions along any deterministic sequence of increasing and absorbing finite volumes Λ_n. These two states will be distinct if there exists $x \in \mathbb{Z}^d$ such that $\mu_\beta^+[\omega][\sigma_x = +1] > 1/2$. Thus, if the translation-invariant, $\mathcal{B}$-measurable event $\{\exists x \in \mathbb{Z}^d : \mu_\beta^+[\omega][\sigma_x = +1] > 1/2\}$ occurs, then these two measures are distinct. But if this event has strictly positive probability, by the ergodic theorem, its probability is 1, and so the two extremal states, $\mu_\beta^\pm$, are then distinct, almost surely. This provides a simple alternative proof of the corollary.

Exercise: Improve the estimates on $\beta(p)$ obtained above. Show in particular that the theorem holds with any $p_0 > 1/3$.

The Peierls approach indicated here does not give optimal results (but has the advantage of clarity and robustness). It is known that $\beta(p)$ is finite if and only if p is larger than the critical value for bond (in $d \geq 3$ plaquette) percolation. This has been proven first by Georgii [123] in $d = 2$ and in more generality in [3, 124]. The latter papers also obtain precise results on the dependence of $\beta(p)$ on p. These results are all based on profound facts from percolation theory, a subject that we will not develop here.

Situations where the random perturbation respects the symmetries of the unperturbed interaction, for any realization of the disorder, is exceptional. In general, the perturbation $H^{(1)}[\omega]$ will break all symmetries of the model for typical ω and thus will render the standard Peierls argument inapplicable. The simplest example of such models is the *random-field Ising model*, whose Hamiltonian is

$$H[\omega](\sigma) \equiv -\sum_{|i-j|=1} \sigma_i \sigma_j - \epsilon \sum_i h_i[\omega] \sigma_i \qquad (6.40)$$

with h_i a family of i.i.d. random variables (say of mean 0 and variance 1). The issue of the RFIM is one of the first occurrences where profound probabilistic thinking has entered the field. I will devote the following chapter to the analysis of this model.

6.5 The Edwards–Anderson model

If in the Hamiltonian of the dilute Ising model we replace the random variables J_{ij} by i.i.d. random variables that are uniformly distributed on the interval $[-1, 1]$, we obtain the *Edwards–Anderson model* of a *spin-glass* [10]. This model has proven to be one of the most elusive and difficult models to analyze, from both the analytical and the numerical points of view. Consequently, the amount of rigorously known facts about this model is frighteningly small. Even on a heuristic level, there are conflicting opinions on the nature of the expected phase structure in various dimensions. I will not discuss this model in any detail (refer to Newman's book [185] for a thorough discussion), but only indicate some basic features.

The basis of the difficulties encountered with this model lies in the fact that it is highly non-trivial to say something sensible about its *ground-states*. The reason is that the couplings take both signs, thus favouring alignment or non-alignment of the spins. Worse, it is clearly impossible to satisfy the demands of all couplings: to see this, consider in dimension two, say, the four sites surrounding one plaquette of the lattice. It is not unusual to find that out of the four couplings around this plaquette, an odd number will be negative, while the remainder are positive. It is then impossible for the spins on the corners to be arranged in such a way that all four couplings have their way: one says that the plaquette is *frustrated*. If the couplings are Bernoulli distributed (± 1 with equal probability), we would find four arrangements contributing equal amounts of energy; one can show that this implies that the ground-states in this case must be infinitely degenerate. In fact, the number of ground-state configurations in a given volume, Λ, is in this case of the order $C^{|\Lambda|}$ [12]. But even in the case of continuous distributions, we encounter plaquettes where four configurations give almost the same contribution to the energy. This allows the possibility that, on a larger scale, there can be numerous spin configurations whose energy is very similar; in particular, ground-state configurations can be very sensitive to boundary conditions and vary dramatically as the volume varies.

As a result, none of the available techniques for analyzing low-temperature phases (Peierls arguments, low-temperature expansions, Pirogov–Sinai theory, etc.) is applicable, and even the most basic questions concerning the low-temperature phases of this model are wide open. It is largely believed that in two dimensions there is a unique Gibbs state at all positive temperatures, while in dimension three and higher, there should be at least two extremal states. There are several schools that predict different pictures in higher dimensions: Fisher and Huse [100, 141] predict the existence of just two extremal states in any dimension, while the school of Parisi [179] suggests a very complex picture based on mean-field theory (see Chapter 11) that would imply the existence of infinitely many extremal Gibbs states in high dimensions. This latter suggestion is mainly based on numerical simulations, which are, however, very difficult to interpret and do not provide unambiguous predictions. Newman and Stein have analyzed a variety of scenarios and checked their compatibility with basic principles. A very recent account summarizing the current state of understanding can be found in [191, 192]. This fascinating problem still awaits new ideas.

7

The random-field Ising model

Quand les physiciens nous demandent la solution d'un problème, ce n'est pas une corvée qu'ils nous imposent, c'est nous au contraire qui leur doivent des remercîments.[1]

Henri Poincaré, La valeur de la science.

The random-field Ising model has been one of the big success stories of mathematical physics and deserves an entire chapter. It will give occasion to learn about many of the more powerful techniques available for the analysis of random systems. The central question heatedly discussed in the 1980s in the physics community was whether the RFIM would show spontaneous magnetization at low temperatures and weak disorder in dimension three, or not. There were conflicting theoretical arguments, and even conflicting interpretations of experiments. Disordered systems, more than others, tend to elude common intuition. The problem was solved at the end of the decade in two rigorous papers by Bricmont and Kupiainen [63] (who proved the existence of a phase transition in $d \geq 3$ for small ϵ) and Aizenman and Wehr [8] (who showed the uniqueness of the Gibbs state in $d = 2$ for all temperatures).

7.1 The Imry–Ma argument

The earliest attempt to address the question of the phase transition in the RFIM goes back to Imry and Ma [142] in 1975. They tried to extend the beautiful and simple Peierls argument to a situation with symmetry breaking randomness. Let us recall that the Peierls argument in its essence relies on the observation that deforming one ground-state, $+1$, in the interior of a contour γ to another ground-state, -1, costs a *surface energy* $2|\gamma|$, while, by symmetry, the 'bulk energies' of the two ground-states are the same. Since the number of contours of a given length L is only of order C^L, the Boltzmann factors, $e^{-2\beta L}$, suppress such deformations sufficiently to make their existence unlikely if β is large enough. What goes wrong with the argument in the RFIM is the fact that the bulk energies of the two ground-states are no longer the same. Indeed, if all σ_i in int γ take the value $+1$, then the random-field term gives a contribution

$$E_{\text{bulk}}(\gamma) = +\epsilon \sum_{i \in \text{int } \gamma} h_i[\omega] \qquad (7.1)$$

---

[1] Approximately: When the physicists ask us for the solution of a problem, it is not a drudgery that they impose on us, on the contrary, it is us who owe them thanks.

while it is equal to minus the same quantity if all σ_i equal -1. Thus deforming the plus state to the minus state within γ produces, in addition to the surface energy term, a bulk term of order $2\epsilon \sum_{i \in \text{int } \gamma} h_i[\omega]$ that can take on any sign. Even when the random fields, h_i, are uniformly bounded, this contribution is bounded *uniformly* only by $2\epsilon |\text{int } \gamma|$ in absolute value and thus can be considerably bigger than the surface term, no matter how small ϵ is, if $|\gamma|$ is sufficiently large. Imry and Ma argued that this uniform bound on $E_{\text{bulk}}(\gamma)$ should not be the relevant quantity to consider. Namely, by the central limit theorem, the 'typical' value of $E_{\text{bulk}}(\gamma)$, for large γ, would be much smaller,

$$E_{\text{bulk}}(\gamma) \sim \pm \epsilon \sqrt{|\text{int } \gamma|} \tag{7.2}$$

Since by the isoperimetric inequality on $\mathbb{Z}^d$ $|\text{int } \gamma| \leq 2d |\gamma|^{\frac{d}{d-1}}$, this means that the typical value of the bulk energy is only $E_{\text{bulk}}(\gamma) \sim \pm \epsilon |\gamma|^{\frac{d}{2(d-1)}}$, which is small compared to $|\gamma|$ if $d > 2$. Otherwise, it is comparable or even larger. This very simple consideration led Imry and Ma to the (*correct* !!) prediction that the RFIM undergoes a phase transition in $d \geq 3$ and does not in $d \leq 2$.

Remark 7.1.1 This argument is meant to work only if $\epsilon \ll 1$; if ϵ is not small, then even in small contours the bulk energy can dominate the surface energy. In particular, it is easy to see that, if $\epsilon > 2d$, and h_i take the values ± 1, then there is a unique random ground-state given by $\sigma_i = \text{sign } h_i$.

It is likely that this argument would have been considered sufficient by the standards of theoretical physics had there not been a more fancy argument, based on field theoretic considerations (due to Parisi and Sourlas [196]), that predicted that the RFIM in dimension d should behave like the Ising model without random field in dimension $d-2$. This *dimensional reduction* argument would then predict the absence of a phase transition in $d = 3$, contrary to the Imry–Ma argument. The two arguments divided the community.[2]

Admittedly, with an alternative option in mind, the Imry–Ma argument looks rather shaky, and anyone would be excused for not trusting it. We will thus distance us a bit from Imry and Ma and try to repeat their reasoning in a more precise way. What we would obviously want to do is to reprove something like Theorem 6.4.1. When trying to re-run the proof of Lemma 4.3.6, all works as before until the last line of (4.52). One obtains instead the two bounds

$$\mu^{+1}_{\text{int } \gamma, \beta}[\sigma_{\gamma^{\text{in}}} = -1] \leq e^{-2\beta|\gamma|} \frac{Z^{-1}_{\text{int } \gamma \setminus \gamma^{\text{in}}, \beta}}{Z^{+1}_{\text{int } \gamma \setminus \gamma^{\text{in}}, \beta}} \tag{7.3}$$

$$\mu^{-1}_{\text{int } \gamma, \beta}[\sigma_{\gamma^{\text{in}}} = +1] \leq e^{-2\beta|\gamma|} \frac{Z^{+1}_{\text{int } \gamma \setminus \gamma^{\text{in}}, \beta}}{Z^{-1}_{\text{int } \gamma \setminus \gamma^{\text{in}}, \beta}}$$

Hence, the analogue of Lemma 4.3.6 becomes:

Lemma 7.1.2 *In the random-field Ising model, for any Gibbs state μ_β,*

$$\mu_\beta[\gamma \in \Gamma(\sigma)] \leq \exp\left(-2\beta|\gamma| + \left|\ln Z^{+1}_{\text{int } \gamma \setminus \gamma^{\text{in}}, \beta} - \ln Z^{-1}_{\text{int } \gamma \setminus \gamma^{\text{in}}, \beta}\right|\right) \tag{7.4}$$

---

[2] And rightly so. Even though in the RFIM dimensional reduction was ultimately shown to make the wrong prediction, in another application of similar reasoning, namely in the problem of critical behaviour of *branched polymers*, Brydges and Imbrie [66] recently proved rigorously that dimensional reduction is correct!

At this point, one may lose all hope when facing the difference of the logarithms of the two partition functions, and one may not even see how to arrive at Imry and Ma's assertion on the 'typical value' of this bulk term.[3] However, the situation is much better than might be feared. The reason is the so-called *concentration of measure* phenomenon that will continue to play an important rôle in the analysis of disordered systems. Roughly, concentration of measure means that in many situations, a Lipschitz continuous function of i.i.d. random variables has fluctuations that are not bigger than those of a corresponding linear function. This phenomenon has been widely investigated over the last 30 years, with culminating results due to M. Talagrand. We refer to [167, 231, 232] for a detailed presentation and references. We will use the following theorem, due to M. Talagrand, whose proof can be found in [232]:

Theorem 7.1.3 *Let $f : [-1, 1]^N \to \mathbb{R}$ be a function whose level sets are convex. Assume that f is Lipschitz continuous with uniform constant C_{Lip}, i.e. for any $X, Y \in [-1, 1]^N$,*

$$|f(X) - f(Y)| \leq C_{\text{Lip}} \|X - Y\|_2 \tag{7.5}$$

Then, if $X_1, \ldots X_N$ are i.i.d. random variables taking values in $[-1, 1]$, and $Z = f(X_1, \ldots, X_N)$, and if $\mathbb{M}_Z$ is a median[4] of Z, then

$$\mathbb{P}[|Z - \mathbb{M}_Z| \geq z] \leq 4 \exp\left(-\frac{z^2}{16 C_{\text{Lip}}^2}\right) \tag{7.6}$$

Remark 7.1.4 In most applications, and in particular when C_{Lip} is small compared to z^2, one can replace the median in (7.6) by the expectation $\mathbb{E}Z$ without harm.

Remark 7.1.5 If X_i are i.i.d. centered Gaussian random variables with variance 1, the conclusion of Theorem 7.1.3 holds even without the assumption of convex level sets, and with the median replaced by the mean, and both constants 4 and 16 replaced by 2 (see [167]). There are many other situations where similar results hold [167, 231].

Remark 7.1.6 Physical quantities satisfying such concentration inequalities are often called *self-averaging*.

Theorem 7.1.3 allows us to prove the following lemma:

Lemma 7.1.7 *Assume that the random fields have a symmetric distribution[5] and are bounded[6] (i.e. $|h_i| \leq 1$), or their distribution is Gaussian. Then there is a constant $C < \infty$ such that, for any $z \geq 0$,*

$$\mathbb{P}\left[\left|\ln Z^{+1}_{\text{int}\,\gamma\setminus\gamma^{\text{in}},\beta} - \ln Z^{-1}_{\text{int}\,\gamma\setminus\gamma^{\text{in}},\beta}\right| > z\right] \leq C \exp\left(-\frac{z^2}{\epsilon^2 \beta^2 C |\text{int}\,\gamma|}\right) \tag{7.7}$$

---

[3] This had been considered to be the truly difficult part of the problem. Chalker in 1983 [77] and Fisher, Fröhlich, and Spencer in 1984 [101] gave a solution of the problem where this difference was ad hoc replaced by the sum over the random fields within int γ. As we will see, however, the real difficulty of the problem lies elsewhere.
[4] A median of a random variable Z is any number such that $\mathbb{P}[Z \geq \mathbb{M}_Z] \geq 1/2$ and $\mathbb{P}[Z \leq \mathbb{M}_Z] \geq 1/2$.
[5] This assumption appears necessary even for the result; otherwise the phase coexistence point could be shifted to some finite value of the external magnetic field.
[6] We make this assumption for convenience; as a matter of fact, essentially the same result holds if we only assume that the h_i have finite exponential moments.

Proof By symmetry of the distribution of h, the two partition functions we consider have, as random variables, the same distribution. In particular,

$$\mathbb{E} \ln Z^{+1}_{\text{int}\,\gamma\setminus\gamma^{\text{in}},\beta} = \mathbb{E} \ln Z^{-1}_{\text{int}\,\gamma\setminus\gamma^{\text{in}},\beta} \tag{7.8}$$

Therefore,

$$\mathbb{P}\big[\big|\ln Z^{+1}_{\text{int}\,\gamma\setminus\gamma^{\text{in}},\beta} - \ln Z^{-1}_{\text{int}\,\gamma\setminus\gamma^{\text{in}},\beta}\big| > z\big] \tag{7.9}$$
$$\leq \mathbb{P}\big[\big\{\big|\ln Z^{+1}_{\text{int}\,\gamma\setminus\gamma^{\text{in}},\beta} - \mathbb{E}\ln Z^{+1}_{\text{int}\,\gamma\setminus\gamma^{\text{in}},\beta}\big| + \big|\mathbb{E}\ln Z^{-1}_{\text{int}\,\gamma\setminus\gamma^{\text{in}},\beta} - \ln Z^{-1}_{\text{int}\,\gamma\setminus\gamma^{\text{in}},\beta}\big|\big\} > z\big]$$
$$\leq 2\mathbb{P}\big[\big|\ln Z^{+1}_{\text{int}\,\gamma\setminus\gamma^{\text{in}},\beta} - \mathbb{E}\ln Z^{+1}_{\text{int}\,\gamma\setminus\gamma^{\text{in}},\beta}\big| > z/2\big]$$

$\ln Z^{+1}_{\text{int}\,\gamma\setminus\gamma^{\text{in}},\beta}$ is a function of the independent random variables h_i, with $i \in \text{int}\,\gamma\setminus\gamma^{\text{in}}$. Moreover, one can check (by differentiation) that it is a convex function. Thus, to use Theorem 7.1.3, we only must verify that $\ln Z$ is Lipschitz continuous and compute the Lipschitz constant. But

$$\big|\ln Z^{+1}_{\text{int}\,\gamma\setminus\gamma^{\text{in}},\beta}[\omega] - \ln Z^{+1}_{\text{int}\,\gamma\setminus\gamma^{\text{in}},\beta}[\omega']\big| \tag{7.10}$$

$$\leq \sup_{\omega''}\bigg|\sum_{i\in\text{int}\,\gamma\setminus\gamma^{\text{in}}}(h_i[\omega]-h_i[\omega'])\frac{\partial \ln Z^{+1}_{\text{int}\,\gamma\setminus\gamma^{\text{in}},\beta}}{\partial h_i}[\omega'']\bigg|$$

$$\leq \epsilon\beta \sup_{i\in\text{int}\,\gamma\setminus\gamma^{\text{in}}}|\mu_{\text{int}\,\gamma\setminus\gamma^{\text{in}},\beta}(\sigma_i)| \sum_{i\in\text{int}\,\gamma\setminus\gamma^{\text{in}}}|h_i[\omega]-h_i[\omega']|$$

$$\leq \epsilon\beta\sqrt{|\text{int}\,\gamma|}\,\|h_{\text{int}\,\gamma}[\omega]-h_{\text{int}\,\gamma}[\omega']\|_2$$

where in the last step we used that the expectation of σ_i is bounded by one and the Cauchy–Schwarz inequality. Theorem 7.1.3 implies (7.7). $\square$

Lemma 7.1.7 implies indeed that, for a given contour γ,

$$\mu_\beta[\gamma \in \Gamma(\sigma)] \leq \exp(-2\beta|\gamma| + \epsilon\beta\sqrt{|\text{int}\,\gamma|}) \tag{7.11}$$

However, the immediate attempt to prove the analogue of Theorem 6.4.1 fails. Namely, we would have to show that

$$\mathbb{P}\big[\exists_{\gamma:\text{int}\,\gamma\ni 0}\big|\ln Z^{+1}_{\text{int}\,\gamma\setminus\gamma^{\text{in}},\beta} - \ln Z^{-1}_{\text{int}\,\gamma\setminus\gamma^{\text{in}},\beta}\big| > \beta|\gamma|\big] \tag{7.12}$$

is small (for small ϵ). The straightforward way to try to prove this is to write

$$\mathbb{P}\big[\exists_{\gamma:\text{int}\,\gamma\ni 0}\big|\ln Z^{+1}_{\text{int}\,\gamma\setminus\gamma^{\text{in}},\beta} - \ln Z^{-1}_{\text{int}\,\gamma\setminus\gamma^{\text{in}},\beta}\big| > \beta|\gamma|\big]$$

$$\leq \sum_{\gamma:\text{int}\,\gamma\ni 0}\mathbb{P}\big[\big|\ln Z^{+1}_{\text{int}\,\gamma\setminus\gamma^{\text{in}},\beta} - \ln Z^{-1}_{\text{int}\,\gamma\setminus\gamma^{\text{in}},\beta}\big| > \beta|\gamma|\big]$$

$$\leq \sum_{\gamma:\text{int}\,\gamma\ni 0}\exp\bigg(-\frac{|\gamma|^2}{C\,\epsilon^2|\text{int}\,\gamma|}\bigg) \tag{7.13}$$

But $\frac{|\gamma|^2}{|\text{int}\,\gamma|}$ can be as small (and is for many γ) as $|\gamma|^{(d-2)/(d-1)}$, and since the number of γ's of given length is of order $C^{|\gamma|}$, the last sum in (7.13) diverges.

Some reflection shows that it is the first inequality in (7.13) that spoiled the estimates. This step would be reasonable if the partition functions for different γ were more or less independent. However, if γ and γ' are very similar, it is clear that this is not the case. A more careful analysis should exploit this fact and hopefully lead to a better bound. Such situations are quite common in probability theory, and in principle there are well-known techniques that go under the name of *chaining* to systematically improve estimates like (7.13). This was done in the papers [77] and [101], however in a model where $\ln Z^{+1}_{\text{int}\,\gamma\setminus\gamma^{\text{in}},\beta} - \ln Z^{-1}_{\text{int}\,\gamma\setminus\gamma^{\text{in}},\beta}$ is ad hoc replaced by $\beta \sum_{i\in\text{int}\,\gamma\setminus\gamma^{\text{in}}} h_i$ (the so-called 'no contours within contours' approximation). In fact, they prove the following:

Proposition 7.1.8 *Assume that there is a finite positive constant C such that, for all $\Lambda, \Lambda' \subset \mathbb{Z}^d$,*

$$\mathbb{P}\big[\big|\ln Z^{+1}_{\Lambda,\beta} - \ln Z^{+1}_{\Lambda',\beta} - \mathbb{E}\big[\ln Z^{+1}_{\Lambda,\beta} - \ln Z^{+1}_{\Lambda',\beta}\big]\big| \geq z\big]$$
$$\leq \exp\left(-\frac{z^2}{C\,\epsilon^2\beta^2|\Lambda\triangle\Lambda'|}\right) \tag{7.14}$$

where $\Lambda\triangle\Lambda'$ denotes the symmetric difference of the two sets Λ and Λ'. Then, if $d \geq 3$, there exist $\epsilon_0 > 0$ and $\beta_0 < \infty$ such that for all $\epsilon \leq \epsilon_0$ and $\beta \geq \beta_0$, for $\mathbb{P}$-almost all $\omega \in \Omega$, there exist at least two extremal infinite-volume Gibbs states μ^+_β and μ^-_β.

Remark 7.1.9 There are good reasons to believe that (7.14) holds, but in spite of multiple efforts, I have not been able to find an easy proof. On a heuristic level, the argument is that the difference appearing in (7.14) should depend very little on the random variables that appear in the intersection of Λ and Λ'. More precisely, when computing the Lipschitz norm, we get, instead of (7.10),

$$\left|\ln Z^{+1}_{\Lambda,\beta}[\omega] - \ln Z^{+1}_{\Lambda,\beta}[\omega'] - \ln Z^{+1}_{\Lambda',\beta}[\omega] + \ln Z^{+1}_{\Lambda',\beta}[\omega']\right| \tag{7.15}$$

$$\leq \sup_{\omega''} \left|\sum_{i\in\Lambda\setminus(\Lambda\cap\Lambda')} (h_i[\omega] - h_i[\omega'])\frac{\partial \ln Z^{+1}_{\Lambda,\beta}}{\partial h_i}[\omega'']\right|$$

$$+ \left|\sum_{i\in\Lambda'\setminus(\Lambda\cap\Lambda')} (h_i[\omega] - h_i[\omega'])\frac{\partial \ln Z^{+1}_{\Lambda',\beta}}{\partial h_i}[\omega'']\right|$$

$$+ \left|\sum_{i\in\Lambda\cap\Lambda'} (h_i[\omega] - h_i[\omega'])\Big(\frac{\partial \ln Z^{+1}_{\Lambda,\beta}}{\partial h_i}[\omega''] - \frac{\partial \ln Z^{+1}_{\Lambda',\beta}}{\partial h_i}[\omega'']\Big)\right|$$

$$\leq \epsilon\beta \left|\sum_{i\in\Lambda\triangle\Lambda'} |h_i[\omega] - h_i[\omega']|\right|$$

$$+ \epsilon\beta \left|\sum_{i\in\Lambda\cap\Lambda'} (h_i[\omega] - h_i[\omega'])(\mu^+_{\beta,\Lambda}[\omega''](\sigma_i) - \mu^+_{\beta,\Lambda'}[\omega''](\sigma_i))\right|$$

$$\leq \epsilon\beta\sqrt{|\Lambda\triangle\Lambda'|}\|h_{\Lambda\triangle\Lambda'}[\omega] - h_{\Lambda\triangle\Lambda'}[\omega']\|_2$$

$$+ \epsilon\beta\sqrt{\sum_{i\in\Lambda\cap\Lambda'}\big(\mu^+_{\beta,\Lambda}[\omega](\sigma_i) - \mu^+_{\beta,\Lambda'}[\omega'](\sigma_i)\big)^2}\|h_{\Lambda\cap\Lambda'}[\omega] - h_{\Lambda\cap\Lambda'}[\omega']\|_2$$

It is natural to believe that the expectation of σ_i with respect to the two measures, $\mu^+_{\beta,\Lambda}$ and $\mu^+_{\beta,\Lambda'}$, will be essentially the same for i well inside the intersection of Λ and Λ', so that it should be possible to bound the coefficient in the last line by $\sqrt{|\Lambda\triangle\Lambda'|}$. If that were the case, the hypothesis of Proposition 7.1.8 would follow from Theorem 7.1.3. Unfortunately, we do not know how to prove such an estimate. The reader should realize that this argument appears nonetheless more convincing and robust than the one given in [101]; the argument there is that (7.14) will hold *if* the $\mu^+_{\beta,\Lambda}[\omega](\sigma_i)$ depend 'weakly' on ω, which is essentially what we are out to prove anyway. On the other hand, smallness of (7.15) can even be expected to hold if the expectation of σ_i depends strongly on the disorder.

Proof To simplify notation, let us set

$$F_\gamma \equiv \ln Z^{+1}_{\operatorname{int}\gamma\setminus\gamma^{\operatorname{in}},\beta} - \mathbb{E}\ln Z^{+1}_{\operatorname{int}\gamma\setminus\gamma^{\operatorname{in}},\beta} \qquad (7.16)$$

The idea behind chaining arguments is to define a sequence of sets Γ_ℓ, $\ell \in \mathbb{N}$ of 'coarse grained' contours and maps $\gamma_\ell : \Gamma_0 \to \Gamma_\ell$, where Γ_0 is the original set of contours. Now write, for $k \in \mathbb{N}$ to be chosen later,

$$\begin{aligned}F_\gamma &= F_{\gamma_k(\gamma)} + (F_{\gamma_{k-1}(\gamma)} - F_{\gamma_k(\gamma)}) \\ &\quad + (F_{\gamma_{k-2}(\gamma)} - F_{\gamma_{k-1}(\gamma)}) + \cdots + (F_\gamma - F_{\gamma_1(\gamma)})\end{aligned} \qquad (7.17)$$

Then we can write

$$\mathbb{P}\left[\sup_{\substack{\gamma:\operatorname{int}\gamma\ni 0\\ |\gamma|=n}}F_\gamma > z\right] \leq \sum_{\ell=1}^{k(n)}\mathbb{P}\left[\sup_{\substack{\gamma:\operatorname{int}\gamma\ni 0\\ |\gamma|=n}}F_{\gamma_\ell(\gamma)}-F_{\gamma_{\ell-1}(\gamma)} > z_\ell\right] + \mathbb{P}\left[\sup_{\substack{\gamma:\operatorname{int}\gamma\ni 0\\ |\gamma|=n}}F_{\gamma_k(\gamma)} > z_{k+1}\right] \qquad (7.18)$$

for any choice of $k = k(n)$ and sequences z_ℓ with $\sum_{\ell=1}^{k+1}z_\ell \leq z$. To estimate the individual probabilities, we just need to count the number, $A_{\ell,n}$, of image points in Γ_ℓ that are reached when mapping all the γ occurring in the supremum (i.e. those of length n and encircling the origin), and use the assumption to get the obvious estimate

$$\mathbb{P}\left[\sup_{\substack{\gamma:\operatorname{int}\gamma\ni 0\\ |\gamma|=n}}F_\gamma > z\right] \qquad (7.19)$$

$$\leq \sum_{\ell=1}^{k(n)} A_{\ell-1,n}A_{\ell,n}\exp\left(-\frac{z_\ell^2}{C\,\epsilon^2\beta^2\,\sup_\gamma|\operatorname{int}\gamma_\ell(\gamma)\Delta\operatorname{int}\gamma_{\ell-1}(\gamma)|}\right)$$

$$+ A_{k(n),n}\exp\left(-\frac{z_{k+1}^2}{C\,\epsilon^2\beta^2\,\sup_\gamma|\operatorname{int}\gamma_k(\gamma)|}\right)$$

We must now make a choice for the sets Γ_ℓ. For this we cover the lattice $\mathbb{Z}^d$ with squares of side-length 2^ℓ centered at the coarser lattice $(2^\ell\mathbb{Z})^d$. The set Γ_ℓ will then be nothing but collections of such squares. Next we need to define the maps γ_ℓ. This is done as follows: Let $V_\ell(\gamma)$ be the set of all cubes, c, of side-length 2^ℓ from the covering introduced above such that

$$|c \cap \operatorname{int}\gamma| \geq 2^{d\ell-1} \qquad (7.20)$$

Then let $\gamma_\ell(\gamma) \equiv \partial V_\ell(\gamma)$ be the set of boundary cubes of $V_\ell(\gamma)$. A simple example is presented in Fig. 7.1. Note that the images $\gamma_\ell(\gamma)$ are in general not connected, but one verifies that the number of connected components cannot exceed *const.* $|\gamma|2^{-(d-1)(\ell-1)}$, and

7.1 The Imry–Ma argument

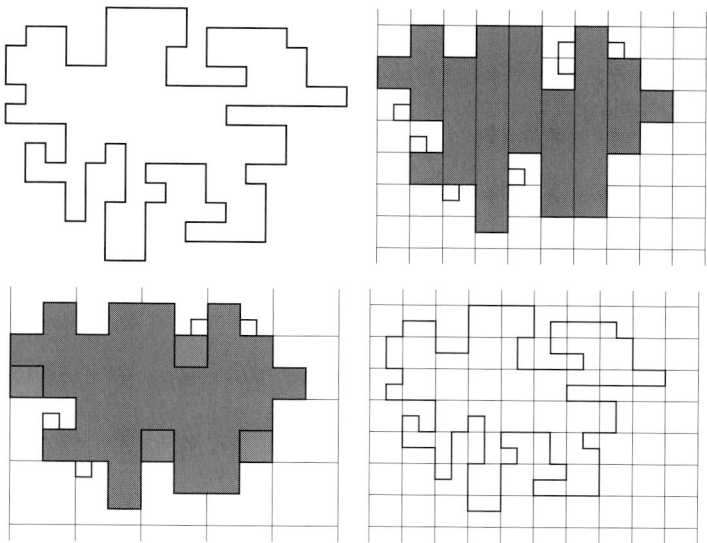

Figure 7.1 Successive coarse graining of a contour.

the maximal distance between any of the connected components is less than $|\gamma|$. This leads, after some tedious counting, to the estimate that

$$A_{\ell,n} \leq \exp\left(\frac{C\ell n}{2^{(d-1)\ell}}\right) \tag{7.21}$$

On the other hand, one readily sees that

$$|\text{int } \gamma_\ell(\gamma) \Delta \text{int } \gamma_{\ell-1}(\gamma)| \leq |\gamma| 2^\ell \tag{7.22}$$

for any γ. Finally, one chooses $k(n)$ such that $2^{k(n)} \sim n^{1/3}$ (i.e. $k(n) \sim \ln n$). Inserting these estimates into (7.19), one concludes that, for small enough ϵ, the sum in (7.20) is bounded by

$$\sum_{\ell=1}^{k(n)} \exp\left(-\frac{z_\ell^2}{C\beta^2 \epsilon^2 2^\ell n} + \frac{C(\ell-1)n}{2^{(d-1)(\ell-1)}}\right) \tag{7.23}$$

and the last on the left of (7.20),

$$A_{k(n),n} \exp\left(-\frac{z_{k(n)+1}^2}{C\epsilon^2 \beta^2 \sup_\gamma |\text{int } \gamma_k(\gamma)|}\right) \tag{7.24}$$

$$\leq \exp\left(C \ln n\, n^{1/3} - \frac{z_{k(n)+1}^2}{C\beta^2 \epsilon^2 n^{d/(d-1)}}\right) \tag{7.25}$$

This allows us to choose $z_\ell = c\beta n \ell^{-2}$ to get a bound of order

$$\mathbb{P}\left[\sup_{\substack{\gamma: \text{int } \gamma \ni 0 \\ |\gamma|=n}} F_\gamma > c\beta n\right] \leq e^{-c' n^{2/3} \epsilon^{-2}} \tag{7.26}$$

and hence

$$\mathbb{P}\left[\sup_{\gamma:\text{int}\,\gamma\ni 0} F_\gamma > c\beta|\gamma|\right] \leq e^{-c'\epsilon^{-2}} \quad (7.27)$$

But from here follows the conclusion of Proposition 7.1.8. □

The existence of two Gibbs states would now follow as in the dilute Ising model if we could verify the hypothesis of Corollary 6.4.2. The only existing full proof of the existence of a phase transition in the RFIM is due to Bricmont and Kupiainen [63] and requires much harder work. We will return to this in Section 7.3.

7.2 Absence of phase transitions: the Aizenman–Wehr method

We have seen that in $d \geq 3$ the random energy that can be gained by flipping spins locally cannot compensate for the surface energy produced in doing so if the disorder is weak and the temperature low. On the other hand, in dimension $d \leq 2$, the Imry–Ma argument predicts that the random bulk energies might outweigh surface terms, and this should imply that the particular realization of the random fields determines locally the orientation of the spins, so that the effects of boundary conditions are not felt in the interior of the system, implying a unique (random) Gibbs state. This argument was made rigorous by Aizenman and Wehr [8, 258] in 1990, using a number of clever and interesting ideas. Roughly, the proof is based on the following reasoning: Consider a volume Λ and two boundary conditions, say all plus spins and all minus spins. Then the difference between the corresponding free energies $f_{\beta,\Lambda^\pm} \equiv \ln Z^\pm_{\beta,\Lambda}$ must always be bounded by *const.* $|\partial\Lambda|$ (just introduce a contour right next to the boundary and proceed as in the Peierls argument). Now get a lower bound on the random fluctuations of that free energy; from the upper bound (7.7) one might guess that these can be of order $C(\beta)\sqrt{|\Lambda|}$, multiplied by a standard Gaussian random variable, say g. If this is so, there is a dilemma: by symmetry, the difference between the two free energies must be as big as the random part, and this implies that $C(\beta)\sqrt{|\Lambda|}g \leq$ *const.* $|\partial\Lambda|$. In $d \leq 2$, this implies that $C(\beta) = 0$. But $C(\beta)$ will be seen to be linked to an *order parameter*, here the *magnetization*, and its vanishing will imply the uniqueness of the Gibbs state. To make this rough argument precise requires, however, a delicate procedure. In what follows I will give the proof of Aizenman and Wehr only for the special case of the RFIM (actually, for any system where FKG inequalities hold).

7.2.1 Translation-covariant states

A key technical idea in [8] is to carry out the argument sketched above in such a way that it gives directly information about infinite-volume states. This will allow the use of ergodicity arguments and this will force us to investigate some covariance properties of random Gibbs measures.

To do so, we equip our probability space $(\Omega, \mathcal{B}, \mathbb{P})$ with some extra structure. First, we define the action, T, of the translation group $\mathbb{Z}^d$ on Ω. We will assume that $\mathbb{P}$ is invariant under this action and that the dynamical system $(\Omega, \mathcal{B}, \mathbb{P}, T)$ is stationary and ergodic. In the random-field model, the action of T is simply

$$(h_{x_1}[T_y\omega], \ldots, h_{x_n}[T_y\omega]) \equiv (h_{x_1+y}[\omega], \ldots, h_{x_n+y}[\omega]) \quad (7.28)$$

and the stationarity and ergodicity assumptions are trivially satisfied if the h_i are i.i.d.

Moreover, we will use that Ω is equipped with an affine structure, i.e. we set $(h_{x_1}[\omega + \omega'], \ldots, h_{x_n}[\omega + \omega']) \equiv (h_{x_1}[\omega] + h_{x_1}[\omega'], \ldots, h_{x_n}[\omega] + h_{x_n}[\omega'])$. We will introduce a subset, $\Omega_0 \subset \Omega$, of random fields that differ from zero only in some finite set, i.e.

$$\Omega_0 \equiv \{\delta\omega \in \Omega : \exists \Lambda \subset \mathbb{Z}^d, \text{ finite}, \forall y \notin \Lambda, h_y[\delta\omega] = 0\} \quad (7.29)$$

We will use the convention to denote elements of Ω_0 by $\delta\omega$.

Definition 7.2.1 A random Gibbs measure μ_β is called covariant if,

(i) for all $x \in \mathbb{Z}^d$, and any continuous function f,

$$\mu_\beta[\omega](T_{-x} f) = \mu_\beta[T_x \omega](f), \text{ a.s.} \quad (7.30)$$

(ii) for all $\delta\omega \in \Omega_0$, for almost all ω and all bounded, continuous f,

$$\mu_\beta[\omega + \delta\omega](f) = \frac{\mu_\beta[\omega]\left(f e^{-\beta(H[\omega+\delta\omega]-H[\omega])}\right)}{\mu_\beta[\omega]\left(e^{-\beta(H[\omega+\delta\omega]-H[\omega])}\right)} \quad (7.31)$$

(Note that $H[\omega + \delta\omega] - H[\omega]$ is a finite sum: if $\delta\omega$ is supported on Λ, then $H[\omega + \delta\omega](\omega) - H[\omega](\sigma) = -\sum_{i \in \Lambda} \sigma_i h_i[\delta\omega]$.)

The properties of covariant random Gibbs measures look natural, but their verification is in general elusive (recall that even the construction of random Gibbs measures was problematic). In the context of the RFIM, we are helped, however, by the FKG inequalities that were discussed in Section 4.3.

Theorem 7.2.2 *Consider the random-field Ising model (7.2) with h_i a stationary and ergodic random field. Then there exist two covariant random Gibbs measures, μ_β^+ and μ_β^-, that satisfy:*

(i) *For almost all ω,*

$$\mu_\beta^\pm[\omega] = \lim_{\Lambda \uparrow \mathbb{Z}^d} \mu_{\beta,\Lambda}^\pm[\omega] \quad (7.32)$$

(ii) *Suppose that, for some β, $\mu_\beta^+ = \mu_\beta^-$. Then, for this value of β, the Gibbs measure for the RFIM model is unique for almost all ω.*

Proof In Section 4.3.3, Corollary 4.3.16, we learned that, due to the FKG inequalities (whose validity remains unaffected by the presence of the random fields), for any $\omega \in \Omega$, we can construct infinite-volume Gibbs states, $\mu_\beta^\pm[\omega]$, as limits of local specifications with constant boundary conditions along arbitrary (ω-independent) increasing and absorbing sequences of finite volumes Λ_n. Thus, the functions $\omega \to \mu_\beta^\pm$ are measurable, since they are limits of measurable functions. It remains to check the covariance properties. Property

(ii) follows immediately from the fact that $\mu_\beta^\pm$ can be represented as a limit of local specifications, and that the formula (7.31) holds trivially for local specifications with Λ large enough to contain the support of $\delta\omega$. Property (i) on the contrary requires the independence of the limit from the chosen sequence Λ_n. Indeed we have

$$\mu_{\beta,\Lambda}^+[\omega](T_{-x}f) = \mu_{\beta,\Lambda+x}^+[T_x\omega](f) \tag{7.33}$$

which implies, by Corollary 4.3.16 that $\mu_\beta^+[\omega](T_{-x}f) = \mu_\beta^+[T_x\omega](f)$ almost surely, as desired. The second assertion of the theorem follows directly from (iv) of Corollary 4.3.16. □

Remark 7.2.3 It is remarkably hard to prove the translation covariance property in the absence of strong results like the FKG inequalities. In fact there are two difficulties. The first is that of the measurability of the limits that we have already discussed above. This can be resolved by the introduction of metastates, and it was precisely in this context that Aizenman and Wehr first applied this idea. The second problem is that, without comparison results between local specifications in different volumes, the relative shift between the function and the volume implicit in (7.33) cannot be removed. A way out of this problem is to construct Gibbs states with *periodic* boundary conditions (i.e. one replaces Λ with a torus, i.e. $(\mathbb{Z} \bmod n)^d$). In that case, one may recover the translation covariance of the limit from translation covariance of the finite-volume measures under the automorphisms of the torus. From the point of view of the general theory as we have presented it so far, this is unsatisfactory. For this reason, we have restricted our exposition of the Aizenman–Wehr method to the RFIM and refer the reader to the original articles for more general results.

7.2.2 Order parameters and generating functions

We learned in Section 4.3.3 that, due to the monotonicity properties implied by the FKG inequalities, we will have a unique Gibbs state, for almost all ω, if and only if the translation-covariant states μ_β^+ and μ_β^- coincide almost surely. Moreover, we know from Proposition 4.3.17 that in the translation-invariant case, uniqueness follows from the vanishing of the total magnetization. We want to extend this result to the random case. We set

$$m^\mu[\omega] \equiv \lim_{\Lambda\uparrow\infty} \frac{1}{|\Lambda|} \sum_{i\in\Lambda} \mu[\omega](\sigma_i) \tag{7.34}$$

provided the limit exists. We will also abuse notation and write $m_\beta^\pm = m^{\mu_\beta^\pm}$. Some simple facts follow from covariance and FKG:

Lemma 7.2.4 *Suppose that μ is a covariant Gibbs state. Then, for almost all ω, the total magnetization $m^\mu[\omega]$ exists and is independent of ω.*

Proof By the covariance of μ,

$$m^\mu[\omega] = \lim_{\Lambda\uparrow\infty} \frac{1}{|\Lambda|} \sum_{i\in\Lambda} \mu[T_{-i}\omega](\sigma_0) \tag{7.35}$$

7.2 The Aizenman–Wehr method

But $\mu_\beta[\omega](\sigma_0)$ is a bounded measurable function of ω; since we assumed that $(\Omega, \mathcal{B}, \mathbb{P}, T)$ is stationary and ergodic, it follows from the ergodic theorem (see, e.g., Appendix A3 of [125] for a good exposition and proofs) that the limit exists, almost surely, and is given by

$$m^\mu = \mathbb{E}\,\mu(\sigma_0) \tag{7.36}$$

□

Lemma 7.2.5 *In the random-field Ising model,*

$$m^+ - m^- = 0 \Leftrightarrow \mu_\beta^+ = \mu_\beta^- \tag{7.37}$$

Proof (7.36) implies that, almost surely,

$$0 = m^+ - m^- = \mathbb{E}(\mu_\beta^+(\sigma_i) - \mu_\beta^-(\sigma_i)) \tag{7.38}$$

and so, since $\mu_\beta^+(\sigma_i) - \mu_\beta^-(\sigma_i) \geq 0$, and there are only countably many sites i, almost surely, for all $i \in \mathbb{Z}^d$, $\mu_\beta^+(\sigma_i) - \mu_\beta^-(\sigma_i) = 0$. □

The order parameters introduced above can be computed as derivatives of certain *generating functions*. We set

$$G_\Lambda^\mu \equiv -\frac{1}{\beta} \ln \mu \left(e^{-\beta \sum_{i \in \Lambda} h_i \sigma_i} \right) \tag{7.39}$$

Note that, if μ is a covariant Gibbs state, then

$$G_\Lambda^{\mu[\omega]} = \frac{1}{\beta} \ln \mu[\omega - \omega_\Lambda] \left(e^{\beta \sum_{i \in \Lambda} h_i \sigma_i} \right) \tag{7.40}$$

Here, $\omega_\Lambda \in \Omega_0$ is defined such that $h_i[\omega_\Lambda] = h_i[\omega]$ if $i \in \Lambda$, and $h_i[\omega_\Lambda] = 0$ if $i \notin \Lambda$. Therefore, for $i \in \Lambda$,

$$\frac{\partial}{\partial h_i} G_\Lambda^{\mu[\omega]} = \frac{\mu[\omega - \omega_\Lambda]\left(\sigma_i e^{\beta \sum_{i \in \Lambda} h_i \sigma_i}\right)}{\mu[\omega - \omega_\Lambda]\left(e^{\beta \sum_{i \in \Lambda} h_i \sigma_i}\right)} = \mu[\omega](\sigma_i) \tag{7.41}$$

where the first equality follows from the fact that $\mu[\omega - \omega_\Lambda]$ is $\mathcal{B}_{\Lambda^c}$-measurable and the second one follows from (7.31). In particular, we get that

$$\mathbb{E}\frac{\partial}{\partial h_i} G_\Lambda^{\mu^\pm} = m^\pm \tag{7.42}$$

Let us now introduce the function

$$F_{\beta,\Lambda} \equiv \mathbb{E}\left[G_\Lambda^{\mu^+} - G_\Lambda^{\mu^-} \big| \mathcal{B}_\Lambda \right] \tag{7.43}$$

Clearly, $\mathbb{E}\frac{\partial}{\partial h_0} F_\Lambda = m^+ - m^-$, and our goal is to prove that this quantity must be zero. The important point is the following a-priori upper bound:

Lemma 7.2.6 *For any value of β, and any volume, Λ,*

$$|F_\Lambda| \leq 2|\partial \Lambda| \tag{7.44}$$

Proof The first step in the proof makes use of (7.40) to express F_Λ in terms of measures that no longer depend on the disorder within Λ. Namely,

$$F_\Lambda = \beta^{-1}\mathbb{E}\left[\ln \frac{\mu_\beta^-[\omega]\left(e^{-\beta\sum_{i\in\Lambda}h_i\sigma_i}\right)}{\mu_\beta^+[\omega]\left(e^{-\beta\sum_{i\in\Lambda}h_i\sigma_i}\right)}\bigg|\mathcal{B}_\Lambda\right] \quad (7.45)$$

$$= \mathbb{E}\left[\ln \frac{\mu_\beta^+[\omega-\omega_\Lambda]\left(e^{\beta\sum_{i\in\Lambda}h_i\sigma_i}\right)}{\mu_\beta^-[\omega-\omega_\Lambda]\left(e^{\beta\sum_{i\in\Lambda}h_i\sigma_i}\right)}\bigg|\mathcal{B}_\Lambda\right]$$

Next, we use the spin-flip symmetry, which implies that $\mu_\beta^+[\omega](f(\sigma)) = \mu_\beta^-[-\omega](f(-\sigma))$, and the symmetry of the distribution of the h_i to show that

$$\mathbb{E}\left[\ln \frac{\mu_\beta^+[\omega-\omega_\Lambda]\left(e^{\beta\sum_{i\in\Lambda}h_i\sigma_i}\right)}{\mu_\beta^-[\omega-\omega_\Lambda]\left(e^{\beta\sum_{i\in\Lambda}h_i\sigma_i}\right)}\bigg|\mathcal{B}_\Lambda\right]$$

$$= \mathbb{E}\left[\ln \frac{\mu_\beta^-[-(\omega-\omega_\Lambda)]\left(e^{-\beta\sum_{i\in\Lambda}h_i\sigma_i}\right)}{\mu_\beta^-[\omega-\omega_\Lambda]\left(e^{\beta\sum_{i\in\Lambda}h_i\sigma_i}\right)}\bigg|\mathcal{B}_\Lambda\right] \quad (7.46)$$

$$= \mathbb{E}\left[\ln \frac{\mu_\beta^-[\omega-\omega_\Lambda]\left(e^{-\beta\sum_{i\in\Lambda}h_i\sigma_i}\right)}{\mu_\beta^-[\omega-\omega_\Lambda]\left(e^{\beta\sum_{i\in\Lambda}h_i\sigma_i}\right)}\bigg|\mathcal{B}_\Lambda\right]$$

We are left with the ratio of two expectations with respect to the same measure. Here we use the DLR equations to compare them:

$$\mu_\beta^-[\omega-\omega_\Lambda]\left(e^{-\beta\sum_{i\in\Lambda}h_i\sigma_i}\right) \quad (7.47)$$

$$= \mathbb{E}_{\sigma_{\Lambda^c}}\mu_\beta^-[\omega-\omega_\Lambda](\sigma_{\Lambda^c})\mu_{\beta,\Lambda}^{\sigma_{\Lambda^c}}\left(e^{-\beta\sum_{i\in\Lambda}h_i\sigma_i}\right)$$

$$= \mathbb{E}_{\sigma_{\Lambda^c}}\frac{\mu_\beta^-[\omega-\omega_\Lambda](\sigma_{\Lambda^c})}{Z_{\beta,\Lambda}^{\sigma_{\Lambda^c}}}\mathbb{E}_{\sigma_\Lambda}e^{\beta\left(\sum_{\substack{i,j\in\Lambda\\|i-j|=1}}\sigma_i\sigma_j+\sum_{\substack{i\in\Lambda,j\in\Lambda^c\\|i-j|=1}}\sigma_i\sigma_j-\sum_{i\in\Lambda}h_i\sigma_i\right)}$$

$$= \mathbb{E}_{\sigma_{\Lambda^c}}\frac{\mu_\beta^-[\omega-\omega_\Lambda](\sigma_{\Lambda^c})}{Z_{\beta,\Lambda}^{\sigma_{\Lambda^c}}}\mathbb{E}_{\sigma_\Lambda}e^{\beta\left(\sum_{\substack{i,j\in\Lambda\\|i-j|=1}}\sigma_i\sigma_j-\sum_{\substack{i\in\Lambda,j\in\Lambda^c\\|i-j|=1}}\sigma_i\sigma_j+\sum_{i\in\Lambda}h_i\sigma_i\right)}$$

$$\leq e^{2\beta|\partial\Lambda|}\mathbb{E}_\sigma\frac{\mu_\beta^-[\omega-\omega_\Lambda](\sigma_{\Lambda^c})}{Z_{\beta,\Lambda}^{\sigma_{\Lambda^c}}}e^{\beta\left(\sum_{\substack{i,j\in\Lambda\\|i-j|=1}}\sigma_i\sigma_j+\sum_{\substack{i\in\Lambda,j\in\Lambda^c\\|i-j|=1}}\sigma_i\sigma_j+\sum_{i\in\Lambda}h_i\sigma_i\right)}$$

$$= \mu_\beta^-[\omega-\omega_\Lambda]\left(e^{\beta\sum_{i\in\Lambda}h_i\sigma_i}\right)e^{2\beta|\partial\Lambda|}$$

Inserting this bound into (7.46) gives the desired estimate. □

Next, we prove a lower bound on the fluctuations of F_Λ, or, more precisely, on its Laplace transform. Namely:

Lemma 7.2.7 *Assume that for some $\epsilon > 0$, the distribution of the random fields h satisfies $\mathbb{E}|h|^{2+\epsilon} < \infty$. Then, for any $t \in \mathbb{R}$, we have that*

$$\liminf_{\Lambda=[-L,L]^d;L\uparrow\infty} \mathbb{E}\exp(tF_\Lambda/\sqrt{|\Lambda|}) \geq \exp\left(\frac{t^2b^2}{2}\right) \quad (7.48)$$

where

$$b^2 \geq \mathbb{E}\left[\mathbb{E}\left[F_\Lambda|\mathcal{B}_0\right]^2\right] \quad (7.49)$$

Remark 7.2.8 It is easy to see that Lemmata 7.2.6 and 7.2.7 contradict each other in $d \leq 2$, *unless* $b = 0$. On the other hand, we will see that $b = 0$ implies $m^+ = m^-$, and thus the uniqueness of the Gibbs state.

Proof The proof of this lemma uses a decomposition of F_Λ as a martingale-difference sequence. That is, we order all the points $i \in \Lambda$ and denote by $\mathcal{B}_{\Lambda,i}$ the sigma-algebra generated by the variables $\{h_j\}_{j \in \Lambda: j \leq i}$. Then we have trivially that

$$F_\Lambda = \sum_{i=1}^{|\Lambda|} \left(\mathbb{E}\left[F_\Lambda | \mathcal{B}_{\Lambda,i}\right] - \mathbb{E}\left[F_\Lambda | \mathcal{B}_{\Lambda,i-1}\right] \right) \equiv \sum_{i=1}^{|\Lambda|} Y_i \tag{7.50}$$

(note that $\mathbb{E} F_\Lambda = 0$!). Using this, we can represent the generating function as

$$\mathbb{E} e^{tF_\Lambda} = \mathbb{E}[\mathbb{E}[\cdots \mathbb{E}[\mathbb{E}[e^{tY_{|\Lambda|}}|\mathcal{B}_{\Lambda,|\Lambda|-1}]e^{tY_{|\Lambda|-1}}|\mathcal{B}_{\Lambda,|\Lambda|-2}]\cdots e^{tY_2}|\mathcal{B}_{\Lambda,1}]e^{tY_1}] \tag{7.51}$$

We want to work up the conditional expectation from the inside out. To do so, we need a lower bound for any of the terms $\mathbb{E}[e^{tY_i}|\mathcal{B}_{\Lambda,i-1}]$. To get such a bound, we use the observation (see [8], Lemma A.2.2) that there exists a continuous function, $g(a)$, with $g(a) \downarrow 0$ as $a \downarrow 0$, such that, for all real x and all $a \geq 0$, $e^x \geq 1 + x + \frac{1}{2}(1 - g(a))x^2 \mathbb{I}_{|x| \leq a}$. Since, moreover, for all $|x| \leq a$, $e^{x^2 e^{-a^2/2}/2} \leq 1 + x^2/2$, it follows that, if $\mathbb{E} X = 0$, then, for $f(a) = 1 - (1 - g(a))e^{-a^2/2}$,

$$\mathbb{E} e^X \geq e^{\frac{1}{2}(1-f(a))\mathbb{E}[X^2 \mathbb{I}_{|X|\leq a}]} \tag{7.52}$$

Using this estimate, we see that

$$\mathbb{E}\left[e^{tY_i}|\mathcal{B}_{\Lambda,i-1}\right] \exp\left(-\frac{t^2}{2}(1 - f(a))\mathbb{E}\left[Y_i^2 \mathbb{I}_{t|Y_i|\leq a}|\mathcal{B}_{\Lambda,i-1}\right]\right) \geq 1 \tag{7.53}$$

Since this quantity is $\mathcal{B}_{\Lambda,i}$-measurable, we can proceed as in (7.51) to see that (we switch to the desired normalization)

$$1 \leq \mathbb{E} e^{tF_\Lambda/\sqrt{|\Lambda|} - \frac{t^2}{2|\Lambda|}(1-f(a))\sum_{i=1}^{|\Lambda|} \mathbb{E}\left[Y_i^2 \mathbb{I}_{t|Y_i|\leq a\sqrt{|\Lambda|}}|\mathcal{B}_{\Lambda,i-1}\right]} \tag{7.54}$$

We will show in a moment that the term

$$V_\Lambda(a) \equiv |\Lambda|^{-1} \sum_{i=1}^{|\Lambda|} \mathbb{E}\left[Y_i^2 \mathbb{I}_{t|Y_i|\leq a\sqrt{|\Lambda|}}|\mathcal{B}_{\Lambda,i-1}\right] \tag{7.55}$$

appearing in the exponent in the last factor of (7.54), converges in probability to a constant, C, independent of $a > 0$. Since, by Lemma 7.2.6, in $d = 2$, $F_\Lambda/\sqrt{|\Lambda|} \leq C'$, this implies easily that

$$\liminf_{\Lambda \uparrow \mathbb{Z}^d} \mathbb{E} e^{tF_\Lambda/\sqrt{|\Lambda|}} \geq e^{t^2 C/2} \tag{7.56}$$

We are left with the task of controlling and identifying the limit of $V_\Lambda(a)$. This will be done by a clever use of the ergodic theorem.

To do so, we introduce new sigma-algebras $\mathcal{B}_i^\leq$, generated by the random variables h_j with $j \leq i$, where $\leq$ refers to the lexicographic ordering. Define

$$W_i \equiv \mathbb{E}\left[G_\Lambda^{\mu^+} - G_\Lambda^{\mu^-}|\mathcal{B}_i^\leq\right] - \mathbb{E}\left[G_\Lambda^{\mu^+} - G_\Lambda^{\mu^-}|\mathcal{B}_{i-1}^\leq\right] \tag{7.57}$$

Using (7.41) one may show that, for all i in Λ, W_i is independent of Λ (the proof uses (7.41) to represent G_Λ^μ in terms of integrals over $\mu(\sigma_i)$, which is independent of Λ). On the other hand, we have the obvious relation that

$$Y_i = \mathbb{E}[W_i | \mathcal{B}_\Lambda] \tag{7.58}$$

We use this first to show that the indicator function in the conditional expectation can be removed, i.e. for all $\epsilon > 0$,

$$\lim_{\Lambda \uparrow \infty} \left[|\Lambda|^{-1} \sum_{i=1}^{|\Lambda|} \mathbb{E}\left[Y_i^2 \mathbb{1}_{t|Y_i| > a\sqrt{|\Lambda|}} | \mathcal{B}_{\Lambda, i-1} \right] > \epsilon \right] = 0 \tag{7.59}$$

To see this, compute the expectation of the left-hand side in probability, and use the Hölder inequality to get

$$\mathbb{E} |\Lambda|^{-1} \sum_{i=1}^{|\Lambda|} \mathbb{E}\left[Y_i^2 \mathbb{1}_{t|Y_i| > a\sqrt{|\Lambda|}} | \mathcal{B}_{\Lambda, i-1} \right] = |\Lambda|^{-1} \sum_{i=1}^{|\Lambda|} \mathbb{E}\left[Y_i^2 \mathbb{1}_{t|Y_i| > a\sqrt{|\Lambda|}} \right]$$

$$\leq |\Lambda|^{-1} \sum_{i=1}^{|\Lambda|} \left(\mathbb{E} Y_i^{2q} \right)^{1/q} (\mathbb{P}[|Y_i| > a\sqrt{|\Lambda|}/t])^{1/p} \tag{7.60}$$

with $1/p + 1/q = 1$. Now, using Jensen's inequality and (7.58), we see that for any $q > 1$ $\mathbb{E} Y_i^{2q} \leq \mathbb{E} W_0^{2q}$ However, using e.g. (7.41), it is easy to see that $|W_0| \leq C|h_0|$, so that, if the $2q$-th moment of h is finite, then $\mathbb{E} W_0^{2q} < \infty$. Using the Chebyshev inequality and the same argument as before, we also conclude that $\mathbb{P}[|Y_i| > a\sqrt{|\Lambda|}/t] \leq \frac{t^2 \mathbb{E} W_0^2}{a^2 |\Lambda|}$, which tends to zero as $\Lambda \uparrow \infty$. We see that (7.60) tends to zero whenever $p < \infty$, for any $a > 0$. By the Chebyshev inequality, this in turn allows us to conclude (7.59).[7]

Next observe that W_i is shift covariant, i.e.

$$W_i[\omega] = W_0[T_{-i}\omega] \tag{7.61}$$

Therefore, by the ergodic theorem, we can conclude that

$$\lim_{\Lambda \uparrow \mathbb{Z}^d} |\Lambda|^{-1} \sum_{i \in \Lambda} \mathbb{E}\left[W_i^2 | \mathcal{B}_{i-1}^{\leq} \right] = \mathbb{E} W_0^2, \text{ in probability} \tag{7.62}$$

Now we will be done if we can show that

$$E\left[Y_i^2 | \mathcal{B}_{\Lambda, i} \right] - E\left[W_i^2 | \mathcal{B}_i^{\leq} \right] \tag{7.63}$$

goes to zero as Λ goes to infinity, in probability. This follows by estimating the expectation of the square of (7.63), some simple estimates using the Cauchy–Schwarz inequality and the fact that for any square integrable function f, $\mathbb{E}[(f - \mathbb{E}[f|\mathcal{B}_\Lambda])^2]$ tends to zero as Λ approaches $\mathbb{Z}^d$.

To arrive at the final conclusion, note that

$$\mathbb{E} W_0^2 \geq \mathbb{E}[(\mathbb{E}[W_0|\mathcal{B}_0])^2] \tag{7.64}$$

(where $\mathcal{B}_0$ is the sigma-algebra generated by the single variable h_0), and $\mathbb{E}[W_0|\mathcal{B}_0] = \mathbb{E}[F_\Lambda|\mathcal{B}_0]$. $\square$

---

[7] In [8], only two moments are required for h. However, the proof given there is in error as it pretends that the function $x^2 \mathbb{1}_{|x|>a}$ is convex, which is manifestly not the case.

Remark 7.2.9 The assertion of the preceding lemma amounts to a central limit theorem; the basic idea of the proof, namely the use of a martingale-difference sequence obtained by successive conditioning, is also the basis of a useful proof of concentration of measure estimates due to Yurinskii [261].

Finally we observe that by (7.41),

$$\frac{\partial}{\partial h_0}\mathbb{E}[F_\Lambda|\mathcal{B}_0] = \mathbb{E}\big[(\mu_\beta^+(\sigma_0) - \mu_\beta^-(\sigma_0))|\mathcal{B}_0\big] \qquad (7.65)$$

Let us denote by $f(h) = \mathbb{E}[F_\Lambda|\mathcal{B}_0]$ (where $h = h_0[\omega]$). Since $1 \geq f'(h) \geq 0$ for all h, $\mathbb{E}f^2 = 0$ implies that $f(h) = 0$ (for $\mathbb{P}$-almost all points) on the support of the distribution of h. But then $f'(h)$ must also vanish on the convex hull of the support of the distribution of h (except if the distribution is concentrated on a single point). Therefore, barring that case $\mathbb{E}[F_\Lambda|\mathcal{B}_0] = 0 \Rightarrow m^+ - m^- = 0$.

Collecting our results we can now prove the following:

Theorem 7.2.10 [8] *In the random-field Ising model with i.i.d. random fields whose distribution is not concentrated on a single point and possesses at least $2 + \epsilon$ finite moments, for some $\epsilon > 0$, if $d \leq 2$, there exists a unique infinite-volume Gibbs state.*

Proof Lemma 7.2.6 implies that for any Λ, $Ee^{tF_{\beta,\Lambda}} \leq e^{t|\partial\Lambda|}$. Combining this with Lemma 7.2.7, we deduce that, if $d \leq 2$, then necessarily $b = 0$. But from what was just shown, this implies $m^+ = m^-$, which in turn implies uniqueness of the Gibbs state. □

With this result we conclude our discussion of the random-field Ising model in $d = 2$. We may stress that Theorem 7.2.10 is in some sense a soft result that gives uniqueness without saying anything more precise about the properties of the Gibbs state. Clearly, there are many interesting questions that could still be asked. How does the Gibbs state at high temperatures distinguish itself from the one at low temperatures, or what does the low-temperature Gibbs state look like in dependence on the strength of the random fields? It is clear that for very low temperatures and very large ϵ, the Gibbs state will be concentrated near configurations $\sigma_i = \text{sign } h_i$. For small ϵ, by contrast, more complicated behaviour is expected. Results of this type are available in $d = 1$ [29], but much less is known in $d = 2$ [258].

7.3 The Bricmont–Kupiainen renormalization group

In 1988 a remarkable article by Bricmont and Kupiainen [63] settled the long-standing dispute on the lower critical dimension of the random-field Ising model through a rigorous mathematical proof of the existence of at least two phases at low temperatures in dimension three and above. Their proof was based on a renormalization group (RG) analysis. In this section we will give an outline of their proof, following mostly the version given in [50], developed for the related problem of interfaces in random environments. The details of the proof are cumbersome, and we will focus here on the structural aspects, omitting the proofs of many of the lemmata that are mainly of combinatorial nature. All the omitted proofs can be found in [50]. A simpler problem where all cluster expansions can be avoided concerns the hierarchical models (see [48, 49, 59]).

The central result of [63] is the following:

Theorem 7.3.1 *Let $d \geq 3$ and assume that the random variables h_x are i.i.d., symmetrically distributed and satisfy $\mathbb{P}[|h_x| > h] \leq \exp(-h^2/\Sigma^2)$ for Σ sufficiently small. Then there exists $\beta_0 < \infty$, $\Sigma_0 > 0$, such that for all $\beta \geq \beta_0$ and $\Sigma \leq \Sigma_0$ for any increasing and absorbing sequence of volumes $\Lambda_n \uparrow \mathbb{Z}^d$, the sequence of measures $\mu^{\pm}_{\Lambda_n,\beta}$ converges to disjoint Gibbs measures $\mu^{\pm}_\beta$, $\mathbb{P}$-almost surely.*

Before entering the details of the proof of this theorem, we explain some of the main ideas and features of the RG approach. The principal idea is to use a low-temperature contour expansion as explained in Chapter 5. As opposed to many deterministic systems, the first (and in some sense *main*) difficulty in most disordered systems is that the ground-state configuration depends in general on the particular realization of the disorder, and, worse, may in principle depend strongly on the shape and size of the finite volume Λ! In dimension $d \geq 3$, we expect, from the arguments given before, that there exist translation-covariant ground-states that more or less look like the plus or the minus configurations, with a few small deviations.

The crucial observation that forms the ideological basis for the RG approach is that while for large volumes Λ we have no a-priori control on the ground-states, for sufficiently small volumes we can give conditions on the random variables h that are fulfilled with large probability under which the ground-state in this volume is actually the same as the one without randomness. Moreover, the size of the regions for which this holds true will depend on the variance of the random variables and increases to infinity as the latter decreases. This allows us to find 'conditioned' ground-states, where the conditioning is on some property of the configuration on this scale, except in some small region of space. Re-summing then over the fluctuations about these conditioned ground-states one obtains a new effective model for the conditions (the coarse-grained variables) with effective random variables that have smaller variance than the previous ones. In this case, this procedure may be iterated, as now conditioned ground-states on a larger scale can be found.

To implement these ideas one has to overcome two major difficulties. The first one is to find a formulation of the model whose form remains invariant under the renormalization group transformation. The second, and more serious, one is that the re-summation procedure as indicated above can only be performed outside a small random region of space, called the *bad region*. While in the first step this may not look too problemematic, in the process of iteration even a very thin region will potentially 'infect' a larger and larger portion of space. This requires us to get some control also in the bad regions and to develop a precise notion of how regions with a certain degree of badness can be reintegrated as 'harmless' on the next scale. For the method to succeed, the bad regions must 'die out' over the scales much faster than new ones are produced.

7.3.1 Renormalization group and contour models

This subsection is intended to serve two purposes. First, we want to describe the principal ideas behind the renormalization group approach for disordered systems in the low-temperature regime. Second, we want to present the particular types of contour models on which the renormalization group will act and to introduce the notation for the latter.

The renormalization group for measure spaces Let us explain what is generally understood by a renormalization group transformation in a statistical mechanics system. We consider a probability space $(\mathcal{S}, \mathcal{F}, \mu)$, where μ is an (infinite volume) Gibbs measure. One may think for the moment of $\mathcal{S}$ as the 'spin'-state over the lattice $\mathbb{Z}^d$, but we shall need more general spaces later. What we shall, however, assume is that $\mathcal{S}$ is associated with the lattice $\mathbb{Z}^d$ in such a way that for any finite subset $\Lambda \subset \mathbb{Z}^d$ there exists a subset $\mathcal{S}_\Lambda \subset \mathcal{S}$ and sub-sigma algebras, $\mathcal{F}_\Lambda$, relative to $\mathcal{S}_\Lambda$ that satisfy $\mathcal{F}_\Lambda \subset \mathcal{F}_{\Lambda'}$ if and only if $\Lambda \subset \Lambda'$. Note that in this case, any increasing and absorbing sequence of finite volumes, $\{\Lambda_n\}_{n \in \mathbb{Z}_+}$, induces a filtration $\{\mathcal{F}_n \equiv \mathcal{F}_{\Lambda_n}\}_{n \in \mathbb{Z}_+}$ of $\mathcal{F}$.

Ideally, a *renormalization group transformation* is a measurable map, $\mathbb{T}$, that maps $\mathbb{Z}^d \to \mathbb{Z}^d$ and $(\mathcal{S}, \mathcal{F}) \to (\mathcal{S}, \mathcal{F})$ in such a way that, for any $\Lambda \subset \mathbb{Z}^d$,

(i) $\mathbb{T}(\Lambda) \subset \Lambda$, and moreover $\exists_{n < \infty} : \mathbb{T}^n(\Lambda) = \{0\}$, where n may depend on Λ.
(ii) $\mathbb{T}(\mathcal{S}_\Lambda) = \mathcal{S}_{\mathbb{T}(\Lambda)}$

The action of $\mathbb{T}$ on space will generally be blocking, e.g. $\mathbb{T}(x) = \mathcal{L}^{-1}x \equiv \text{int}(x/L)$. The action on $\mathcal{S}$ has to be compatible with this blocking but needs to be defined carefully.

Having the action of $\mathbb{T}$ on the measure space $(\mathcal{S}, \mathcal{F})$ we get a canonical action on measures via

$$(\mathbb{T}\mu)(\mathcal{A}') = \mu(\mathbb{T}^{-1}(\mathcal{A}')) \qquad (7.66)$$

for any Borel-set $\mathcal{A} \in \mathbb{T}(\mathcal{F})$. The fundamental relation of the renormalization group allows us to decompose the measure μ into a conditional expectation and the renormalized measure on the condition, i.e. for any Borel-set $\mathcal{A} \in \mathcal{F}$ we have

$$\mu(\mathcal{A}) = \int_{\mathbb{T}(\mathcal{S})} \mu\left(\mathcal{A} | \mathbb{T} = \omega'\right) (\mathbb{T}\mu)(d\omega') \qquad (7.67)$$

A renormalization group transformation is useful if this decomposition has the virtue that the measure $\mathbb{T}\mu$ is 'simpler' than the measure μ itself, and if the conditioned expectations are easier to control at least on a subspace of large measure with respect to $\mathbb{T}\mu$.

So far, we have not made reference to the specific situation in random systems. In such a situation the specific choice of the renormalization group transformation has to be adapted to the particular realization of the disorder, i.e. will itself have to be a – complicated – random function. In particular, the renormalization group transformation cannot be simply iterated, since after each step the properties of the new measure have to be taken into account when specifying the new map. We will even allow the underlying spaces $\mathcal{S}$ to be random and to change under the application of the renormalization group map.

A final aspect that should be kept in mind is that the renormalized measures (or even their local specifications) can only *in principle* be computed exactly, while in practice our knowledge is limited to certain bounds.

Contour models The concept of *contours* has already been introduced in the context of low-temperature expansions in Chapter 5. The idea is that the support of a contour indicates where a configuration deviates from a ground-state configuration. In our situation, the true ground-states are not known, but we will proceed *as if* the constant configurations were ground-states. The trick introduced by Bricmont and Kupiainen is to correct for this sloppiness by incorporating into the support also those parts of space where the disorder is so big that this assumption is questionable, the so-called *bad regions*. Section 7.3.2 will pinpoint this idea by dealing exclusively with the ground-state problem.

Definition 7.3.2 A contour, Γ, is a pair $(\underline{\Gamma}, \sigma)$, where $\underline{\Gamma}$ is a subset of $\mathbb{Z}^d$, called the support of Γ, and $\sigma \equiv \sigma(\Gamma) : \mathbb{Z}^d \to \{-1, 1\}$ is a map that is constant on connected components of $\underline{\Gamma}^c$.

In the sequel, $\mathcal{S}$ shall denote the space of all contours. Also, $\mathcal{S}_\Lambda$ will denote the space of contours in the finite volume Λ.

We will also need spaces of contours satisfying some constraints. To explain this, we must introduce some notation. Let D be a subset of $\mathbb{Z}^d$. We denote by $\mathcal{S}(D)$ all contours whose support contains D, i.e.

$$\mathcal{S}(D) \equiv \{\Gamma \in \mathcal{S} | D \subset \underline{\Gamma}\} \tag{7.68}$$

As we have indicated above, a renormalization group transformation may depend on the realization of the disorder, and in particular on a bad region D. The bad regions will be affected by the renormalization, so that we will have to construct maps, $\mathbb{T}_D$, that map the spaces $\mathcal{S}(D)$ into $\mathcal{S}(D')$ for suitably computed D'. The resulting structure will then be a measurable map, $\mathbb{T}_D : (\mathcal{S}(D), \mathcal{F}(D)) \to (\mathcal{S}(D'), \mathcal{F}(D'))$, that can be lifted to the measure μ such that, for any $\mathcal{A} \in \mathcal{F}(D')$,

$$(\mathbb{T}_D \mu)(\mathcal{A}) = \mu(\mathbb{T}_D^{-1}(\mathcal{A})) \tag{7.69}$$

We want to iterate this procedure. As a first step, let us rewrite the original RFIM as a contour model.

The RFIM as a contour model We need to introduce some more notation. We always use the metric $d(x, y) = \max_{i=1}^{d} |x_i - y_i|$ for points in $\mathbb{Z}^d$. We call a set $A \subset \mathbb{Z}^d$ connected if and only if, for all $x \in A$, $d(x, A\backslash\{x\}) \leq 1$. A maximal connected subset of a set A will be called a connected component of A. We write $\overline{A}$ for the set of points whose distance from A is not bigger than 1, and we write $\partial A \equiv \overline{A}\backslash A$ and call ∂A the boundary of A. A further important notion is the *interior* of a set A, int A. It is defined as follows: For any set $A \subset \mathbb{Z}^d$, let $\hat{A} \subset \mathbb{R}^d$ denote the set in $\mathbb{R}^d$ obtained by embedding A into $\mathbb{R}^d$ and surrounding each of its points by the unit cube in d dimensions. Then the complement of $\hat{A}$ may have finite connected components. Their union with $\hat{A}$ is called int $\hat{A}$, and the intersection of this set with $\mathbb{Z}^d$ is defined to be int A.

Important operations will consist of the cutting and gluing of contours. First, for any contour Γ we may decompose its support, $\underline{\Gamma}$, into connected components, $\underline{\gamma}_i$, in the sense described above. Note that a contour is uniquely described by specifying its support $\underline{\Gamma}$, the values of σ on the support and the values σ takes on each of the connected components of the boundary of the support. This makes it possible to associate with each connected component $\underline{\gamma}_i$ of the support a contour, γ_i, by furnishing the additional information of the signs on $\underline{\gamma}_i$ and on the connected components of $\partial \underline{\gamma}_i$. We will call a contour with connected support a *connected contour*. In the same spirit we call a connected contour γ_i obtained from a connected component of the support of a contour Γ a *connected component* of Γ. A collection, $\{\gamma_1, \ldots, \gamma_n\}$, of connected contours is called *compatible* if there exists a contour,

7.3 The Bricmont–Kupiainen renormalization group

Γ, such that $\gamma_1, \ldots, \gamma_n$ are the connected components of Γ. This contour will also be called $(\gamma_1, \ldots, \gamma_n)$.

We will also use a notion of *weak connectedness*: a set $\mathcal{A} \subset \mathbb{Z}^d$ is *weakly connected* if int $\mathcal{A}$ is connected. All the notions of the previous paragraph then find their weak analogues.

Defining

$$E_s(\Gamma) = \frac{1}{2} \sum_{\substack{x,y \in \overline{\Gamma} \\ |x-y|=1}} (\sigma_x(\Gamma) - \sigma_y(\Gamma))^2 \qquad (7.70)$$

we could write

$$H(\sigma) = E_s(\Gamma) + (h, \sigma(\Gamma)) \qquad (7.71)$$

with $\underline{\Gamma}$ defined for a given function σ as the set of x that possess a nearest neighbour, y, for which $\sigma_y \neq \sigma_x$. Then the term $E_s(\Gamma)$ could be written as a sum over connected components, $E_s(\Gamma) = \sum_i E_s(\gamma_i)$. This would be reasonable if the constant configurations were indeed ground-states. But the field terms may deform the ground-states. What we want to do is to indicate where such deformations may have occurred in space. To implement this, we allow only h_x that are small enough to remain in the field term. For a fixed $\delta > 0$, to be chosen later, we set

$$S_x \equiv h_x \mathbb{1}_{|h_x|<\delta} \qquad (7.72)$$

For fields that are not small, we introduce a *control field* that keeps rough track of their size,

$$N_x \equiv \delta^{-1} \mathbb{1}_{|h_x|\geq\delta} |h_x| \qquad (7.73)$$

The prefactor δ^{-1} is such that non-zero control fields have minimal size one. The region D, the *bad region*, is then defined as

$$D \equiv D(N) \equiv \{x \in \mathbb{Z}^d | N_x > 0\} \qquad (7.74)$$

The bad region will always be considered part of the contour of a configuration, irrespective of the signs. We define the mass of a contour Γ as

$$\mu(\Gamma) \equiv \rho(\Gamma) e^{-\beta(S,\sigma(\Gamma))} \mathbb{1}_{\underline{\Gamma}=\{x|\exists y:|x-y|=1:\sigma_y(\Gamma)\neq\sigma_x(\Gamma)\}\cup D(\Gamma)} \qquad (7.75)$$

where

$$\rho(\Gamma) = e^{-\beta(E_s(\Gamma)+(h,\sigma(\Gamma))_{D(\Gamma)\cap\underline{\Gamma}})} \qquad (7.76)$$

The important fact is that $\rho(\Gamma)$ factors over the connected components of Γ, i.e. if $\Gamma = (\gamma_1, \ldots, \gamma_n)$, then

$$\rho(\Gamma) = \prod_{i=1}^{n} \rho(\gamma_i) \qquad (7.77)$$

Note that (7.75) implies a one-to-one relation between spin-configurations and contours with non-zero weight.

We would wish that the form of the measures on the contours would remain in this form under renormalization, i.e. activities factorizing over connected components plus a small-field contribution. Unfortunately, except in the case of zero temperature, things will get a bit more complicated. In general, the renormalization will introduce *non-local interactions* between connected components of supports as well as a (small) *non-local random field* $\{S_C\}$ indexed by the connected subsets C of $\mathbb{Z}^d$. We will also introduce the notations

$$V_\pm(\Gamma) \equiv \{x \in \mathbb{Z}^d | \sigma_x(\Gamma) = \pm\} \tag{7.78}$$

and

$$(S, V(\Gamma)) \equiv \sum_{C \subset V_+(\Gamma)} S_C^+ + \sum_{C \subset V_-(\Gamma)} S_C^- \tag{7.79}$$

where sums over C here and henceforth are over connected sets, and the superscript $\pm$ on S refers to whether C is contained in the plus or the minus phase. If C is a single site, $C = x$, we set $S_x^\pm = \pm S_x$. The final structure of the contour measures will be the following:

$$\mu(\Gamma) = \frac{1}{Z_{\beta,\Lambda}} e^{-\beta(S, V(\Gamma))} \sum_{\mathbb{Z}^d \supset G \supset \underline{\Gamma}} \rho(\Gamma, G) \tag{7.80}$$

where the activities, $\rho(\Gamma, G)$, factor over connected components of G.

The functions S, the activities ρ and the fields N will be the parameters on which the action of the renormalization group will finally be controlled.

Renormalization of contours We will now define the action of the renormalization group on contours. This cannot yet be done completely, since, as indicated above, the renormalization group map will depend on the bad regions, basically through the fields N_x. These details will be filled in later.

The renormalization group transformation consists of three steps:

(1) Summation of small connected components of contours
(2) Blocking of the remaining large contours
(3) Dressing of the supports by the new bad region

Note that step (3) is to some extent cosmetic and requires knowledge of the renormalized bad regions. We note that this causes no problem, as the bad regions may already be computed after step (1).

Let us now give a brief description of the individual steps:

STEP 1: We would like to sum in this step over all those classes of contours for which we can get a convergent expansion in spite of the random fields. In practice, we restrict ourselves to a much smaller, but sufficiently large, class of contours. We define a connected component as 'small' if it is geometrically small (in the sense that $d(\gamma_i) < L$), and if its support does not intersect the bad region, with the exception of a suitably defined 'harmless' subset of the bad region. This latter point is important since it will allow us to forget about this harmless part in the next stage of the iteration and this will assure that the successive bad regions become sparser and sparser. Precise definitions are given in Section 7.3.2.

A contour that contains no small connected component is called large, and we denote by $\mathcal{S}^l(D)$ the subspace of large contours. The first step of RG transformation is the canonical projection from $\mathcal{S}(D)$ to $\mathcal{S}^l(D)$, i.e. to any contour in $\mathcal{S}$ we associated the large contour composed of only the large components of Γ.

STEP 2: In this step the large contours are mapped to a coarse-grained lattice. We choose the simplest action of $\mathbb{T}$ on $\mathbb{Z}^d$, namely $(\mathbb{T}x)_i = \mathcal{L}^{-1} \equiv \text{int}(x_i/L)$. We will denote by $\mathcal{L}x$ the set of all points y such that $\mathcal{L}^{-1}y = x$. The action of $\mathcal{L}^{-1}$ on spin configurations is defined as averaging, i.e.

$$(\mathcal{L}^{-1}\sigma)_y = \text{sign} \sum_{x \in \mathcal{L}y} \sigma_x \qquad (7.81)$$

With this definition the action of $\mathcal{L}^{-1}$ on large contours is

$$\mathcal{L}^{-1}\Gamma \equiv (\mathcal{L}^{-1}\underline{\Gamma}, \mathcal{L}^{-1}\sigma) \qquad (7.82)$$

STEP 3: The action of $\mathbb{T}$ given by (2.19) does not yet give a contour in $\mathcal{S}(D')$. Thus, the last step in the RG transformation consists of enlarging the supports of the contours by the newly created bad regions, which requires that we compute those. This will in fact be the most subtle and important part of the entire renormalization program. Given a new region D', the effect on the contours is to replace $\underline{\mathcal{L}^{-1}\Gamma}$ by $\underline{\mathcal{L}^{-1}\Gamma} \cup D'(\mathcal{L}^{-1}\Gamma)$, so that finally the full RG transformation on the contours can be written as

$$\mathbb{T}_D(\Gamma) \equiv (D'(\mathcal{L}^{-1}\Gamma'(\Gamma)) \cup \mathcal{L}^{-1}\underline{\Gamma}'(\Gamma), \mathcal{L}^{-1}h(\Gamma'(\Gamma))) \qquad (7.83)$$

7.3.2 The ground-states

The crucial new feature in the analysis of the low-temperature phase of the RFIM lies in the fact that even the analysis of the properties of the ground-state becomes highly non-trivial. We therefore present the analysis of the ground-states first.

Formalism and setup To simplify things, we will only show that, with probability larger than $1/2$ (and in fact tending to 1 if $\Sigma \downarrow 0$), the spin at the origin is $+1$ in any ground-state configuration with $+$ boundary conditions. More precisely, define

$$\mathcal{G}_\Lambda^{(\Gamma)} \equiv \left\{ \Gamma^* \in \mathcal{S} \middle| \Gamma^*_{\Lambda^c} = \Gamma_{\Lambda^c} \wedge H_\Lambda(\Gamma^*) = \inf_{\Gamma': \Gamma'_{\Lambda^c} = \Gamma_{\Lambda^c}} H_\Lambda(\Gamma') \right\} \qquad (7.84)$$

Here Γ_{Λ^c} denotes the restriction of Γ to Λ^c. We want to show that for a suitable sequence of cubes Λ_n,

$$\mathbb{P}\left[\liminf_{n \uparrow \infty} \min_{\Gamma \in \mathcal{G}_{\Lambda_n}^{\emptyset,+1}} \sigma_0(\Gamma) = +1\right] > 1/2 \qquad (7.85)$$

Let us introduce the abbreviation $\mathcal{S}_n \equiv \mathcal{S}_{\Lambda_n}$. The analysis of ground-states via the renormalization group method then consists of the following inductive procedure. Let $\mathbb{T}$ be a map $\mathbb{T} : \mathcal{S}_n \to \mathcal{S}_{n-1}$. Then clearly

$$\inf_{\Gamma \in \mathcal{S}_n} H_{\Lambda_n} = \inf_{\tilde{\Gamma} \in \mathcal{S}_{n-1}} \left(\inf_{\Gamma \in \mathbb{T}^{-1}\tilde{\Gamma}} H_{\Lambda_n}(\Gamma) \right) \qquad (7.86)$$

which suggests the definition

$$(\mathbb{T}H_{\Lambda_{n-1}})(\tilde{\Gamma}) \equiv \inf_{\Gamma \in \mathbb{T}^{-1}\tilde{\Gamma}} H_{\Lambda_n}(\Gamma) \qquad (7.87)$$

Here $\mathbb{T}^{-1}\tilde{\Gamma}$ denotes the set of pre-images of $\tilde{\Gamma}$ in $\mathcal{S}_n$. Defining $\mathbb{T}\mathcal{G}^{(0)}_{\Lambda_{n-1}}$ to be the set of ground-states with respect to the energy function $\mathbb{T}H$, we have that

$$\mathcal{G}^{\emptyset,+1}_{\Lambda_n} = \left\{ \Gamma^* \,\middle|\, H_{\Lambda_n}(\Gamma^*) = \inf_{\Gamma \in \mathbb{T}^{-1}(\mathbb{T}\mathcal{G}^{\emptyset,+1}_{\Lambda_{n-1}})} H_{\Lambda_n}(\Gamma) \right\} \tag{7.88}$$

that is, if we can determine the ground-states with respect to $\mathbb{T}H$ in the smaller volume Λ_{n-1}, then we have to search for the true ground-state only within the inverse image of this set.

We will now give a precise description of the class of admissible energy functions. The original energy function describing the RFIM was already introduced. To describe the general class of models that will appear in the RG process, we begin with the 'control' fields N. We let $\{N_x\}_{x \in \Lambda_n}$ be a family of non-negative real numbers. They will later be assumed to be a random field satisfying certain specific probabilistic assumptions. Given N, $D(N)$ is defined in (7.74). We denote by $\mathcal{S}_n(D) \subset \mathcal{S}_n$ the space

$$\mathcal{S}_n(D) \equiv \{\Gamma \in \mathcal{S}_n | D(N) \subset \underline{\Gamma}\} \tag{7.89}$$

Definition 7.3.3 An N-bounded contour energy, ϵ, of level k is a map, $\epsilon : \mathcal{S}_n(D) \to \mathbb{R}$, such that:

(i) If $\gamma_1, \ldots, \gamma_m$ are the connected components of Γ, then

$$\epsilon(\Gamma) = \sum_{i=1}^{m} \epsilon(\gamma_i) \tag{7.90}$$

(ii) If γ is a connected contour in $\mathcal{S}_n(D)$, then

$$\epsilon(\gamma) \geq E_s(\gamma) + L^{-(d-2)k}|\underline{\gamma} \setminus \overline{D}(\gamma)| - (N, V(\gamma) \cap \underline{\gamma}) \tag{7.91}$$

where $E_s(\gamma)$ is defined in (7.70).

(iii) Let $C \subset D$ be connected and γ be the connected component of a contour $\Gamma \subset \mathcal{S}_n(D)$. Then

$$\epsilon(\gamma) \leq \sum_{x \in C} N_x \tag{7.92}$$

An N-bounded energy function of level k is a map $H_{\Lambda_n} : \mathcal{S}_n \to \mathbb{R}$ of the form

$$H_{\Lambda_n}(\Gamma) = \epsilon(\Gamma) + (S, V(\Gamma)) \tag{7.93}$$

where S_x are bounded random fields (see (7.72)) and ϵ is a N-bounded contour energy of level k.

Remark 7.3.4 The appearance of the dimension- and k-dependent constant in the lower bound (7.91) is due to the fact that in the RG process no uniform constant suppressing supports of contours outside the bad region is maintained.

We now define the notion of a proper RG transformation.

Definition 7.3.5 For a given control field N, a proper renormalization group transformation, $\mathbb{T}^{(N)}$, is a map from $\mathcal{S}_n(D(N))$ into $\mathcal{S}_{n-1}(D(N'))$ such that, if H_{Λ_n} is of the form (7.93) with ϵ an N-bounded contour energy of level k, then $H'_{\Lambda_{n-1}} \equiv \mathbb{T}^{(N)} H_{\Lambda_n}$ is of the form

$$H'_{\Lambda_{n-1}}(\Gamma) = \epsilon'(\Gamma) + (S', V(\Gamma)) \tag{7.94}$$

where ϵ' is an N'-bounded contour energy of level $k + 1$, and S' is a new bounded random field and N' is a new control field.

In order to make use of an RG transformation, it is crucial to study the action of the RG on the random and control fields. As both are random fields, this control will be probabilistic. We must therefore specify more precisely the corresponding assumptions.

Recall that the energy functions H are random functions on a probability space $(\Omega, \mathcal{B}, \mathbb{P})$ and that H_{Λ_n} is assumed to be $\mathcal{B}_{\Lambda_n}$-measurable (this is evident, e.g., in the original model, where H_{Λ_n} is a function of the stochastic sequences h_x with $x \in \Lambda_n$ only, and $\mathcal{B}_{\Lambda_n}$ is the sigma-algebra generated by these sequences). The renormalized energy functions are still random variables on this same probability space. It is useful to consider an action of the RG map on the sigma-algebras and to introduce $\mathcal{B}^{(k)} = \mathbb{T}^k \mathcal{B}$, where $\mathbb{T}\mathcal{B}^{(k)}_\Lambda \subset \mathcal{B}^{(k-1)}_{\mathcal{L}\Lambda}$, such that after k iterations of the RG the resulting energy function is $\mathcal{B}^{(k)}$-measurable. Naturally, $\mathcal{B}^{(k)}$ is endowed with a filtration with respect to the renormalized lattice. In the general step we will drop the reference to the level in the specification of this sigma-algebra and write simply $\mathcal{B}$. We need to maintain certain locality properties that we state as follows:

(i) The stochastic sequences $\{N_x\}$ and $\{S_x\}$ are measurable with respect to the sigma-algebras $\mathcal{B}_{\bar{x}}$.
(ii) For connected contours $\gamma \in \mathcal{S}_n(D)$, $\epsilon(\gamma)$ is measurable with respect to $\mathcal{B}_{\bar{\gamma}}$.
(iii) The distribution of $\{S_x\}_{x \in \tilde{\Lambda}_n}$ is jointly symmetric, and the distribution of the contour energies $\{\epsilon(\gamma)\}_{\gamma \subset \tilde{\Lambda}_n}$ is symmetric under a global spin flip.

Finally, we need assumptions on the smallness of the disorder. Here the S-fields are centered and bounded, i.e.

(iv) $|S_x| \leq \delta$, for δ small enough (for instance $\delta = \frac{1}{8L}$ will work).
(v) $$\mathbb{P}\left[S_y \geq \epsilon\right] \leq \exp\left(-\frac{\epsilon^2}{2\Sigma^2}\right) \tag{7.95}$$

The control fields N_x should also satisfy bounds like (7.95), but actually the situation there is more complicated. Notice that in the original model the N-fields as defined in (7.73) satisfy bounds $\mathbb{P}(N_x > z) \leq 2\exp(-\frac{z^2}{2\Sigma^2})$, and 'moreover' the smallest non-zero value they take is δ. The precise formulation of the conditions on N are postponed to the end of this section.

Absorption of small contours The first part of the RG map consists of the re-summing of 'small contours'. These can be defined as connected components of small size (on scale L) with support outside the bad regions. The definition of the bad regions excludes the existence of such small components in a ground-state contour. Actually, there is even a large portion of the bad region that may be removed if we are willing to allow for the appearance of 'flat' small contours, i.e contours with non-empty supports but constant sign even on their support. It is crucial to take advantage of this fact. The following definition describes this 'harmless' part of the bad region.

Definition 7.3.6 Let D_i denote the $L^{1/2}$-connected[8] components of D. Such a connected component is called small, on level k, if

(i) $|D_i| < L^{(1-\alpha)/2}$
(ii) $d(D_i) \leq L/4$
(iii) $\sum_{y \in D_i} N_y < LL^{-(d-2)k} \Sigma^2$

Here $\alpha > 0$ is a constant that will be fixed later and $\Sigma^2 \equiv \Sigma_0^2$ refers to the variance of the original random fields, not to those at level k. Define

$$\mathcal{D} \equiv \bigcup_{D_i \text{ small}} D_i \tag{7.96}$$

Remark 7.3.7 The definition of $\mathcal{D}$ is 'local': If we consider a point x and a set $E \subset \Lambda_n$ containing x, then the event $\{E \text{ is a component of } \mathcal{D}\}$ depends only on $N_{x'}$-fields such that $d(x, x') \leq L/3$.

Definition 7.3.8 A connected contour $\gamma \in \mathcal{S}_n(D)$ is called small, if and only if

(i) $d(\gamma) < L$, and
(ii) $(D \backslash \mathcal{D}) \cap \text{int } \underline{\gamma} = \emptyset$

A contour Γ is called small if and only if the maximal connected component of each weakly connected component is small. A contour that is not small is called *large*. We denote by $\mathcal{S}_n^s(D)$ the set of small contours and by $\mathcal{S}_n^l(D)$ the set of large contours.

Remark 7.3.9 Notice that $\mathcal{S}_n^l(D) \subset \mathcal{S}_n(D \backslash \mathcal{D})$, but in general it is *not* a subset of $\mathcal{S}_n(D)$!

Definition 7.3.10 The map $T_1 : \mathcal{S}_n(D) \to \mathcal{S}_n^l(D)$ is the canonical projection, i.e. if $\Gamma = (\gamma_1, \ldots, \gamma_r, \gamma_{r+1}, \ldots, \gamma_q)$ with γ_i large for $i = 1, \ldots, t$ and small for $i = r+1, \ldots, q$, then

$$T_1(\Gamma) \equiv \Gamma^l \equiv (\gamma_1, \ldots, \gamma_r) \tag{7.97}$$

To give a precise description of the conditioned ground-states under the projection T_1, we need to define the following sets. First let $\overline{\mathcal{D}}_i$ denote the *ordinary* connected components of $\overline{\mathcal{D}}$ (in contrast to the definition of D_i!). Given a contour $\Gamma^l \in \mathcal{S}_n^l(D)$ we write $\mathcal{B}_i(\Gamma^l) \equiv \overline{\mathcal{D}}_i \backslash \underline{\Gamma^l}$ for all those components such that $\overline{\mathcal{D}}_i \subset V_\pm(\Gamma^l) \backslash \underline{\Gamma^l}$. Let $\mathcal{B}(\Gamma^l) \equiv \bigcup_i \mathcal{B}_i(\Gamma^l) = \overline{\mathcal{D}}(\Gamma^l) \backslash \underline{\Gamma^l}$. Finally we set $\mathcal{D}_i = \overline{\mathcal{D}}_i \cap \mathcal{D}$. Note that the $\mathcal{D}_i$ need not be connected.

Let $\mathcal{G}_{\Gamma^l,1}$ be the set of contours in $\mathcal{S}_n(D)$ that minimize H_n under the condition $T_1 \Gamma = \Gamma^l$. Then:

Lemma 7.3.11 *Let* $\Gamma^l \in \mathcal{S}_\Lambda^l(D)$. *Then, for any* $\Gamma \in \mathcal{G}_{1,\Gamma^l}$:

(i) $\underline{\Gamma} \backslash \underline{\Gamma^l} \subset \mathcal{B}(\Gamma^l)$, *and*
(ii) *for all* x, $\sigma_x(\Gamma) \equiv \sigma_x(\Gamma^l)$.

Remark 7.3.12 This lemma is the crucial result of the first step of the RG transformation. It makes manifest that fluctuations on length scale L can only arise due to 'large fields in the bad regions'. Since this statement will hold in each iteration of the RG, it shows that the spins are constant outside the bad regions.

---

[8] It should be clear what is meant by $L^{1/2}$-connectedness: a set A is called $L^{1/2}$-connected if there exists a path in A with steps of length less than or equal to $L^{1/2}$ joining each point in A.

The next lemma gives a formula for the renormalized energy function under T_1. We set

$$\epsilon^{\pm}(\mathcal{B}_i(\Gamma^l)) \equiv \inf_{\gamma : \mathcal{D}_i \subset \underline{\gamma} \subset \mathcal{B}_i(\Gamma^l), \gamma = (\underline{\gamma}, \sigma_x \equiv \pm)} \epsilon(\gamma) \tag{7.98}$$

Note that γ here is not necessarily connected.

Lemma 7.3.13 *For any $\Gamma^l \in \mathcal{S}_n^l(D)$, let*

$$(T_1 H_n)(\Gamma^l) \equiv \inf_{\Gamma \in \mathcal{S}_n(D) : T_1(\Gamma) = \Gamma^l} H_n(\Gamma) \tag{7.99}$$

Then

$$(T_1 H_n)(\Gamma^l) - H_n(\Gamma^l) = \sum_i \epsilon^{\pm}(\mathcal{B}_i(\Gamma^l)) \tag{7.100}$$

where the sign $\pm$ is such that $\mathcal{B}_i(\Gamma^l) \subset V_{\pm}(\Gamma^l)$.

Note that in the expression $H_n(\Gamma^l)$, we view Γ^l as a contour in $\mathcal{S}_n(D\backslash\mathcal{D})$; that is, the contributions to the energy in the regions $\mathcal{D}\backslash\underline{\Gamma^l}$ are ignored.

We will skip the proof, which is essentially book-keeping and using the isoperimetric inequality

$$E_s(\gamma) \geq \frac{d}{L} \sum_{x \in \text{int}(\underline{\gamma})} (\sigma_x(\gamma) - \sigma_\gamma)^2 \tag{7.101}$$

where γ is a weakly connected contour such that $d(\text{int } \underline{\gamma}) \leq L$ and σ_γ denotes the sign of γ on $\partial \text{int } \underline{\gamma}$.

Remark 7.3.14 The proof of Lemma 7.3.13 requires a smallness condition on Σ^2 with respect to L, which is the reason for the constant $1/8L$ in (iii) of Definition 7.3.6.

From the preceding lemmata, and Definition 7.3.6, we obtain the following uniform bounds on the $\epsilon^{\pm}$:

Lemma 7.3.15 *For any Γ^l, and any component $\mathcal{B}_i(\Gamma^l)$*

$$|\epsilon^{\pm}(\mathcal{B}_i(\Gamma^l))| \leq L L^{-(d-2)k} \Sigma^2 \tag{7.102}$$

Here we see the rationale for the definition of the harmless part of the large field region, namely that the ground-state contours supported in them only introduce an extremely small correction to the energy, which can, as we will see in the next step, be absorbed locally in the small fields.

The blocking We now want to map the configuration space $\mathcal{S}_n$ to $\mathcal{S}_{n-1}$. The corresponding operator, T_2, will be chosen as $T_2 \equiv \mathcal{L}^{-1}$, with $\mathcal{L}^{-1}$ defined in (7.81) and (7.82). We will use the name $\mathcal{L}^{-1}$ when referring to the purely geometric action of T_2. Notice that $\mathcal{L}^{-1}$ is naturally a map from $\mathcal{S}_n(D\backslash\mathcal{D})$ into $\mathcal{S}_{n-1}(\mathcal{L}^{-1}(D\backslash\mathcal{D}))$, where $\mathcal{L}^{-1}(D\backslash\mathcal{D})$ is defined as the union of the sets $\mathcal{L}^{-1}(D\backslash\mathcal{D})$. We must construct the induced action of this map on the energy functions and on the random fields S and N. Consider first the small fields. Recall that we wanted to absorb the contributions of the small contours into the renormalized small fields. This would be trivial if the $\mathcal{B}_i(\Gamma^l)$ did not depend on Γ^l. To take this effect into account,

we write
$$\epsilon^{\pm}(\mathcal{B}_i(\Gamma^l)) = \epsilon^{\pm}(\overline{\mathcal{D}_i}) + (\epsilon^{\pm}(\mathcal{B}_i(\Gamma^l)) - \epsilon^{\pm}(\overline{\mathcal{D}_i})) \qquad (7.103)$$

and add the first term to the small fields while the second is non-zero only for $\mathcal{D}_i$ that touch the contours of Γ^l and will later be absorbed in the new contour energies. Thus we define the (preliminary) new small fields by

$$\tilde{S}'_y \equiv L^{-(d-1-\alpha)} \left(\sum_{x \in \mathcal{L}y} S_x + \sum_{i : \overline{\mathcal{D}_i} \cap \mathcal{L}y \neq \emptyset} \frac{\epsilon^{\pm}(\overline{\mathcal{D}_i})}{|\mathcal{L}^{-1}\overline{\mathcal{D}_i}|} \right) \qquad (7.104)$$

The prefactor in this definition anticipates the scaling factor of the surface energy term under blocking. Note here that the $\tilde{S}'_y$ satisfy the locality conditions (i): $\tilde{S}'_y$ and $\tilde{S}'_{y'}$ are independent stochastic sequences if $|y - y'| > 1$, since the $\overline{\mathcal{D}_i}$ cannot extend over distances larger than L.

The (preliminary) new control field is defined as
$$\tilde{N}'_y \equiv L^{-(d-1-\alpha)} \sum_{x \in \mathcal{L}y \setminus \mathcal{D}} N_x \qquad (7.105)$$

Note here that the summation over x excludes the regions $\mathcal{D}$, as the contributions there are dealt with elsewhere. This is crucial, as otherwise the regions with positive $\tilde{N}'$ would grow, rather than shrink, in the RG process.

The induced energy function $T_2 T_1 H_n$ on $\mathcal{S}_{n-1}(\mathcal{L}^{-1}(D \setminus \mathcal{D}))$ is
$$(T_2 T_1 H_n)(\Gamma') \equiv \inf_{\Gamma^l : \mathcal{L}^{-1}\Gamma^l = \Gamma'} (T_1 H_n)(\Gamma^l) \qquad (7.106)$$

The following lemma states that this energy function is essentially of the same form as H_n:

Lemma 7.3.16 For any $\Gamma' \in \mathcal{S}_{n-1}(\mathcal{L}^{-1}(D \setminus \mathcal{D}))$ we have

$$(T_2 T_1 H_n)(\Gamma') = L^{d-1-\alpha} \left(\sum_{i=1}^q \bar{\epsilon}(\gamma'_i) + (\tilde{S}', V(\Gamma')) \right) \qquad (7.107)$$

where the γ'_i are the connected components of Γ', and $\bar{\epsilon}$ satisfies the lower bound

$$\bar{\epsilon}(\gamma') \geq c_1 L^\alpha E_s(\gamma') + c_2 L^\alpha L L^{-(d-2)(k+1)} |\underline{\gamma'} \setminus \overline{\tilde{D}'}(\gamma')| - (\tilde{N}', V(\gamma') \cap \underline{\gamma'}) \qquad (7.108)$$

where $\tilde{D}' \equiv D(\tilde{N}')$ is the preliminary bad field region. Moreover, for flat contours of the form $\gamma' = (C, \sigma_y \equiv s)$ with $C \subset \tilde{D}'$ connected, we have the upper bound

$$\bar{\epsilon}(\gamma') \leq (\tilde{N}', V(\gamma') \cap C) \qquad (7.109)$$

We will again skip the details of the proof, which is largely a matter of book-keeping, i.e. suitably distributing the various terms to the new energy functionals and the new field terms. To obtain the desired estimates on the energy terms, we need the following isoperimetric inequalities:

Lemma 7.3.17 Let $\sigma' = \text{sign} \sum_{x \in \mathcal{L}0} \sigma_x$. Then

$$\sum_{<x,y>:x,y \in \mathcal{L}0} |\sigma_x - \sigma_y| \geq \frac{1}{L} \sum_{x \in \mathcal{L}0} |\sigma_x - \sigma'| \qquad (7.110)$$

Lemma 7.3.18 *Let* $\Gamma \in \mathcal{L}^{-1}\gamma'$. *Then*

$$E_s(\Gamma) \geq \frac{L^{d-1}}{d+1} E_s(\gamma') \tag{7.111}$$

The most tricky part is to obtain the term proportional to $|\gamma'\setminus\overline{\tilde{D}'(\gamma')}|$, where $\tilde{D}'$ is the bad region associated with the new control field. The problem is that the original estimate is only in the volume of $\underline{\Gamma}^l$ *outside* the bad region, while the new estimate involves the new bad region, which is smaller than the image of the bad region under $\mathcal{L}^{-1}$ since the harmless part, $\mathcal{D}$, has been excluded in the definition of the $\tilde{N}'$. In fact, the geometric constraints in the definition of $\mathcal{D}$ were essentially made in order to get the desired estimate.

Final shape-up The hard part of the RG transformation is now done. However, not all of the properties of the original model are shared by the renormalized quantities; in particular, the renormalized weak field $\tilde{S}'$ is not centered and it may have become too large. Both defects are, however, easily rectified. We define

$$S'_y \equiv \tilde{S}'_y \mathbb{1}_{|\tilde{S}'_y| < \delta} - \mathbb{E}\left(\tilde{S}'_y \mathbb{1}_{|\tilde{S}'_y| < \delta}\right) \tag{7.112}$$

What is left, i.e. the large part of the small field, is taken account of through the redefined control field. We define the final renormalized control field by

$$N'_y \equiv L^{-(d-1-\alpha)} \sum_{x \in \mathcal{L}y \setminus \mathcal{D}} N_x + |\tilde{S}'_y| \mathbb{1}_{|\tilde{S}'_y| > \delta} \tag{7.113}$$

Given N', we may now define $D' \equiv D(N')$ as (7.74). Then let T_3 (given N') be the map from $\mathcal{S}_{n-1}$ to $\mathcal{S}_{n-i}(D')$ defined through

$$T_3(\Gamma) = (\sigma(\Gamma), \underline{\Gamma} \cup D'(\Gamma)) \tag{7.114}$$

We define the contour energies

$$\epsilon'(\gamma'') \equiv \inf_{\substack{\gamma':\sigma(\gamma')=\sigma(\gamma'') \\ \gamma' \subset \gamma''}} \overline{\epsilon}(\gamma') + \sum_{y \in \gamma''} \tilde{S}'_y \sigma_y(\gamma'') \mathbb{1}_{|\tilde{S}'_y| \geq \frac{1}{16L}}$$

$$+ \sum_{\substack{y \in \gamma'' \\ d(y, \Lambda^c_{n-1})=1}} \mathbb{E}\left[S'_y \mathbb{1}_{|\tilde{S}'_y| < \frac{1}{16L}}\right] \sigma_y(\gamma'') \tag{7.115}$$

Notice that the terms in the second line of (7.115) form a boundary term that is due to the fact that the renormalized fields have mean zero if they are at least at a distance 2 from the boundary.

The final form of the renormalization group map is then given through the following:

Lemma 7.3.19 *For any* $\Gamma' \in \mathcal{S}_{\Lambda'}(N')$ *we have*

$$T_3 T_2 T_1 E(\Gamma') = L^{d-1-\alpha}(\epsilon'(\Gamma') + (S', V(\Gamma'))) \tag{7.116}$$

where ϵ' is an N'-bounded contour energy of level $k+1$.

Proof The form of the renormalized energy follows from the construction. The N'-boundedness of ϵ' is essentially a consequence of Lemma 7.3.16. The only problem is the boundary terms in the second line of (7.114). But these can again be compensated for by giving away a small fraction of the interaction energy. □

This concludes the construction of the entire RG transformation. We may summarize the results of the previous three subsections in the following:

Proposition 7.3.20 *Let $\mathbb{T}^{(N)} \equiv T_3 T_2 T_1 : \mathcal{S}_n(D(N)) \to \mathcal{S}_{n-1}(D(N'))$ with T_1, T_2 and T_3 defined above; let N' and S' and ϵ' be defined as above and define $H'_{n-1} \equiv L^{-(d-1-\alpha)}(\mathbb{T}^{(N)} H_n)$ through*

$$H'_{n-1}(\Gamma) = \epsilon'(\Gamma) + (S', V(\Gamma)) \tag{7.117}$$

If H_n is an N-bounded energy function of level k, then H'_{n-1} is an N'-bounded energy function of level $k+1$.

This proposition allows us to control the flow of the RG transformation on the energies through its action on the random fields S and N. What is now left to do is to study the evolution of the probability distributions of these random fields under the RG map.

Probabilistic estimates Our task is now to control the action of the RG transformation on the random fields S and N, i.e. given the probability distribution of these random fields, we must compute the distribution of the renormalized random fields S' and N' as defined through (7.104), (7.105), (7.112), and (7.113). Of course, we only compute certain bounds on these distributions.

Let us begin with the small fields. In the k-th level of iteration, the distributions of the random fields are governed by a parameter Σ_k^2 (essentially the variance of S_x^k) that decreases exponentially fast to zero with k. We will set

$$\Sigma_k^2 \equiv L^{-(d-2-\eta)k} \Sigma^2 \tag{7.118}$$

where η may be chosen as $\eta \equiv 4\alpha$. We denote by $S^{(k)}$ the small random field obtained from S after k iterations of the RG map $\mathbb{T}$ (where the action of $\mathbb{T}$ on S is defined through (7.112) and (7.104)).

Proposition 7.3.21 *Let $d \geq 3$. Assume that the initial S satisfy assumptions (i), (iii), and (v) (with Σ^2 sufficiently small). Then, for all $k \in \mathbb{N}$ and for all $\epsilon \geq 0$,*

$$\mathbb{P}\left[S_y^{(k)} \geq \epsilon\right] \leq \exp\left(-\frac{\epsilon^2}{2\Sigma_k^2}\right) \tag{7.119}$$

with Σ_k defined through (7.118), and $S^{(k)}$ satisfies assumptions (i),(iii), (iv), and (v).

Proof The renormalized small fields are sums of the old ones, where by assumption the old random variables are independent if their distance is larger than 1. This allows us to represent the sum $\sum_{x \in \mathcal{L}_y} S_x$ as a sum of 2^d sums of independent random variables. Now note that successive use of Hölder's inequality implies that

$$\mathbb{E} e^{t \sum_{i=1}^k Z_i} \leq \prod_{i=1}^k \left(\mathbb{E} e^{tk Z_i}\right)^{1/k} \tag{7.120}$$

Thus the estimates of the Laplace transforms of S' can be reduced to those of i.i.d. random variables satisfying Gaussian tail estimates. The proposition follows thus from standard computations using the exponential Chebyshev inequality. Details can be found in [63] or [50]. □

Next we turn to the distribution of the control fields. We denote by $N_x^{(k)}$ the fields obtained after k iterations of the RG transformation from a starting field $N^{(0)}$, where the iterative steps are defined by equations (7.105) and (7.113). We denote by $D^{(k)}$ and $\mathcal{D}^{(k)}$ the bad regions and harmless bad regions in the k-th RG step. What we need to prove for the control fields are two types of results: first, they must be large only with very small probability; second, and more important, they must be equal to zero with larger and larger probability, as k increases. This second fact implies that the 'bad regions' become smaller and smaller in each iteration of the RG group. The proof of this second fact must take into account the absorption of parts of the bad regions, the $\mathcal{D}$, in each step. What is happening is that once a large field has been scaled down sufficiently, it will drop to zero, since it finds itself in the region $\mathcal{D}$. Due to the complications arising from interactions between neighbouring blocks, this is not quite true, as the field really drops to zero only if the fields at neighbouring sites are small, too. This is being taken into account by considering an upper bound on the control field that is essentially the sum of the original N over small blocks. We define

$$\bar{N}_y^{(0)} = N_y^{(0)} \tag{7.121}$$
$$\bar{N}_y^{(k+1)} = L^{-(d-1-\alpha)} \sum_{x \in \mathcal{L}\{y\} \cap D^{(k+1)} \setminus \mathcal{D}^{(k)}} \bar{N}_x^{(k)} + \left|\tilde{S}_y^{(k+1)}\right| \mathbb{1}_{|\tilde{S}_y^{(k+1)}| > \delta}$$

The fields $\bar{N}$ bound the original N from above, but also, in an appropriate sense, from below. Namely:

Lemma 7.3.22 *The fields $\bar{N}^{(k)}$ defined in (7.121) satisfy*

$$\bar{N}_x^{(k)} \geq N_x^{(k)} \tag{7.122}$$

Moreover, if M is an arbitrary subset of $\mathbb{Z}^d$ and if $K \subset \mathbb{Z}^d$ denotes the union of the connected components of $D^{(k)}$ that intersect M, then

$$\sum_{x \in M} \bar{N}_x^{(k)} \leq C_1^k \sum_{x \in \overline{\overline{M}} \cap K} N_x^{(k)} \tag{7.123}$$

Proof The lower bound (7.122) is obvious. The upper bound is proven by induction. Assume (7.123) for k. To show that it then also holds for $k+1$, we need to show that

$$\sum_{y \in M} \sum_{x \in \mathcal{L}\{y\} \cap \mathcal{D}^{(k+1)} \setminus \mathcal{D}^{(k)}} \bar{N}_x^{(k)} \leq C_1^k \sum_{y \in \overline{M} \cap K} \sum_{x \in \mathcal{L}y \setminus \mathcal{D}^{(k)}} N_x^{(k)}$$
$$= C_1^{k+1} \sum_{x \in \mathcal{L}(\overline{\overline{M}} \cap K) \setminus \mathcal{D}^{(k)}} N_x^{(k)} \tag{7.124}$$

Now, quite obviously,

$$\sum_{y \in M} \sum_{x \in \mathcal{L}\{y\} \cap \mathcal{D}^{(k+1)} \setminus \mathcal{D}^{(k)}} \bar{N}_x^{(k)} \leq |\overline{\{0\}}| \sum_{x \in \mathcal{L}(\overline{M} \cup K) \setminus \mathcal{D}^{(k)}} \bar{N}_x^{(k)} \tag{7.125}$$

where $|\overline{\{0\}}| = 3^d$ takes into account the maximal possible over-counting due to the double sum over y and x. The restriction of the sum over x to the image of K is justified, since all other x must either lie in $D^{(k)}$ or give a zero contribution. Using the induction hypothesis and the definition of $\bar{N}^{(k)}$, a simple calculation shows now that

$$\sum_{x \in \mathcal{L}(\overline{M} \cup K) \setminus \mathcal{D}^{(k)}} \bar{N}_x^{(k)} \leq \left(1 + C_1^k\right) \sum_{x \in \mathcal{L}(\overline{M} \cup K) \setminus \mathcal{D}^{(k)}} N_x^{(k)} \tag{7.126}$$

from which we get (7.124) if C_1 is chosen to be $2 \cdot 3^d$. □

Remark 7.3.23 The bound (7.123) is relevant in the estimates for the finite-temperature case only.

The main properties of the control fields are given by the following:

Proposition 7.3.24 *Let $f_d(z) \equiv z^2 \mathbb{I}_{z \geq 1} + z^{\frac{d-2}{d-1}} \mathbb{I}_{u<1}$. Then*

$$\mathbb{P}\left[L^{-(d-3/2)k} \Sigma > \bar{N}_y^{(k)} > 0\right] = 0 \tag{7.127}$$

and, for $z \geq L^{-(d-3/2)k} \Sigma$

$$\mathbb{P}\left[\bar{N}_y^{(k)} \geq z\right] \leq \exp\left(-\frac{f_d(z)}{4(16L)^{d/d-1} \Sigma_k^2}\right) \tag{7.128}$$

Proof The proof of this proposition will be by induction over k. Note that it is trivially verified for $k = 0$. Thus we assume (7.127) and (7.128) for k.

Let us first show that (7.127) holds for $k+1$. The event under consideration cannot occur if $|\tilde{S}_y^{(k+1)}| > \delta$. Therefore, unless $\bar{N}_y^{(k+1)} = 0$, the site y must lie within $\tilde{D}^{(k+1)}$. But this implies that

$$\mathcal{L}y \cap \left(D^{(k)} \setminus \mathcal{D}^{(k)}\right) \neq \emptyset \tag{7.129}$$

and hence there must exist an $L^{\frac{1}{2}}$-connected component $D_i \subset D^{(k)}$ intersecting $\mathcal{L}y$ that violates one of the conditions of 'smallness' from Definition 7.3.6. Assume first that *only* condition (iii) is violated. In this case, D_i is so small that it is contained in $\mathcal{L}\bar{y}$ and therefore contributes a term larger than $L^{-(d-2)(k+1)+\alpha} \Sigma^2$ to $\bar{N}_y$, and since $\Sigma^2 \sim 1/L$, this already exceeds $L^{-(d-3/2)(k+1)} \Sigma$. Thus, either condition (i) or (ii) must be violated. In both cases, this implies that the number of sites in D_i exceeds $L^{(1-\alpha)/2}$. Each site in D_i contributes at least the minimal non-zero value of $N_x^{(k)}$, which by inductive assumption is $L^{-(d-3/2)k} \Sigma$. Therefore

$$\bar{N}_y^{(k+1)} \geq L^{-(d-1-\alpha)} \sum_{x \in D_i \cap \mathcal{L}y} L^{-(d-3/2)k} \Sigma$$

$$\geq L^{(1-\alpha)/2} L^{-(d-1-\alpha)} L^{-(d-3/2)k} \Sigma$$

$$\geq L^{-(d-3/2)(k+1)} \Sigma \tag{7.130}$$

But this proves (7.127).

To complete the proof of (7.128) we need a property of the function f_d. Before stating it, let us point out that it is crucial to have the function $f_d(z)$, rather than simply z^2; namely, our goal is to show that $\bar{N}_x^{(k)}$ is non-zero with very small probability, which is true if $f_d(L^{-(d-3/2)k} \Sigma) \frac{\delta^2}{\Sigma_k^2}$ is large and grows with k. This is true if f_d, for small values of its argument, cannot decay too fast!

Lemma 7.3.25 *The function f_d defined in Proposition 7.3.24 satisfies*

$$\sum_{x \in \mathcal{L}\bar{y}} f_d\left(\bar{N}_x^{(k)}\right) \geq L^{d-2-3\alpha} f_d\left(L^{-(d-1-\alpha)} \sum_{x \in \mathcal{L}\bar{y}} \bar{N}_x^{(k)}\right) \tag{7.131}$$

Proof See [50]. □

7.3 The Bricmont–Kupiainen renormalization group

We are now ready to prove (7.128) for $k + 1$. Obviously,

$$\mathbb{P}\left[\bar{N}_y^{(k+1)} \geq z\right] \leq \mathbb{P}\left[L^{-(d-1-\alpha)} \sum_{x \in \mathcal{L}\bar{y}\setminus\mathcal{D}} \bar{N}_x^{(k)} \geq z/2\right] + \mathbb{P}\left[|\tilde{S}_y^{(k)}| 1\!\!1_{\tilde{S}_y^{(k)} > \delta} > z/2\right] \quad (7.132)$$

Let us consider the first term in (7.132). By Lemma 7.3.25,

$$\mathbb{P}\left[L^{-(d-1-\alpha)} \sum_{x \in \mathcal{L}\bar{y}\setminus\mathcal{D}} \bar{N}_x^{(k)} \geq z/2\right]$$

$$= \mathbb{P}\left[f_d\left(L^{-(d-1-\alpha)} \sum_{x \in \mathcal{L}\bar{y}\setminus\mathcal{D}} \bar{N}_x^{(k)}\right) \geq f_d(z/2)\right]$$

$$\leq \mathbb{P}\left[\sum_{x \in \mathcal{L}\bar{y}} f_d\left(\bar{N}_x^{(k)}\right) \geq L^{d-2-3\alpha} f_d(z)\right] \quad (7.133)$$

The variables $f_d(N_x^{(k)})$ are essentially exponentially distributed in their tails. We can bound their Laplace transform by

$$\mathbb{E}\left(e^{t f_d\left(N_x^{(k)}\right)}\right) \leq \mathbb{P}\left[N_x^{(k)} = 0\right] + t \int_{m_k}^{\infty} e^{tf} e^{-f\alpha_k} df$$

$$\leq 1 + t \frac{e^{(t-\alpha_k) f_0}}{\alpha_k - t} \quad (7.134)$$

where we have set $m_k \equiv f_d(L^{-(d-3/2)k}\Sigma)$ (the minimal non-zero value $f_d(N^{(k)})$ can take) and $\alpha_k \equiv 1/4(16L)^d \Sigma_k^2$. We will bound the Laplace transform *uniformly* for all $t \leq t^* \equiv (1-\epsilon)\alpha_k$, for some small $\epsilon > 0$. Since $\gamma_k \gg (1-\epsilon)\alpha_k f_0$ (check!), we get in this range of parameters

$$\mathbb{E}\left(e^{t f_d\left(N_x^{(k)}\right)}\right) \leq 1 + \frac{1-\epsilon}{\epsilon} e^{-\epsilon m_k \alpha_k} \quad (7.135)$$

Using the independence of well-separated $\bar{N}_x^{(k)}$, we find

$$\mathbb{P}\left[\sum_{x \in \mathcal{L}\bar{y}} f_d(\bar{N}_x^{(k)}) \geq L^{d-2-3\alpha} f_d(z/2)\right]$$

$$\leq e^{-L^{d-2-3\alpha} f_d(z/2) \frac{t^*}{5d}} \mathbb{E}\left(e^{\frac{t^*}{5d} \sum_{x \in \mathcal{L}\bar{y}} f_d\left(\bar{N}_x^{(k)}\right)}\right)$$

$$\leq e^{-L^{d-2-3\alpha} f_d(z) \frac{\alpha_k(1-\epsilon)}{5d}} \left[\mathbb{E}\left(e^{t^* f_d\left(\bar{N}_x^{(k)}\right)}\right)\right]^{\frac{L^d}{5d}}$$

$$\leq e^{-L^{d-2-3\alpha} f_d(z/2) \frac{\alpha_k(1-\epsilon)}{5d}} \left[1 + \frac{1-\epsilon}{\epsilon} e^{-\epsilon m_k \alpha_k}\right]^{\frac{L^d}{5d}} \quad (7.136)$$

The last factor in (7.136) is close to 1 and may be absorbed in a constant in the exponent, as we only want a bound for $z \geq L^{-(d-3/2)(k+1)}\Sigma$. Moreover, $f_d(z/2) \geq f(z)/2$, and, for L

large enough,

$$L^{d-2-3\alpha} f_d(z/2) \frac{\alpha_k(1-\epsilon)}{5^d} \gg L^{-d-2-\eta}\alpha_k$$
$$= \frac{1}{4(16L)^{d/d-1}\Sigma_{k+1}^2} \quad (7.137)$$

This gives a bound of the desired form for the first term in (7.132). The bound on the second term follows easily from the estimates of Proposition 7.3.21. The proof of Proposition 7.3.24 is now finished. □

Control of the ground-states From our construction it follows that, in a ground-state configuration, the spin at $x \in \mathbb{Z}^d$ will take on the value $+1$, if in no iteration of the renormalization group, the point x will fall into the bad set, D. But Proposition 7.3.24 implies that this is quite likely to be the case. More precisely, we get the following:

Corollary 7.3.26 *Let $d \geq 3$, Σ^2 small enough. Then there exists a constant c' (of order unity) such that for any $x \in \mathbb{Z}^d$*

$$\mathbb{P}[\exists_{k \geq 0} : N_{\mathcal{L}^{-k}x}^{(k)} \neq 0] \leq \exp\left(-\frac{\delta^2}{c'\Sigma^{2-\frac{d-2}{d-1}}}\right) \quad (7.138)$$

Proof The supremum in (7.138) is bounded from above by $\bar{N}_x^{(k)}$. Moreover, $\bar{N}_x^{(k)}$ is either zero or larger than $L^{-(d-3/2)k}\Sigma$. Therefore

$$\mathbb{P}[\exists_{k \geq 0} : N_{\mathcal{L}^{-k}x} \neq 0] \leq \mathbb{P}[\exists_{k \geq 0} \bar{N}_{\mathcal{L}^{-k}x}^{(k)} \neq 0] \quad (7.139)$$
$$\leq \sum_{k=0}^{\infty} \exp\left(-L^{\left(\frac{d-2}{2(d-1)}-\eta\right)k} \frac{\delta^2}{a\Sigma^{2-\frac{d-2}{d-1}}}\right)$$

which gives (7.138) for a suitable constant c'. □

Let us denote by $D^{(k)}$, $\mathcal{D}^{(k)}$ the bad regions and 'harmless' bad regions in the k-th level. Set further

$$\Delta^{(k)} \equiv \bigcup_{i=0}^{k} \mathcal{L}^i \overline{\text{int } \mathcal{D}^{(i)}} \quad (7.140)$$

One may keep in mind that the sets $D^{(k)}$ depend in principle on the finite volume in which we are working; however, this dependence is quite weak and only occurs near the boundary. We therefore suppress this dependence in our notations.

In this terminology Corollary 7.3.26 states that even $\Delta^{(\infty)}$ is a very sparse set. This statement has an immediate implication for the ground-states, via the following:

Proposition 7.3.27 *Let $\Lambda_n \equiv \mathcal{L}^n 0$, and let $\mathcal{G}_{\Lambda_n}^{(0)}$ be defined through (7.84). Then for any $\Gamma^* \in \mathcal{G}_{\Lambda_n}^{(0)}$,*

$$\underline{\Gamma}^* \subset \Delta^{(n-1)} \cup \mathcal{L}^n D^{(n)} \quad (7.141)$$

Proof Let γ_i^* denote the maximal weakly connected components of Γ^*. It is clear that for all these components $\sigma_{\partial \text{int } \gamma_i^*} = +1$. Let $\tilde{\gamma}_i^*$ denote the 'outer' connected component of γ_i^*, i.e. the outer connected component of γ_i^* is the connected component with the property that the interior of its support contains all the supports of the connected components of γ_i^*. If $\tilde{\gamma}_i^*$ is 'small' (in the sense of Definition 7.3.8, since it occurs in a ground-state, by Lemma 7.3.11), it is 'flat' (i.e. $\sigma_x(\tilde{\gamma}_i^*) \equiv 0$) and its support is contained in $\overline{\mathcal{D}}$ (in the first step this set is even empty). Then all the other connected components of γ_i^* are also small, so that γ_i^* is flat and its support is contained in $\overline{\mathcal{D}}$. Thus $\underline{\Gamma^*} \subset \text{int } \underline{\Gamma^{*,l}} \cup \overline{\mathcal{D}^{(0)}}$. On the other hand, $\underline{\Gamma^{*,l}} \subset \mathcal{L}(\mathbb{T}\Gamma^*)$; again the support of the small components of $\mathbb{T}\Gamma^*$ will be contained, by the same argument, in the closure of the small parts of the new bad regions, and so $\underline{\mathbb{T}\Gamma^{*,s}} \subset \overline{\mathcal{D}^{(1)}}$, while $\underline{\mathbb{T}\Gamma^{*,l}} \subset \mathcal{L}\text{int}\,(\mathbb{T}^2\Gamma^{*,l})$. This may be iterated as long as the renormalized contours still have non-empty supports; in the worst case, after n steps, we are left with $\mathbb{T}^n\Gamma^*$, whose support consists at most of the single point 0, and this only if 0 is in the n-th level bad set $D^{(n)}$. But this proves the proposition. $\square$

The task of the next section will be to carry over these results to the finite-temperature case and the Gibbs measures.

7.3.3 The Gibbs states at finite temperature

In this section we repeat the construction and analysis of the renormalization maps for the finite-temperature Gibbs measures. The steps will follow closely those of the previous section and we will be able to make use of many of the results obtained there. The probabilistic analysis will mostly carry over. The difficulties here lie in the technicalities of the various expansions that we will have to use.

Setup and inductive assumptions Just as in Section 7.3.2, an object of crucial importance will be the control field N_x. Given such a field, the bad region $D \equiv D(N)$ is defined exactly as in (7.77).

Analogously to Definition 7.3.3 we now define an N-bounded contour measure:

Definition 7.3.28 An N-bounded contour measure is a probability measure on $\mathcal{S}_n(D)$ of the form

$$\mu(\Gamma) = \frac{1}{Z} e^{-\beta(S, V(\Gamma))} \sum_{\Lambda_n \supset G \supset \Gamma} \rho(\Gamma, G) \qquad (7.142)$$

where

(i) S is a non-local small random field, that is, a map that assigns to each connected (non-empty) set $C \subset \Lambda_n$ and sign $\pm$ a real number $S_C^\pm$ such that

$$\left|S_C^\pm\right| \leq e^{-\tilde{b}|C|}, \text{ if}|C| > 1 \qquad (7.143)$$

and for sets made of a single point x,

$$|S_x| \leq \delta \qquad (7.144)$$

(ii) $\rho(\Gamma, G)$ are positive activities factorizing over connected components of G, i.e. if $(G_1, \ldots, G_l)$ are the connected components of G and if Γ_i denotes the contour made from those connected components of Γ whose supports are contained in G_i, then

$$\rho(\Gamma, G) = \prod_{i=1}^{l} \rho(\Gamma_i, G_i) \tag{7.145}$$

where it is understood that $\rho(\Gamma, G) = 0$ if $\underline{\Gamma} = \emptyset$. They satisfy the upper bound

$$0 \leq \rho(\Gamma, G) \leq e^{-\beta E_s(\Gamma) - \tilde{b}|G \setminus \overline{D(\Gamma)}| + \beta B(N, V(\Gamma) \cap \underline{\Gamma}) + A|G \cap D(\Gamma)|} \tag{7.146}$$

Let $C \subset D$ be connected and $\gamma = (C, \sigma_x(\Gamma) \equiv s)$ be a connected component of a contour $\Gamma \subset \mathcal{S}_n(D)$. Then

$$\rho(\gamma, C) \geq e^{-\beta(N, V(\gamma) \cap C)} \tag{7.147}$$

Z is the partition function that turns μ into a probability measure.

Here β and $\tilde{b}$ are parameters ('temperatures') that will be renormalized in the course of the iterations. In the k-th level, they will be shown to behave as $\beta^{(k)} = L^{(d-1-\alpha)k}$ and $\tilde{b}^{(k)} = L^{(1-\alpha)k}$. B and A are further k-dependent constants. B will actually be chosen close to 1, i.e. with $B = 1$ in level $k = 0$ we can show that in all levels $1 \leq B \leq 2$. A is close to zero; in fact $A \sim e^{-\tilde{b}^{(k)}}$. These constants are in fact quite irrelevant, but cannot be completely avoided for technical reasons. We have suppressed the dependence of μ and ρ on their parameters to lighten the notation.

The probabilistic assumptions are completely analogous to those in Section 7.3.2 and we will not restate them; all quantities depending on sets C are supposed to be measurable with respect to $\mathcal{B}_{\overline{C}}$.

The definition of a proper RG transformation will now be adapted to this setup.

Definition 7.3.29 For a given control field N, a proper renormalization group transformation, $\mathbb{T}^{(N)}$, is a map from $\mathcal{S}_n(D(N))$ into $\mathcal{S}_{n-1}(D(N'))$ such that, if μ is an N-bounded contour measure on $\mathcal{S}_n(D(N))$ with 'temperatures' β and $\tilde{b}$ and small field S (of level k), then $\mu'_{\Lambda_{n-1}} \equiv \mathbb{T}^{(N)}\mu_{\Lambda_n}$ is an N'-bounded contour measure on $\mathcal{S}_{n-1}(D(N'))$ for some control field N', with temperatures β' and $\tilde{b}'$ and small field S' (of level $k+1$).

Absorption of small contours The construction of the map T_1 on the level of contours proceeds now exactly as before, i.e. Definition 7.3.6 defines the harmless large field region, Definition 7.3.8 the 'small' contours, and Definition 7.3.10 the map T_1. What we have to do is to control the induced action of T_1 on the contour measures. Let us for convenience denote by $\hat{\mu} \equiv Z\mu$ the non-normalized measures; this only simplifies notations since T_1 leaves the partition functions invariant (i.e. $T_1\mu = \frac{1}{Z}T_1\hat{\mu}$).

We have, for any $\Gamma' \in \mathcal{S}_N^l(D)$,

$$(T_1\hat{\mu})(\Gamma') \equiv \sum_{\Gamma: T_1(\Gamma) = \Gamma'} \hat{\mu}(\Gamma)$$

$$= \sum_{\Gamma: T_1(\Gamma) = \Gamma'} e^{-\beta(S, V(\Gamma))} \sum_{G \supset \Gamma} \rho(\Gamma, G) \tag{7.148}$$

Now we write

$$(S, V(\Gamma)) = (S, V(\Gamma')) + [(S, V(\Gamma)) - (S, V(\Gamma'))] \tag{7.149}$$

7.3 The Bricmont–Kupiainen renormalization group

Here the first term is what we would like to have; the second reads explicitly

$$[(S, V(\Gamma)) - (S, V(\Gamma'))] = \left[\sum_x S_x \sigma_x(\Gamma) - S_x \sigma(\Gamma')\right]$$

$$+ \sum_{\pm} \left[\sum_{\substack{C \subset V_{\pm}(\Gamma) \\ C \cap \text{int}\,\underline{\Gamma}^s \neq \emptyset}} S_C^{\pm} - \sum_{\substack{C \subset V_{\pm}(\Gamma') \\ C \cap \text{int}\,\underline{\Gamma}^s \neq \emptyset}} S_C^{\pm}\right]$$

$$\equiv \delta S_{\text{loc}}(\Gamma, \Gamma') + \delta S_{\text{nl}}(\Gamma, \Gamma') \quad (7.150)$$

where we used the suggestive notation $\underline{\Gamma}^s \equiv \underline{\Gamma}\backslash\underline{\Gamma}'$. Note that all sets C are assumed to have a volume of at least 2 and to be connected. The conditions on C to intersect $\underline{\Gamma}^s$ just make manifest that otherwise the two contributions cancel. Thus all these unwanted terms are attached to the supports of the 'small' components of Γ. Thus, the local piece, δS_{loc}, poses no particular problem. The non-local piece, however, may join up 'small' and 'large' components, which spoils the factorization properties of ρ. To overcome this difficulty, we apply a cluster expansion. It is useful to introduce the notation

$$\tilde{\sigma}_{\Gamma,\Gamma'}(C) \equiv \sum_{\pm} S_C^{\pm} \left(\mathbb{I}_{C \subset V_{\pm}(\Gamma)} - \mathbb{I}_{C \subset V_{\pm}(\Gamma')}\right) \quad (7.151)$$

so that

$$\delta S_{\text{nl}}(\Gamma, \Gamma') = \sum_{C \cap \text{int}\,\underline{\Gamma}^s \neq \emptyset} \tilde{\sigma}_{\Gamma,\Gamma'}(C) \quad (7.152)$$

Unfortunately the $\tilde{\sigma}_{\Gamma,\Gamma'}(C)$ have arbitrary signs. Therefore, expanding $\exp(-\beta \delta S_{nl})$ directly would produce a polymer system with possibly negative activities. However, by assumption,

$$|\tilde{\sigma}_{\Gamma,\Gamma'}(C)| \leq 2 \max_{\pm} |S_C^{\pm}| \leq 2e^{-\tilde{b}|C|} \equiv f(C) \quad (7.153)$$

Therefore, $\tilde{\sigma}_{\Gamma,\Gamma'}(C) - f(C) \leq 0$ and setting

$$F(\text{int}\,\underline{\Gamma}^s) \equiv \sum_{C \cap \text{int}\,\underline{\Gamma}^s \neq \emptyset} f(C) \quad (7.154)$$

we get

$$e^{-\beta \delta S_{\text{nl}}(\Gamma,\Gamma')} = e^{-\beta F(\text{int}\,\underline{\Gamma}^s)} e^{\beta \sum_{C \cap \text{int}\,\underline{\Gamma}^s \neq \emptyset}(f(C) - \tilde{\sigma}_{\Gamma,\Gamma'}(C))} \quad (7.155)$$

where the second exponential could be expanded in a sum over positive activities. The first exponential does not factor over connected components. However, it is dominated by such a term, and the remainder may be added to the Σ-terms. This follows from the next lemma.

Lemma 7.3.30 *Let $A \subset \mathbb{Z}^d$ and let $(A_1, \ldots, A_l)$ be its connected components. Let $F(A)$ be as defined in (7.155) and set*

$$\delta F(A) \equiv F(A) - \sum_{i=1}^{l} F(A_i) \quad (7.156)$$

Then

$$\delta F(A) = -\sum_{C \cap A \neq \emptyset} k(A, C) f(C) \quad (7.157)$$

where

$$0 \leq k(A,C)f(C) \leq e^{-\tilde{b}(1-\kappa)|C|} \tag{7.158}$$

for $\kappa = \tilde{b}^{-1}$

Proof The sum $\sum_{i=1}^{l} F(A_i)$ counts all C that intersect k connected components of A exactly k times, whereas in $F(A)$ such a C appears only once. Thus, (7.157) holds with $k(A,C) = \#\{A_i : A_i \cap C \neq \emptyset\} - 1$. Furthermore, if C intersects k components, then certainly $|C| \geq k$, from which the upper bound in (7.158) follows. $\square$

Now we can write the non-local terms in their final form:

Lemma 7.3.31 *Let $\delta S_{nl}(\Gamma, \Gamma')$ be defined as in (7.150). Then*

$$e^{-\beta \delta S_{nl}(\Gamma,\Gamma')} = r(\underline{\Gamma}^s) \sum_{l=0}^{\infty} \frac{1}{l!} \sum_{\substack{C_1,\ldots,C_l \\ C_i \cap \text{int}\,\underline{\Gamma}^s \neq \emptyset \\ C_i \neq C_j}} \prod_{i=1}^{l} \phi_{\Gamma,\Gamma'}(C_i)$$

$$\equiv r(\underline{\Gamma}^s) \sum_{\mathcal{C}: \mathcal{C} \cap \text{int}\,\underline{\Gamma}^s \neq \emptyset} \phi_{\Gamma,\Gamma'}(\mathcal{C}) \tag{7.159}$$

where $\phi_{\Gamma,\Gamma'}(C)$ satisfies

$$0 \leq \phi_{\Gamma,\Gamma'}(C) \leq e^{-\tilde{b}|C|/2} \tag{7.160}$$

$r(\underline{\Gamma}^s)$ is a non-random positive activity factoring over connected components of int $\underline{\Gamma}^s$; *for a weakly connected component γ^s,*

$$1 \geq r(\underline{\gamma}^s) \equiv e^{-\beta F(\text{int}\,\underline{\gamma}^s)} \geq e^{-\beta|\text{int}\,\underline{\gamma}^s|e^{-a\tilde{b}}} \tag{7.161}$$

with some constant $0 < a < 1$.

Proof Define for $|C| \geq 2$

$$\sigma_{\Gamma,\Gamma'}(C) \equiv \tilde{\sigma}_{\Gamma,\Gamma'}(C) - f(C)(k(\text{int}\,\underline{\Gamma}^s, C) + 1) \tag{7.162}$$

Then we may write

$$e^{-\beta \sum_{C \cap \text{int}\,\underline{\Gamma}^s \neq \emptyset} \sigma_{\Gamma,\Gamma'}(C)} = \prod_{C \cap \text{int}\,\underline{\Gamma}^s \neq \emptyset} \left(e^{-\beta \sigma_{\Gamma,\Gamma'}(C)} - 1 + 1\right) \tag{7.163}$$

$$= \sum_{l=0}^{\infty} \sum_{\substack{C_1,\ldots,C_l \\ C_i \cap \text{int}\,\underline{\Gamma}^s \neq \emptyset \\ C_i \neq C_j}} \prod_{i=1}^{l} \left(e^{-\beta \sigma_{\Gamma,\Gamma'}(C_i)} - 1\right)$$

which gives (7.159). But since $|\sigma_{\Gamma,\Gamma'}(C)| \leq 2e^{-\tilde{b}(1-\kappa)|C|}$ by (7.158) and the assumption on S_C, (7.160) follows if only $2\beta \leq e^{\tilde{b}(1-2\kappa)/2}$. Given the behaviour of β and $\tilde{b}$ as described in the remark after Definition 7.3.29, if this relation holds for the initial values of the parameters, then it will continue to hold after the application of the renormalization group map for the new values of the parameters. The initial choice will be $\tilde{b} = \beta/L$, and with this relation we must only choose β large enough, e.g. $\beta \geq L(\ln L)^2$ will do.

The properties of $r(\underline{\Gamma}^s)$ follow from Lemma 7.3.30. These activities depend only on the geometry of the support of Γ^s and are otherwise non-random. $\square$

7.3 The Bricmont–Kupiainen renormalization group

Next we write

$$
\begin{aligned}
(T_1\hat{\mu})(\Gamma^l) &= \mathrm{e}^{-\beta(S,V(\Gamma^l))} \sum_{\Gamma:T_1(\Gamma)=\Gamma^l} r(\underline{\Gamma}^s) \sum_{G\supset\underline{\Gamma}} \rho(\Gamma,G)\mathrm{e}^{-\beta\delta S_{\mathrm{loc}}(\Gamma,\Gamma^l)} \sum_{\mathcal{C}:\mathcal{C}\cap\mathrm{int}\,\underline{\Gamma}^s\ne\emptyset} \phi_{\Gamma,\Gamma^l}(\mathcal{C}) \\
&= \mathrm{e}^{-\beta(S,V(\Gamma^l))} \sum_{\Gamma:T_1(\Gamma)=\Gamma^l} \sum_{K\supset\underline{\Gamma}} \sum_{\underline{\Gamma}\subset G\subset K} \sum_{\substack{\mathcal{C}\subset K \\ \mathcal{C}\cap\mathrm{int}\,\underline{\Gamma}^s\ne\emptyset \\ \mathcal{C}\cup G=K}} r(\underline{\Gamma}^s)\rho(\Gamma,G)\mathrm{e}^{-\beta\delta S_{\mathrm{loc}}(\Gamma,\Gamma^l)} \phi_{\Gamma,\Gamma^l}(\mathcal{C})
\end{aligned}
$$
(7.164)

We decompose the set K into its connected components and call K_1 the union of those components that contain components of $\underline{\Gamma}^l$. We set $K_2 = K\setminus K_1$. Everything factorizes over these two sets, including the sum over Γ (the possible small contours that can be inserted into Γ^l are independent from each other in these sets). We make this explicit by writing

$$
\begin{aligned}
(T_1\hat{\mu})(\Gamma^l) &= \mathrm{e}^{-\beta(S,V(\Gamma^l))} \sum_{K_1\supset\underline{\Gamma}^l} \sum_{\Gamma_1:T_1(\Gamma_1)=\Gamma^l} \sum_{\underline{\Gamma_1}\subset G_1\subset K_1} \sum_{\substack{\mathcal{C}_1\subset K_1 \\ \mathcal{C}_1\cap\mathrm{int}\,\underline{\Gamma}_1^s\ne\emptyset \\ \mathcal{C}_1\cup G_1=K_1}} \\
&\quad \times r(\underline{\Gamma}_1^s)\rho(\Gamma_1,G_1)\mathrm{e}^{-\beta\delta S_{\mathrm{loc}}(\Gamma_1,\Gamma^l)} \phi_{\Gamma_1,\Gamma^l}(\mathcal{C}_1) \\
&\quad \times \sum_{K_2:K_2\cap\overline{K_1}=\emptyset} \sum_{\Gamma_2:T_1(\Gamma_2)=\Gamma^l} \sum_{\underline{\Gamma_2}\subset G_2\subset K_2} \sum_{\substack{\mathcal{C}_2\subset K_2 \\ \mathcal{C}_2\cap\mathrm{int}\,\underline{\Gamma}_2^s\ne\emptyset \\ \mathcal{C}_2\cup G_2=K_2}} \\
&\quad \times r(\underline{\Gamma}_2^s)\rho(\Gamma_2,G_2)\mathrm{e}^{-\beta\delta S_{\mathrm{loc}}(\Gamma_2,\Gamma^l)} \phi_{\Gamma_2,\Gamma^l}(\mathcal{C}_2) \\
&\equiv \mathrm{e}^{-\beta(S,V(\Gamma^l))} \sum_{K_1\supset\underline{\Gamma}^l} \hat{\rho}(\Gamma^l,K_1) \sum_{K_2:K_2\cap\overline{K_1}=\emptyset} \tilde{\rho}(\Gamma^l,K_2)
\end{aligned}
$$
(7.165)

Here, the contours Γ_1 and Γ_2 are understood to have small components with supports only within the sets K_1 and K_2, respectively. Also, the set K_2 must contain $D(\Gamma^l)\cap K_1^c$. The final form of (7.165) is almost the original one, except for the sum over K_2. This latter will give rise to an additional *non-local* field term, as we will now explain.

The sum over K_2 can be factored over the connected components of K_1^c. In these components, $\tilde{\rho}$ depends on Γ^l only through the (constant) value of the spin $\sigma(\Gamma^l)$ in this component. Let Y denote such a connected component. We have

Lemma 7.3.32 *Let $\tilde{\rho}$ be defined as in (7.165). Then*

$$
\sum_{D\cap Y\subset K\subset Y} \tilde{\rho}(\Gamma^l,K) = \mathrm{e}^{-\beta\sum_{C\subset Y}\bar{\psi}_C - \beta\sum_{C\subset Y,\overline{C}\cap Y^c\ne\emptyset}\psi_C^s(Y)} \prod_i \tilde{\rho}'(\mathcal{B}_i^Y)
$$
(7.166)

where the $\mathcal{B}_i^Y$ denote the connected components of the set $\mathcal{B}^Y \equiv \overline{\mathcal{D}}\cap Y = \mathcal{B}(\Gamma^l)\cap Y$ in Y. The sum over C is over connected sets such that $C\setminus\overline{D}\ne\emptyset$. The fields $\bar{\psi}_C$ are independent of Y and Γ^l. Moreover, there exists a strictly positive constant $1 > g > 0$ such that

$$
|\bar{\psi}_C| \le \mathrm{e}^{-g\bar{b}|C\setminus\overline{D}|}
$$
(7.167)

and

$$
|\psi_C^s(Y)| \le \mathrm{e}^{-g\bar{b}|C\setminus\overline{D}|}
$$
(7.168)

and a constant $C_1 > 0$ such that

$$\left|\frac{1}{\beta}\ln\left(\tilde{\rho}'(\mathcal{B}_i^Y)\right)\right| \leq B\sum_{x\in\mathcal{D}_i} N_x + \frac{C_1}{\beta}|\mathcal{B}_i^Y| \qquad (7.169)$$

Proof Naturally, the form (7.166) will be obtained through a Mayer-expansion, considering the connected components of K as polymers subjected to a hard-core interaction. A complication arises from the fact that these polymers must contain the set $\mathcal{D} \cap Y$. Thus we define the set $\mathcal{G}(Y)$ of permissible polymers through

$$\mathcal{G}(Y) = \left\{K \subset Y, \text{conn.}, K \cap \mathcal{B}(\Gamma^l) = \cup_{\mathcal{B}_i^Y \cap K \neq \emptyset} \mathcal{B}_i^Y\right\} \qquad (7.170)$$

That is, any polymer in this set will *contain* all the connected components of $\overline{\mathcal{D}} \cap Y$ it intersects. For such polymers we define the activities

$$\tilde{\rho}'(K) = \sum_{\substack{\tilde{K}:K\cap\mathcal{D}\subset\tilde{K}\\ \tilde{K}\setminus\mathcal{B}^Y=K\setminus\mathcal{B}^Y}} \sum_{\Gamma:T_1(\Gamma)=(\emptyset,\pm)} \sum_{\underline{\Gamma}\subset G\subset\tilde{K}} \sum_{\substack{\mathcal{C}\subset\tilde{K}\\ \mathcal{C}\cap\text{int}\underline{\Gamma}^s\neq\emptyset\lor\mathcal{C}=\emptyset\\ \mathcal{C}\cup G=\tilde{K}}} e^{-\beta\sum_{x\in\underline{\Gamma}}[S_x\sigma_x(\Gamma)\mp S_x)]} r(\underline{\Gamma})\rho(\Gamma, G)\phi_\Gamma(\mathcal{C})$$

$$(7.171)$$

Note that by summing over $\tilde{K}$ we collect all polymers that differ only within $\mathcal{B}^Y$. Thus we get

$$\sum_{\mathcal{D}\cap Y\subset K\subset Y} \tilde{\rho}(\Gamma^l, K) = \sum_{N=0}^{\infty} \frac{1}{N!} \sum_{\substack{K_1,\ldots,K_N:K_i\in\mathcal{G}(Y)\\ \cup_{i=1}^N K_i\supset\mathcal{B}(\Gamma^l)}} \prod_{i=1}^{N} \tilde{\rho}'(K_i) \prod_{1\leq i<j\leq N} \mathbb{1}_{\overline{K_i\cap K_j=\emptyset}} \qquad (7.172)$$

Next we have to extract the contributions of those polymers that can occur in the ground-states. We set

$$\bar{\rho}(K) = \frac{\tilde{\rho}'(K)}{\prod_{\mathcal{B}_i^Y\subset K} \tilde{\rho}'(\mathcal{B}_i^Y)} \qquad (7.173)$$

Then $\bar{\rho}(\pm, \mathcal{B}_i^Y) = 1$, i.e. the $\mathcal{B}_i^Y$ play the role of the empty polymer. This procedure allows us to remove the restriction $\cup_{i=1}^N K_i \supset \mathcal{B}(\Gamma^l)$ in the following way. Set

$$\mathcal{G}'(Y) = \mathcal{G}(Y) \setminus \{\mathcal{B}_i^Y, i \in \mathbb{Z}\} \qquad (7.174)$$

Each polymer K_i is either in $\mathcal{G}'(Y)$ or is one of the $\mathcal{B}_i^Y$. Moreover, once all the $K_i \in \mathcal{G}'(Y)$ are chosen, the hard-core interaction plus the constraint $\cup_{i=1}^N K_i \supset \mathcal{B}(\Gamma^l)$ fix the remaining K_i uniquely up to permutations. Since their activities $\bar{\rho}$ are equal to one, the entire sum over these polymers outside $\mathcal{G}'(Y)$ just contributes a factor 1. Therefore

$$\sum_{\mathcal{D}\cap Y\subset K\subset Y} \tilde{\rho}(\Gamma^l, K) = \prod_{\mathcal{B}_i^Y\subset Y} \tilde{\rho}'(\mathcal{B}_i^Y) \sum_{N=0}^{\infty} \frac{1}{N!} \sum_{\substack{K_1,\ldots,K_N\\ K_i\in\mathcal{G}'(Y)}} \prod_{i=1}^{N} \bar{\rho}(K_i) \prod_{1\leq i<j\leq N} \mathbb{1}_{\overline{K_i\cap K_j=\emptyset}}$$

$$(7.175)$$

This is now (up to the prefactor) the standard form of a polymer partition function with hard-core interaction (see Section 5.2). It can be exponentiated and yields the estimates of the lemma provided we get the bound

$$\bar{\rho}(\pm, K) \leq e^{-c_2\tilde{b}|K\setminus\overline{\mathcal{D}}|} \qquad (7.176)$$

on the activities. We skip the tedious details of these estimates, which can be found in [50]. □

Next we need to control the activities $\hat{\rho}(\Gamma^l, K)$. Our aim is to show that they satisfy bounds similar to the original ρ. As this is similar to the proof of Lemma 7.3.32 we can skip the details.

We can now write the expression for $T_1\hat{\mu}$ in the following pleasant form:

$$(T_1\hat{\mu})(\Gamma^l) = e^{-\beta(S, V(\Gamma^l))} \sum_{K \supset \underline{\Gamma}^l} \hat{\rho}'(\Gamma^l, K) \prod_{\pm, i : \overline{\mathcal{D}_i} \subset V_\pm(\Gamma^l) \cap \overline{K}} \tilde{\rho}'(\overline{\mathcal{D}_i})$$
$$\times \prod_{\pm, i : \mathcal{D}_i \subset V_\pm(\Gamma^l) \cap \overline{K}^c} \tilde{\rho}'(\overline{\mathcal{D}_i} \cap \overline{K}^c)$$
$$\times \exp\left(-\beta(\bar{\psi}, V(\Gamma^l) \cap \overline{K}^c) - \beta \sum_{\substack{\pm, C : C \subset V_\pm(\Gamma^l) \cap \overline{K}^c \\ C \cap \overline{K} \neq \emptyset}} \psi_C^s(K)\right) \quad (7.177)$$

Here the $\tilde{\rho}'(\overline{\mathcal{D}_i})$ are independent of the contour and K and can be exponentiated to yield a non-local field. For the activities we have the following bounds.

Lemma 7.3.33

$$0 \leq \hat{\rho}'(\Gamma^l, K) \leq e^{-\beta E_s(\Gamma^l) - \frac{g\bar{b}}{2}|K \setminus \overline{D(\Gamma^l)}| + \beta B(N, V(\Gamma^l) \cap \underline{\Gamma}^l)} 25^{|K|} \quad (7.178)$$

For contours $\Gamma^l = (C, h_x \equiv h)$, with $C \subset D \setminus \mathcal{D}$ connected, we have moreover

$$\hat{\rho}'(\Gamma^l, \underline{\Gamma}^l) \geq e^{-\beta B(N, V(\Gamma^l) \cap \underline{\Gamma}^l)} \quad (7.179)$$

Proof Notice that $\hat{\rho}'(\Gamma^l, \underline{\Gamma}^l) = \tilde{\rho}(\Gamma^l, \underline{\Gamma}^l) = \rho(\Gamma^l, \underline{\Gamma}^l)$ so that (7.179) follows from the assumptions on ρ. The upper bound (7.178) is proven in the same way as the upper bound on $\tilde{\rho}$, since small contours can be summed over in each connected component of the complement of $\underline{\Gamma}^l$ in K. □

The blocking We now turn to the main step of the RG transformation, the blocking. As before, nothing changes as far as the action of $\mathbb{T}$ on contours is concerned and all we have to do is to study the effect on the contour measures.

First we exponentiate all terms in (7.177) that give rise to the new random fields. We set

$$z_C \equiv \sum_i \mathbb{1}_{C = \overline{\mathcal{D}_i}} \left(-\frac{1}{\beta} \ln(\tilde{\rho}'(\overline{\mathcal{D}_i}))\right) \quad (7.180)$$

Setting now

$$\tilde{S}_C^\pm \equiv S_C^\pm + z_C + \bar{\psi}_C \quad (7.181)$$

and noticing that

$$(\bar{\psi}, V(\Gamma^l) \cap \overline{K}^c) = (\bar{\psi}, V(\Gamma^l)) - \sum_{\substack{\pm, C \subset V_\pm(\Gamma^l) \\ C \cap \overline{K} \neq \emptyset}} \bar{\psi}_C \quad (7.182)$$

We have

$$(T_l\hat{\mu})(\Gamma^l) = e^{-\beta(\tilde{S}, V(\Gamma^l))} \sum_{K \supset \underline{\Gamma^l}} \hat{\rho}'(\Gamma^l, K) \left(\prod_{\substack{\mathcal{D}_i \subset V_\pm(\Gamma^l) \cap \overline{K}^c \\ \overline{\mathcal{D}_i} \not\subset \overline{K}^c}} \frac{\tilde{\rho}'(\overline{\mathcal{D}_i} \cap \overline{K}^c)}{\tilde{\rho}'(\overline{\mathcal{D}_i})} \right)$$

$$\times \exp \left(\beta \sum_{\substack{\pm, C \subset V_\pm(\Gamma^l) \\ C \cap \overline{K} \neq \emptyset}} \bar{\psi}_C - \beta \sum_{\substack{\pm, C: C \subset V_\pm(\Gamma^l) \cap \overline{K}^c \\ \overline{C} \cap \overline{K} \neq \emptyset}} \psi_C^s(K) \right) \quad (7.183)$$

where the random field and the activity-like contributions are almost well separated. We first prepare the field term for blocking. For given $\Gamma' \subset S_{n-1}(\mathcal{L}^{-1}D)$, we can split the term into three parts:

$$(\tilde{S}, V_\pm(\Gamma^l)) = L^{d-1-\alpha}(\tilde{S}', V_\pm(\Gamma')) + \delta\tilde{S}_{\text{loc}}(\Gamma^l, \Gamma') + \delta\tilde{S}_{\text{nl}}(\Gamma^l, \Gamma') \quad (7.184)$$

where for single points y

$$\tilde{S}'^{\pm}_y \equiv L^{-(d-1-\alpha)} \left(\sum_{x \in \mathcal{L}y} \tilde{S}_x \sigma_x(\Gamma') + \sum_{\substack{C \subset V_\pm(\Gamma'): C \cap \mathcal{L}y \neq \emptyset \\ d(C) < L/4 \vee C \subset \mathcal{L}y}} \frac{\tilde{S}^{\pm}_C}{|\mathcal{L}^{-1}(C)|} \right) \quad (7.185)$$

and for $|C'| > 1$

$$\tilde{S}'_{C'} \equiv L^{-(d-1-\alpha)} \sum_{\substack{C: \mathcal{L}^{-1}(C) = C' \\ d(C) \geq L/4}} \tilde{S}_C \quad (7.186)$$

Equations (7.185) and (7.186) are the analogues of (7.104) and almost the final definitions of the renormalized 'small random fields'. Furthermore,

$$\delta\tilde{S}_{\text{loc}}(\Gamma^l, \Gamma') \equiv \sum_{y \in \Lambda_{n-1}} \left[\sum_{x \in \mathcal{L}y} \left(\tilde{S}_x \sigma_x(\Gamma^l) - \tilde{S}_x \sigma_{\mathcal{L}^{-1}x}(\Gamma') \right) \right.$$

$$\left. + \sum_{\substack{\pm, C: C \cap \mathcal{L}y \neq \emptyset \\ d(C) < L/4 \vee C \subset \mathcal{L}y}} \tilde{S}^{\pm}_C \left[\mathbb{I}_{C \subset V_\sigma(\Gamma^l)} - \frac{\mathbb{I}_{\sigma_y(\Gamma') = \pm}}{|\mathcal{L}^{-1}C|} \right] \right] \quad (7.187)$$

and

$$\delta\tilde{S}_{\text{nl}}(\Gamma^l, \Gamma') \equiv \sum_{\substack{\pm, C: C \subset \Lambda_{n-1} \\ d(C) \geq L/4 \wedge |\mathcal{L}^{-1}C| \geq 2}} \tilde{S}^{\pm}_C \left[\mathbb{I}_{C \subset V_\pm(\Gamma^l)} - \mathbb{I}_{\mathcal{L}^{-1}C \subset V_\pm(\Gamma')} \right]$$

$$\equiv \sum_{\substack{C: C \subset \Lambda_n \\ d(C) \geq L/4 \wedge |\mathcal{L}^{-1}C| \geq 2}} \tilde{s}_{\Gamma^l, \Gamma'}(C) \quad (7.188)$$

The point here is that the contributions from $\delta\tilde{S}_{\text{loc}}$ will factor over the connected components of the blocked K, while the non-local $\delta\tilde{S}_{\text{nl}}$ can be expanded and gives only very small contributions, due to the minimal size condition on the C occurring in it.

7.3 The Bricmont–Kupiainen renormalization group

In a similar way we decompose the exponent on the last line of (7.183). Here it is convenient to slightly enlarge the supports of the ψ and to define

$$\tilde{\psi}_{\Gamma^l,K}(\tilde{C}) \equiv - \sum_{\substack{\pm, C \subset V_\pm(\Gamma^l), \overline{C}=\tilde{C} \\ C \cap K \neq \emptyset}} \bar{\psi}_C + \sum_{\substack{\pm, C: C \subset V_\pm(\Gamma^l) \cap \overline{K}^c, \overline{C}=\tilde{C} \\ \overline{C} \cap K \neq \emptyset}} \psi_C^s(K) \qquad (7.189)$$

This has the advantage that now $\tilde{\psi}_{\Gamma^l,K}(C) = 0$ if $C \cap K = \emptyset$. We then decompose

$$-\sum_{\substack{\pm, C \subset V_\pm(\Gamma^l) \\ C \cap K \neq \emptyset}} \bar{\psi}_C + \sum_{\substack{\pm, C: C \subset V_\pm(\Gamma^l) \cap \overline{K}^c \\ \overline{C} \cap K \neq \emptyset}} \psi_C^s(K) = \sum_{C: C \cap K \neq \emptyset} \tilde{\psi}_{\Gamma^l,K}(C)$$

$$= \sum_{y \in \Lambda_{n-1}} \sum_{\substack{C \cap \mathcal{L} y \neq \emptyset \\ C \cap K \neq \emptyset \\ d(C) < L/4 \vee C \subset \mathcal{L} y}} \frac{\tilde{\psi}_{\Gamma^l,K}(C)}{|\mathcal{L}^{-1}C|} + \sum_{\substack{C \cap K \neq \emptyset \\ d(C) \geq L/4 \wedge |\mathcal{L}^{-1}C| \geq 2}} \tilde{\psi}_{\Gamma^l,K}(C)$$

$$\equiv \delta \psi_{\text{loc}}(\Gamma^l, K) + \delta \psi_{\text{nl}}(\Gamma^l, K) \qquad (7.190)$$

In all of the non-local terms only sets C give a contribution for which $C \cap \mathcal{L}(\mathcal{L}^{-1}K) \neq \emptyset$, $d(C) \geq L/4$ and $|\mathcal{L}^{-1}C| \geq 2$. Moreover, for connected C with $d(C) > L/4$ we have that $|C| \leq const.|C \setminus \overline{D}|$ and hence (7.167) implies $|\tilde{\psi}_{\Gamma^l,K}(C)| \leq e^{-const.\tilde{\beta}|C|}$. The inductive hypothesis yields a similar estimate for $\tilde{S}$ so that in fact

$$\left|\tilde{s}_{\Gamma^l,\Gamma'}(C) + \tilde{\psi}_{\Gamma^l,K}(C)\right| \leq e^{-const.'\tilde{b}|C|} \equiv \tilde{f}(C) \qquad (7.191)$$

In analogy to Lemma 7.3.31 we can therefore expand these contributions to get

$$e^{-\beta(\delta \tilde{S}_{\text{nl}}(\Gamma^l,\Gamma')+\delta \psi_{\text{nl}}(\Gamma^l,K))} = R(K) \sum_{l=0}^{\infty} \frac{1}{l!} \sum_{\substack{C_1,...,C_l: C_i \neq C_j \\ C_i \cap \mathcal{L}(\mathcal{L}^{-1}K) \neq \emptyset \\ d(C) \geq L/4 \wedge |\mathcal{L}^{-1}C| \geq 2}} \prod \Phi_{\Gamma',\Gamma^l,K}(C_i)$$

$$\equiv \sum_{\substack{\mathcal{C}: C \cap \mathcal{L}(\mathcal{L}^{-1}K) \neq \emptyset \\ d(C) \geq L/4 \wedge |\mathcal{L}^{-1}C| \geq 2}} \Phi_{\Gamma',\Gamma^l,K}(\mathcal{C}) \qquad (7.192)$$

where the activities Φ satisfy

$$0 \leq \Phi_{\Gamma',\Gamma^l,K}(C) \leq e^{-g''\tilde{b}|C|} \qquad (7.193)$$

and $R(K)$ are non-random activities factoring over connected components of $\mathcal{L}(\mathcal{L}^{-1}K)$, satisfying, for a connected component,

$$1 \geq R(K) \equiv \exp\left(-\sum_{\substack{C \cap \mathcal{L}(\mathcal{L}^{-1}K) \neq \emptyset \\ d(C) \geq L/4 \wedge |\mathcal{L}^{-1}C| \geq 2}} \tilde{f}(C)\right) \geq e^{-|\mathcal{L}(\mathcal{L}^{-1}K)|e^{-\frac{L}{4}\tilde{b}''}} \qquad (7.194)$$

Note that in these bounds the terms $\overline{D}$ no longer appear.

With these preparations we can now write down the blocked contour measures in the form

$$(\mathbb{T}T_1\hat{\mu})(\Gamma') = e^{-\beta L^{d-1-\alpha}(\tilde{S}',V(\Gamma'))} \sum_{G' \supset \Gamma'} \rho'(\Gamma', G') \qquad (7.195)$$

where

$$\rho'(\Gamma', G') \equiv \sum_{G' \supset K' \supset \underline{\Gamma}'} \sum_{\mathcal{C}': \mathcal{C}' \cup K' = G'} \sum_{\Gamma^l: T_2(\Gamma^l) = \Gamma'} \sum_{\substack{K \supset \underline{\Gamma}^l \\ \mathcal{L}^{-1} K = K'}} \sum_{\substack{\mathcal{C}: \mathcal{L}^{-1} \mathcal{C} = \mathcal{C}' \\ d(\mathcal{C}) \geq L/4 \wedge |\mathcal{L}^{-1} \mathcal{C}| \geq 2}}$$

$$\times \mathrm{e}^{-\beta \left(\delta \tilde{S}_{\mathrm{loc}}(\Gamma', \Gamma^l) + \delta \psi_{\mathrm{loc}}(\Gamma^l, K) \right)} R(K) \hat{\rho}'(\Gamma^l, K) \Phi_{\Gamma', \Gamma^l, K}(\mathcal{C})$$

$$\times \prod_{\substack{\mathcal{D}_i \subset V_{\pm}(\Gamma^l) \cap \overline{K}^c \\ \overline{\mathcal{D}_i} \not\subset \overline{K}^c}} \frac{\tilde{\rho}'(\overline{\mathcal{D}_i} \cap \overline{K}^c)}{\tilde{\rho}'(\overline{\mathcal{D}_i})} \qquad (7.196)$$

Notice that by construction the C occurring in the local fields $\delta \tilde{S}$ and $\delta \psi$ cannot connect disconnected components of G', and therefore $\rho'(\Gamma', G')$ factorizes over connected components of G'. The main task that is left is to prove that ρ' yields an N'-bounded contour measure for a suitably defined N'. As in Section 7.3.2, we define the preliminary new control field by

$$\tilde{N}'_y \equiv L^{-(d-1-\alpha)} \sum_{x \in \mathcal{L}y \setminus \mathcal{D}} N_x \qquad (7.197)$$

where N has been defined already in (7.142). We will now prove the following:

Lemma 7.3.34 *Let $\tilde{N}'$ be defined as in (7.197) and set $\hat{D}' \equiv D(\tilde{N}')$. Then the activities ρ' defined in (7.196) factor over connected components of G' and, for any connected G',*

$$0 \leq \rho'(\Gamma,' G')$$
$$\leq \mathrm{e}^{-c_1 L^{d-1} \beta E_s(\Gamma') - c_2 L \tilde{b} |G' \setminus \overline{\hat{D}'(\Gamma')}| + L^{d-1-\alpha} \beta B(\tilde{N}', V(\Gamma') \cap G') + C_3 L^d |G'|} \qquad (7.198)$$

for some positive constants c_1, c_2, C_3. For $\Gamma' = (C, h_y \equiv h)$, with $C \subset \hat{D}'$ connected,

$$\rho'(\Gamma', \underline{\Gamma}') \geq \mathrm{e}^{-L^{d-1-\alpha} \beta B(\tilde{N}', V(\Gamma')) - \mathrm{e}^{\mathrm{const.} \tilde{b}} |C|} \qquad (7.199)$$

Proof We will skip the cumbersome, but fairly straightforward, proofs of these estimates. □

Final tidying Just as in Section 7.3.2 we must make some final changes to the definition of the small and control fields and to the definition of the contours to recover the exact form of N'-bounded contour models. We will also take care of the entropy terms that were created in the estimates in Lemma 7.3.34.

The definition of the local small fields (7.112) and the control fields (7.113) remain unchanged. The non-local small fields will be left unaltered, i.e. we simply set $S'^{\pm}_{C'} \equiv \tilde{S}'^{\pm}_{C'}$. The centering has no effect on the contour measures, as the effect cancels with the partition functions (which are *not* invariant under this last part of the RG map), except for some boundary effects that can be easily dealt with as in Section 7.3.2. The final result is then the following:

Proposition 7.3.35 *Let $\mathbb{T}^{(N)} \equiv T_3 T_2 T_1 : \mathcal{S}_n(D(N)) \to \mathcal{S}_{n-1}(D(N'))$ with $T_1, T_2,$ and T_3 defined above; let N' and S' and ρ' be defined as above and let μ be an N-bounded contour measure at temperatures β and $\tilde{b}$ of level k. Then $\mu' \equiv \mathbb{T}\mu$ is an N'-bounded contour measure with temperatures $\beta' = L^{d-1-\alpha} \beta$ and $\tilde{b}' = L^{1-\alpha} \tilde{b}$ of level $k+1$, for suitably chosen $\alpha > 0$.*

Proof This is again tedious book-keeping and will be skipped. □

Remark 7.3.36 Let us briefly summarize where we stand now. Equation (7.142) provides a form of contour measures that remains invariant under renormalization. The specific form of the bounds on the activities is not so important, but they have three main features: the term $E_s(\Gamma)$ (in our case) weighs the renormalized configurations; the term $|G \setminus \overline{D(\Gamma)}|$ suppresses 'bad histories', i.e. contours that are images of 'unlikely' original configurations; and finally, the control field terms allow deviations from ground-states in exceptional regions; the probabilistic estimates must then ensure that such regions become less and less prominent.

Proof (of the main theorem) From the definition of the renormalized small fields and control fields it is clear that the probabilistic estimates carried out in Section 7.3.2 apply unaltered at small temperatures provided the hypothesis of Proposition 7.3.35 holds, i.e. if the RG program can be carried through. We will now show how these estimates can be used to prove Theorem 7.3.1. The main idea here is that contours are suppressed outside the union of all the bad regions in all hierarchies and that this latter set is, by the estimates on the control fields, very sparse. Moreover, the randomness essentially only produces local deformations that are very weakly correlated over larger distances, and thus finite-volume measures with plus and minus boundary conditions (whose existence follows from the FKG inequalities) will remain distinct.

Let us now assume that β is large enough, Σ small enough, and the parameters L, α, and η chosen such that the preceding results are all valid. We denote by $\mu_\Lambda \equiv \mu_{\Lambda,\beta}$ the finite-volume measure in Λ with plus boundary conditions.

A key point needed to prove Theorem 7.3.1 is that:

Lemma 7.3.37 *Under the assumptions of Theorem 7.3.1,*

$$\mathbb{P}\left[\lim_{M\uparrow\infty} \mu^+_{\mathcal{L}^M 0,\beta}(\sigma_0 = +1) > 1/2\right] > 0 \qquad (7.200)$$

Given Lemma 7.3.37, Theorem 7.3.1 follows from the monotonicity properties of the Gibbs measures and ergodicity as in Corollary 6.4.2. □

Proof (of Lemma 7.3.37) Let us introduce, for any contour $\Gamma \subset \mathcal{S}_M(D)$, the notation γ_0 for the unique weakly connected component of Γ whose interior contains the origin. If no such component exists, γ_0 is understood to be the empty set. Then

$$\mu_{\mathcal{L}^M 0}(\sigma_0 = -1) = \sum_{G \subset \mathcal{L}^M 0, G \ni 0, \vee G = \emptyset} \mu_{\mathcal{L}^M 0}(\text{int } \underline{\gamma_0} = G)\, \mu_{\mathcal{L}^M 0}(\sigma_0 = -1 | \text{int } \underline{\gamma_0} = G)$$

$$\leq \sum_{k=1}^{M} \sum_{\substack{G \subset \mathcal{L}^M 0, G \ni 0 \\ G \subset \mathcal{L}^k 0,\, G \not\subset \mathcal{L}^{k-1} 0}} \mu_{\mathcal{L}^M 0}(\text{int } \underline{\gamma_0} = G) \qquad (7.201)$$

The final estimate in (7.201) can be rewritten in the form

$$\mu_{\mathcal{L}^M 0}(\sigma_0 = -1) \leq \sum_{k=1}^{M} \alpha_M^{(k-1)} \qquad (7.202)$$

where
$$\alpha_M^{(k)} \equiv \mu_{\mathcal{L}^M 0}(\text{int } \underline{\gamma_0} \not\subset \mathcal{L}^k 0) \tag{7.203}$$

We must prove that $\alpha_M^{(k)}$ decays rapidly with k; a crude estimate on $s_M^{(k)}$ will then suffice. The estimate on $\alpha_M^{(k)}$ is the Peierls-type estimate we alluded to before. It will tell us that it is indeed unlikely that a connected component with large support encircles the origin. Of course, such an estimate has to be conditioned on the environments. The precise form is:

Lemma 7.3.38 *Let $0 \leq k \leq M - 1$ and let $F_{l,M} \subset \mathcal{A}$ denote the event*
$$F_{l,M} \equiv \left\{ d\left(D^{(l)}, 0\right) \leq \frac{L}{2} \right\} \tag{7.204}$$

Then there exists a constant $b > 0$ such that
$$\{\alpha_M^{(k)} \geq e^{-b\tilde{b}^{(k)}}\} \subset \bigcup_{l=k}^{M} F_{l,M} \tag{7.205}$$

The proof of this lemma will be postponed. Assuming Lemma 7.3.38, it is easy to prove Lemma 7.3.37.

Note first that the events $F_{l,M}$ are independent of M (recall that $D^{(k)}$ depends on the finite volume only near the boundary). Therefore,
$$\sum_{k=0}^{\infty} \mathbb{P}\left[\sup_{M \geq k} \alpha_M^{(k)} \geq e^{-b\tilde{b}^{(k)}}\right] \leq \sum_{k=0}^{\infty} \mathbb{P}[F_{k,\infty}] + \sum_{k=0}^{\infty} \mathbb{P}[F_{k,k}] \tag{7.206}$$

Moreover, the probabilities of the events $F_{l,M}$ satisfy
$$\mathbb{P}[F_{k,M}] \leq L^d \exp\left(-L^{\left(\frac{d-2}{2(d-1)} - \eta\right)k} \frac{\delta^2}{a\Sigma^{2 - \frac{d-2}{d-1}}}\right) \tag{7.207}$$

and are estimated as in Corollary 7.3.26. Since $\sum_{k=1}^{\infty} e^{-b\tilde{b}^{(k-1)}} \delta_k < C e^{-b\tilde{b}} \delta_1$,
$$\sum_{k=1}^{\infty} \sup_{M \geq k} \alpha_M^{(k-1)} < \infty, \quad \mathbb{P}\text{- a.s.} \tag{7.208}$$

and choosing β large enough, this quantity is in fact smaller than $\frac{1}{2}$ with positive probability. This proves Lemma 7.3.37. $\square$

Proof (of Lemma 7.3.38) For simplicity let us fix $\Lambda \equiv \mathcal{L}^M 0$ and let us write $\mu^{(k)} \equiv \mathbb{T}^k \mu_{\Lambda,\beta}$ for the renormalized measures. The key observation allowing the use of the RG in this estimate is that, if Γ is such that $\gamma_0(\Gamma) \not\subset \mathcal{L}^k 0$, then int $\mathbb{T}^k(\Gamma) \ni 0$ (simply because a connected component of such a size cannot have become 'small' in only $k - 1$ RG steps). But this implies that
$$\mu(\gamma_0 \not\subset \mathcal{L}^k 0) \leq \mu^{(k)}(\text{int } \underline{\Gamma} \ni 0) \tag{7.209}$$

To analyze the right-hand side of this bound, we decompose the event int $\underline{\Gamma} \ni 0$ according to the decomposition of contours into small and large parts: either 0 is contained in the interior of the support of Γ^l, or else it is in the interior of the support of Γ^s and *not* in that

7.3 The Bricmont–Kupiainen renormalization group

of Γ^l. That is,

$$\mu^{(k)}(\text{int }\underline{\Gamma} \ni 0) \leq \mu^{(k)}(\text{int }\underline{\Gamma}^l \ni 0) + \mu^{(k)}(\text{int }\underline{\Gamma}^s \ni 0, \text{ int }\underline{\Gamma}^l \not\ni 0) \tag{7.210}$$

If int $\underline{\Gamma}^l \ni 0$, then int $\underline{\mathbb{T}\Gamma} \ni 0$, which allows us to push the estimation of the first term in (7.210) into the next hierarchy; the second term concerns an event that is sufficiently 'local' to be estimated, as we will see. Iterating this procedure, we arrive at the bound

$$\mu(\gamma_0 \not\subset \mathcal{L}^k 0) \leq \sum_{l=k}^{M-1} \mu^{(l)}(\text{int }\underline{\Gamma}^s \ni 0, \text{ int }\underline{\Gamma}^l \not\ni 0) + \mu^{(M)}(\underline{\Gamma} \ni 0) \tag{7.211}$$

The last term in (7.211) concerns a single-site measure and will be very easy to estimate. To bound the other terms, we have to deal with the non-locality of the contour measures. To do so, we introduce the non-normalized measure

$$\nu(\Gamma) \equiv \frac{1}{Z} e^{-\beta(S, V(\Gamma))} \sum_{G \supset \Gamma} \rho(\Gamma, G) \mathbb{I}_{G \ni 0} \tag{7.212}$$

For all G contributing to ν (i.e. containing the origin) we write $G_0 \equiv G_0(G)$ for the connected component of G that contains the origin. We then define further

$$\nu_s(\Gamma) \equiv \frac{1}{Z} e^{-\beta(S, V(\Gamma))} \sum_{G \supset \Gamma} \rho(\Gamma, G) \mathbb{I}_{G \ni 0} \mathbb{I}_{G_0 \cap \underline{\Gamma}^l = \emptyset} \tag{7.213}$$

and

$$\nu_l(\Gamma) \equiv \frac{1}{Z} e^{-\beta(S, V(\Gamma))} \sum_{G \supset \Gamma} \rho(\Gamma, G) \mathbb{I}_{G \ni 0} \mathbb{I}_{G_0 \cap \underline{\Gamma}^l \neq \emptyset} \tag{7.214}$$

Of course, $\nu = \nu_s + \nu_l$. Let us further set

$$m_s \equiv \frac{1}{Z} \sum_{\Gamma} e^{-\beta(S, V(\Gamma))} \sum_{G \supset \Gamma} \rho(\Gamma, G) \mathbb{I}_{\text{int }\underline{\Gamma}^s \ni 0} \mathbb{I}_{\text{int }\underline{\Gamma}^l \not\ni 0} \mathbb{I}_{g_0 \cap \underline{\Gamma}^l = \emptyset} \tag{7.215}$$

and

$$m_l \equiv \frac{1}{Z} \sum_{\Gamma} e^{-\beta(S, V(\Gamma))} \sum_{G \supset \Gamma} \rho(\Gamma, G) \mathbb{I}_{\text{int }\underline{\Gamma}^s \ni 0} \mathbb{I}_{\text{int }\underline{\Gamma}^l \not\ni 0} \mathbb{I}_{g_0 \cap \underline{\Gamma}^l \neq \emptyset} \tag{7.216}$$

where $g_0 \equiv g_0(G, \Gamma)$ denotes the connected component of G that contains the maximal connected component of Γ_s whose interior contains the origin. (Note that, in general, $g_0 \neq G_0$.) The point here is that

$$\mu(\text{int }\underline{\Gamma}^s \ni 0, \text{ int }\underline{\Gamma}^l \not\ni 0) = m_s + m_l \tag{7.217}$$

We will shortly see that we can easily estimate m_s. On the other hand, the estimation of m_l can be pushed to the next RG level. Namely,

$$\sum_{\Gamma: T(\Gamma) = \Gamma'} \nu_l(\Gamma) \leq \nu'(\Gamma') \tag{7.218}$$

and

$$m_l \leq \nu'(\mathcal{S}) \tag{7.219}$$

To see why (7.218) holds, consider just the first two steps of the RG procedure. The point is that the G_0 contributing to ν_l, as they contain the support of a large component of Γ, are never summed over in the first RG step. In the second step (the blocking) they contribute to terms in which G' is such that $\mathcal{L}G' \supset G \ni 0$, and in particular $G' \ni 0$. Therefore

$$Z \sum_{\Gamma: T_2 T_1(\Gamma) = \tilde{\Gamma}'} \nu_l(\Gamma) \leq e^{-\beta'(\tilde{S}', V(\Gamma'))} \sum_{G' \supset \tilde{\Gamma}'} \rho'(\tilde{\Gamma}', G') \mathbb{1}_{G' \ni 0} \qquad (7.220)$$

In the third step, finally, the number of terms on the right can only be increased, while the constant produced by centering small fields cancels against the corresponding change of the partition function. This then yields (7.218).

Equation (7.219) is understood in much the same way. The set γ_0 is not summed away in the first step. But g_0 contains a small connected component γ_0 whose interior contains the origin. By the geometric smallness of these components, $\mathcal{L}^{-1}\Gamma_0 = \{0\}$ and so $\mathcal{L}^{-1}G_0 \ni 0$, implying (7.219).

Iterating these two relations, we get, in analogy to (7.211),

$$\nu^{(l+1)}(\mathbb{I}) \leq \sum_{j=l+1}^{M-1} \nu_s^{(j)}(\mathbb{I}) + \nu^{(M)}(\mathbb{I}) \qquad (7.221)$$

where the superscripts refer to the RG level. Combining all this, we get

$$\mu_{\mathcal{L}^M 0}(\gamma_0 \not\subset \mathcal{L}^k 0) \qquad (7.222)$$

$$\leq \sum_{l=k}^{M-1} \left[m_{s,\mathcal{L}^M 0}^{(l)} + \sum_{j=l+1}^{M-1} \nu_{s,\mathcal{L}^M 0}^{(j)}(\mathbb{I}) + \nu_{\mathcal{L}^M 0}^{(M)}(\mathbb{I}) \right] + \mu_{\mathcal{L}^M 0}^{(M)}(\Gamma \ni 0)$$

$$= \sum_{l=k}^{M-1} m_{s,\mathcal{L}^M 0}^{(l)} + \sum_{j=k+1}^{M-1} (j-k)\nu_{s,\mathcal{L}^M 0}^{(j)}(\mathbb{I}) + (M-k)\nu_{\mathcal{L}^M 0}^{(M)}(\mathbb{I}) + \mu_{\mathcal{L}^M 0}^{(M)}(\Gamma \ni 0)$$

All the terms appearing in this final bound can be estimated without recourse to further renormalization. The result is:

Lemma 7.3.39 *Let $F_{l,M} \subset \mathcal{A}$ be defined as in Lemma 7.3.38. Then there exists a positive constant $\bar{b} > 0$ such that*

$$\left\{ m_{s,\mathcal{L}^M 0}^{(l)} \geq e^{-\bar{b}\bar{b}^{(l)}} \right\} \subset F_{l,M},$$

$$\left\{ \nu_{s,\mathcal{L}^M 0}^{(l)}(1) \geq e^{-\bar{b}\bar{b}^{(l)}} \right\} \subset F_{l,M} \qquad (7.223)$$

$$\left\{ \nu_{\mathcal{L}^M 0}^{(M)}(1) \geq e^{-\bar{b}\bar{b}^{(M)}} \right\} \subset F_{M,M},$$

$$\left\{ \mu_{\mathcal{L}^M 0}^{(M)}(\Gamma \ni 0) \geq e^{-\bar{b}\bar{b}^{(M)}} \right\} \subset F_{M,M} \qquad (7.224)$$

Proof Relations (7.224) are easy to verify as they refer to systems with a single lattice site. The proof of the two relations (7.223) is similar. We explain only for the first one. We

7.3 The Bricmont–Kupiainen renormalization group

suppress the index l in our notation:

$$m_s = \frac{1}{Z} \sum_{\substack{\gamma_0 \text{ small} \\ \text{int}\,\gamma_0 \ni 0}} \sum_{\substack{G_0 \supset \gamma_0 \\ G_0 \text{ conn.}}} \sum_{\substack{\Gamma_0^s:\Gamma_0^s \subset G_0 \\ \gamma_0 \subset \Gamma_0^s}} \rho(\Gamma_0^s, G_0)$$

$$\times \sum_{G \cap \overline{G_0} = \emptyset} \sum_{\substack{\Gamma:\underline{\Gamma} \subset G \\ \text{int}\,\underline{\Gamma} \not\supset G_0}} \rho(\Gamma, G) e^{-\beta(S, V(\Gamma \cup \Gamma_0^s))} \quad (7.225)$$

Note that the second line almost reconstitutes a partition function outside the region G_0, except for the constraint on the support of Γ and the fact that the field term is not the correct one. This latter problem can be repaired by noting that

$$(S, V(\Gamma \cup \Gamma_0^s)) = (S, V(\Gamma) \setminus G_0) + \sum_{\pm} \sum_{\substack{C \subset V_\pm(\Gamma \cup \Gamma_0^s) \\ C \cap G_0 \neq \emptyset}} S_C^\pm \quad (7.226)$$

The second term on the right consists of a local term (i.e. involving only C consisting of a single site x) that depends only on Γ_0^s, and the non-local one, which as in the previous instances is very small, namely

$$\left| \sum_{\pm} \sum_{\substack{C \subset V_\pm(\Gamma \cup \Gamma_0^s),\ |C| \geq 2 \\ C \cap G_0 \neq \emptyset}} S_C^\pm \right| \leq const. |G_0| e^{-\tilde{b}} \quad (7.227)$$

Thus we get the upper bound

$$m_s \leq \sum_{\substack{\gamma_0 \text{ small} \\ \text{int}\,\gamma_0 \ni 0}} \sum_{\substack{G_0 \supset \gamma_0 \\ G_0 \text{ conn.}}} \sum_{\substack{\Gamma_0^s:\Gamma_0^s \subset G_0 \\ \gamma_0 \subset \Gamma_0^s}} \rho(\Gamma_0^s, G_0) e^{-\beta(S_{loc}, V(\Gamma_0^s) \cap G_0)} e^{const.|G_0| e^{-\tilde{b}}}$$

$$\times \frac{1}{Z} \sum_{G \cap \overline{G_0} = \emptyset} \sum_{\substack{\Gamma:\underline{\Gamma} \subset G \\ \text{int}\,\underline{\Gamma} \not\supset G_0}} \rho(\Gamma, G) e^{-\beta(S, V(\Gamma) \setminus G_0)} \quad (7.228)$$

The last line has the desired form. A slight problem is that the contours contributing to the denominator are not (in general) allowed to have empty support in G_0, as the support of any Γ must contain $D(\Gamma)$. However, G_0 is necessarily such that $\mathcal{D} \cap G_0 \subset \mathcal{D}$, since otherwise G_0 would have to contain support from large contours. Thus, for given G_0, we may bound the partition function from below by summing only over contours that within G_0 have $\sigma_x(\Gamma) \equiv +1$, and the support of those in G_0 is exactly given by $\mathcal{D} \cap G_0$. Treating the small-field term as above gives the lower bound on the partition function

$$Z \geq \prod_{i: \mathcal{D}_i \subset G_0} \rho(\mathcal{D}_i, \mathcal{D}_i) e^{-\beta \sum_{x \in G_0} S_x} e^{const.|G_0| e^{-\tilde{b}}}$$

$$\times \sum_{G \cap \overline{G_0} = \emptyset} \sum_{\substack{\Gamma:\underline{\Gamma} \subset G \\ \text{int}\,\underline{\Gamma} \not\supset G_0}} \rho(\Gamma, G) e^{-\beta(S, V(\Gamma) \setminus G_0)} \quad (7.229)$$

and so

$$m_s \leq \frac{1}{Z} \sum_{\substack{\gamma_0 \text{ small} \\ \text{int } \underline{\gamma_0} \ni 0}} \sum_{\substack{G_0 \supset \gamma_0 \\ G_0 \text{ conn.}}} \sum_{\substack{\Gamma_0^s : \Gamma_0^s \subset G_0 \\ \gamma_0 \subset \Gamma_0^s}} e^{2\,const.\,|G_0|e^{-\tilde{b}}} \qquad (7.230)$$

$$\times e^{-\beta(S_{\text{loc}}, V(\Gamma_0^s) \cap G_0) + \beta \sum_{x \in G_C} S_x \sigma_x} \frac{\rho(\Gamma_0^s, G_0)}{\prod_{i:\mathcal{D}_i \subset G_0} \rho(\mathcal{D}_i, \mathcal{D}_i)}$$

Here the ρ's appearing in the denominator are exactly those for which we have lower bounds. Note that for this reason we could not deal directly with expressions in which G_0 is allowed to contain large components of Γ. The estimation of the sums in (7.230) is now performed as in the absorption of small contours. Γ_0^s with non-constant spins give essentially no contribution, and due to the separatedness of the components $\mathcal{D}_i$, and the smallness of the total control field on one such component, the main contribution comes from the term where Γ_0^s has support in only one component $\mathcal{D}_i$. If there is such a component that surrounds 0, this could give a contribution of order one. But on $F_{l,M}$ this is excluded, so that G_0 cannot be contained in $\mathcal{D}$ and therefore

$$m_{s, \mathcal{L}^M 0}^{(l)} \leq \text{const.}\, e^{-b\tilde{b}^{(l)}} \qquad (7.231)$$

as claimed. □

From Lemma 7.3.39 and the bound (7.222), Lemma 7.3.38 follows immediately. □

Part III

Disordered systems: mean-field models

8

Disordered mean-field models

> Les mathématiques ont un triple but. Elles doivent fournir un instrument pour l'étude de la nature. Mais ce n'est pas tout: elles ont un but philosophique et, j'ose le dire, un but esthétique.[1]
>
> *Henri Poincaré, La valeur de la science.*

In the previous chapters we have seen that with considerable work it is possible to study some simple aspects of random perturbations of the Ising model. On the other hand, models that are genuinely random and promise to bring some new features to light are at the moment not seriously accessible to rigorous analysis. In such situations it it natural to turn to simplified models, and a natural reflex in statistical mechanics is to turn to the *mean-field approximation*. In the case of disordered systems, this has turned out, quite surprisingly, to open a Pandora's box.

The common feature of mean-field models is that the spatial structure of the lattice $\mathbb{Z}^d$ is abandoned in favour of a simpler setting, where sites are indexed by the natural numbers and all spins are supposed to interact with each other, irrespective of their distance. The prototype of all such models is the *Curie–Weiss model*, that we studied in Section 3.5.

We now turn to the question of what should be the natural class of *disordered* mean-field models to study? This question requires some thought, as there are at least two natural classes that propose themselves. Recall that we could represent the Hamiltonian of the Curie–Weiss model through equation (3.30) as a function of the total *empirical magnetization*. This fact was instrumental for the treatment of the model with large deviation methods and is the reason why it is fairly easy to solve. One may thus consider as proper mean-field models only models that share this property, i.e. whose Hamiltonian is a function of macroscopic variables.

One can then seek to introduce macroscopic variables that depend on some quenched disorder variables. The simplest natural example of this kind would be the *random-field Curie–Weiss model*, which we introduced at the end of Section 6.3. Its Hamiltonian can be represented in the form

$$H_N[\omega](\sigma) - \frac{N}{2}(m_N(\sigma))^2 - N\delta n_N[\omega](\sigma)$$

---

[1] Approximately: Mathematics has a threefold purpose. It must provide an instrument for the study of nature. But this is not all: it has a philosophical purpose, and, I daresay, an aesthetic purpose.

where
$$n_N[\omega](\sigma) \equiv \frac{1}{N} \sum_{i=1}^{N} h_i[\omega] \sigma_i \tag{8.1}$$

is a second, random, macroscopic variable. We will discuss more complicated and more interesting generalizations along this line in the context of the so-called *Hopfield models* in Chapter 12.

While the mean-field analogue of the random-field model fits naturally in this class, models with random pair interactions are less naturally incorporated. In fact, the naive analog of the Curie–Weiss Hamiltonian with random couplings would be

$$H_N[\omega](\sigma) = -\frac{1}{2N} \sum_{1 \leq i,j \leq N} J_{ij}[\omega] \sigma_i \sigma_j \tag{8.2}$$

for, say, J_{ij} some family of i.i.d. random variables. This Hamiltonian cannot be written as a function of some macroscopic variables. The properties of this model depend strongly on the choice of the random variables J_{ij}. The main interest in this model concerns the case when the random couplings have mean zero. In this case, we will see shortly that the normalization factor, N^{-1}, is actually inappropriate and must be replaced by $N^{-1/2}$, to obtain an interesting model. Namely, we certainly want the free energy to be an extensive quantity, i.e. to be of order N. This means that, for typical realizations of the disorder, there must be at least some spin configurations σ for which $H_N(\sigma) \sim CN$, for some $C > 0$. Thus, we must estimate $\mathbb{P}[\max_\sigma H_N(\sigma) \geq CN]$. But

$$\mathbb{P}\Big[\max_\sigma H_N(\sigma) \geq CN\Big] \leq \sum_{\sigma \in \mathcal{S}_N} \mathbb{P}[H_N(\sigma) \geq CN] \tag{8.3}$$

$$= \sum_{\sigma \in \mathcal{S}_N} \inf_{t \geq 0} e^{-tCN} \mathbb{E} e^{t \frac{1}{2N} \sum_{i,j \in \Lambda_N \times \Lambda_N} J_{ij}[\omega] \sigma_i \sigma_j}$$

$$= \sum_{\sigma \in \mathcal{S}_N} \inf_{t \geq 0} e^{-tCN} \prod_{i,j \in \Lambda_N \times \Lambda_N} \mathbb{E} e^{t \frac{1}{2N} J_{ij}[\omega] \sigma_i \sigma_j}$$

where we assumed that the exponential moments of J_{ij} exist. A standard estimate then shows that, for some constant c, $\mathbb{E} e^{t \frac{1}{2N} J_{ij}[\omega] \sigma_i \sigma_j} \leq e^{c \frac{t^2}{2N^2}}$, and so

$$\mathbb{P}\Big[\max_\sigma H_N(\sigma) \geq CN\Big] \leq 2^N \inf_{t \geq 0} e^{-tCN} e^{ct^2/2} \leq 2^N e^{-\frac{C^2 N^2}{2c}} \tag{8.4}$$

which tends to zero with N. Thus, our Hamiltonian is never of order N, but at best of order $\sqrt{N}$. The proper Hamiltonian for what is called the *Sherrington–Kirkpatrick model* (or SK model) is thus

$$H_N^{SK} \equiv -\frac{1}{\sqrt{2N}} \sum_{i,j \in \Lambda_N \times \Lambda_N} J_{ij} \sigma_i \sigma_j \tag{8.5}$$

where the random variables $J_{ij} = J_{ji}$ are i.i.d. for $i \leq j$ with mean zero (or at most $J_0 N^{-1/2}$) and variance normalized to one for $i \neq j$ and to two for $i = j$.[2] In its original, and mostly considered, form, the distribution is moreover taken to be Gaussian. Note that $\sum_{ij} |N^{-1/2} J_{ij} \sigma_i \sigma_j| \sim N^{3/2}$, and that competing signs play a major role.

---

[2] This choice is for notational convenience. Of course the self-couplings J_{ii} have no physical relevance whatsoever.

This model was introduced by Sherrington and Kirkpatrick in 1976 [221] as an attempt to furnish a simple, solvable mean-field model for the then newly discovered class of materials called *spin-glasses*. However, it turned out that the innocent looking modifications made to create a spin-glass model that looks similar to the Curie–Weiss model had thoroughly destroyed the simplifying properties that made the latter so easily solvable, and that a model with an enormously complex structure had been invented. Using highly innovative ideas based on ad hoc mathematical structures, Parisi (see [179]) produced in the mid-1980s a heuristic framework that explained the properties of the model. Only very recently, these predictions have to some extent been rigorously justified through work of F. Guerra [133] and M. Talagrand [240], which we will explain later.

It will be useful to introduce a different point of view on the SK model, which allows us to put it in a wider context. This point of view consists of regarding the Hamiltonian (8.5) as a *Gaussian random process*[3] indexed by the set $\mathcal{S}_N$, i.e. by the N-dimensional hypercube. We will restrict our attention to the case when the J_{ij} are centered Gaussian random variables. In this case, $H_N(\sigma)$ is in fact a centered Gaussian random process which is fully characterized by its covariance function

$$\text{cov}(H_N(\sigma), H_N(\sigma')) = \frac{1}{2N} \sum_{1 \leq i,j,l,k \leq N} \mathbb{E} J_{ij} J_{kl} \sigma_i \sigma_j \sigma'_k \sigma'_l \qquad (8.6)$$

$$= \frac{1}{N} \sum_{1 \leq i,j \leq N} \sigma_i \sigma'_i \sigma_j \sigma'_j = N R_N(\sigma, \sigma')^2$$

where $R_N(\sigma, \sigma') \equiv N^{-1} \sum_{i=1}^{N} \sigma_i \sigma'_i$ is usually called the *overlap* between the two configurations σ and σ'. It is useful to recall that the overlap is closely related to the *Hamming distance* $d_{\text{HAM}}(\sigma, \sigma') \equiv \#(i \leq N : \sigma_i \neq \sigma'_i)$, namely $R_N(\sigma, \sigma') = (1 - 2N^{-1} d_{\text{HAM}}(\sigma, \sigma'))$.

Seen this way, the SK model is a particular example of a class of models whose Hamiltonians are centered Gaussian random process on the hypercube with covariance depending only on $R_N(\sigma, \sigma')$,

$$\text{cov}(H_N(\sigma), H_N(\sigma')) = N \xi(R_N(\sigma, \sigma')) \qquad (8.7)$$

normalized such that $\xi(1) = 1$. A class of examples considered in the literature are the so-called p-spin SK models, which are obtained by choosing $\xi(x) = |x|^p$. They enjoy the property that they may be represented in a form similar to the SK Hamiltonian, except that the two-spin interaction must be replaced by a p-spin one:

$$H_N^{p-SK}(\sigma) = \frac{-1}{\sqrt{N^{p-1}}} \sum_{1 \leq i_1, \ldots, i_p \leq N} J_{i_1 \cdots i_p} \sigma_{i_1} \cdots \sigma_{i_p} \qquad (8.8)$$

with $J_{i_1,\ldots,i_p}$ i.i.d. standard normal random variables.[4] As we will see later, the difficulties in studying the statistical mechanics of these models is closely linked to the understanding of

---

[3] The choice of Gaussian couplings and hence Gaussian processes may appear too restrictive and, from a physical point of view, poorly motivated. It turns out, however, that the Gaussian nature of the processes considered is not really important, and that a large class of models have the same asymptotics as the corresponding Gaussian ones (at least on the level of the free energy) [72]. It is, however, a good idea to start with the simplest situation.

[4] Sometimes the terms where some indices in the sum (8.8) coincide are omitted. This can be included in our framework by making ξ explicitly N-dependent in a suitable way. Although this has an effect, for instance, on the fluctuations of the free energy (see [56]), for our present purposes this is not relevant and we choose the form with the simplest expression for the covariance.

the extremal properties of the corresponding random processes. While Gaussian processes have been heavily analyzed in the mathematical literature (see, e.g., [1, 167]), the known results were not enough to recover the heuristic results obtained in the physics literature. This is one reason why this particular field of mean-field spin-glass models has considerable intrinsic interest for mathematics.

The class of models we have just introduced depends on the particular choice of the Hamming distance as the metric on the hypercube. It is only natural to think of further classes of models that can be obtained by other choices of metric. A particularly important alternative choice is the *lexicographic distance*: Given two sequences σ and τ, we look at the first value of the index i for which the sequences differ, i.e. $\sigma_i \neq \tau_i$. Naturally, if this value is N, then $\sigma = \tau$ and thus their distance is zero, while, if $i = 1$, then we consider them maximally apart. The quantity

$$d_N(\sigma, \tau) \equiv N^{-1} \left(\min(i : \sigma_i \neq \tau_i) - 1 \right) \tag{8.9}$$

is thus analogous to the overlap $R_N(\sigma, \tau)$. The corresponding Gaussian processes are then characterized by covariances given by

$$\text{cov}(H_N(\sigma), H_N(\tau)) = N A(d_N(\sigma, \tau)) \tag{8.10}$$

where A can be chosen to be any non-decreasing function on $[0, 1]$, and can be thought of as a probability distribution function. The choice of the lexicographic distance entails some peculiar features. First, this distance is an *ultrametric*, i.e. for any three configurations σ, τ, ρ,

$$d_N(\sigma, \tau) = \min \left(d_N(\sigma, \rho), d_N(\tau, \rho) \right) \tag{8.11}$$

This fact will be seen to have remarkable consequences, that make the Gibbs measures of these models fully analyzable, even though they will show as much complexity as those of the SK models. Moreover, a clever comparison between the two types of processes is instrumental in the analysis of the SK models themselves, as will be explained later. I will therefore devote a considerable amount of attention to the analysis of these models before returning to the study of the SK models.

Finally, let us mention that there is no need to stop here. One may invent even larger classes of models, with covariances depending, e.g., on several 'distances'. An interesting class of examples of such a type was investigated by Bolthausen and Kistler [32].

Let us conclude this introductory chapter with some comments on the infinite-volume limit in mean-field models. The fact that the finite-volume Hamiltonians in mean-field models generally depend on N in an explicit parametric way shows that they cannot be considered as restrictions of some infinite-volume Hamiltonian. Similarly, the finite-volume Gibbs measures will not be conditional distributions of the infinite-volume measures, i.e. the infinite-volume Gibbs states cannot be defined as solutions of the DLR equations. Thus, in mean-field models, we will define infinite-volume Gibbs states as limits of sequences of finite-volume Gibbs states. The proper characterization and description of infinite-volume Gibbs states, and in particular the choice of an appropriate topology, in the case of disordered mean-field models will be, as we shall see, a non-trivial and interesting problem in itself, and will take considerable space in the following chapters.

9

The random energy model

> D'ailleurs, une science uniquement faite en vue des applications est impossible; les vérités ne sont fécondés que si elles sont enchaînées les unes aux autres. Si l'on s'attache seulement à celles dont on attend un résultat immédiat, les anneaux intermédiaires manqueront, et il n'y aura plus de chaîne.[1]
>
> Henri Poincaré, *La valeur de la science.*

Both classes of Gaussian models we have just introduced have, fortunately, an extreme common member. It corresponds to the case when the Gaussian process is just an i.i.d. field. The corresponding model is known as the *random energy model* or *REM*.

The REM was introduced by Derrida [85, 86] in 1980 and can be considered as the ultimate toy model of a disordered system. Little is left of the structure of interacting spins, but we will still be able to gain a lot of insight into the peculiarities of disordered systems by studying this simple system. Early rigorous results on the REM were obtained in [95, 97, 113, 194].

The REM is a model with state space $\mathcal{S}_N = \{-1, +1\}^N$. The Hamiltonian is given by

$$H_N(\sigma) = -\sqrt{N} X_\sigma \qquad (9.1)$$

where $X_\sigma, \sigma \in \mathcal{S}_N$, are 2^N i.i.d. standard normal random variables.

9.1 Ground-state energy and free energy

The first and as we will see crucial information we require concerns the value of the maximum of the variables X_σ, i.e. the ground-state energy. For i.i.d. random variables, this is of course not very hard.

Lemma 9.1.1 *The family of random variables introduced above satisfies*

$$\lim_{N \uparrow \infty} \max_{\sigma \in \mathcal{S}_N} N^{-1/2} X_\sigma = \sqrt{2 \ln 2} \qquad (9.2)$$

both almost surely and in mean.

---

[1] Approximately: Moreover, a science made only in view of applications is impossible; the truths are fertilized only if they are linked with one another. If one cares only for those from which one expects an immediate result, the intermediary links will be missing, and there is no longer a chain.

Proof Since everything is independent,

$$\mathbb{P}\left[\max_{\sigma \in \mathcal{S}_N} X_\sigma \leq u\right] = \left(1 - \frac{1}{\sqrt{2\pi}} \int_u^\infty e^{-x^2/2} dx\right)^{2^N} \quad (9.3)$$

and we just need to know how to estimate the integral appearing here. This is something we should get used to quickly, as it will occur all over the place. It will always be done using the fact that, for $u > 0$,

$$\frac{1}{u} e^{-u^2/2}(1 - 2u^{-2}) \leq \int_u^\infty e^{-x^2/2} dx \leq \frac{1}{u} e^{-u^2/2} \quad (9.4)$$

Exercise: Prove these two bounds!

We see that for our probability to converge neither to zero nor to one, u must be chosen in such a way that the integral is of order 2^{-N}. With the help of the bounds (9.4), one can show with a little computation that, if we define $u_N(x)$ by

$$\frac{2^N}{\sqrt{2\pi}} \int_{u_N(x)}^\infty e^{-z^2/2} dz = e^{-x} \quad (9.5)$$

then (for $x > -\ln N / \ln 2$)

$$u_N(x) = \sqrt{2N \ln 2} + \frac{x}{\sqrt{2N \ln 2}} - \frac{\ln(N \ln 2) + \ln 4\pi}{2\sqrt{2N \ln 2}} + o(1/\sqrt{N}) \quad (9.6)$$

Thus

$$\mathbb{P}\left[\max_{\sigma \in \mathcal{S}_N} X_\sigma \leq u_N(x)\right] = (1 - 2^{-N} e^{-x})^{2^N} \to e^{-e^{-x}} \quad (9.7)$$

In other words, the random variable $u_N^{-1}(\max_{\sigma \in \mathcal{S}_N} X_\sigma)$ converges in distribution to a random variable with double-exponential distribution (this is the most classic result of *extreme value statistics*, see [162]). The assertion of the lemma is now a simple corollary of this fact. □

Next we turn to the analysis of the partition function. In this model, the partition function is just the sum of i.i.d. random variables, i.e.

$$Z_{\beta,N} \equiv 2^{-N} \sum_{\sigma \in \mathcal{S}_N} e^{\beta \sqrt{N} X_\sigma} \quad (9.8)$$

It is convenient to introduce the quantity

$$\Phi_{\beta,N} \equiv \frac{1}{N} \ln Z_{\beta,N} \quad (9.9)$$

which is related to the specific free energy via $f_{\beta,N} \equiv -\beta^{-1} \Phi_{\beta,N}$, and then to try to find its limit as $N \uparrow \infty$.[2] In our simple model we expect of course to be able to compute this limit exactly. A first guess would be that a *law of large numbers* might hold, implying that $Z_{\beta,N} \sim \mathbb{E} Z_{\beta,N}$, and hence

$$\lim_{N \uparrow \infty} \Phi_{\beta,N} = \lim_{N \uparrow \infty} \frac{1}{N} \ln \mathbb{E} Z_{\beta,N} = \frac{\beta^2}{2}, \text{ a.s.} \quad (9.10)$$

---

[2] The problem of the existence of such limits in general will be discussed later.

9.1 Ground-state energy and free energy

It turns out that this is indeed true, but only for small enough values of β, and that there is a critical value β_c associated with a breakdown of the law of large numbers. The analysis of this problem will allow us to compute the free energy exactly.

Theorem 9.1.2 *In the REM,*

$$\lim_{N\uparrow\infty} \mathbb{E}\Phi_{\beta,N} = \begin{cases} \frac{\beta^2}{2}, & \text{for } \beta \leq \beta_c \\ \frac{\beta_c^2}{2} + (\beta - \beta_c)\beta_c, & \text{for } \beta \geq \beta_c \end{cases} \qquad (9.11)$$

where $\beta_c = \sqrt{2\ln 2}$.

Proof We will not give the most efficient proof of this result, but one that introduces useful ideas that can be applied to other models. It uses what we call the method of truncated second moments, which was introduced in the context of spin-glasses by M. Talagrand [231, 233, 236].

We will first derive an upper bound for $\mathbb{E}\Phi_{\beta,N}$. Note first that by Jensen's inequality, $\mathbb{E}\ln Z \leq \ln \mathbb{E}Z$, and thus

$$\mathbb{E}\Phi_{\beta,N} \leq \frac{\beta^2}{2} \qquad (9.12)$$

On the other hand we have that

$$\mathbb{E}\frac{d}{d\beta}\Phi_{\beta,N} = N^{-1/2}\mathbb{E}\frac{\mathbb{E}_\sigma X_\sigma e^{\beta\sqrt{N}X_\sigma}}{Z_{\beta,N}} \qquad (9.13)$$

$$\leq N^{-1/2}\mathbb{E}\max_{\sigma\in\mathcal{S}_N} X_\sigma \leq \beta\sqrt{2\ln 2}(1 + C/N)$$

for some constant C. Combining (9.12) and (9.13), we deduce that

$$\mathbb{E}\Phi_{\beta,N} \leq \inf_{\beta_0 \geq 0} \begin{cases} \frac{\beta^2}{2}, & \text{for } \beta \leq \beta_0 \\ \frac{\beta_0^2}{2} + (\beta - \beta_0)\sqrt{2\ln 2}(1 + C/N), & \text{for } \beta \geq \beta_0 \end{cases} \qquad (9.14)$$

It is easy to see that the infimum is realized (ignore the C/N correction) for $\beta_0 = \sqrt{2\ln 2}$. This shows that the right-hand side of (9.11) is an upper bound.

It remains to show the corresponding lower bound. Note that, since $\frac{d^2}{d\beta^2}\Phi_{\beta,N} \geq 0$, the slope of $\Phi_{\beta,N}$ is non-decreasing, so that the theorem will be proven if we can show that $\Phi_{\beta,N} \to \beta^2/2$ for all $\beta < \sqrt{2\ln 2}$, i.e. that the law of large numbers holds up to this value of β. A natural idea to prove this is to estimate the variance of the partition function.[3] Naively, one would compute

$$\mathbb{E}Z_{\beta,N}^2 = \mathbb{E}_\sigma \mathbb{E}_{\sigma'}\mathbb{E}e^{\beta\sqrt{N}(X_\sigma + X_{\sigma'})}$$

$$= 2^{-2N}\left(\sum_{\sigma\neq\sigma'} e^{N\beta^2} + \sum_\sigma e^{2N\beta^2}\right) \qquad (9.15)$$

$$= e^{N\beta^2}\left[(1 - 2^{-N}) + 2^{-N}e^{N\beta^2}\right]$$

---

[3] This idea can be traced to Aizenman, Lebowitz, and Ruelle [5], and, later, to Comets and Neveu [81], who used it in the proofs of a central limit theorem for the free energy in the SK model.

where all we used is that, for $\sigma \neq \sigma'$, X_σ and $X_{\sigma'}$ are independent. The second term in the square brackets is exponentially small if and only if $\beta^2 < \ln 2$. For such values of β we have that

$$\mathbb{P}\left[\left|\ln \frac{Z_{\beta,N}}{\mathbb{E}Z_{\beta,N}}\right| > \epsilon N\right] = \mathbb{P}\left[\frac{Z_{\beta,N}}{\mathbb{E}Z_{\beta,N}} < e^{-\epsilon N} \text{ or } \frac{Z_{\beta,N}}{\mathbb{E}Z_{\beta,N}} > e^{\epsilon N}\right]$$

$$\leq \mathbb{P}\left[\left(\frac{Z_{\beta,N}}{\mathbb{E}Z_{\beta,N}} - 1\right)^2 > (1 - e^{-\epsilon N})^2\right]$$

$$\leq \frac{\mathbb{E}Z_{\beta,N}^2/(\mathbb{E}Z_{\beta,N})^2 - 1}{(1 - e^{-\epsilon N})^2}$$

$$\leq \frac{2^{-N} + 2^{-N}e^{N\beta^2}}{(1 - e^{-\epsilon N})^2} \qquad (9.16)$$

which is more than enough to get (9.10). But of course this does not correspond to the critical value of β claimed in the proposition! Some reflection shows that the point here is that when computing $\mathbb{E}e^{\beta\sqrt{N}2X_\sigma}$, the dominant contribution comes from the part of the distribution of X_σ where $X_\sigma \sim 2\beta\sqrt{N}$, whereas in the evaluation of $\mathbb{E}Z_{\beta,N}$ the values of X_σ where $X_\sigma \sim \beta\sqrt{N}$ give the dominant contribution. Thus one is led to realize that instead of the second moment of Z one should compute a truncated version of it, namely, for $c \geq 0$,

$$\tilde{Z}_{\beta,N}(c) \equiv \mathbb{E}_\sigma e^{\beta\sqrt{N}X_\sigma} \mathbb{I}_{X_\sigma < c\sqrt{N}} \qquad (9.17)$$

An elementary computation using (9.4) shows that, if $c > \beta$, then

$$\mathbb{E}\tilde{Z}_{\beta,N}(c) = e^{\frac{\beta^2 N}{2}} \left(1 - \frac{e^{-N\beta^2/2}}{\sqrt{2\pi N}(c-\beta)}(1 + O(1/N))\right) \qquad (9.18)$$

so that such a truncation essentially does not influence the mean partition function. Now compute the mean of the square of the truncated partition function (neglecting irrelevant $O(1/N)$ errors):

$$\mathbb{E}\tilde{Z}_{\beta,N}^2(c) = (1 - 2^{-N})[\mathbb{E}\tilde{Z}_{\beta,N}(c)]^2 + 2^{-N}\mathbb{E}e^{\beta\sqrt{N}2X_\sigma} \mathbb{I}_{X_\sigma < c\sqrt{N}}) \qquad (9.19)$$

where

$$\mathbb{E}e^{2\beta\sqrt{N}X_\sigma} \mathbb{I}_{X_\sigma < c\sqrt{N}} = \begin{cases} e^{2\beta^2 N}, & \text{if } 2\beta < c \\ 2^{-N}\frac{e^{2c\beta N - \frac{c^2 N}{2}}}{(2\beta-c)\sqrt{2\pi N}}, & \text{otherwise} \end{cases} \qquad (9.20)$$

Combined with (9.18) this implies that, for $c/2 < \beta < c$,

$$\frac{2^{-N}\mathbb{E}e^{2\beta\sqrt{N}X_\sigma} \mathbb{I}_{X_\sigma < c\sqrt{N}}}{(\mathbb{E}\tilde{Z}_{N,\beta})^2} = \frac{e^{-N(c-\beta)^2 - N(2\ln 2 - c^2)/2}}{(2\beta - c)\sqrt{N}} \qquad (9.21)$$

Therefore, for all $c < \sqrt{2\ln 2}$, and all $\beta < c$,

$$\mathbb{E}\left[\frac{\tilde{Z}_{\beta,N}(c) - \mathbb{E}\tilde{Z}_{\beta,N}(c)}{\mathbb{E}\tilde{Z}_{\beta,N}(c)}\right]^2 \leq e^{-Ng(c,\beta)} \qquad (9.22)$$

with $g(c, \beta) > 0$. Thus Chebyshev's inequality implies that

$$\mathbb{P}[|\tilde{Z}_{\beta,N}(c) - \mathbb{E}\tilde{Z}_{\beta,N}(c)| > \delta \mathbb{E}\tilde{Z}_{\beta,N}(c)] \leq \delta^{-2} e^{-Ng(c,\beta)} \tag{9.23}$$

and so, in particular,

$$\lim_{N \uparrow \infty} \frac{1}{N} \mathbb{E} \ln \tilde{Z}_{\beta,N}(c) = \lim_{N \uparrow \infty} \frac{1}{N} \ln \mathbb{E} \tilde{Z}_{\beta,N}(c) \tag{9.24}$$

for all $\beta < c < \sqrt{2 \ln 2} = \beta_c$. But this implies that for all $\beta < \beta_c$, we can chose c such that

$$\lim_{N \uparrow \infty} \frac{1}{N} \ln \mathbb{E} Z_{\beta,N} \geq \lim_{N \uparrow \infty} \frac{1}{N} \ln \mathbb{E} \tilde{Z}_{\beta,N}(c) = \frac{\beta^2}{2} \tag{9.25}$$

This proves the theorem. □

9.2 Fluctuations and limit theorems

In the previous section we went to some length to compute the limit of the free energy. However, computing the free energy is not quite enough to get a full understanding of a model, and in particular the Gibbs states. The limit of the free energy has been seen to be a non-random quantity. A question of central importance is to understand how and on what level the randomness shows up in the corrections to the limiting behaviour. We will see later that the knowledge gained here is sufficient to derive detailed information about the Gibbs measures. The results of this section were obtained with I. Kurkova and M. Löwe in [56]. For earlier results see also [97, 113, 194].

Theorem 9.2.1 *The partition function of the REM has the following fluctuations:*

(i) *If $\beta < \sqrt{\ln 2/2}$, then*

$$e^{\frac{N}{2}(\ln 2 - \beta^2)} \ln \frac{Z_{\beta,N}}{\mathbb{E} Z_{\beta,N}} \xrightarrow{\mathcal{D}} \mathcal{N}(0, 1). \tag{9.26}$$

(ii) *If $\beta = \sqrt{\ln 2/2}$, then*

$$e^{\frac{N}{2}(\ln 2 - \beta^2)} \ln \frac{Z_{\beta,N}}{\mathbb{E} Z_{\beta,N}} \xrightarrow{\mathcal{D}} \mathcal{N}(0, 1/2) \tag{9.27}$$

(iii) *Let $\alpha \equiv \beta/\sqrt{2 \ln 2}$. If $\sqrt{\ln 2/2} < \beta < \sqrt{2 \ln 2}$, then*

$$e^{\frac{N}{2}(\sqrt{2 \ln 2}-\beta)^2 + \frac{\alpha}{2}[\ln(N \ln 2) + \ln 4\pi]} \ln \frac{Z_{\beta,N}}{\mathbb{E} Z_{\beta,N}}$$

$$\xrightarrow{\mathcal{D}} \int_{-\infty}^{\infty} e^{\alpha z} (\mathcal{P}(dz) - e^{-z} dz), \tag{9.28}$$

where $\mathcal{P}$ denotes the Poisson point process[4] on $\mathbb{R}$ with intensity measure $e^{-x} dx$.

(iv) *If $\beta = \sqrt{2 \ln 2}$, then*

$$e^{\frac{1}{2}[\ln(N \ln 2) + \ln 4\pi]} \left(\frac{Z_{\beta,N}}{\mathbb{E} Z_{\beta,N}} - \frac{1}{2} + \frac{\ln(N \ln 2) + \ln 4\pi}{4\sqrt{\pi N \ln 2}} \right)$$

$$\xrightarrow{\mathcal{D}} \int_{-\infty}^{0} e^z (\mathcal{P}(dz) - e^{-z} dz) + \int_{0}^{\infty} e^z \mathcal{P}(dz) \tag{9.29}$$

---

[4] For a thorough exposition on point processes and their connection to extreme value theory, see in particular [210].

(v) If $\beta > \sqrt{2\ln 2}$, then

$$e^{-N[\beta\sqrt{2\ln 2}-\ln 2]+\frac{\alpha}{2}[\ln(N\ln 2)+\ln 4\pi]}Z_{\beta,N} \xrightarrow{\mathcal{D}} \int_{-\infty}^{\infty} e^{\alpha z}\mathcal{P}(dz) \quad (9.30)$$

and

$$N(\Phi_{\beta,N} - \mathbb{E}\Phi_{\beta,N}) \xrightarrow{\mathcal{D}} \ln\int_{-\infty}^{\infty} e^{\alpha z}\mathcal{P}(dz) - \mathbb{E}\ln\int_{-\infty}^{\infty} e^{\alpha z}\mathcal{P}(dz). \quad (9.31)$$

Remark 9.2.2 Note that expressions like $\int_{-\infty}^{0} e^{z}(\mathcal{P}(dz) - e^{-z}dz)$ are always understood as $\lim_{y\downarrow-\infty}\int_{y}^{0} e^{z}(\mathcal{P}(dz) - e^{-z}dz)$. We will see that all the functionals of the Poisson point process that appear are almost surely finite random variables. Theorem 9.2.1 can be extended to non-Gaussian random variables. This has been done by Ben Arous, Bogachev, and Molchanov [11] for a large class of distributions with super-exponential tails. It also has interesting applications well beyond spin-glass theory.

Proof We will prove these results step by step. We first prove (i). Since the partition function is a sum of i.i.d. random variables, our first guess would be to look for a central limit theorem, and in fact in this parameter regime the result follows from the standard CLT for triangular arrays. Let us first write

$$\ln\frac{Z_{\beta,N}}{\mathbb{E}Z_{\beta,N}} = \ln\left(1 + \frac{Z_{\beta,N} - \mathbb{E}Z_{\beta,N}}{\mathbb{E}Z_{\beta,N}}\right) \quad (9.32)$$

We will show that the second term in the logarithm properly normalized will converge to a normal random variable. To see this, write

$$\frac{Z_{\beta,N} - \mathbb{E}Z_{\beta,N}}{\mathbb{E}Z_{\beta,N}} = 2^{-N}\sum_{\sigma\in\mathcal{S}_N}\left(e^{\beta\sqrt{N}X_\sigma - N\beta^2/2} - 1\right) \quad (9.33)$$

If we set

$$\mathcal{Y}_N(\sigma) \equiv \frac{e^{\beta\sqrt{N}X_\sigma - N\beta^2/2} - 1}{e^{N\beta^2/2}\sqrt{1 - e^{-N\beta^2}}} \quad (9.34)$$

then $\mathcal{Y}_N(\sigma)$ has mean zero and variance one, and

$$\frac{Z_{\beta,N} - \mathbb{E}Z_{\beta,N}}{\mathbb{E}Z_{\beta,N}} = e^{-\frac{N}{2}(\ln 2 - \beta^2)}\sqrt{1 - e^{-N\beta^2}}\frac{1}{2^{N/2}}\sum_{\sigma\in\mathcal{S}_N}\mathcal{Y}_N(\sigma) \quad (9.35)$$

By the CLT for triangular arrays (see [222]), it follows readily that

$$\frac{1}{2^{N/2}}\sum_{\sigma\in\mathcal{S}_N}\mathcal{Y}_N(\sigma) \xrightarrow{\mathcal{D}} \mathcal{N}(0,1) \quad (9.36)$$

if the Lindeberg condition holds, that is, if for any $\epsilon > 0$,

$$\lim_{N\uparrow\infty}\mathbb{E}\mathcal{Y}_N^2(\sigma)\mathbb{I}_{\{|\mathcal{Y}_N(\sigma)|\geq \epsilon 2^{N/2}\}} = 0 \quad (9.37)$$

But

$$\mathbb{E}\mathcal{Y}_N^2(\sigma)\mathbb{I}_{\{|\mathcal{Y}_N(\sigma)|\geq \epsilon 2^{N/2}\}}$$

$$= \frac{e^{-2N\beta^2}}{\sqrt{2\pi}(1-e^{-N\beta^2})} \int_{\sqrt{N}\left(\frac{\ln 2}{2\beta}+\beta\right)+\frac{\ln \epsilon}{\sqrt{N}\beta}+o\left(\frac{1}{\sqrt{N}}\right)}^{\infty} e^{2\sqrt{N}\beta z-\frac{z^2}{2}}dz + o(1)$$

$$= \frac{1}{\sqrt{2\pi}(1-e^{-N\beta^2})} \int_{\sqrt{N}\left(\frac{\ln 2}{2\beta}-\beta\right)+\frac{\ln \epsilon}{\sqrt{N}\beta}+o\left(\frac{1}{\sqrt{N}}\right)}^{\infty} e^{-\frac{z^2}{2}}dz + o(1) \quad (9.38)$$

It is easy to check that the latter integral converges to zero if and only if $\beta^2 < \ln 2/2$. Using the fact that $\ln(1+x) = x + o(x)$ as $x \to 0$, the assertion (i) follows immediately.

Since the Lindeberg condition clearly fails for $2\beta^2 \geq \ln 2$, it is clear that we cannot expect a simple CLT beyond this regime. Such failure of a CLT is always a problem related to *heavy tails*, and results from the fact that extremal events begin to influence the fluctuations of the sum. It appears therefore reasonable to separate from the sum the terms where X_σ is anomalously large. For Gaussian random variables it is well known that the right scale of separation is given by $u_N(x)$ defined in (9.5). The key to most of what follows relies on the famous result on the convergence of the extreme value process to a Poisson point process (for a proof see, e.g., [162]):

Theorem 9.2.3 *Let $\mathcal{P}_N$ be point process on $\mathbb{R}$ given by*

$$\mathcal{P}_N \equiv \sum_{\sigma \in \mathcal{S}_N} \delta_{u_N^{-1}(X_\sigma)} \quad (9.39)$$

Then $\mathcal{P}_N$ converges weakly to a Poisson point process on $\mathbb{R}$ with intensity measure $e^{-x}dx$.

Let us define

$$\mathbb{Z}_{N,\beta}^x \equiv \mathbb{E}_\sigma e^{\beta\sqrt{N}X_\sigma}\mathbb{I}_{\{X_\sigma \leq u_N(x)\}} \quad (9.40)$$

We may write

$$\frac{Z_{\beta,N}}{\mathbb{E}Z_{\beta,N}} = 1 + \frac{\mathbb{Z}_{\beta,N}^x - \mathbb{E}\mathbb{Z}_{\beta,N}^x}{\mathbb{E}Z_{\beta,N}} + \frac{Z_{\beta,N} - \mathbb{Z}_{\beta,N}^x - \mathbb{E}(Z_{\beta,N} - \mathbb{Z}_{\beta,N}^x)}{\mathbb{E}Z_{\beta,N}} \quad (9.41)$$

Let us first consider the last summand. We introduce the random variable

$$\mathcal{W}_N(x) = \frac{Z_{\beta,N} - \mathbb{Z}_{\beta,N}^x}{\mathbb{E}Z_{\beta,N}} = e^{-N(\ln 2 + \beta^2/2)} \sum_{\sigma \in \mathcal{S}_N} e^{\beta\sqrt{N}X_\sigma}\mathbb{I}_{\{X_\sigma > u_N(x)\}} \quad (9.42)$$

It is convenient to rewrite this as (we ignore the sub-leading corrections to $u_N(x)$ and only keep the explicit part of (9.6))

$$\mathcal{W}_N(x) = e^{-N(\ln 2 + \beta^2/2)} \sum_{\sigma \in \mathcal{S}_N} e^{\beta\sqrt{N}u_N\left(u_N^{-1}(X_\sigma)\right)}\mathbb{I}_{\{u_N^{-1}(X_\sigma)>x\}}$$

$$= \frac{1}{C(\beta,N)} \sum_{\sigma \in \mathcal{S}_N} e^{\alpha u_N^{-1}(X_\sigma)}\mathbb{I}_{\{u_N^{-1}(X_\sigma)>x\}} \quad (9.43)$$

where
$$C(\beta, N) \equiv e^{\frac{N}{2}(\sqrt{2\ln 2}-\beta)^2 + \frac{\alpha}{2}[\ln(N\ln 2) + \ln 4\pi]} \tag{9.44}$$

Clearly, the weak convergence of $\mathcal{P}_N$ to $\mathcal{P}$ implies convergence in law of the right-hand side of (9.43), provided that $e^{\alpha x}$ is integrable on $[x, \infty)$ with respect to the Poisson point process with intensity e^{-x}. This is, in fact, never a problem: the Poisson point process has almost surely support on a finite set, and therefore $e^{\alpha x}$ is always almost surely integrable. Note, however, that for $\beta \geq \sqrt{2\ln 2}$ the mean of the integral is infinite, indicating the passage to the low-temperature regime. Note also that the variance of the integral is finite exactly if $\alpha < 1/2$, i.e. $\beta^2 < \ln 2/2$, i.e. when the CLT holds. On the other hand, the mean of the integral diverges if $x \downarrow \infty$; note that the points of the Poisson point process accumulate at minus infinity, and thus the integral itself diverges as well. The following lemma provides the first step of the proof of part (iii) of Theorem 9.2.1:

Lemma 9.2.4 *Let $\mathcal{W}_N(x)$, α be defined as above, and let $\mathcal{P}$ be the Poisson point process with intensity measure $e^{-z}dz$. Then*

$$C(\beta, N)\mathcal{W}_N(x) \xrightarrow{\mathcal{D}} \int_x^\infty e^{\alpha z} \mathcal{P}(dz) \tag{9.45}$$

We now need to turn to the remaining term,

$$\frac{Z_{\beta,N}^x - \mathbb{E}Z_{\beta,N}^x}{\mathbb{E}Z_{\beta,N}} \equiv \mathcal{V}_N(x) \tag{9.46}$$

One might first hope that this term upon proper scaling would converge to a Gaussian; however, one can easily check that this is not the case. In fact, it is not hard to compute all moments of this term:

Lemma 9.2.5 *Let $\mathcal{V}_N(x)$ be defined by (9.46). Then for $\alpha > 1/2$ and any integer $k \geq 2$*

$$\lim_{N\uparrow+\infty} C(N, \beta)^k \mathbb{E}[\mathcal{V}_N(x)]^k$$
$$= \sum_{i=1}^k \frac{1}{i!} \sum_{\substack{\ell_1 \geq 2, \cdots, \ell_i \geq 2 \\ \sum_j \ell_j = k}} \frac{k!}{\ell_1! \cdots \ell_i!} \frac{e^{(k\alpha - i)x}}{(\ell_1\alpha - 1)\cdots(\ell_i\alpha - 1)} \tag{9.47}$$

For $\alpha = 1/2$, we have for k even

$$\lim_{N\uparrow+\infty} \mathbb{E}[\mathcal{V}_N(x)]^k e^{kN(\sqrt{2\ln 2}-\beta)^2/2} = \frac{k!}{(k/2)! \, 2^k} = \frac{(k-1)!!}{2^{k/2}} \tag{9.48}$$

and for k odd

$$\lim_{N\uparrow+\infty} \mathbb{E}[\mathcal{V}_N(x)]^k e^{kN(\sqrt{2\ln 2}-\beta)^2/2} = 0 \tag{9.49}$$

(which are the moments of the normal distribution with variance $1/2$).

Proof To prove the lemma we need to compute the moments of the random variables $T_N(\sigma) \equiv 2^{-N} e^{\beta\sqrt{N}X_\sigma} \mathbb{I}_{\{X_\sigma \leq u_N(x)\}}$. They are computed in the same way as the moments of the truncated partition function in the preceding section, using completion of the square in the exponent and the asymptotic estimates (9.4). We get:

(i) If $k\beta < \sqrt{2\ln 2}$,
$$\mathbb{E}T_N^k(\sigma) \sim 2^{-kN}e^{k^2\beta^2 N/2} \qquad (9.50)$$

(ii) if $k\beta = \sqrt{2\ln 2}$,
$$\mathbb{E}[T_N(\sigma)]^k \sim \frac{2^{-kN}e^{k^2\beta^2 N/2}}{2} = \frac{2^{-N}e^{(k\alpha-1)}}{2}e^{k[(\beta\sqrt{2\ln 2}-\ln 2)N]} \qquad (9.51)$$

(iii) if $k\beta > \sqrt{2\ln 2}$,
$$\mathbb{E}[T_N(\sigma)]^k \sim \frac{2^{-N}e^{x(k\alpha-1)}}{k\alpha - 1}e^{k[(\beta\sqrt{2\ln 2}N-\ln 2) - \frac{\alpha}{2}[\ln(N\ln 2)+\ln 4\pi]]} \qquad$$

Now,
$$(\mathbb{E}Z_{\beta,N})^k \mathbb{E}[\mathcal{V}_N(x)]^k = \mathbb{E}\left(\sum_{\sigma \in \mathcal{S}_N}[T_N(\sigma) - \mathbb{E}T_N(\sigma)]\right)^k$$

$$= \sum_{\sigma_1,\ldots,\sigma_k \in \mathcal{S}_N} \mathbb{E}\prod_{i=1}^{k}[T_N(\sigma_i) - \mathbb{E}T_N(\sigma_i)]$$

$$= \sum_{i=1}^{k}\sum_{\substack{\ell_1,\ldots,\ell_i \geq 2 \\ \sum_j \ell_j = k}} \frac{k!}{\ell_1! \cdots \ell_i!}\binom{2^N}{i}\mathbb{E}[T_N(\sigma)$$
$$- \mathbb{E}T_N(\sigma)]^{\ell_1} \cdots \mathbb{E}[T_N(\sigma) - \mathbb{E}T_N(\sigma)]^{\ell_i} \qquad (9.52)$$

Note that, for $k\beta \geq \sqrt{2\ln 2}$, $\mathbb{E}T_N(\sigma)$ is essentially of the form $2^{-N}C_N^k$; thus, for $\ell \geq 2$ and $\beta \geq \sqrt{\ln 2/2}$, and for $\ell \geq 3$ and $\beta = \sqrt{\ln 2/2}$,

$$\mathbb{E}[T_N(\sigma) - \mathbb{E}T_N(\sigma)]^{\ell} = \sum_{j=1}^{\ell}(-1)^j\binom{\ell}{j}\mathbb{E}T_N(\sigma)^{\ell-j}[\mathbb{E}T_N(\sigma)]^j \qquad (9.53)$$

$$\sim \mathbb{E}T_N(\sigma)^{\ell} \sim \frac{2^{-N}e^{x(\ell\alpha-1)}}{k\alpha - 1}\left[2^{-N}e^{N\beta\sqrt{2\ln 2} - \frac{\alpha}{2}[\ln(N\ln 2)+\ln 4\pi]}\right]^{\ell}$$

Inserting this result into (9.52) gives the assertion (9.47).

For $\beta = \sqrt{\ln 2/2}$ and $\ell = 2$, we get

$$\mathbb{E}[\tilde{T}_N(\sigma) - \mathbb{E}\tilde{T}_N(\sigma)]^2 \sim \frac{2^{-N}e^{-x}}{2}[2^{-N}e^{N\beta\sqrt{2\ln 2}}e^{\alpha x}]^2 \qquad (9.54)$$

Inserting this formula into (9.52) we see that the term with $\ell_1, \ldots, \ell_i = 2, i = k/2$ brings the main contribution to the sum, and all others are of smaller order, because of the polynomial terms $e^{-\ell\frac{\alpha}{2}\ln(N\ln 2)}$ in (9.53). This implies (9.48) and (9.49) and the lemma is proven. $\square$

A standard consequence of Lemma 9.2.5 is the weak convergence of the normalized version of $\mathcal{V}_N(x)$:

Corollary 9.2.6 For $\sqrt{\ln 2/2} < \beta$,

$$C(\beta, N)\mathcal{V}_N(x) \xrightarrow{\mathcal{D}} \mathcal{V}(x, \alpha) \qquad (9.55)$$

where $\mathcal{V}(x, \alpha)$ is the random variable with mean zero and k-th moments given by the right-hand side of (9.47). For $\beta = \sqrt{\ln 2/2}$, for all $x \in \mathbb{R}$

$$e^{\frac{N}{2}(\sqrt{2\ln 2}-\beta)^2}\mathcal{V}_N(x) \xrightarrow{\mathcal{D}} \mathcal{N}(0, 1/2) \qquad (9.56)$$

Equation (9.56) together with Lemma 9.2.4 implies that, in the case $\beta = \sqrt{\ln 2/2}$, the contribution from $\mathcal{W}(x)$ vanishes, and the one from $\mathcal{V}(x)$ yields the Gaussian claimed in (ii) of Theorem 9.2.1 by taking $x \uparrow +\infty$. The next proposition will imply (iii) of Theorem 9.2.1.

Proposition 9.2.7 Let $\sqrt{\ln 2/2} < \beta < \sqrt{2\ln 2}$. Then, for $x \in \mathbb{R}$ chosen arbitrarily,

$$e^{\frac{N}{2}(\sqrt{2\ln 2}-\beta)^2 + \frac{\alpha}{2}[\ln(N\ln 2)+\ln 4\pi]} \ln \frac{Z_{\beta,N}}{\mathbb{E}Z_{\beta,N}}$$

$$\xrightarrow{\mathcal{D}} \mathcal{V}(x, \alpha) + \int_x^\infty e^{\alpha z}\mathcal{P}(dz) - \int_x^\infty e^{\alpha z}e^{-z}dz \qquad (9.57)$$

where $\mathcal{V}(x, \alpha)$ and $\mathcal{P}$ are independent random variables.

Proof Equation (9.57) would be immediate from Lemma 9.2.4 and Corollary 9.2.6 if $\mathcal{W}_N(x)$ and $\mathcal{V}_N(x)$ were independent. However, while this is not true, they are not far from independent. To see this, note that if we condition on the number, $n_N(x)$, of variables X_σ that exceed $u_N(x)$, then the decomposition in (9.41) is independent. On the other hand, one readily verifies that Corollary 9.2.6 also holds under the conditional law $\mathbb{P}[\cdot|n_N(x) = n]$, for any finite n, with the same right-hand side $\mathcal{V}(x, \alpha)$. But this implies that the limit can be written as the sum of two independent random variables, as desired. □

When $\alpha > 1/2$, since $\mathbb{E}\mathcal{V}(x, \alpha)^2 = e^{x(2\alpha-1)}/(2\alpha - 1)$, $\mathcal{V}(x, \alpha)$ tends to zero in distribution, as $x \downarrow -\infty$. Therefore,

$$\mathcal{V}(x, \alpha) \xrightarrow{\mathcal{D}} \lim_{y\uparrow+\infty} \int_{-y}^x e^{\alpha z}\mathcal{P}(dz) - \int_{-y}^x e^{\alpha z}e^{-z}dz \qquad (9.58)$$

which means that we can give sense to the Poisson integral $\int_{-\infty}^\infty e^{\alpha z}(\mathcal{P}(dz) - e^{-z}dz)$. From here we get (iii) of Theorem 9.2.1.

We now turn to the proof of parts (iv) and (v) of Theorem 9.2.1. We will see that the computations above almost suffice to conclude the low-temperature case as well. With the notations from above, we write

$$Z_{\beta,N} = Z_{\beta,N}^x + (Z_{\beta,N} - Z_{\beta,N}^x) \qquad (9.59)$$

Clearly, for $\beta \geq \sqrt{2\ln 2}$,

$$Z_{\beta,N} - Z_{\beta,N}^x = e^{N[\beta\sqrt{2\ln 2}-\ln 2]-\frac{\alpha}{2}[\ln(N\ln 2)+\ln 4\pi]} \sum_{\sigma \in \mathcal{S}_N} \mathbb{1}_{\{u_N^{-1}(\sigma)>x\}}e^{\alpha u_N^{-1}(X_\sigma)} \qquad (9.60)$$

so that, for any $x \in \mathbb{R}$,

$$(Z_{\beta,N} - Z_{\beta,N}^x)e^{-N[\beta\sqrt{2\ln 2}-\ln 2]+\frac{\alpha}{2}[\ln(N\ln 2)+\ln 4\pi]} \xrightarrow{\mathcal{D}} \int_x^\infty e^{\alpha z}\mathcal{P}(dz) \qquad (9.61)$$

9.3 The Gibbs measure

Let us first treat the case $\beta > \sqrt{2\ln 2}$. By (9.52), we have

$$e^{-N[\beta\sqrt{2\ln 2} - \ln 2] + \frac{\alpha}{2}[\ln(N\ln 2) + \ln 4\pi]} \mathbb{E} Z^x_{\beta, N} \sim \frac{2^{-N} e^{x(\alpha - 1)}}{\alpha - 1} \qquad (9.62)$$

which tends to zero, as $x \downarrow -\infty$. Hence, with the normalization of (9.61), the contribution from $Z^x_{\beta, N}$ converges to zero in probability. The assertion (v) of the theorem is now immediate.

Let us finally consider the case $\beta = \sqrt{2\ln 2}$. Proceeding as in (9.52),

$$\mathbb{E} Z^0_{\beta, N} = \frac{2^N}{\sqrt{2\pi}} \int_{-\infty}^{u_N(0) - \sqrt{2N\ln 2}} e^{-z^2/2} dz$$

$$= 2^N \left(\frac{1}{2} - \frac{\ln(N\ln 2) + \ln 4\pi}{4\sqrt{N\pi \ln 2}} + O\left(\frac{(\ln N)^2}{N}\right) \right) \qquad (9.63)$$

We use the decomposition

$$Z_{\beta, N} = Z_{\beta, N} - Z^0_{\beta, N} + \mathbb{E} Z^0_{\beta, N} + (Z^0_{\beta, N} - \mathbb{E} Z^0_{\beta, N}) \qquad (9.64)$$

By (9.63), $\mathbb{E} Z^0_{\beta, N} / \mathbb{E} Z_{\beta, N} \sim 1/2$. By (9.43), we see easily that

$$\frac{Z_{\beta, N} - Z^0_{\beta, N}}{\mathbb{E} Z_{\beta, N}} = \mathcal{W}_N(x) \to 0, \text{ a.s.} \qquad (9.65)$$

even though $\mathbb{E} \mathcal{W}_N(0) = 1/2$! Thus the more precise statement consists of saying that

$$e^{\frac{1}{2}[\ln(N\ln 2) + \ln 4\pi]} \mathcal{W}_N(0) \xrightarrow{\mathcal{D}} \int_0^\infty e^z \mathcal{P}(dz) \qquad (9.66)$$

Note that of course the limiting variable has infinite mean, but is almost surely finite. Finally, by Corollary 9.2.6,

$$e^{\frac{1}{2}[\ln(N\ln 2) + \ln 4\pi]} \frac{Z^0_{\beta, N} - \mathbb{E} Z^0_{\beta, N}}{\mathbb{E} Z_{\beta, N}} \xrightarrow{\mathcal{D}} \mathcal{V}(0, 1) \qquad (9.67)$$

The same arguments as those given after Proposition 9.2.7 allow us to identify $\mathcal{V}(0, 1)$ with the centered Poisson integral $\int_{-\infty}^0 e^z (\mathcal{P}(dz) - e^{-z} dz)$. This implies (9.30). Equation (9.31) is an immediate corollary. This concludes the proof of Theorem 9.2.1. □

9.3 The Gibbs measure

With our preparation on the fluctuations of the free energy, we have accumulated enough understanding about the partition function that we can deal with the Gibbs measures. Clearly, there are a number of ways of trying to describe the asymptotics of the Gibbs measures. Recalling the general discussion on random Gibbs measures from Part II, it should be clear that we are seeking a result on the convergence in distribution of random measures. To be able to state such a result, we have to introduce a topology on the spin configuration space that makes it uniformly compact. The first natural candidate would seem to be the product topology. However, given what we already know about the partition function, this topology does not appear ideally adapted to give adequate information. Recall that at low temperatures, the partition function was dominated by a 'few' spin configurations with exceptionally large energy. This is a feature that should remain visible in a limit theorem. A nice way to do this consists in mapping the hypercube to the interval $(0, 1]$ via[5]

$$\mathcal{S}_N \ni \sigma \to r_N(\sigma) \equiv 1 - \sum_{i=1}^{N}(1 - \sigma_i)2^{-i-1} \in (0, 1] \tag{9.68}$$

Define the pure point measure $\tilde{\mu}_{\beta,N}$ on $(0, 1]$ by

$$\tilde{\mu}_{\beta,N} \equiv \sum_{\sigma \in \mathcal{S}_N} \delta_{r_N(\sigma)} \mu_{\beta,N}(\sigma) \tag{9.69}$$

Our results will be expressed in terms of the convergence of these measures. It will be understood in the sequel that the space of measures on $(0, 1]$ is equipped with the topology of weak convergence, and all convergence results hold with respect to this topology.

As the diligent reader will have expected, in the high-temperature phase the limit is the same as for $\beta = 0$, namely:

Theorem 9.3.1 *If $\beta \leq \sqrt{2 \ln 2}$, then*

$$\tilde{\mu}_{\beta,N} \to \lambda, \text{ a.s.} \tag{9.70}$$

where λ denotes the Lebesgue measure on $[0, 1]$.

Proof Note that we have to prove that for any finite collection of intervals $\Delta_1, \ldots, \Delta_k \subset (0, 1]$, the family of random variables $\{\tilde{\mu}_{\beta,N}(\Delta_1), \ldots, \tilde{\mu}_{\beta,N}(\Delta_k)\}$ converges jointly almost surely to $\{\frac{1}{2}|\Delta_1|, \ldots, \frac{1}{2}|\Delta_k|\}$. Our strategy is to first get very sharp estimates for a family of special intervals.

Below we always assume that $N \geq n$. We denote by Π_n the canonical projection from $\mathcal{S}_N$ to $\mathcal{S}_n$. To simplify notation, we will often write $\sigma_n \equiv \Pi_n \sigma$ when no confusion can arise. For $\sigma \in \mathcal{S}_N$, set

$$a_n(\sigma) \equiv r_n(\Pi_n \sigma) \tag{9.71}$$

and

$$\Delta_n(\sigma) \equiv (a_n(\sigma) - 2^{-n}, a_n(\sigma)] \tag{9.72}$$

Note that the union of all these intervals forms a disjoint covering of $(-1, 1]$. Obviously, these intervals are constructed in such a way that

$$\tilde{\mu}_{\beta,N}(\Delta_n(\sigma)) = \mu_{\beta,N}(\{\sigma' \in \mathcal{S}_N : \Pi_n(\sigma') = \Pi_n(\sigma)\}) \tag{9.73}$$

The first step in the proof consists in showing that the masses of all the intervals $\Delta_n(\sigma)$ are remarkably well approximated by their uniform mass.

Lemma 9.3.2 *Set $\beta' \equiv \sqrt{\frac{N}{N-n}}\beta$. For any $\sigma \in \mathcal{S}_n$:*

(i) *If $\beta' \leq \sqrt{\frac{\ln 2}{2}}$, then*

$$|\tilde{\mu}_{\beta,N}(\Delta_n(\sigma)) - 2^{-n}| \leq 2^{-n} e^{-(N-n)(\ln 2 - \beta'^2)} Y_{N-n} \tag{9.74}$$

where Y_N has bounded variance, as $N \uparrow \infty$.

---

[5] The choice of the map r_N is slightly different from the (more straightforward) one in my earlier notes [39]. The reason is purely cosmetic: it avoids an atom at the point 0, so that the point processes constructed below live on the half-open interval $(0, 1]$.

(ii) If $\sqrt{\frac{\ln 2}{2}} < \beta' < \sqrt{2\ln 2}$, then

$$|\tilde{\mu}_{\beta,N}(\Delta_n(\sigma)) - 2^{-n}| \leq 2^{-n} e^{-(N-n)(\sqrt{2\ln 2} - \beta')^2/2 - \alpha \ln(N-n)/2} Y_{N-n} \qquad (9.75)$$

where Y_N is a random variable with bounded mean absolute value.

(iii) If $\beta = \sqrt{2\ln 2}$, then, for any n fixed,

$$|\tilde{\mu}_{\beta,N}(\Delta_n(\sigma)) - 2^{-n}| \to 0, \text{ in probability} \qquad (9.76)$$

Remark 9.3.3 Note that in the sub-critical case, the results imply convergence to the uniform product measure on $\mathcal{S}$ in a *very strong sense*. In particular, the base-size of the cylinders considered (i.e. n) can grow proportionally to N, *even if almost sure convergence is required to hold uniformly for all cylinders!* However, one should not be deceived by this fact: even though seen from the cylinder masses the Gibbs measures look like the uniform measure, seen from the point of view of individual spin configurations the picture is quite different. In fact, the measure concentrates on an *exponentially* small fraction of the full hypercube, namely those $O(\exp(N(\ln 2 - \beta^2/2)))$ vertices that have energy $\sim \beta N$. The seemingly paradoxical effect results from the fact that this set is still exponentially large, as long as $\beta < \sqrt{2\ln 2}$, and is very uniformly dispersed over $\mathcal{S}_N$. The weaker result in the critical case is not artificial. In fact, it is not true that almost sure convergence will hold, as can be seen from Theorem 1 in [113]. One should of course anticipate some signature of the phase transition at the critical point.

Proof The proof of this lemma is a simple application of the first three points in Theorem 9.2.1. Just note that the partial partition functions

$$Z_{\beta,N}(\sigma_n) \equiv \mathbb{E}_{\sigma'} e^{\beta\sqrt{N} X_{\sigma'}} \mathbb{I}_{\Pi_n(\sigma') = \sigma_n} \qquad (9.77)$$

are independent and have the same distribution as $2^{-n} Z_{\beta', N-n}$. But

$$\tilde{\mu}_{\beta,N}(\Delta_n(\sigma_n)) = \frac{Z_{\beta,N}(\sigma_n)}{\sum_{\sigma'_n \in \Sigma_n} Z_{\beta,N}(\sigma'_n)] + Z_{\beta,N}(\sigma_n)} \qquad (9.78)$$

Note that all 2^n random variables $Z_{\beta,N}(\sigma_n)$ are independent, and we have excellent control on the speed of convergence to their mean value. The assertions of the lemma follow easily. □

Once we have the excellent approximation of the measure on all of the intervals $\Delta_n(\sigma)$, it is not difficult to approximate more general collections of intervals by unions of these intervals and to use the same arguments as before to show that their masses converge to their length, which proves the theorem. □

The behaviour of the measure at low temperatures is much more interesting. Let us introduce the Poisson point process $\mathcal{R}$ on the strip $(0, 1] \times \mathbb{R}$ with intensity measure $\frac{1}{2} dy \times e^{-x} dx$. If (Y_k, X_k) denote the atoms of this process, define a new point process $\mathcal{M}_\alpha$ on $(0, 1] \times (0, 1]$ whose atoms are (Y_k, w_k), where

$$w_k \equiv \frac{e^{\alpha X_k}}{\int \mathcal{R}(dy, dx) e^{\alpha x}} \qquad (9.79)$$

for $\alpha > 1$. With this notation we have that:

Theorem 9.3.4 *If $\beta > \sqrt{2 \ln 2}$, with $\alpha = \beta/\sqrt{2 \ln 2}$, then*

$$\tilde{\mu}_{\beta,N} \xrightarrow{D} \tilde{\mu}_{\beta} \equiv \int_{(0,1]\times(0,1]} \mathcal{M}_{\alpha}(dy, dw)\delta_y w \qquad (9.80)$$

Proof With $u_N(x)$ defined in (9.6), we define the point process $\mathcal{R}_N$ on $(0,1] \times \mathbb{R}$ by

$$\mathcal{R}_N \equiv \sum_{\sigma \in \mathcal{S}_N} \delta_{(r_N(\sigma), u_N^{-1}(X_\sigma))} \qquad (9.81)$$

A standard result of extreme value theory (see [162], Theorem 5.7.2) is easily adapted to yield that

$$\mathcal{R}_N \xrightarrow{D} \mathcal{R}, \text{ as } N \uparrow \infty \qquad (9.82)$$

where the convergence is in the sense of weak convergence on the space of sigma-finite measures endowed with the (metrizable) topology of vague convergence. Note that

$$\mu_{\beta,N}(\sigma) = \frac{e^{\alpha u_N^{-1}(X_\sigma)}}{\sum_\sigma e^{\alpha u_N^{-1}(X_\sigma)}} = \frac{e^{\alpha u_N^{-1}(X_\sigma)}}{\int \mathcal{R}_N(dy, dx) e^{\alpha x}} \qquad (9.83)$$

Since $\int \mathcal{R}_N(dy, dx) e^{\alpha x} < \infty$ almost surely, we can define the point process

$$\mathcal{M}_{\alpha,N} \equiv \sum_{\sigma \in \mathcal{S}_N} \delta_{\left(r_N(\sigma), \frac{\exp(\alpha u_N^{-1}(X_\sigma))}{\int \mathcal{R}_N(dy, dx) \exp(\alpha x)}\right)} \qquad (9.84)$$

on $(0,1] \times (0,1]$. Then

$$\tilde{\mu}_{\beta,N} = \int \mathcal{M}_{\alpha,N}(dy, dw)\delta_y w \qquad (9.85)$$

The only non-trivial point in the convergence proof is to show that the contribution to the partition functions in the denominator from atoms with $u_N(X_\sigma) < x$ vanishes as $x \downarrow -\infty$. But this is precisely what we have shown to be the case in the proof of part (v) of Theorem 9.2.1. Standard arguments then imply that first $\mathcal{M}_{\alpha,N} \xrightarrow{D} \mathcal{M}_\alpha$, and consequently, (9.80). □

Remark 9.3.5 In [216], Ruelle introduced a process $\mathcal{W}_\alpha$ that is nothing but the marginal of $\mathcal{M}_\alpha$ on the 'masses', i.e. on the second variable, as an asymptotic description of the distribution of the masses of the Gibbs measure of the REM in the infinite-volume limit. Our result implies in particular that indeed $\sum_{\sigma \in \mathcal{S}_N} \delta_{\mu_{\beta,N}(\sigma)} \xrightarrow{D} \mathcal{W}_\alpha$ if $\alpha > 1$. Neveu in [182] gave a sketch of the proof of this fact. Note that Theorem 9.3.4 contains in particular the convergence of the Gibbs measure in the product topology on $\mathcal{S}_N$, since cylinders correspond to certain subintervals of $(0, 1]$. The formulation of Theorem 9.3.4 is very much in the spirit of the metastate approach to random Gibbs measures. The limiting measure is a measure on a continuous space, and each point measure on this set may appear as a 'pure state'. The 'metastate', i.e. the law of the random measure $\tilde{\mu}_\beta$, is a probability distribution, concentrated on the countable convex combinations of pure states, randomly chosen by a Poisson point process from an uncountable collection, while the coefficients of the convex combination are again random variables and are selected via another point process. The only aspect of metastates that is missing here is that we have not 'conditioned on the disorder'. The point

9.3 The Gibbs measure

is, however, that there is no natural filtration of the disorder space compatible with, say, the product topology, and thus in this model we have no natural urge to 'fix the disorder locally'; note, however, that it is possible to represent the i.i.d. family X_σ as a sum of 'local' couplings, i.e. let J_Δ, for any $\Delta \subset \mathbb{N}$ be i.i.d. standard normal variables. Then we can represent $X_\sigma = 2^{-N/2} \sum_{\Delta \subset \{1,\dots,N\}} \sigma_\Delta J_\Delta$; obviously these variables become independent of any of the J_Δ, with Δ fixed, so that conditioning on them would not change the metastate.

Let us discuss the properties of the limiting process $\tilde{\mu}_\beta$. It is easy to see that, with probability one, the support of $\tilde{\mu}_\beta$ is the entire interval $(0, 1]$. But its mass is concentrated on a countable set, i.e. the measure is pure point. To see this, consider the rectangle $A_\epsilon \equiv (\ln \epsilon, \infty) \times (0, 1]$. The process $\mathcal{R}$ restricted to this set has finite total intensity given by ϵ^{-1}, i.e. the number of atoms in that set is a Poissonian random variable with parameter ϵ^{-1}. If we remove the projection of these finitely many random points from $(0, 1]$, the remaining mass is given by

$$\int_{(0,1]\times(-\infty,\ln\epsilon)} \mathcal{R}(dy, dx) \frac{e^{\alpha x}}{\int \mathcal{P}(dx') e^{\alpha x'}} = \int_{-\infty}^{\ln \epsilon} \mathcal{P}(dx) \frac{e^{\alpha x}}{\int \mathcal{P}(dx') e^{\alpha x'}} \quad (9.86)$$

We want to get a lower bound in probability on the denominator. The simplest possible bound is obtained by estimating the probability of the integral by the contribution of the largest atom, which of course follows the double-exponential distribution. Thus

$$\mathbb{P}\left[\int \mathcal{P}(dx) e^{\alpha x} \leq Z\right] \leq e^{-e^{-\ln Z/\alpha}} = e^{-Z^{-\frac{1}{\alpha}}} \quad (9.87)$$

Setting $\Omega_Z \equiv \{\mathcal{P} : \int \mathcal{P}(dx) e^{\alpha x} \leq Z\}$, we conclude that, for $\alpha > 1$,

$$\mathbb{P}\left[\int_{-\infty}^{\ln \epsilon} \mathcal{P}(dx) \frac{e^{\alpha x}}{\int \mathcal{P}(dx') e^{\alpha x'}} > \gamma\right] \quad (9.88)$$

$$\leq \mathbb{P}\left[\int_{-\infty}^{\ln \epsilon} \mathcal{P}(dx) \frac{e^{\alpha x}}{\int \mathcal{P}(dx') e^{\alpha x'}} > \gamma, \Omega_Z^c\right] + \mathbb{P}[\Omega_Z]$$

$$\leq \mathbb{P}\left[\int_{-\infty}^{\ln \epsilon} \mathcal{P}(dx) e^{\alpha x} > \gamma Z, \Omega_Z^c\right] + \mathbb{P}[\Omega_Z]$$

$$\leq \mathbb{P}\left[\int_{-\infty}^{\ln \epsilon} \mathcal{P}(dx) e^{\alpha x} > \gamma Z\right] + \mathbb{P}[\Omega_Z]$$

$$\leq \frac{\mathbb{E} \int_{-\infty}^{\ln \epsilon} \mathcal{P}(dx) e^{\alpha x}}{\gamma} + \mathbb{P}[\Omega_Z] \leq \frac{\epsilon^{\alpha-1}}{(\alpha - 1)\gamma Z} + e^{-Z^{-\frac{1}{\alpha}}}$$

Obviously, for any positive γ it is possible to choose Z as a function of ϵ in such a way that the right-hand side tends to zero. But this implies that, with probability one, all of the mass of the measure $\tilde{\mu}_\beta$ is carried by a countable set, implying that $\tilde{\mu}_\beta$ is pure point. A more refined analysis is given in [156].

So we see that the phase transition in the REM expresses itself via a change of the properties of the infinite-volume Gibbs measure mapped to the interval from the Lebesgue measure at high temperatures to a random dense pure point measure at low temperatures.

9.4 The replica overlap

While the random measure description of the phase transition in the REM yields an elegant description of the thermodynamic limit, the projection to the unit interval loses the geometric structure of the space $\mathcal{S}_N$. The description of this geometry is a central issue that we will continue to pursue. We would like to describe 'where' in $\mathcal{S}_N$ the mass of the Gibbs measure is located. In a situation where no particular reference configuration exists, a natural possibility is to compare two independent copies of spin configurations drawn from the same Gibbs distribution to each other. To make this precise, recall the function $r_N : \mathcal{S}_N \times \mathcal{S}_N \to (0, 1]$ defined in (9.68). We are interested in the probability distribution of $R_N(\sigma, \sigma')$ under the product measure $\mu_{\beta,N} \otimes \mu_{\beta,N}$, i.e. define a probability measure, $\psi_{\beta,N}$, on $[-1, 1]$ by

$$\psi_{\beta,N}[\omega](\mathrm{d}z) \equiv \mu_{\beta,N}[\omega] \otimes \mu_{\beta,N}[\omega](R_N(\sigma, \sigma') \in \mathrm{d}z) \tag{9.89}$$

As we will see later, the analysis of the replica overlap is a crucial tool for studying the Gibbs measures of more complicated models. The following exposition is intended to give a first introduction to this approach:

Theorem 9.4.1

(i) For all $\beta < \sqrt{2 \ln 2}$,

$$\lim_{N \uparrow \infty} \psi_{\beta,N} = \delta_0, \text{ a.s.} \tag{9.90}$$

(ii) For all $\beta > \sqrt{2 \ln 2}$,

$$\psi_{\beta,N} \xrightarrow{\mathcal{D}} \delta_0 \left(1 - \int \mathcal{W}_\alpha(\mathrm{d}w) w^2\right) + \delta_1 \int \mathcal{W}_\alpha(\mathrm{d}w) w^2 \tag{9.91}$$

Proof We write for any $\Delta \subset [-1, 1]$

$$\psi_{\beta,N}(\Delta) = Z_{\beta,N}^{-2} \mathbb{E}_\sigma \mathbb{E}_{\sigma'} \sum_{\substack{t \in \Delta \\ R_N(\sigma,\sigma')=t}} e^{\beta\sqrt{N}(X_\sigma + X_{\sigma'})} \tag{9.92}$$

First, the denominator is bounded from below by $[\tilde{Z}_{\beta,N}(c)]^2$, and by (9.23), with probability of order $\delta^{-2} \exp(-Ng(c, \beta))$, this in turn is larger than $(1 - \delta)^2 [\mathbb{E}\tilde{Z}_{\beta,N}(c)]^2$. First let $\beta < \sqrt{2 \ln 2}$. Assume initially that $\Delta \subset (0, 1) \cup [-1, 0)$. We conclude that

$$\mathbb{E}\psi_{\beta,N}(\Delta) \leq \frac{1}{(1-\delta)^2} \mathbb{E}_\sigma \mathbb{E}_{\sigma'} \sum_{\substack{t \in \Delta \\ R_N(\sigma,\sigma')=t}} 1 + \delta^{-2} e^{-g(c,\beta)N} \tag{9.93}$$

$$\sim \frac{1}{\sqrt{2\pi N}} \frac{1}{(1-\delta)^2} \sum_{t \in \Delta} \frac{2e^{-NI(t)}}{\sqrt{1-t^2}} + \delta^{-2} e^{-g(c,\beta)N}$$

for any $\beta < c < \sqrt{2 \ln 2}$, where $I : [-1, 1] \to \mathbb{R}$ denotes the Cramèr entropy function defined in (3.38). Here we used that, if $(1 - t)N = 2\ell$, $\ell = 0, \ldots, N$, then

$$\mathbb{E}_\sigma \mathbb{E}_{\sigma'} \mathbb{I}_{R_N(\sigma,\sigma')=t} = 2^{-N} \binom{N}{\ell} \tag{9.94}$$

and the approximation of the binomial coefficient given in (3.37) and (3.39). Under our assumptions on Δ, we see immediately from this representation that the right-hand side of (9.93) is clearly exponentially small in N. It remains to consider the mass at the point 1, i.e.

$$\psi_{\beta,N}(1) = Z_{\beta,N}^{-2} \mathbb{E}_\sigma 2^{-N} e^{2\beta\sqrt{N}X_\sigma} \qquad (9.95)$$

But we can split

$$\mathbb{E}_\sigma e^{2\beta\sqrt{N}X_\sigma} = Z_{2\beta,N}^x + (Z_{\beta,N} - Z_{2\beta,N} - Z_{2\beta,N}^x) \qquad (9.96)$$

For the first, we use that

$$\mathbb{E} Z_{2\beta,N}^x \leq 2^{-N} e^{2\beta N\sqrt{2\ln 2}} \qquad (9.97)$$

and for the second, we use that it is

$$e^{N 2\beta\sqrt{2\ln 2} e^\alpha [\ln(N \ln 2) + 4\pi]} \sum_\sigma e^{2\alpha u_N^{-1}(x_\sigma)} \qquad (9.98)$$

Both terms are exponentially smaller than $2^N e^{\beta^2 N}$, and thus the mass of $\psi_{\beta,N}$ at 1 also vanishes. This proves (9.90).

Now let $\beta > \sqrt{2\ln 2}$. We use the truncation introduced in Section 9.2. Note first that for any interval Δ

$$\left| \psi_{\beta,N}(\Delta) - Z_{\beta,N}^{-2} \mathbb{E}_\sigma \mathbb{E}_{\sigma'} \sum_{\substack{t \in \Delta \\ R_N(\sigma,\sigma')=t}} \mathbb{I}_{X_\sigma, X_{\sigma'} \geq u_N(x)} e^{\beta\sqrt{N}(X_\sigma + X_{\sigma'})} \right| \leq \frac{2 Z_{\beta,N}^x}{Z_{\beta,N}} \qquad (9.99)$$

We have already seen in the proof of Theorem 9.2.1 (see (9.62)) that the right-hand side of (9.99) tends to zero in probability, as first $N \uparrow \infty$ and then $x \downarrow -\infty$. On the other hand, for $t \neq 1$,

$$\mathbb{P}\left[\exists_{\sigma,\sigma_\cdot: R_N(\sigma,\sigma')=t} : X_\sigma > u_N(x) \wedge X_{\sigma'} > u_N(x)\right] \qquad (9.100)$$

$$\leq \mathbb{E}_{\sigma\sigma'} \mathbb{I}_{R_N(\sigma,\sigma')=t} 2^{2N} \mathbb{P}[X_\sigma > u_N(x)]^2 = \frac{2 e^{-I(t)N} e^{-2x}}{\sqrt{2\pi N}\sqrt{1-t^2}}$$

by the definition of $u_N(x)$ (see (9.5)). This implies again that any interval $\Delta \subset [-1, 1) \cup [-1, 0)$ has zero mass. To conclude the proof it is enough to compute $\psi_{\beta,N}(1)$. Clearly

$$\psi_{\beta,N}(1) = \frac{2^{-N} \mathbb{Z}_{2\beta,N}}{Z_{\beta,N}^2} \qquad (9.101)$$

By (v) of Theorem 9.2.1, one sees easily that

$$\psi_{\beta,N}(1) \xrightarrow{\mathcal{D}} \frac{\int e^{2\alpha z} \mathcal{P}(dz)}{\left(\int e^{\alpha z} \mathcal{P}(dz)\right)^2} \qquad (9.102)$$

Expressing the left-hand side of (9.102) in terms of the point process $\mathcal{W}_\alpha$, defined in (9.79), yields the expression for the mass of the atom at 1; since the only other atom is at zero, the assertion (ii) follows from the fact that $\psi_{\beta,N}$ is a probability measure. This concludes the proof. □

9.5 Multi-overlaps and Ghirlanda–Guerra relations

The distribution of the replica overlap apparently does not contain all the information on the Gibbs state we have acquired so far. It will be instructive to look at the joint distribution of k independent copies of the spin variables. Interestingly, there are some very strong general principles that allow us to relate multiple overlaps in terms of the two-spin overlaps. These identities were discovered by Ghirlanda and Guerra [127] in 1998. They have proven to be of extreme importance, and we will keep encountering them in the following chapters. Similar relations were derived by Aizenman and Contucci [4].

We begin with the following simple observation:

Proposition 9.5.1 *For any value of β,*

$$\mathbb{E}\frac{d}{d\beta}\Phi_{\beta,N} = \beta(1 - \mathbb{E}\psi_{\beta,N}(1)) \qquad (9.103)$$

Proof Obviously,

$$\mathbb{E}\frac{d}{d\beta}\Phi_{\beta,N} = N^{-1}\mathbb{E}\frac{\mathbb{E}_\sigma \sqrt{N} X_\sigma e^{\beta\sqrt{N}X_\sigma}}{\mathbb{E}_\sigma e^{\beta\sqrt{N}X_\sigma}} \qquad (9.104)$$

We now use for the first time the following so-called *Gaussian integration by parts formula*:

Lemma 9.5.2 *If X is a standard normal variable, and g any differentiable function of at most polynomial growth, then*

$$\mathbb{E}[Xg(X)] = \mathbb{E}g'(X) \qquad (9.105)$$

Proof Left as an exercise (or see, e.g., Appendix A of [239]). □

Using this identity in the right-hand side of (9.104) with respect to the average over X_σ, we get immediately that

$$\mathbb{E}\frac{\mathbb{E}_\sigma \sqrt{N}X_\sigma e^{\beta\sqrt{N}X_\sigma}}{\mathbb{E}_\sigma e^{\beta\sqrt{N}X_\sigma}} = N\beta\mathbb{E}\left(1 - \frac{2^{-N}\mathbb{E}_\sigma e^{2\beta\sqrt{N}X_\sigma}}{(\mathbb{E}_\sigma e^{\beta\sqrt{N}X_\sigma})^2}\right)$$

$$= N\beta\mathbb{E}\left(1 - \mu_{\beta,N}^{\otimes 2}\left(\mathbb{1}_{\sigma_1=\sigma_2}\right)\right) \qquad (9.106)$$

which is the assertion of the proposition. □

In exactly the same way one can prove the following generalization:

Lemma 9.5.3 *Let $h : \mathcal{S}_N^n \to \mathbb{R}$ be any bounded function of n spins. Then*

$$\frac{1}{\sqrt{N}}\mathbb{E}\mu_{\beta,N}^{\otimes n}(X_{\sigma^k}h(\sigma^1,\ldots,\sigma^n)) \qquad (9.107)$$

$$= \beta\mathbb{E}\mu_{\beta,N}^{\otimes n+1}\left(h(\sigma^1,\ldots,\sigma^n)\left(\sum_{l=1}^n \mathbb{1}_{\sigma^k=\sigma^l} - n\mathbb{1}_{\sigma^k=\sigma^{n+1}}\right)\right)$$

Proof Left as an exercise. □

The strength of Lemma 9.5.3 comes out when combined with a factorization result that in turn is a consequence of self-averaging.

Lemma 9.5.4 *Let h be as in the previous lemma. Then*

$$\frac{1}{\sqrt{N}}\left|\mathbb{E}\mu_{\beta,N}^{\otimes n}(X_{\sigma^k}h(\sigma^1,\ldots,\sigma^n)) - \mathbb{E}\mu_{\beta,N}(X_{\sigma^k})\mathbb{E}\mu_{\beta,N}^{\otimes n}(h(\sigma^1,\ldots,\sigma^n))\right| = \delta_N(u) \quad (9.108)$$

where $\delta_N(\beta)$ here and in the sequel stands for constants that satisfy, for any $a > 0$,

$$\lim_{N\uparrow\infty}\int_{-a}^{a}|C(N,\beta+x)|dx = 0 \quad (9.109)$$

Proof Let us write

$$\left(\mathbb{E}\mu_{\beta,N}^{\otimes n}(X_{\sigma^k}h(\sigma^1,\ldots,\sigma^n)) - \mathbb{E}\mu_{\beta,N}(X_{\sigma^k})\mathbb{E}\mu_{\beta,N}^{\otimes n}(h(\sigma^1,\ldots,\sigma^n))\right)^2$$
$$= \left(\mathbb{E}\mu_{\beta,N}^{\otimes n}\left((X_{\sigma^k} - \mathbb{E}\mu_{\beta,N}^{\otimes n}X_{\sigma^k})h(\sigma^1,\ldots,\sigma^n)\right)\right)^2$$
$$\leq \mathbb{E}\mu_{\beta,N}^{\otimes n}\left(X_{\sigma^k} - \mathbb{E}\mu_{\beta,N}^{\otimes n}X_{\sigma^k}\right)^2\mathbb{E}\mu_{\beta,N}^{\otimes n}(h(\sigma^1,\ldots,\sigma^n))^2 \quad (9.110)$$

where the last inequality is the Cauchy–Schwarz inequality applied to the joint expectation with respect to the Gibbs measure and the disorder. Obviously the first factor in the last line is equal to

$$\mathbb{E}\left(\mu_{\beta,N}(X_\sigma^2) - [\mu_{\beta,N}(X_\sigma)]^2\right) + \mathbb{E}(\mu_{\beta,N}(X_\sigma) - \mathbb{E}\mu_{\beta,N}(X_\sigma))^2$$
$$= \beta^{-2}\mathbb{E}\frac{d^2}{d\beta^2}\Phi_{\beta,N} + N\beta^{-2}\mathbb{E}\left(\frac{d}{d\beta}\Phi_{\beta,N} - \mathbb{E}\frac{d}{d\beta}\Phi_{\beta,N}\right)^2 \quad (9.111)$$

We know that $\Phi_{\beta,N}$ converges as $N\uparrow\infty$ and that the limit is infinitely differentiable for all $\beta \geq 0$, except at $\beta = \sqrt{2\ln 2}$; moreover, $-\Phi_{\beta,N}$ is convex in β. Now, for any $\delta > 0$,

$$\limsup_{N\uparrow\infty}\int_{-\delta}^{\delta}dx\,\mathbb{E}\frac{d^2}{d\beta^2}\Phi_{\beta+x,N} \quad (9.112)$$
$$= \lim_{N\uparrow\infty}[(\beta+\delta)\mathbb{E}\mu_{\beta+\delta,N}(Y_\sigma/\sqrt{N}) - (\beta-\delta)\mathbb{E}\mu_{\beta-\delta,N}(Y_\sigma/\sqrt{N})] < \infty$$

which will vanish when divided by N. To see that the coefficient of N of the second term gives a vanishing contribution, we use the following lemma on self-averaging the derivatives of convex functions:

Lemma 9.5.5 *Let $f_N(x)$ be a family of random, real-valued functions on some open set, $U \subset \mathbb{R}$, satisfying the following hypotheses:*

(i) *For any N, f_N is convex (or concave).*
(ii) *There exists a sequence $C_N \downarrow 0$ such that, for all $x \in U$, $\mathbb{E}(f_N(x) - \mathbb{E}f_N(x))^2 \leq C_N$.*
(iii) *There exists a sequence $C'_N \downarrow 0$ and a function $f(x)$ such that, for any $x \in U$, $|\mathbb{E}f_N(x) - f(x)| \leq C'_N$.*

Set $g_N(x) = \mathbb{E}f_N(x)$. Then, for any $x \in U$ such that $f''(x) < \infty$,

$$\lim_{N\uparrow\infty}\mathbb{E}\left(f'_N(x) - g'_N(x)\right)^2 = 0 \quad (9.113)$$

Proof Convexity implies that, for any $x \in U$ and $y > 0$ such that $x \pm y \in U$,

$$[f_N(x) - f_N(x-y)]/y \leq f'(x) \leq [f(x+y) - f(x)]/y \quad (9.114)$$

The same holds of course for g_N. Hence

$$[f_N(x) - f_N(x - y) - g_N(x + y) - g_N(x)]/y \le f'_N(x) - g'_N(x) \quad (9.115)$$
$$\le [f_N(x + y) - f_N(x) - g_N(x) + g_N(x - y)]/y$$

Using (ii), and the inequality $(a + b)^2 \le 2a^2 + 2b^2$, this implies that

$$\left(f'_N(x) - g'_N(x)\right)^2 \le 2\left(g_N(x + y) - g_N(x) - g_N(x) + g_N(x - y)\right)^2/y^2 + 4C_N/y^2 \quad (9.116)$$

Now we use (iii) to replace g_N by f to get

$$\left(f'_N(x) - g'_N(x)\right)^2 \le 2y^2([f(x + y) - f(x) - f(x) + f(x - y)]/y^2)^2 \\ + 4C_N/y^2 + 32C'_N/y^2 \quad (9.117)$$

But, by assumption,

$$\lim_{y \downarrow 0}[f(x + y) - f(x) - f(x) + f(x - y)]/y^2 = f''(x) < \infty \quad (9.118)$$

Thus, if we chose a sequence $y_N = (\max(C_N, C'_N))^{1/4}$, we get that

$$\left(f'_N(x) - g'_N(x)\right)^2 \le \left(\max(C_N, C'_N)\right)^{1/2} [f''(x) + \delta_N + 32] \quad (9.119)$$

where $\delta_N \downarrow 0$. This implies the assertion of the lemma. □

Remark 9.5.6 The assumption that $f''(x)$ is finite holds in general for all but a countable set of points, since $\int_{-a}^{a} f''(x + y)dy = f'(x + a) - f'(x - a)$ is finite (for all but a countable set of values a), by general results on convex functions (see Section 25 of [211]). The proof above is inspired by Lemma 2.12.4 of [240].

In Theorem 9.2.1 we have more than established that the variance of $\Phi_{\beta,N}$ tends to zero, and hence we can apply Lemma 9.5.5 with $f_N(x) = -\Phi_{N,x}$. The assertion of the lemma then follows. □

Exercise: Give an alternative proof of Lemma 9.5.4 in the special case of the REM for all $\beta > \sqrt{2 \ln 2}$, using the previous results on the convergence of the Gibbs measures.

If we combine Proposition 9.5.1, Lemma 9.5.3, and Lemma 9.5.4 we arrive immediately at:

Proposition 9.5.7 *For any bounded function $h : S_N^n \to \mathbb{R}$,*

$$\left| \mathbb{E}\mu_{\beta,N}^{\otimes n+1}(h(\sigma^1, \ldots, \sigma^n) \mathbb{I}_{\sigma^k = \sigma^{n+1}}) \right. \quad (9.120)$$
$$\left. - \frac{1}{n} \mathbb{E}\mu_{\beta,N}^{\otimes n+1}\left(h(\sigma^1, \ldots, \sigma^n) \left(\sum_{l \ne k} \mathbb{I}_{\sigma^l = \sigma^k} + \mathbb{E}\mu_{\beta,N}^{\otimes 2}(\mathbb{I}_{\sigma_1 = \sigma_2}) \right) \right) \right| = \delta_N(\beta)$$

Together with the fact that the product Gibbs measures are concentrated on the sets where the overlaps take values 0 and 1 (9.120) allows us to compute the distribution of all higher overlaps in terms of the two-replica overlap. For example, if we put

$$A_n \equiv \lim_{N\uparrow\infty} \mathbb{E}\mu_{\beta,N}^{\otimes n}(\mathbb{1}_{\sigma^1=\sigma^2=\cdots=\sigma^n}) \tag{9.121}$$

then (9.120) with $h = \mathbb{1}_{\sigma^1=\sigma^2=\cdots=\sigma^n}$ provides the recursion

$$A_{n+1} = \frac{n-1}{n}A_n + \frac{1}{n}A_n A_2 = A_n\left(1 - \frac{1-A_2}{n}\right)$$

$$= \prod_{k=2}^{n}\left(1 - \frac{1-A_2}{k}\right)A_2 = \frac{\Gamma(n+A_2)}{\Gamma(n+1)\Gamma(A_2)} \tag{9.122}$$

Note that we can alternatively use Theorem 9.3.1 to compute, for the non-trivial case $\beta > \sqrt{2\ln 2}$,

$$\lim_{N\uparrow\infty} \mu_{\beta,N}^{\otimes n}(\mathbb{1}_{\sigma^1=\sigma^2=\cdots=\sigma^n}) = \int \mathcal{W}_\alpha(dw)w^n \tag{9.123}$$

so that (9.122) implies a formula for the mean of the n-th moments of $\mathcal{W}_\alpha$,

$$\mathbb{E}\int \mathcal{W}_\alpha(dw)w^n = \frac{\Gamma(n+A_2)}{\Gamma(n+1)\Gamma(A_2)} \tag{9.124}$$

where $A_2 = \mathbb{E}\int \mathcal{W}_\alpha(dw)w^2$. This result has been obtained by a direct computation by Ruelle ([216], Corollary 2.2).

Clearly, we can generalize this result by considering more complicated events

$$A_{n_1,\ldots,n_r} \equiv \lim_{N\uparrow\infty} \mathbb{E}\mu_{\beta,N}^{\otimes(n_1+\cdots+n_r)}\left(\mathbb{1}_{\sigma_1=\cdots=\sigma_{n_1}}\mathbb{1}_{\sigma_{n_1+1}=\cdots=\sigma_{n_2}}\cdots\mathbb{1}_{\sigma_{n_{r-1}+1}=\cdots=\sigma_{n_r}}\right) \tag{9.125}$$

The Ghirlanda–Guerra relations provide, for any $1 \leq i \leq r$ and $n_i \geq 2$, the recursion

$$A_{n_1,\ldots,n_r+1} = \frac{1}{n_1+\cdots+n_r}\left((n_r-1)A_{n_1,\ldots,n_r} + \sum_{j=1}^{r-1} n_j A_{n_1,\ldots,n_j+n_r,\ldots,n_{r-1}} + A_2 A_{n_1,\ldots,n_r}\right) \tag{9.126}$$

It is easy to see that these recursions can all be solved in terms of A_2: We have seen that this is the case for $r = 1$. Assume that it is true for all $p \leq r-1$ and all $n_i \geq 2$. Then (9.126), for $n_r = 1$, allows us to compute all $A_{n_1,\ldots,n_{r-1},2}$ in terms of A_2 and $A_{n_1,\ldots,n_{r-1}}$, and thus entirely in terms of A_2, by the hypothesis. From here we can use (9.126) to successively compute all other $A_{n_1,\ldots,n_{r-1},n_r}$, $n_r > 2$ as well.

The significance of this observation comes from the fact that the collection of these quantities determines completely Ruelle's process $\mathcal{W}_\alpha$. Namely, in analogy to (9.123), we have that

$$A_{n_1\ldots n_r} = \mathbb{E}\int \mathcal{W}_\alpha(dw_1)w_1^{n_1}\cdots\int \mathcal{W}_\alpha(dw_r)w_r^{n_r} \tag{9.127}$$

which provide a sufficiently large class of functions to determine the law of the marginal of $\mathcal{W}$. These observations are due to Talagrand [240]. While in the REM they appear as a curiosity, their analogues will be invaluable in more complicated models, as we shall see in the next chapter.

10

Derrida's generalized random energy models

> En un mot, pour tirer la loi de l'expérience, il faut généraliser; c'est une nécessité qui s'impose à l'observateur le plus circonspect.[1]
> Henri Poincaré, *La valeur de la science*.

We will now turn to the investigation of the second class of Gaussian models we mentioned in Chapter 8, namely Gaussian processes whose covariance is a function of the lexicographic distance on the hypercube (see (8.10)). B. Derrida introduced these models in the case where A is a step function with finitely many jumps as a natural generalization of the REM and called it the *generalized random energy model* (GREM) [87, 114, 115, 116].

We split the presentation into two parts. In the first one, we describe the standard GREM, using only explicit computations. The second part is devoted to the general case, where we return to the use of the Ghirlanda–Guerra identities that we have already encountered in the REM. This chapter is based on collaborations with I. Kurkova [51, 52, 53].

10.1 The standard GREM and Poisson cascades

A key element in the analysis of the REM was the theory of convergence to Poisson processes of the extreme value statistics of (i.i.d.) random variables. In the GREM, analogous results will be needed in the correlated case.

We assume that A is the distribution function of a measure that is supported on a finite number, n, of points $x_1, \ldots, x_n \in [0, 1]$, as shown in Fig. 10.1. In that case we denote the mass of the atoms x_i by a_i, and we set

$$\ln \alpha_i = (x_i - x_{i-1}) \ln 2, \quad i = 1, \ldots, n \tag{10.1}$$

where $x_0 \equiv 0$. We normalize in such a way that $\sum_{i=1}^n a_i = 1$ and $\prod_{i=1}^n \alpha_i = 2$.

It is very useful that there is an explicit representation of the corresponding process X_σ. We will write $\sigma = \sigma_1 \sigma_2 \cdots \sigma_n$ where $\sigma_i \in \mathcal{S}_{N \ln \alpha_i / \ln 2}$. Usually we will assume that $x_1 > 0$, $x_n = 1$, and all $a_i > 0$.

Then the Gaussian process X_σ can be constructed from independent standard Gaussian random variables $X_{\sigma_1}, X_{\sigma_1 \sigma_2}, \ldots, X_{\sigma_1 \cdots \sigma_2}$, where $\sigma_i \in \{-1, 1\}^{N \ln \alpha_i / \ln 2}$, as

$$X_\sigma \equiv \sqrt{a_1} X_{\sigma_1} + \sqrt{a_2} X_{\sigma_1 \sigma_2} + \cdots + \sqrt{a_n} X_{\sigma_1 \sigma_2 \ldots \sigma_n}, \quad \text{if } \sigma = \sigma_1 \sigma_2 \cdots \sigma_n \tag{10.2}$$

---

[1] Approximately: In one word, to draw the rule from experience, one must generalize; this is a necessity that imposes itself on the most circumspect observer.

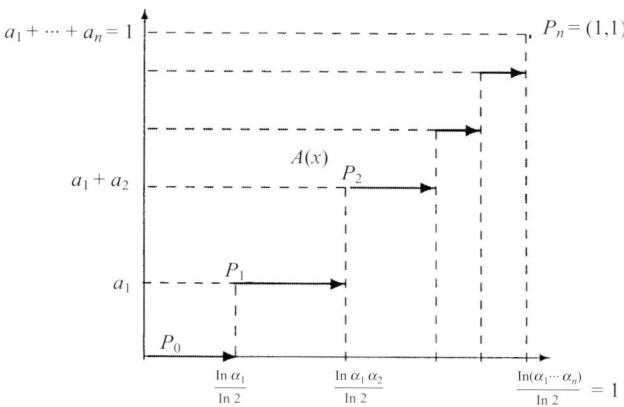

Figure 10.1 The function $A(x)$.

10.1.1 Poisson cascades and extremal processes

Our first concern is understanding the structure of the extremes of such processes. The key ideas are easiest understood in the case where $n = 2$. Let us consider the set, S_x, of σ_1 for which $X_{\sigma_1} \sim \sqrt{a_1 2N \ln \alpha_1 x}$. We know that the cardinality of this set is rather precisely $\alpha_1^{N(1-x)}$ if $x < 1$. Now all the $\alpha_1^{N(1-x)} \alpha_2^N = 2^N \alpha_1^{-xN}$ random variables $X_{\sigma_1 \sigma_2}$ with $\sigma_1 \in S_x$ are independent, so that we know that their maximum is roughly $\sqrt{2a_2 N(\ln 2 - x \ln \alpha_1)}$. Hence, the maximum of the X_σ with $\sigma_1 \in S_x$ is

$$\sqrt{a_1 2N \ln a_1 x} + \sqrt{2a_2 N(\ln 2 - x \ln \alpha_1)} \tag{10.3}$$

Finally, to determine the overall maximum, it suffices to find the value of x that maximizes this quantity, which turns out to be given by $x^* = \frac{a_1 \ln 2}{\ln \alpha_1}$, provided the constraint $\frac{a_1 \ln 2}{\ln \alpha_1} < 1$ is satisfied. In that case we also find that

$$\sqrt{a_1 2N \ln a_1 x} + \sqrt{2a_2 N(\ln 2 - x^* \ln \alpha_1)} = \sqrt{2 \ln 2} \tag{10.4}$$

i.e. the same value as in the REM. On the other hand, if $\frac{a_1 \ln 2}{\ln \alpha_1} > 1$, the maximum is realized by selecting the largest values in the first generation, corresponding to $x = 1$, and then for each of them the extremal members of the corresponding second generation. The value of the maximum is then (roughly)

$$\sqrt{a_1 2N \ln a_1} + \sqrt{2a_2 N \ln \alpha_2} \leq \sqrt{2 \ln 2} \tag{10.5}$$

where equality holds only in the borderline case $\frac{a_1 \ln 2}{\ln \alpha_1} = 1$, which requires more care. The condition $\frac{a_1 \ln 2}{\ln \alpha_1} < 1$ has a nice interpretation: it simply means that the function $A(x) < x$, for all $x \in (0, 1)$.

In terms of the point processes, the above considerations suggest the following picture (which actually holds true): If $\frac{a_1 \ln 2}{\ln \alpha_1} < 1$, the point process

$$\sum_{\sigma S_N} \delta_{u_N^{-1}(X_\sigma)} \to \mathcal{P} \tag{10.6}$$

exactly as in the REM, while in the opposite case this process would surely converge to zero. On the other hand, we can construct (in both cases) another point process,

$$\sum_{\sigma=\sigma_1\sigma_2\in\{-1,+1\}^N} \delta_{\sqrt{a_1}u_{\ln\alpha_1,N}^{-1}(X_{\sigma_1})+\sqrt{a_2}u_{\ln\alpha_2,N}^{-1}(X_{\sigma_1\sigma_2})} \tag{10.7}$$

where we set

$$u_{\alpha,N}(x) \equiv u_{N\ln\alpha/\ln 2}(x) \tag{10.8}$$

This point process will converge to a process obtained from a *Poisson cascade*: The process

$$\sum_{\sigma_1\in\{-1,+1\}^{\ln\alpha_1 N}} \delta_{u_{\alpha_1,N}^{-1}(X_{\sigma_1})} \tag{10.9}$$

converges to a Poisson point process, and, for any σ_1, so do the point processes

$$\sum_{\sigma_2\in\{-1,+1\}^{\ln\alpha_2 N}} \delta_{u_{\alpha_2,N}^{-1}(X_{\sigma_1\sigma_2})} \tag{10.10}$$

Then the two-dimensional point process

$$\sum_{\sigma=\sigma_1\sigma_2\in\{-1,+1\}^N} \delta_{(u_{\alpha_1,N}^{-1}(X_{\sigma_1}),u_{\alpha_2,N}^{-1}(X_{\sigma_1\sigma_2}))} \tag{10.11}$$

converges to a *Poisson cascade* in $\mathbb{R}^2$: we place the Poisson process (always with intensity measure $e^{-x}dx$) on $\mathbb{R}$, and then, for each atom, we place an independent PPP on the line orthogonal to the first line that passes through that atom. Adding up the atoms of these processes with the right weight yields the limit of the process defined in (10.7). Now this second point process does not yield the extremal process, as long as the first one exists, i.e. as long as the process (10.6) does not converge to zero. Interestingly, when we reach the borderline, the process (10.6) converges to the PPP with intensity $Ke^{-x}dx$ with $0 < K < 1$, while the cascade process yields points that differ from those of this process only in the subleading order.

Having understood the particular case of two levels, it is not difficult to figure out the general situation, which we will now describe more precisely.

The first and most important result tells us which Poisson point processes we can construct.

Theorem 10.1.1 *Let* $0 < a_i < 1$, $\alpha_i > 1$, $i = 1, 2, \ldots, n$ *with* $\sum_{i=1}^n a_i = 1$. *Set* $\bar{\alpha} \equiv \prod_{i=1}^n \alpha_i$. *Then the point process*

$$\sum_{\sigma=\sigma_1\cdots\sigma_n\in\{-1,+1\}^{N\ln\bar{\alpha}/\ln 2}} \delta_{u_{\bar{\alpha},N}^{-1}(\sqrt{a_1}X_{\sigma_1}+\sqrt{a_2}X_{\sigma_1\sigma_2}+\cdots+\sqrt{a_n}X_{\sigma_1\sigma_2\cdots\sigma_n})} \tag{10.12}$$

converges weakly to the Poisson point process $\mathcal{P}$ *on* $\mathbb{R}$ *with intensity measure* $Ke^{-x}dx$, $K \in \mathbb{R}$, *if and only if, for all* $i = 2, 3, \cdots, n$,

$$a_i + a_{i+1} + \cdots + a_n \geq \ln(\alpha_i\alpha_{i+1}\cdots\alpha_n)/\ln\bar{\alpha} \tag{10.13}$$

Furthermore, if all inequalities in (10.13) are strict, then the constant $K = 1$. *If some of them are equalities, then* $0 < K < 1$.

Remark 10.1.2 An explicit formula for K can be found in [52].

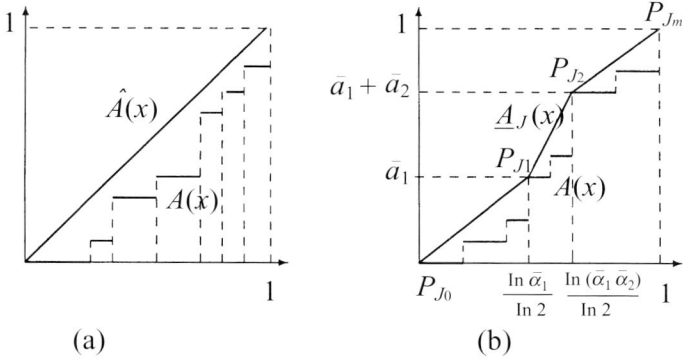

Figure 10.2 The conditions (10.13) (*a*) and (10.17) (*b*).

Remark 10.1.3 The conditions (10.13) can be expressed as $A(x) \leq x$ for all $x \in (0, 1)$. See Fig. 10.2(*a*).

Theorem 10.1.4 *Let* $\alpha_i \geq 1$, *and set* $\bar{\alpha} \equiv \prod_{i=1}^{k} \alpha_i$. *Let* $Y_{\sigma_1}, Y_{\sigma_1 \sigma_2}, \ldots, Y_{\sigma_1 \ldots \sigma_k}$ *be identically distributed random variables such that the vectors* $(Y_{\sigma_1})_{\sigma_1 \in \{-1,1\}^{N \ln \alpha_1 / \ln \bar{\alpha}}}$, $(Y_{\sigma_1 \sigma_2})_{\sigma_2 \in \{-1,1\}^{N \ln \alpha_2 / \ln \bar{\alpha}}}, \ldots, (Y_{\sigma_1 \sigma_2 \ldots \sigma_k})_{\sigma_k \in \{-1,1\}^{N \ln \alpha_k / \ln \bar{\alpha}}}$ *are independent. Let* $v_{N,1}(x)$, $\ldots, v_{N,k}(x)$ *be functions on* $\mathbb{R}$ *such that the following point processes*

$$\sum_{\sigma_1} \delta_{v_{N,1}(Y_{\sigma_1})} \to \mathcal{P}_1$$

$$\sum_{\sigma_2} \delta_{v_{N,2}(Y_{\sigma_1 \sigma_2})} \to \mathcal{P}_2, \quad \forall \sigma_1$$

$$\ldots$$

$$\sum_{\sigma_k} \delta_{v_{N,k}(Y_{\sigma_1 \sigma_2 \ldots \sigma_k})} \to \mathcal{P}_k, \quad \forall \sigma_1 \cdots \sigma_{k-1} \quad (10.14)$$

converge weakly to Poisson point processes, $\mathcal{P}_1, \ldots, \mathcal{P}_k$, *on* $\mathbb{R}$ *with intensity measures* $K_1 e^{-x} dx, \ldots, K_k e^{-x} dx$, *for some constants* $K_1, \ldots, K_k$. *Then the point processes on* $\mathbb{R}^k$,

$$\mathcal{P}_N^{(k)} \equiv \sum_{\sigma_1} \delta_{v_{N,1}(Y_{\sigma_1})} \sum_{\sigma_2} \delta_{v_{N,2}(Y_{\sigma_1 \sigma_2})} \cdots \sum_{\sigma_k} \delta_{v_{N,k}(Y_{\sigma_1 \sigma_2 \cdots \sigma_k})} \to \mathcal{P}^{(k)} \quad (10.15)$$

converge weakly to point processes $\mathcal{P}^{(k)}$ *on* $\mathbb{R}^k$, *called Poisson cascades with k levels.*

Poisson cascades are best understood in terms of the following iterative construction. If $k = 1$, it is just a Poisson point process on $\mathbb{R}$ with intensity measure $K_1 e^{-x} dx$. To construct $\mathcal{P}^{(2)}$ on $\mathbb{R}^2$, we place the process $\mathcal{P}^{(1)}$ for $k = 1$ on the axis of the first coordinate and through each of its points draw a straight line parallel to the axis of the second coordinate. Then we put on each of these lines independently a Poisson point process with intensity measure $K_2 e^{-x} dx$. These points on $\mathbb{R}^2$ form the process $\mathcal{P}^{(2)}$. This procedure is now simply iterated k times.

Theorems 10.1.1 and 10.1.4 combined give a first important result that establishes which different point processes may be constructed in the GREM.

Theorem 10.1.5 Let $\alpha_i \geq 1$, $0 < a_i < 1$, such that $\prod_{i=1}^n \alpha_i = 2$, $\sum_{i=1}^n a_i = 1$. Let $J_1, J_2, \ldots, J_m \in \mathbb{N}$ be the indices such that $0 = J_0 < J_1 < J_2 < \cdots < J_m = n$. We denote by $\bar{a}_l \equiv \sum_{i=J_{l-1}+1}^{J_l} a_i$, $\bar{\alpha}_l \equiv \prod_{i=J_{l-1}+1}^{J_l} \alpha_i$, $l = 1, 2, \ldots, m$, and set

$$\bar{X}^{\sigma_1 \cdots \sigma_{J_{l-1}}}_{\sigma_{J_{l-1}+1} \sigma_{J_{l-1}+2} \cdots \sigma_{J_l}} \equiv \frac{1}{\sqrt{\bar{a}_l}} \sum_{i=1}^{J_l - J_{l-1}} \sqrt{a_{J_{l-1}+i}} X_{\sigma_1 \cdots \sigma_{J_{l-1}+i}} \quad (10.16)$$

Assume that the partition $J_1, J_2, \ldots, J_m$ satisfies the following condition: for all $l = 1, 2, \ldots, m$ and all k such that $J_{l-1} + 2 \leq k \leq J_l$

$$(a_k + a_{k+1} \cdots + a_{J_l-1} + a_{J_l})/\bar{a}_l \geq \ln(\alpha_k \alpha_{k+1} \cdots \alpha_{J_l-1} \alpha_{J_l})/\ln(\bar{\alpha}_l). \quad (10.17)$$

Then the point process

$$\mathcal{P}_N^{(m)} \equiv \sum_{\sigma_1 \ldots \sigma_{J_1}} \delta_{u_{\bar{\alpha}_1,N}^{-1}(\bar{X}_{\sigma_1 \cdots \sigma_{J_1}})} \sum_{\sigma_{J_1+1} \ldots \sigma_{J_2}} \delta_{u_{\bar{\alpha}_2,N}^{-1}(\bar{X}^{\sigma_1 \cdots \sigma_{J_1}}_{\sigma_{J_1+1} \cdots \sigma_{J_2}})} \cdots$$
$$\cdots \sum_{\sigma_{J_{m-1}+1} \ldots \sigma_{J_m}} \delta_{u_{\bar{\alpha}_m,N}^{-1}(\bar{X}^{\sigma_1 \cdots \sigma_{J_{m-1}}}_{\sigma_{J_{m-1}+1} \cdots \sigma_{J_m}})} \quad (10.18)$$

converges weakly to the process $\mathcal{P}^{(m)}$ *on* $\mathbb{R}^m$ *defined in Theorem 10.1.4 with constants* $K_1, \ldots, K_m$. *The constant*

$$K_l = 1 \quad (10.19)$$

if all $J_l - J_{l-1} - 1$ inequalities in (10.17) for $k = J_{l-1} + 2, \ldots, J_l$ are strict. Otherwise $0 < K_l < 1$.

Remark 10.1.6 An explicit expression for K_l is given in [52].

Remark 10.1.7 The conditions on the choice of the partitions can be expressed in a simple geometric way: Replace the function A by the function A_J obtained by joining the sequence of straight line segments going from $(x_{J_i}, A(x_{J_i}))$ to $(x_{J_{i=1}}, A(x_{J_{i+1}}))$, $i = 0, \ldots, m - 1$. Then a partiton, J, is admissible if $A(x) \leq A_J(x)$, for all $x \in [0, 1]$. This condition allows, in general, several possible partitions, and therefore the corollary yields a *family* of convergent point processes associated with the GREM. See Fig. 10.2(*b*).

Having constructed all possible point processes, we now find the extremal process by choosing the one that yields the largest values. It is easy to see that this is achieved if as many intermediate hierarchies as possible are grouped together. In terms of the geometrical construction just described, this means that we must choose the partition J in such a way that the function A_J has no convex pieces, i.e. that A_J is the *concave hull*, $\bar{A}$, of the function A (see Fig. 10.3). (The concave hull, $\bar{A}$, of a function A is the smallest concave function such that $\bar{A}(x) \geq A(x)$, for all x in the domain considered.) Algorithmically, this is achieved by setting $J_0 \equiv 0$, and

$$J_l \equiv \min\{J > J_{l-1} : A_{J_{l-1}+1,J} > A_{J+1,k}, \forall k \geq J + 1\} \quad (10.20)$$

where $A_{j,k} \equiv \sum_{i=j}^k a_i/(2\ln(\prod_{i=j}^k \alpha_i))$.

10.1 The standard GREM and Poisson cascades

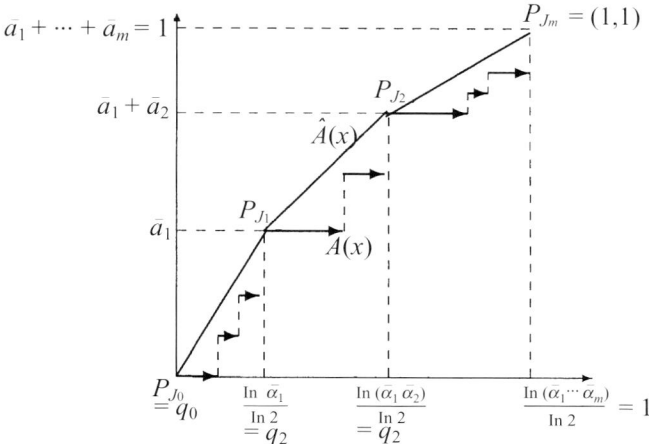

Figure 10.3 The concave hull of $A(x)$.

Set $\gamma_l \equiv \sqrt{\bar{a}_l}/\sqrt{2 \ln \bar{\alpha}_l}$, $l = 1, 2, \ldots, m$. Clearly, by (10.20), $\gamma_1 > \gamma_2 > \cdots > \gamma_m$. Define the function $U_{J,N}$ by

$$U_{J,N}(x) \equiv \sum_{l=1}^{m} \left(\sqrt{2N\bar{a}_l \ln \bar{\alpha}_l} - N^{-1/2} \gamma_l (\ln(N(\ln \bar{\alpha}_l)) + \ln 4\pi)/2 \right) + N^{-1/2} x \tag{10.21}$$

and the point process

$$\mathcal{E}_N \equiv \sum_{\sigma \in \{-1,1\}^N} \delta_{U_{J,N}^{-1}(X_\sigma)} \tag{10.22}$$

Theorem 10.1.8

(i) The point process $\mathcal{E}_N$ converges weakly, as $N \uparrow \infty$, to the point process on $\mathbb{R}$

$$\mathcal{E} \equiv \int_{\mathbb{R}^m} \mathcal{P}^{(m)}(dx_1, \ldots, dx_m) \delta_{\sum_{l=1}^{m} \gamma_l x_l} \tag{10.23}$$

where $\mathcal{P}^{(m)}$ is the Poisson cascade introduced in Theorem 10.1.5 corresponding to the partition $J_1, \ldots, J_m$ given by (10.20).

(ii) $\mathcal{E}$ exists, since $\gamma_1 > \cdots > \gamma_m$. It is the cluster point process on $\mathbb{R}$ containing an a.s. finite number of points in any interval $[b, \infty)$, $b \in \mathbb{R}$. The probability that there exists at least one point of $\mathcal{E}$ in the interval $[b, \infty)$ is decreasing exponentially, as $b \uparrow \infty$.

Remark 10.1.9 If some of the inequalities in the condition (10.17) hold with equality, there are several point processes that to leading order are equally large. The degeneracy is lifted in favour of the process with the longest increments in J_i through the $\ln N$ term in the functions $u_{\alpha,N}$ defined in (10.8). If one considers what happens here as a function of the parameters, one sees that the process with fewer levels 'dies out', and the process with extra levels takes over. While the values of the extremes are in leading order the same at the coexistence point, as the inequality gets strictly violated these values now begin to drop substantially.

The formal proofs of these theorems can be found in [52]. I will not reproduce them here, since they are tedious rather than difficult. Not much is involved beyond those results used in the REM together with the observations made above in the two-level case. Only the computations of the constants K in the degenerate cases require a considerable effort.

10.1.2 Convergence of the partition function

We will now turn to the study of the Gibbs measures. Technically, the main step in the proof will be to show that the infinite-volume limit of the properly rescaled partition function can be expressed as a certain functional of Poisson cascade processes, as suggested by Ruelle [216].

For *any* sequence of indices $0 < J_1 < \cdots < J_m = n$ that satisfies the conditions (10.17), the partition function can be written as:

$$Z_{\beta,N} = e^{\sum_{j=1}^{m} \left(\beta N \sqrt{2\bar{a}_j \ln \bar{\alpha}_j} - \beta \gamma_j [\ln(N \ln \bar{\alpha}_j) + \ln 4\pi]/2 \right)}$$
$$\times \mathbb{E}_{\sigma_1 \cdots \sigma_{J_1}} e^{\beta \gamma_1 u_{\bar{\alpha}_1,N}^{-1}(\bar{X}_{\sigma_1 \cdots \sigma_{J_1}})} \quad (10.24)$$
$$\cdots \mathbb{E}_{\sigma_{J_{m-1}+1} \cdots \sigma_{J_m}} e^{\beta \gamma_m u_{\bar{\alpha}_m,N}^{-1}(\bar{X}_{\sigma_{J_{m-1}+1} \cdots \sigma_{J_m}}^{\sigma_1 \cdots \sigma_{J_{m-1}}})}$$

Clearly, not all of these representations can be useful, i.e the sums in the second line should converge to a finite random variable. For this to happen, from what we learned in the REM, each of the sums should be at 'low temperature', meaning here that $\beta \gamma_\ell > 1$. Moreover, we should expect that there is a relation to the maximum process; in fact, this will follow from the condition that $\gamma_i > \gamma_{i+1}$, for all i that appear. Thus we will have to choose the partition J that yields the extremal process, and we have to cut the representation (10.24) at some temperature-dependent level, $J_{l(\beta)}$, and treat the remaining hierarchies as *high-temperature* REM's, i.e. replace them by their mean value. The level $l(\beta)$ is determined by

$$l(\beta) \equiv \max\{l \geq 1 : \beta \gamma_l > 1\} \quad (10.25)$$

and $l(\beta) \equiv 0$ if $\beta \gamma_1 \leq 1$.

From these considerations it is now very simple to compute the asymptotics of the partition function. The resulting formula for the free energy was first found in [71]:

Theorem 10.1.10 [71] *With the notation introduced above,*

$$\lim_{N \to \infty} \Phi_{\beta,N} = \beta \sum_{i=1}^{l(\beta)} \sqrt{2\bar{a}_i \ln \bar{\alpha}_i} + \sum_{i=J_{l(\beta)}+1}^{n} \beta^2 a_i/2, \text{ a.s.} \quad (10.26)$$

The condition that for $\beta \leq \beta_c$, $l(\beta) = 0$, defines the critical temperature, $\beta_c = 1/\gamma_1$.

The more precise asymptotics of the partition function follow:

Theorem 10.1.11 *Let $J_1, J_2, \ldots, J_m \in \mathbb{N}$ be the sequence of indices defined by (10.20) and $l(\beta)$ defined by (10.25). Then, with the notations introduced above,*

$$e^{-\beta \sum_{j=1}^{l(\beta)} \left(N \sqrt{2\bar{a}_j \ln \bar{\alpha}_j} - \gamma_j [\ln(N \ln \bar{\alpha}_j) + \ln 4\pi]/2 \right) - N\beta^2 \sum_{i=J_{l(\beta)}+1}^{n} a_i/2} Z_{\beta,N}$$
$$\xrightarrow{\mathcal{D}} C(\beta) \int_{\mathbb{R}^{l(\beta)}} e^{\beta \sum_{i=1}^{l(\beta)} \gamma_i x_i} \mathcal{P}^{(l(\beta))}(\mathrm{d}x_1 \cdots \mathrm{d}x_{l(\beta)}) \quad (10.27)$$

This integral is over the process $\mathcal{P}^{(l(\beta))}$ on $\mathbb{R}^{l(\beta)}$ from Theorem 10.1.4 with constants K_j from Theorem 10.1.5. The constant $C(\beta)$ satisfies

$$C(\beta) = 1, \text{ if } \beta\gamma_{l(\beta)+1} < 1 \tag{10.28}$$

and $0 < C(\beta) < 1$, otherwise.

Remark 10.1.12 An explicit formula for $C(\beta)$ is given in [52].

The integrals over the Poisson cascades appearing in Theorem 10.1.11 are to be understood as

$$\int_{\mathbb{R}^m} e^{\beta\gamma_1 x_1 + \cdots + \beta\gamma_m x_m} \mathcal{P}^{(m)}(dx_1 \cdots dx_m) \tag{10.29}$$

$$\equiv \lim_{x \downarrow -\infty} \int_{\substack{(x_1,\ldots,x_m) \in \mathbb{R}^m: \\ \exists i, 1 \leq i \leq m: \gamma_1 x_1 + \cdots + \gamma_i x_i > (\gamma_1 + \cdots + \gamma_i)x}} e^{\beta\gamma_1 x_1 + \cdots + \beta\gamma_m x_m} \mathcal{P}^{(m)}(dx_1 \cdots dx_m)$$

The existence of these limits requires the conditions on the γ_i mentioned before, and thus can be seen as responsible for the selection of the partition J and the cutoff level $l(\beta)$. Namely [52]:

Proposition 10.1.13 *Assume that $\gamma_1 > \gamma_2 > \ldots > \gamma_m > 0$ and $\beta\gamma_m > 1$. Then:*

(i) *For any $a \in \mathbb{R}$ the process $\mathcal{P}^{(m)}$ contains almost surely a finite number of points $(x_1, \ldots, x_m)$ such that $\gamma_1 x_1 + \cdots + \gamma_m x_m > a$.*
(ii) *The limit in (10.29) exists and is finite almost surely*

10.1.3 The Gibbs measures

We will now turn to the asymptotic description of the Gibbs measures in the GREM. In the REM one considers Ruelle's process of the Gibbs masses, obtained as the limit of the process $\mathcal{W}_N \equiv \sum_\sigma \delta_{\mu_{\beta,N}(\sigma)}$. Together with the information that the replica overlap in the REM can take on only the values 0 and 1 in the limit, this process describes fully the structure of the Gibbs measure: namely, if one is interested in capturing an arbitrary large fraction $1 - p$ of the total mass, then it suffices to consider for some $\epsilon = \epsilon(p) > 0$ the atoms of the process $\mathcal{W}$ with mass larger than ϵ, and to place them at random on a set of orthogonal vectors on the infinite-dimensional unit sphere.

In the GREM, this picture is insufficient, since the overlap distribution may now take on values that are different from 0 and 1. Thus, the points carrying the masses described by the process $\mathcal{W}$ are distributed in a more complicated way in space. Ruelle took this fact into account when defining the 'probability cascades' in his version of the GREM. We will define the corresponding objects in the context of the GREM and prove their convergence to Ruelle's cascades.

The overlap distribution As in the REM, we define the overlap distribution, i.e. the random probability distribution

$$\tilde{\psi}_{\beta,N}(q) \equiv \mu_{\beta,N}^{\otimes 2}(R_N(\sigma,\sigma') \leq q) \tag{10.30}$$

where $R_N(\sigma, \sigma')$ is the overlap defined in (8.6). In the context of the GREM, it appears natural to introduce the distribution of d_N,

$$\psi_{\beta,N}(q) \equiv \mu_{\beta,N}^{\otimes 2}(d_N(\sigma, \sigma') \leq q) \tag{10.31}$$

An interesting and important result will be the fact that these two notions coincide in the thermodynamic limit, implying that the Gibbs measures concentrate on sets where between any two points the two distance measures coincide.

With the notation introduced in Theorem 10.1.11, according to the partition $J_0, \ldots, J_m$ defined in (10.20), let us set

$$q_l \equiv \sum_{n=1}^{l} \frac{\ln \bar{\alpha}_n}{\ln 2} \tag{10.32}$$

and

$$q_{\max}(\beta) \equiv \sum_{n=1}^{l(\beta)} \frac{\ln \bar{\alpha}_n}{\ln 2} \tag{10.33}$$

We will see that the measure $\psi_{\beta,N}$ converges to a limiting random measure ψ_β with support on the set $\{0, q_1, \ldots, q_{l(\beta)}\}$.

Point processes of masses and Ruelle's probability cascades We will introduce a number of point processes that appear to be good candidates for a more detailed description of the Gibbs measure.

Let us introduce the sets

$$B_l(\sigma) \equiv \{\sigma' \in \mathcal{S}_N : d_N(\sigma, \sigma') \geq q_l\}, \quad l = 1, 2, \ldots, l(\beta) \tag{10.34}$$

We define point processes $\mathcal{W}_{\beta,N}^{(m)}$ on $(0, 1]^m$ given by

$$\mathcal{W}_{\beta,N}^{(m)} = \sum_\sigma \delta_{(\mu_{\beta,N}(B_1(\sigma)),\ldots,\mu_{\beta,N}(B_m(\sigma)))} \frac{\mu_{\beta,N}(\sigma)}{\mu_{\beta,N}(B_m(\sigma))} \tag{10.35}$$

as well as their projection on the last coordinate,

$$\mathcal{R}_{\beta,N}^{(m)} \equiv \sum_\sigma \delta_{\mu_{\beta,N}(B_m(\sigma))} \frac{\mu_{\beta,N}(\sigma)}{\mu_{\beta,N}(B_m(\sigma))} \tag{10.36}$$

It is easy to see that the processes $\mathcal{W}_{\beta,N}^{(m)}$ satisfy

$$\mathcal{W}_{\beta,N}^{(m)}(dw_1, \ldots, dw_m) = \int_0^1 \mathcal{W}_{\beta,N}^{m+1}(dw_1, \ldots, dw_m, dw_{m+1}) \frac{w_{m+1}}{w_m} \tag{10.37}$$

where the integration is of course over the last coordinate w_{m+1}. Note that these processes will in general not all converge, but will do so only when for some σ, $\mu_\beta(B_m(\sigma))$ is strictly positive. From our experience with the Gibbs measure, it is clear that this will be the case precisely when $m \leq l(\beta)$. In fact, we prove the following:

Theorem 10.1.14 *If $m \leq l(\beta)$, the point process $\mathcal{W}_{\beta,N}^{(m)}$ on $(0, 1]^m$ converges weakly to the point process $\mathcal{W}_\beta^{(m)}$ whose atoms, $w(i)$, are given in terms of the atoms, $(x_1(i), \ldots, x_m(i))$,* of the point process $\mathcal{P}^{(m)}$ by

$$(w_1(i), \ldots, w_m(i)) = \left(\frac{\int \mathcal{P}^{(m)}(dy)\delta(y_1 - x_1(i))e^{\beta(\gamma,y)}}{\int \mathcal{P}^{(m)}(dy)e^{\beta(\gamma,y)}}, \ldots \right. \tag{10.38}$$

$$\left. \ldots, \frac{\int \mathcal{P}^{(m)}(dy)\delta(y_1 - x_1(i))\ldots\delta(y_m - x_m(i))e^{\beta(\gamma,y)}}{\int \mathcal{P}^{(m)}(dy)e^{\beta(\gamma,y)}} \right)$$

and the point processes $\mathcal{R}_{\beta,N}^{(m)}$ converge to the point process $\mathcal{R}_{\beta}^{(m)}$, whose atoms are the last components of the atoms in (10.38).

Of course the most complete object we can reasonably study is the process $\hat{\mathcal{W}}_\beta \equiv \mathcal{W}_\beta^{l(\beta)}$. Analogously, we will set $\hat{\mathcal{R}}_\beta \equiv \mathcal{R}_\beta^{l(\beta)}$.

The point processes $\hat{\mathcal{W}}_\beta$ take values on vectors whose components form decreasing sequences in $(0, 1]$. Moreover, these atoms are naturally clustered in a hierarchical way. These processes were introduced by Ruelle [216] and called *probability cascades*.

There is an intimate relation between the distance distributions ψ_β and these point processes.

Theorem 10.1.15 *With the notation introduced above, we have that:*

(i) *The random distribution functions $\psi_{\beta,N}$ and $\tilde{\psi}_{\beta,N}$ converge, in distribution and in mean, to the same random distribution function ψ_β.*
(ii) *ψ_β is a step function with jumps at the values $\{0, q_1, \ldots, q_{l(\beta)}\}$. For any $q \in [q_{i-1}, q_i)$*

$$\psi_\beta(q) = \int \hat{\mathcal{W}}_\beta(dw_1, \ldots, dw_{l(\beta)}) w_{l(\beta)}(1 - w_i), \quad i = 1, \ldots, l(\beta); \tag{10.39}$$

$$\psi_\beta(q) = 1, \quad \text{for } q \geq q_{l(\beta)}$$

The proof of (ii) is quite straightforward and similar to the analogous statement in the REM (Theorem 9.4.1). The proof of the more interesting result (i) is tedious and can be found in [52].

10.2 Models with continuous hierarchies: the CREM

We now turn to the case when A is an arbitrary probability distribution function, and in particular when it contains a continuous part. The analysis of the standard GREM seemed to rely very crucially on the fact that the number of hierarchies, n, was fixed finite when we took the limit $N \uparrow \infty$. Treating the case when $n \sim N$ with the same methods seems hopeless. Nonetheless, there is a very encouraging observation: the formula for the free energy, (10.26), can be written in a way that makes sense for arbitrary A, namely,

$$\lim_{N \uparrow \infty} \Phi_{\beta,N} = \Phi_\beta = \sqrt{2 \ln 2} \beta \int_0^{x_\beta} \sqrt{\bar{a}(x)} dx + \frac{\beta^2}{2}(1 - \bar{A}(x_\beta)) \tag{10.40}$$

where

$$x_\beta \equiv \sup\left(x \,\Big|\, \bar{a}(x) > \frac{2 \ln 2}{\beta^2} \right) \tag{10.41}$$

The mean of the limiting distance distribution function

$$\psi_\beta(x) \equiv \lim_{N \uparrow \infty} \mu_{\beta,N}^{\otimes 2}(d_N(\sigma, \sigma') \leq x) \tag{10.42}$$

as given in Gardner and Derrida [115], or Proposition 1.11 of [52], can be written as

$$\mathbb{E}\psi_\beta(x) = \begin{cases} \beta^{-1}\sqrt{2\ln 2}/\sqrt{\bar{a}(x)}, & \text{if } x < x_\beta \\ 1, & \text{if } x \geq x_\beta \end{cases} \tag{10.43}$$

Here $\bar{A}$ denotes the concave hull of the function $A(x)$, and $\bar{a}$ the right-derivative of this function, $\bar{a}(x) \equiv \lim_{\epsilon \downarrow 0} \epsilon^{-1}(\bar{A}(x+\epsilon) - \bar{A}(x))$, which exists for all $x \in (0, 1]$. To avoid complications that are not of interest here, we will assume that A is a piecewise smooth function. We will draw advantage from the fact that any such function can be approximated arbitrarily well by non-decreasing step functions, for which we have computed everything explicitly. The task is to show that the corresponding processes can be used to capture precisely the asymptotic properties of the processes of interest.

The next important step is to reconsider the representation of the Gibbs measures, which in the standard case were represented by Ruelle's probability cascades, i.e. as point processes on $(0, 1]^m$. However, the vectors $(\mu_{\beta,N}(B_1(\sigma)), \ldots, \mu_{\beta,N}(B_m(\sigma)))$ should be thought of as the values of the mass distribution

$$m_\sigma(x) \equiv \mu_{\beta,N}(d_N(\sigma, \sigma') > x) \tag{10.44}$$

at the values q_i, for which the limiting distribution has jumps. Seen in this light, it seems most appropriate to introduce the following objects, which we have called *empirical distance distribution functions*

$$\mathcal{K}_{\beta,N} \equiv \sum_{\sigma \in \mathcal{S}_N} \mu_{\beta,N}(\sigma) \delta_{m_\sigma(\cdot)} \tag{10.45}$$

as a random measure on $\mathcal{M}_1([0, 1])$, the set of probability measures on $[0, 1]$. Note that the first moment of $\mathcal{K}_{\beta,N}$ is related to the distance distribution function, $\psi_{\beta,N}$, via

$$\int \mathcal{K}_{\beta,N}(\mathrm{d}m) = 1 - \psi_{\beta,N} \tag{10.46}$$

In the case of the GREM with finitely many levels, the results stated above imply readily the convergence of $\mathcal{K}_{\beta,N}$ to a measure that is concentrated on atomic distributions. The values, q_i, of the atoms are controlled by the point process $\mathcal{W}_\beta^{(m)}$, and whence are given, finally, in terms of the processes $\mathcal{P}^{(m)}$. This approach clearly fails in the continuous case. However, the *Ghirlanda–Guerra* identities [4, 127] provide an alternative approach to the construction of the infinite-volume limit. In the CREM this idea is a crucial tool.

10.2.1 Free energy

The basis of all results is control of the convergence of the free energy. The idea here is to consider sequences, A_n, of piecewise constant distributions functions that converge to A in the sup-norm, to use the fact that, for any such A_n, the formula of the free energy is known, and to let $n \uparrow \infty$, after the thermodynamic limit. This might not tell us anything about the thermodynamic limit of the free energy corresponding to A, were it not

10.2 Models with continuous hierarchies: the CREM

for the following fundamental comparison lemma, due to Kahane, which we take from Theorem 3.11 of [167]:

Lemma 10.2.1 *Let X and Y be two independent n-dimensional Gaussian vectors. Let D_1 and D_2 be subsets of $\{1, \ldots, n\} \times \{1, \ldots, n\}$. Assume that*

$$\begin{aligned} \mathbb{E} X_i X_j &\leq \mathbb{E} Y_i Y_j, & \text{if} \quad (i,j) \in D_1 \\ \mathbb{E} X_i X_j &\geq \mathbb{E} Y_i Y_j, & \text{if} \quad (i,j) \in D_2 \\ \mathbb{E} X_i X_j &= \mathbb{E} Y_i Y_j, & \text{if} \quad (i,j) \notin D_1 \cup D_2 \end{aligned} \quad (10.47)$$

Let f be a function on $\mathbb{R}^n$ such that its second derivatives satisfy

$$\frac{\partial^2}{\partial x_i \partial x_j} f(x) \geq 0, \quad \text{if} \quad (i,j) \in D_1$$

$$\frac{\partial^2}{\partial x_i \partial x_j} f(x) \leq 0, \quad \text{if} \quad (i,j) \in D_2 \quad (10.48)$$

Then

$$\mathbb{E} f(X) \leq \mathbb{E} f(Y) \quad (10.49)$$

Remark 10.2.2 This theorem, and even more its proof, has played a crucial rôle in recent work on mean-field spin-glasses, and we will encounter it several times in the sequel. For this reason we give its (simple) proof.

Proof The first step of the proof consists of writing

$$f(X) - f(Y) = \int_0^1 dt \frac{d}{dt} f(X^t) \quad (10.50)$$

where we define the interpolating process

$$X^t \equiv \sqrt{t} X + \sqrt{1-t} Y \quad (10.51)$$

Next observe that

$$\frac{d}{dt} f(X^t) = \frac{1}{2} \sum_{i=1}^n \frac{\partial}{\partial x_i} f(X^t) \left(t^{-1/2} X_i - (1-t)^{-1/2} Y_i \right) \quad (10.52)$$

Finally, we use the generalization of the Gaussian integration by parts formula (9.105) to the multivariate setting:

Lemma 10.2.3 *Let $X_i, i \in \{1, \ldots, n\}$ be a multivariate Gaussian process, and let $g : \mathbb{R}^n \to \mathbb{R}$ be a differentiable function of at most polynomial growth. Then*

$$\mathbb{E} g(X) X_i = \sum_{j=1}^n \mathbb{E}(X_i X_j) \mathbb{E} \frac{\partial}{\partial x_j} g(X) \quad (10.53)$$

(See, e.g., Appendix A of [239] for a simple proof.)

Applied to the mean of the left-hand side of (10.52) this yields

$$\mathbb{E} f(X) - \mathbb{E} f(Y) = \frac{1}{2} \sum_{i,j} \int_{0,1} (\mathbb{E} X_i X_j - \mathbb{E} Y_i Y_j) \mathbb{E} \frac{\partial^2}{\partial x_j \partial x_i} f(X^t) \quad (10.54)$$

from which the assertion of the theorem can be read off. □

Theorem 10.2.4 Let X_σ be a centered Gaussian process on $\mathcal{S}_N$ with covariance given by (8.10). Then

$$\lim_{N\uparrow\infty} N^{-1}\mathbb{E}\ln Z_{\beta,N} = \sqrt{2\ln 2}\beta \int_0^{x_\beta} \sqrt{\bar{a}(x)}dx + \frac{\beta^2}{2}(1-\bar{A}(x_\beta)) \qquad (10.55)$$

where x_β is defined in (10.42). The critical temperature, β_c, is given by

$$\beta_c = \sqrt{\frac{2\ln 2}{\lim_{x\downarrow 0}\bar{a}(x)}} \qquad (10.56)$$

Proof We may easily check that the function $\ln\sum_\sigma e^{\beta\sqrt{N}X_\sigma}$ satisfies the hypothesis of Lemma 10.2.1 with $D_1 = \{\sigma,\sigma'\in\mathcal{S}_N^2 : \sigma\neq\sigma'\}$ and $D_2 = \emptyset$. Thus taking $A_n^\pm(1) = A(1)$ and $A_n^\pm(x)$ smaller, respectively, larger than A elsewhere, we can indeed construct upper and lower bounds for Φ_β that converge to the same limit when $n\uparrow\infty$. □

A further crucial observation is that the distance distribution function can be expressed as a derivative of the free energy (see the case of the REM) as a function of the covariance. In the continuous case, this will be a little more subtle and will require the introduction of some temporal structure into our Gaussian process X_σ; we define the Gaussian process $X_\sigma(t), \sigma\in\mathcal{S}_N, t\in[0,1]$, with covariance

$$\text{cov}(X_\sigma(t), X_{\sigma'}(s)) = A(t\wedge s\wedge [d_N(\sigma,\sigma')]) \qquad (10.57)$$

Note that $X_\sigma(t)$ is a martingale in the variable t and that its increments with respect to t are independent. It is also useful to realize that we may represent $X_\sigma(t)$ as

$$X_\sigma(t) = Y_\sigma(A(t)) \qquad (10.58)$$

where $Y_\sigma(t)$ is a continuous Gaussian martingale with covariance

$$\text{cov}(Y_\sigma(t), Y_{\sigma'}(s)) = t\wedge s\wedge A(d_N(\sigma,\sigma')) \qquad (10.59)$$

Observe that there is the following integration by parts formula:

Lemma 10.2.5 *For any $t\in(0,1]$ and $\epsilon > 0$,*

$$\mathbb{E}\mathbb{E}_\sigma \frac{(X_\sigma(t+\epsilon)-X_\sigma(t))e^{\beta\sqrt{N}X_\sigma}}{\mathbb{E}_{\sigma'}e^{\beta\sqrt{N}X_{\sigma'}}}$$

$$= \beta\sqrt{N}\int_t^{t+\epsilon} dA(s)\mathbb{E}\mu_{\beta,N}^{\otimes 2}[d_N(\sigma,\sigma')\leq s] \qquad (10.60)$$

Proof Let us introduce the infinitesimal increments, $dY_\sigma(t)$, of the process $Y_\sigma(t)$. Clearly

$$\mathbb{E}dY_\sigma(t)dY_{\sigma'}(s) = dt\, ds\, \delta(s-t)\mathbb{I}_{A(d_N(\sigma,\sigma'))>t} \qquad (10.61)$$

The proof makes use of the infinitesimal version of the Gaussian integration by parts formula (10.53),

$$\mathbb{E}dY_\sigma(t)f\left(\int dY_{\sigma'}(s)\right) = \mathbb{E}f'\left(\int dY_{\sigma'}(s)\right)\int \mathbb{E}dY_\sigma(t)dY_{\sigma'}(s)$$

$$= \mathbb{E}f'(Y_{\sigma'})\mathbb{I}_{A(d_N(\sigma,\sigma'))>t}dt \qquad (10.62)$$

where f is any differentiable function of moderate growth. Using (10.62) in (10.61), we get

$$\mathbb{E}\mathbb{E}_\sigma \frac{dY_\sigma(t)e^{\beta\sqrt{N}Y_\sigma}}{\mathbb{E}_{\sigma'}e^{\beta\sqrt{N}Y_{\sigma'}}} \qquad (10.63)$$

$$= dt\,\mathbb{E}\mathbb{E}_\sigma \mathbb{E}_{\sigma'} \frac{\left(1 - \mathbb{1}_{A(d_N(\sigma,\sigma'))>t}\right) e^{\beta\sqrt{N}(X_\sigma + X_{\sigma'})}}{\left(\mathbb{E}_{\sigma'}e^{\beta\sqrt{N}X_{\sigma'}}\right)^2} \beta\sqrt{N}$$

Thus

$$\mathbb{E}\mathbb{E}_\sigma \frac{dx_\sigma(t)e^{\beta\sqrt{N}X_\sigma}}{\mathbb{E}_{\sigma'}e^{\beta\sqrt{N}X_{\sigma'}}} = dA(t)\mathbb{E}\mu_{\beta,N}^{\otimes 2}\left(\mathbb{1}_{A(d_N(\sigma,\sigma'))\leq A(t)}\right)\beta\sqrt{N} \qquad (10.64)$$

Integrating, and using that, if t is a point of increase of A, $A(d_N(\sigma,\sigma')) \leq A(t)$ if and only if $d_N(\sigma,\sigma') \leq t$, yields the lemma. $\square$

Next we want to express the right-hand side of (10.60) as a derivative of the free energy. To that end consider, for $t \in [0,1]$ and $\epsilon > 0$ fixed, the random process

$$X_\sigma^u \equiv X_\sigma + u[X_\sigma(t+\epsilon) - X_\sigma(t)] \qquad (10.65)$$

Clearly

$$\text{cov}(X_\sigma^u, X_{\sigma'}^u) = A^u(d_N(\sigma,\sigma')) \qquad (10.66)$$

where

$$A^u(x) = \begin{cases} A(x), & \text{if } x \leq t \\ A(x) + (2u+u^2)(A(x) - A(t)), & \text{if } t < x \leq t+\epsilon \\ A(x) + (2u+u^2)(A(t+\epsilon) - A(t)), & \text{if } x > t+\epsilon \end{cases} \qquad (10.67)$$

Let us denote the partition function corresponding to the process with covariance A^u by $Z_{\beta,N}^u$, etc. Clearly we have that

$$\beta\sqrt{N}\mathbb{E}\mathbb{E}_\sigma \frac{(X_\sigma(t+\epsilon) - X_\sigma(t))e^{\beta\sqrt{N}X_\sigma}}{\mathbb{E}_{\sigma'}e^{\beta\sqrt{N}X_{\sigma'}}} = \frac{d}{du}\left(\mathbb{E}\ln Z_{\beta,N}^u\right)_{u=0} \qquad (10.68)$$

This yields:

Lemma 10.2.6 *With the notation introduced above we have, for any $t \in (0,1]$ and any $\epsilon > 0$, that*

$$\beta^{-2}N^{-1}\frac{d}{du}\left(\mathbb{E}\ln Z_{\beta,N}^u\right)_{u=0} = \int_t^{t+\epsilon} dA(s)\mathbb{E}\mu_{\beta,N}^{\otimes 2}[d_N(\sigma,\sigma') \leq s] \qquad (10.69)$$

This allows us to obtain an explicit formula for the distance distribution function.

Theorem 10.2.7 *Under the assumptions of Theorem 10.2.4,*

$$\lim_{N\uparrow\infty}\mathbb{E}\mu_{\beta,N}^{\otimes 2}(d_N(\sigma,\sigma') \leq x) = \begin{cases} \beta^{-1}\sqrt{2\ln 2}/\sqrt{a(x)}, & \text{if } x < x_\beta \\ 1, & \text{if } x \geq x_\beta \end{cases} \qquad (10.70)$$

Proof Observe that $\Phi^u_{\beta,N} \equiv N^{-1}\mathbb{E}\ln Z^u_{\beta,N}$ is a convex function of u. A trivial extension of Theorem 10.2.4 shows that $\Phi^u_{\beta,N}$ converges to the function Φ^u_β given by the expression (10.56) when A is replaced by A^u. By convexity, this implies that $\lim_{N\uparrow\infty}\frac{d}{du}\Phi_{\beta,N} = \frac{d}{du}\Phi_\beta$ at all points, u, where Φ^u_β is differentiable. Thus we only have to compute this derivative. We can write

$$\Phi^u_\beta - \Phi_\beta = \sqrt{2\ln 2}\beta \int_0^{x_\beta} dx\left(\sqrt{\bar{a}^u(x)} - \sqrt{\bar{a}(x)}\right)$$

$$+ \sqrt{2\ln 2}\beta \int_{x_\beta}^{x^u_\beta} dx\sqrt{\bar{a}^u(x)} + \frac{\beta^2}{2}\left(\bar{A}^u(x_\beta) - \bar{A}^u(x^u_\beta)\right)$$

$$+ \frac{\beta^2}{2}((\bar{A}^u(1) - \bar{A}^u(x_\beta)) - (\bar{A}(1) - \bar{A}(x_\beta))) \tag{10.71}$$

If $x_\beta < t$, (10.71) simplifies to

$$\Phi^u_\beta - \Phi_\beta = \frac{\beta^2}{2}((\bar{A}^u(1) - \bar{A}^u(x_\beta)) - (\bar{A}(1) - \bar{A}(x_\beta))) \tag{10.72}$$

$$= \frac{\beta^2}{2}(\bar{A}^u(1) - \bar{A}(1)) = (2u + u^2)\frac{\beta^2}{2}[A(t+\epsilon) - A(t)]$$

This is due to the fact that $\bar{A}$ cannot be linear in a neighbourhood of x_β, while on the other hand $\bar{A}^u(x) = \bar{A}(x)$ up to a point $z(u)$ that is either of order $t - O(u)$ (if the function $\bar{A}$ is strictly convex in a left-neighbourhood of t), or equals the lower boundary of the region containing t where $\bar{A}$ is linear. In both cases $x_\beta < z(u)$ if u is small enough. Hence,

$$\frac{d}{du}\left(\Phi^u_\beta\right)_{u=0} = \beta^2[A(t+\epsilon) - A(t)] \tag{10.73}$$

Inserting this into (10.69) and letting ϵ tend to zero, we obtain that, for $t > x_\beta$, $\lim_{N\uparrow\infty}\mathbb{E}\mu^{\otimes 2}_{N,\beta}[d_N(\sigma,\sigma') \leq t] = 1$.

If $x_\beta > t$ (and consequently $x_b > t + \epsilon$, for $\epsilon > 0$ small enough), we must distinguish two cases: (a) A is strictly convex in a left-neighbourhood of $t + \epsilon$, then $\bar{A}(x) = A(x)$ for $x \in [t, t+\epsilon]$; (b) $\bar{A}$ is linear in a left-neighbourhood of $t + \epsilon$.

(a) In this case the function $A^u(x)$ is not convex in a neighbourhood of t, since, for $x < t$, $a^u(x) = a(x)$, and for $x > t$, $a^u(x) = a(x)(1 + 2u + u^2)$. To construct its convex hull, $\bar{A}^u(x)$, one should find the points, $z_1(u), z_2(u), z_1(u) < t < z_2(u) \leq t + \epsilon$, such that the straight line passing through $A(z_1(u))$ and $A^u(z_2(u))$ is tangent to $A(x)$ at $x = z_1(u)$ and to $A^u(x)$ at $x = z_2(u)$. In other words, $a(z_1(u)) = a(z_2(u))(1 + 2u + u^2)$ and $A(z_2(u)) + (2u + u^2)(A(z_2(u)) - A(t)) = A(z_1(u)) + (z_2(u) - z_1(u))a(z_1(u))$. Then $\bar{A}^u(x)$ coincides with this straight line, for $x \in [z_1(u), z_2(u)]$, while $\bar{A}^u(x) = \bar{A}(x)$, for $x \in [0, z_1(u))$, $\bar{A}^u(x) = A^u(x)$, for $x \in (z_2(u), t+\epsilon]$, and $\bar{A}^u(x) = \bar{A}(x) + (2u + u^2)(A(t+\epsilon) - A(t))$,

10.2 Models with continuous hierarchies: the CREM

for $x \in (t+\epsilon, 1]$. Then the last terms in (10.71) are zero and

$$\Phi^u_\beta - \Phi_\beta = \sqrt{2\ln 2}\beta \int_0^{x_\beta} dx \left(\sqrt{\bar{a}^u(x)} - \sqrt{\bar{a}(x)}\right) \tag{10.74}$$

$$= \sqrt{2\ln 2}\beta \int_{z_1(u)}^{t} dx \left(\sqrt{a(z_1(u))} - \sqrt{a(x)}\right)$$

$$+ \sqrt{2\ln 2}\beta \int_t^{z_2(u)} dx \left(\sqrt{a(z_2(u))(1+2u+u^2)} - \sqrt{a(x)(1+2u+u^2)}\right)$$

$$+ \sqrt{2\ln 2}\beta \int_t^{t+\epsilon} dx \left(\sqrt{a(x)(1+2u+u^2)} - \sqrt{a(x)}\right)$$

Note that the straight line tangent to $A(x)$ at the point $x = z(u) < t$ such that $a(z(u)) = a(t)(1+2u+u^2)$ does not cross $A^u(x)$, for $x > z(u)$. Then $z(u) < z_1(u) < t$. Since $z(u) = t + O(u)$, it follows that $z_1(u) = t + O(u)$. It follows that $a(z_1(u)) = a(t) + O(u)$, $a(z_2(u)) = a(t) + O(u)$, and finally $z_2(u) = t + O(u)$. Then the integrand in the first term satisfies

$$0 \leq \sqrt{a(z_1(u))} - \sqrt{a(x)} \leq \sqrt{a(z_1(u))} - \sqrt{a(t)} = O(u) \tag{10.75}$$

and in the the second one

$$0 \leq \sqrt{a(z_2(u))(1+2u+u^2)} - \sqrt{a(x)(1+2u+u^2)}$$
$$\leq \sqrt{a(z_2(u))(1+2u+u^2)} - \sqrt{a(t)} = O(u) \tag{10.76}$$

Since $z_2(u) - z_1(u) = O(u)$, the integrals over both of these terms are of order $O(u^2)$ and do not contribute to the derivative of Φ^u_β. The integral of the first term can be written as

$$\left(\sqrt{1+2u+u^2} - 1\right) \int_t^{t+\epsilon} \sqrt{a(x)} dx = u(1+o(1)) \int_t^{t+\epsilon} \frac{1}{\sqrt{\bar{a}(x)}} dA(x) \tag{10.77}$$

Therefore, by (10.69), we get that

$$\sqrt{2\ln 2}\beta^{-1} \int_t^{t+\epsilon} dA(x) \frac{1}{\sqrt{\bar{a}(x)}}$$
$$= \int_t^{t+\epsilon} dA(x) \lim_{N\uparrow\infty} \mathbb{E}\mu^{\otimes 2}_{N,\beta}[d_N(\sigma, \sigma') \leq x] \tag{10.78}$$

Since this is true for any $\epsilon > 0$, (10.70) follows.

(b) We now consider the case when $\bar{A}$ is linear to the left of $t + \epsilon$. The deformation of $\bar{A}$ now extends further down to the beginning of the linear piece of $\bar{A}$. Assume that $\bar{A}$ is linear on the interval $(y, z] \supset (t, t+\epsilon]$. Then, for small enough u, the slope of $\bar{A}^u$ will differ from that of $\bar{A}$ only in $(y, z]$. Moreover, $\bar{A}^u(y) = \bar{A}(y)$ and $\bar{A}^u(z) = \bar{A}(z) + (2u+u^2)(A(t+\epsilon) - A(t))$. Let

$$z^* \equiv \sup_{x \in (y,z]} \{\bar{A}^u(x) - \bar{A}(x) < (2u+u^2)(A(t+\epsilon) - A(t))\} \tag{10.79}$$

Obviously, $x_\beta \notin (y, z)$. If $t \leq x_\beta$, we get that

$$\beta^{-1}(2\ln 2)^{-1/2}\left(\Phi_\beta^u - \Phi_\beta\right)$$

$$= \int_y^{z^*} dx \left(\sqrt{\bar{a}(y) + (2u + u^2)(A(t+\epsilon) - A(t))/(z^* - y)} - \sqrt{\bar{a}(y)}\right)$$

$$= \sqrt{\bar{a}(y)} \int_y^{z^*} dx \left(\sqrt{1 + (2u + u^2)(A(t+\epsilon) - A(t))/(\bar{a}(y)(z^* - y))} - 1\right)$$

$$= (\bar{a}(y))^{-1/2} u(A(t+\epsilon) - A(t)) + O(u^2) \tag{10.80}$$

Again (10.70) follows now from (10.69). □

Remark 10.2.8 It is clear from the above consideration that we may repeat the same computation with $u < 0$ to find the left-derivative of Φ_β^u at zero. The results coincide, except when $t = x_\beta$. Similarly, one shows that the second derivative of Φ_β^u is finite in a neighbourhood of zero whenever $t \neq x_\beta$.

10.2.2 The empirical distance distribution

The key object we will use to describe the geometry of the Gibbs measures in the continuous case (and which is the appropriate generalization of the point process of masses of Ruelle) is the *empirical distance distribution* defined in (10.45). This object is a *random probability measure* on the space of mass distributions. It has a very appealing physical interpretation: it tells, for a fixed realization of the disorder, with what probability an observer who is distributed with the Gibbs distribution will see a given distribution of mass around him/herself.

Let us say a word more on the interpretation of these processes. Recall that $\int \mathcal{K}_{\beta,N}(dm) m(q) = 1 - \psi_{\beta,N}(q)$. Thus, $\mathcal{K}_{\beta,N}$ will be asymptotically concentrated on distributions for which $m(q) = 0$ if $q \geq x_\beta$. In other words, the smallest blocks

$$\Sigma_\sigma(t) \equiv \{\sigma' \in \mathcal{S}_N | d_N(\sigma, \sigma') > t\} \tag{10.81}$$

around any point σ that have positive mass are of size $2^{(1-x_\beta)N}$. Since the mass distribution around any point within such a 'massive' block is identical, such a block contributes with mass $\mu_{\beta,N}(\sigma)$ with a Dirac measure on the empirical mass distribution around itself.

While in the discrete case, the convergence of these processes could be proven directly, it is a priori not clear how this could be achieved in the general case. But recall that, in the REM, instead of constructing the limiting processes directly, the Ghirlanda–Guerra identities provide an alternative way. We will see that this path is still open in the CREM.

Recall that $\mathcal{K}_{\beta,N}$ are random probability distributions on $\mathcal{M}_1([0, 1])$, the space of probability distributions on $[0, 1]$, and as such random elements of $\mathcal{M}_1(\mathcal{M}_1([0, 1]))$. As such, their laws are elements of the space $\mathcal{M}_1(\mathcal{M}_1(\mathcal{M}_1([0, 1])))$. Equipping $\mathcal{M}_1([0, 1])$ and $\mathcal{M}_1(\mathcal{M}_1([0, 1]))$ with the topologies of weak convergence, there is no obstacle to defining weak convergence of our objects. Just note that the continuous functions of a measure, $m \in \mathcal{M}_1([0, 1])$, can be approximated arbitrarily well by monomials in finite collections of integrals with respect to m of indicator functions of (disjoint) intervals, $\Delta_1, \ldots, \Delta_l \subset [0, 1]$, and that in turn continuous functions of a measure, $\mathcal{W} \in \mathcal{M}_1(\mathcal{M}_1([0, 1]))$, can be approximated by polynomials in a collection of integrals of such functions. Thus, if we show that for any collection, $\Delta_{ij} \subset [0, 1]$, and integers, $r_{ij}, i = 1, \ldots, \ell, j = 1, \ldots, k_i$,

$$\mathbb{E}\left(\int \mathcal{K}_{\beta,N}(dm_1) m_1(\Delta_{11})^{r_{11}} \cdots m_1(\Delta_{1j_1})^{r_{1j_1}} \cdots \right.$$
$$\left. \cdots \int \mathcal{K}_{\beta,N}(dm_\ell) m_\ell(\Delta_{\ell 1})^{r_{\ell 1}} \cdots m_\ell(\Delta_{\ell j_\ell})^{r_{\ell j_\ell}}\right) \quad (10.82)$$

converges, then the point process $\mathcal{K}_{N,\beta}$ converges in distribution to a limit $\mathcal{K}_\beta$. We will investigate this process in some detail in Section 10.3.

10.2.3 Multi-overlap distributions

Our task is now to prove the Ghirlanda–Guerra identities in the general case.

Theorem 10.2.9 *For any $n \in \mathbb{N}$ and any $x \in [0, 1] \setminus x_\beta$,*

$$\left|\mathbb{E}\mu_{\beta,N}^{\otimes n+1}\left(h(\sigma^1, \ldots, \sigma^n)\mathbb{I}_{d_N(\sigma^k, \sigma^{n+1}) > x}\right) - \frac{1}{n}\mathbb{E}\mu_{\beta,N}^{\otimes n+1}\left(h(\sigma^1, \ldots, \sigma^n)\right.\right.$$
$$\left.\left. \times \left[\sum_{l \neq k}^{n} \mathbb{I}_{d_N(\sigma^k, \sigma^l) > x} + \mathbb{E}\mu_{\beta,N}^{\otimes 2}(\mathbb{I}_{d_N(\sigma^1, \sigma^2) > x})\right]\right)\right| = \delta_N(u) \quad (10.83)$$

where the Gibbs measures are considered to depend on u through the deformed functions A^u, and $\delta_N(u)$ is as in Lemma 9.5.4.

Proof One of the pillars of the Ghirlanda–Guerra identities is concentration of measure for the free energy, that holds here also:

Lemma 10.2.10 *For any β and for any $\epsilon \geq 0$,*

$$\mathbb{P}[|\ln Z_{\beta,N} - \mathbb{E}\ln Z_{\beta,N}| > \epsilon] \leq 2\exp\left(-\frac{\epsilon^2}{2\beta^2 N}\right) \quad (10.84)$$

Proof This follows, e.g., from the standard Gaussian concentration of measure theorem (see [167]) and the representation of $\ln Z_{\beta,N}$ as a Lipshitz function of $2^{N+1} - 2$ independent standard Gaussian random variables with Lipshitz constant $\beta\sqrt{N}$. □

As a first step, we need the generalization of Lemma 10.2.5:

Lemma 10.2.11 *For any $t \in (0, 1]$ and $\epsilon > 0$, and $h : \mathcal{S}_N^n \to \mathbb{R}$ any bounded function of n spin-configurations:*

$$\frac{1}{\sqrt{N}}\mathbb{E}\mu_{\beta,N}^{\otimes n}\left((X_{\sigma^k}(t+\epsilon) - X_{\sigma^k}(t))h(\sigma^1, \ldots, \sigma^n)\right)$$
$$= \beta \int_t^{t+\epsilon} dA(s) \mathbb{E}\mu_{\beta,N}^{\otimes n+1} \quad (10.85)$$
$$\times \left(h(\sigma^1, \ldots, \sigma^n)\left[\sum_{l=1}^{n} \mathbb{I}_{d_N(\sigma^k, \sigma^l) > s} - n\mathbb{I}_{d_N(\sigma^k, \sigma^{n+1}) > s}\right]\right)$$

Proof Exactly analogous to the proof of Lemma 10.2.5. □

The more important step is the following:

Lemma 10.2.12 *Let h be as in the previous lemma. Then*

$$\frac{1}{\sqrt{N}}\left|\mathbb{E}\mu_{\beta,N}^{\otimes n}((X_{\sigma^k}(t+\epsilon) - X_{\sigma^k}(t))h(\sigma^1,\ldots,\sigma^n)) \right.\tag{10.86}$$

$$\left. - \mathbb{E}\mu_{\beta,N}(X_{\sigma^k}(t+\epsilon) - X_{\sigma^k}(t))\,\mathbb{E}\mu_{\beta,N}^{\otimes n}(h(\sigma^1,\ldots,\sigma^n))\right| = \delta_N(u)$$

Proof Let us write

$$\left(\mathbb{E}\mu_{\beta,N}^{\otimes n}\left((X_{\sigma^k}(t+\epsilon) - X_{\sigma^k}(t))h(\sigma^1,\ldots,\sigma^n)\right)\right.\tag{10.87}$$

$$\left. - \mathbb{E}\mu_{\beta,N}(X_{\sigma^k}(t+\epsilon) - X_{\sigma^k}(t))\,\mathbb{E}\mu_{\beta,N}^{\otimes n}\left(h(\sigma^1,\ldots,\sigma^n)\right)\right)^2$$

$$= \left(\mathbb{E}\mu_{\beta,N}^{\otimes n}\left(((X_{\sigma^k}(t+\epsilon) - X_{\sigma^k}(t))\right.\right.$$

$$\left.\left.- \mathbb{E}\mu_{\beta,N}^{\otimes n}(X_{\sigma^k}(t+\epsilon) - X_{\sigma^k}(t)))h(\sigma^1,\ldots,\sigma^n)\right)\right)^2$$

$$\le \mathbb{E}\mu_{\beta,N}^{\otimes n}\left((X_{\sigma^k}(t+\epsilon) - X_{\sigma^k}(t))\right.$$

$$\left.- \mathbb{E}\mu_{\beta,N}^{\otimes n}(X_{\sigma^k}(t+\epsilon) - X_{\sigma^k}(t)))^2\,\mathbb{E}\mu_{\beta,N}^{\otimes n}\left(h(\sigma^1,\ldots,\sigma^n)\right)^2$$

where the last inequality is the Cauchy–Schwarz inequality applied to the joint expectation with respect to the Gibbs measure and the disorder. Obviously the first factor in the last line is equal to

$$\mathbb{E}\mu_{\beta,N}\left((X_{\sigma^k}(t+\epsilon) - X_{\sigma^k}(t)) - \mu_{\beta,N}(X_{\sigma^k}(t+\epsilon) - X_{\sigma^k}(t))\right)^2$$

$$+ \mathbb{E}\left(\mu_{\beta,N}(X_{\sigma^k}(t+\epsilon) - X_{\sigma^k}(t)) - \mathbb{E}\mu_{\beta,N}(X_{\sigma^k}(t+\epsilon) - X_{\sigma^k}(t))\right)^2$$

$$= -\beta^{-2}\mathbb{E}\frac{d^2}{du^2}\Phi_{\beta,N}^{u=0} + N\beta^{-2}\mathbb{E}\left(\frac{d}{du}\Phi_{\beta,N}^{u=0} - \mathbb{E}\frac{d}{du}\Phi_{\beta,N}^{u=0}\right)^2 \tag{10.88}$$

where we used the same notation as in the proof of Theorem 10.2.4. We know that $\Phi_{\beta,N}^u$ converges as $N \uparrow \infty$, except possibly when $x_\beta = t$; moreover, $\Phi_{\beta,N}^u$ is convex in the variable u. Then a standard result of convex analysis (see [211], Theorem 25.7) implies that the derivative exists for almost all values of u, and, hence, the second derivative integrated over a small interval is bounded. Thus, the first term in (10.88) will vanish when divided by N. To see that the coefficient of N of the second term gives a vanishing contribution, we use Lemma 9.5.5. Hence the result of the lemma is proven. □

The theorem now follows easily by using (10.86) on the left-hand side of (10.85), and expressing the resulting term with the help of (10.60). Noting that the result holds for any $\epsilon > 0$ then yields (10.83). □

Following [127], we now define the family of measures $\mathbb{Q}_N^{(n)}$ on the space $[0,1]^{n(n-1)/2}$,

$$\mathbb{Q}_{\beta,N}^{(n)}(\underline{d} \in \mathcal{A}) \equiv \mathbb{E}\mu_{N,\beta}^{\otimes n}[\underline{d}_N \in \mathcal{A}] \tag{10.89}$$

where $\underline{d}_N$ denotes the vector of replica distances whose components are $d_N(\sigma^l, \sigma^k)$,

10.2 Models with continuous hierarchies: the CREM

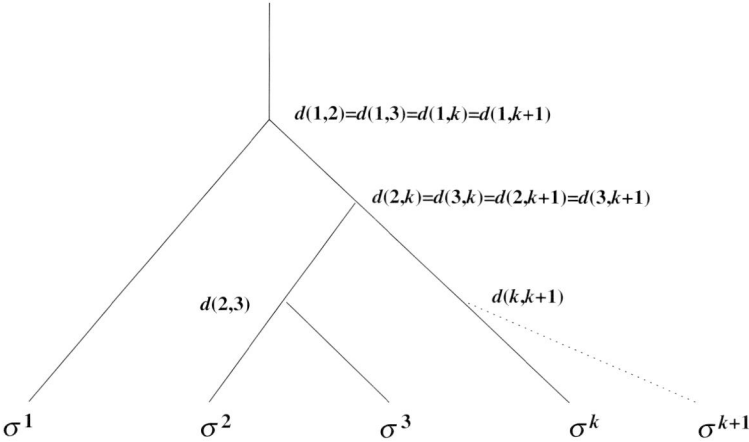

Figure 10.4 Ultrametric relation between distances.

$1 \leq l < k \leq n$. Denote by $\mathcal{B}_k$ the sigma-algebra generated by the first $k(k-1)/2$ coordinates, and let A be a Borel set in $[0, 1]$.

Theorem 10.2.13 *The family of measures $\mathbb{Q}_{\beta,N}^{(n)}$ converge to limiting measures $\mathbb{Q}_\beta^{(n)}$ for all finite n, as $N \uparrow \infty$. Moreover, these measures are uniquely determined by the distance distribution functions ψ_β. They satisfy the identities*

$$\mathbb{Q}_\beta^{(n+1)}\left(d_{k,n+1} \in A | \mathcal{B}_n\right) = \frac{1}{n}\mathbb{Q}_\beta^{(2)}(A) + \frac{1}{n}\sum_{l \neq k}^{n} \mathbb{Q}_\beta^{(n)}\left(d_{k,l} \in A | \mathcal{B}_n\right) \quad (10.90)$$

for any Borel set A. As a consequence, the random measure $\mathcal{K}_{\beta,N}$ converges in distribution to the random measure $\mathcal{K}_\beta$, whose generalized moments are given by $\mathbb{Q}_\beta$. Here convergence is understood as convergence of finite-dimensional marginals integrated over deformations of A.

Proof Choosing h as the indicator function of any desired event in $\mathcal{B}_k$, one sees that (10.83) implies (10.90). This level is simply determined by the condition that the levels beyond $J_{l(\beta)}$, considered as REMs, are in the high-temperature phase, i.e. that $\beta \gamma_{l(\beta)+1} \leq 1$ (which by concavity implies that $\beta \gamma_{l(\beta)+j} \leq 1$, for all $i \geq 1$). The limit as $N \uparrow \infty$, the family of measures $\mathbb{Q}_{\beta,N}^{(n)}$, is entirely determined by the two-replica distribution function. While this may not appear obvious, it follows when taking into account the ultrametric property of the function d_N. This is most easily seen by realizing that the prescription of the mutual distances between k spin configurations amounts to prescribing a tree (start all k configurations at the origin and continue on top of each other as long as the coordinates coincide, then branch off). This is depicted in Fig. 10.4. To determine the full tree of $k + 1$ configurations, it is sufficient to know the overlap of configuration $\sigma^{(k+1)}$ with the configuration it has maximal overlap with, since then all overlaps with all other configurations are determined. But the corresponding probabilities can be computed recursively via (10.90).

We have already seen that $\mathbb{Q}_{\beta,N}^{(2)} = \mathbb{E}\psi_{\beta,N}$ converges. Therefore the relation (10.90) implies the convergence of all distributions $\mathbb{Q}_{\beta,N}^{(n)}$, and proves that the relation (10.90) holds for the limiting measures.

It is clear that all expressions of the form (10.82) can be expressed in terms of the measures $\mathbb{Q}_{\beta,N}^{(k)}$ for k sufficiently large (we leave this as an exercise for the reader to write down). Thus all limit points of sequences of distributions of the measures $\mathcal{K}_{\beta,N}$ must coincide. By compactness of the space $\mathcal{M}_1(\mathcal{M}_1(\mathcal{M}_1([0,1])))$, this implies the convergence of the process $\mathcal{K}_{\beta,N}$ to a limit $\mathcal{K}_\beta$. □

A remarkable feature emerges again if we are only interested in the marginal process $K_\beta(t)$ for fixed t. This process is a simple point process on $[0, 1]$ and is fully determined in terms of the moments

$$\mathbb{E}\left(\int K_{\beta,N}(t)(\mathrm{d}x)x^{r_1}\cdots\int K_{\beta,N}(t)(\mathrm{d}x)x^{r_j}\right)$$
$$= \mathbb{E}\mu_{\beta,N}^{\otimes r_1+\cdots+r_j+j}\left(\mathbb{I}_{d_N(\sigma^1,\sigma^{j+1})>t}\cdots\mathbb{I}_{\ldots,d_N(\sigma^1,\sigma^{j+r_1})>t}\cdots\right.$$
$$\left.\cdots\mathbb{I}_{d_N(\sigma^j,\sigma^{j+r_1+\cdots+r_{j-1}+1})>t}\cdots\mathbb{I}_{d_N(\sigma^j,\sigma^{j+r_1+\cdots+r_j})>t}\right) \quad (10.91)$$

This restricted family of moments satisfies via the Ghirlanda–Guerra identities exactly the same recursion as in the case of the REM. This implies:

Theorem 10.2.14 *Assume that t is such that $\mathbb{E}\mu_\beta^{\otimes 2}(d_N(\sigma,\sigma')<t)=1/\alpha>0$. Then the random measure $K_\beta(t)$ is a Poisson–Dirichlet process (see, e.g., [216, 236]) with parameter α.*

In fact much more is true. We can consider the processes on arbitrary finite-dimensional marginals, i.e.

$$K_{\beta,N}(t_1,\ldots,t_m) \equiv \sum_{\sigma\in\mathcal{S}_N}\mu_{\beta,N}(\sigma)\delta_{m_\sigma(t_1),\ldots,m_\sigma(t_m)} \quad (10.92)$$

for $0<t_1<\cdots<t_m<1$. The point is that this process is entirely determined by the expressions (10.82) with the Δ_{ij} all of the form $(t_i,1]$ for t_i in the fixed set of values $t_1,\ldots,t_m$. This in turn implies that the process is determined by the multi-replica distribution functions $\mathbb{Q}_{\beta,N}^{(n)}$ restricted to the discrete set of events $\{d_N(\sigma^i,\sigma^j)>t_k\}$. Since these numbers are totally determined through the Ghirlanda–Guerra identities, they are identical to those obtained in a GREM with m levels, i.e. a function A having steps at the values t_i, whose two-replica distribution function takes the same values as as that of the model with continuous A at the points t_i and is constant between those values. In fact:

Theorem 10.2.15 *Let $0<t_1<\cdots<t_k\leq q_{max}(\beta)$ be points of increase of $\mathbb{E}\psi_\beta$. Consider a GREM with k levels and parameters α_i, a_i and temperature $\tilde{\beta}$ that satisfy $\ln\alpha_i/\ln 2 = t_i - t_{i-1}$, $\tilde{\beta}^{-1}\sqrt{2\ln\alpha_i/a_i}=\mathbb{E}\psi_\beta(t_i)$. Then*

$$\lim_{N\uparrow\infty}\mathcal{K}_{\beta,N}(t_1,\ldots,t_k) = \mathcal{W}_{\tilde{\beta}}^{(k)} \quad (10.93)$$

Thus, if the t_i are chosen in such a way that for all of them $\mathbb{E}\psi_\beta(t_i)>0$, then we can construct an explicit representation of the limiting marginal process $\mathcal{K}_\beta(t_1,\ldots,t_m)$ in terms of a Poisson cascade process via the corresponding formulae in the associated m-level

GREM. In this sense we obtain an explicit description of the limiting mass distribution function $\mathcal{K}_\beta$.

10.3 Continuous state branching and coalescent processes

Having observed that the empirical distance distributions in the CREM converge to some limit, we will now conclude this analysis by identifying this limit in terms of the *genealogical structure* associated with a particular *continuous state branching process* introduced in 1992 by Neveu [182], shortly after Ruelle's paper [216] appeared. Neveu noted that the Poisson cascades of Ruelle are naturally embedded in this rich object.[2] Following a much later paper by Bolthausen and Sznitman [33], where it was explained how the results of the replica theory of spin-glasses can be interpreted in terms of a *coalescent process* (now known as the *Bolthausen–Sznitman coalescent*), Bertoin and Le Gall [25] finally gave a precise and complete form of the relation between continuous state branching processes, Ruelle's GREM, and the Bolthausen–Sznitman coalescent. The exposition below follows [51], which takes a look at this issue from the perspective of $\mathcal{K}_\beta$.

10.3.1 Genealogy of flows of probability measures

In the REM we considered the image of the Gibbs measure on the unit interval through the map $r_N : \mathcal{S}_N \to (0, 1]$. The apparent disadvantage of this procedure, if applied to the GREM or CREM, is that the limiting measure does not capture the complex geometry of the Gibbs measure. In fact, they always look essentially as they do in the REM. It is instructive to explain why this is the case. On the hypercube we are interested in the masses of sets $\{\sigma' : d_N(\sigma, \sigma') > t\}$. If we map such sets to the unit interval via r_N, we obtain intervals $(r_{[Nt]} - 2^{-[tN]}, r_{[Nt]}]$ of length $2^{-[tN]}$. In fact there is no difficulty in expressing, e.g., $\mathcal{K}_{\beta,N}$ for N fixed in terms of defined quantities with respect to the image measure on the hypercube. However, the construction involves masses of intervals of exponentially small size (in N), and in the limit all these intervals of different size are mapped to a single point. So what should one do in the limit when N is infinite? Obviously, we cannot analyze the structure by looking at intervals of size $2^{-t\infty}$. Thus, we need to remember the masses of intervals of such sizes on all scales. A natural construction leads to the notion of a *flow of probability measures*.

Definition 10.3.1 A two-parameter family of measures with probability distribution functions $S^{(s,t)}$ on $[0, 1]$, $s \leq t \in I \subset \mathbb{R}$, is called a *flow of compatible probability measures* on I if and only if for any collection $t_1 \leq t_2 \leq \cdots \leq t_n \subset I$

$$S^{(t_1,t_n)} = S^{(t_{n-1},t_n)} \circ S^{(t_{n-2},t_{n-1})} \circ \cdots \circ S^{(t_2,t_3)} \circ S^{(t_1,t_2)} \tag{10.94}$$

holds.

With such a flow we can associate the notion of a genealogy. We say that each point $a \in [0, 1]$ is an individual in generation s and its image $S^{(s,t)}(a) \in [0, 1]$ is its offspring in

---

[2] Unfortunately, these observations are only contained in an internal report that has never been published.

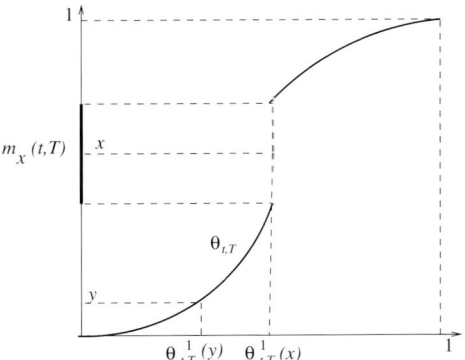

Figure 10.5 Family structure of a population at time T according to relationships at time t.

generation t. Let us define for any distribution function $\Theta(x)$ its inverse function

$$\Theta^{-1}(x) = \inf\{a \mid \Theta(a) \geq x\} \qquad (10.95)$$

Then each individual $x \in [0, 1]$ in generation t has an ancestor a in generation s which is $a = (S^{(s,t)})^{-1}(x)$.

Given an individual $x \in [0, 1]$ in generation t, let us look for individuals x' having the same ancestor as x in generation s (see Fig. 10.5):

$$m_x(s, t) \equiv \{x' : (S^{(s,t)})^{-1}(x') = (S^{(s,t)})^{-1}(x)\} \qquad (10.96)$$

Two situations are possible. If $S^{(s,t)}$ is continuous at $a = (S^{(s,t)})^{-1}(x)$, then any individual $x' \neq x$ has a different ancestor from the one of x, i.e. $(S^{(s,t)})^{-1}(x') \neq (S^{(s,t)})^{-1}(x)$, $m_x(s, t) = \{x\}$. If $S^{(s,t)}$ jumps at $a = (S^{(s,t)})^{-1}(x)$, then by definition $S^{s,t}(a) > x$, any individual $x' \in (\lim_{\epsilon \downarrow 0} S^{(s,t)}(a - \epsilon), S^{(s,t)}(a)]$ has the same ancestor as x in generation s, i.e. $(S^{(s,t)})^{-1}(x') = a$. Hence, the family (10.96) of the individual x having the same ancestor as x in generation s is the interval:

$$m_x(s, t) = \lim_{\epsilon \downarrow 0} \left(S^{(s,t)}((S^{(s,t)})^{-1}(x) - \epsilon), S^{(s,t)} \circ (S^{(s,t)})^{-1}(x) \right] \qquad (10.97)$$

We are mainly interested in the nontrivial case when the functions $S^{(s,t)}$ have lots of jumps.

The next lemma justifies our terminology. It says that any individual having an ancestor in common with x in generation s necessarily has an ancestor in common with x in any generation $s' < s$. In other words, if we partition the interval $[0, 1]$ into families $m_x(s', t)$ having the same ancestor in generation s', then the partition into families $m_x(s, t)$ having the same ancestor in generation $s > s'$ is a refinement of the previous one.

Lemma 10.3.2 *Let $S^{(s,t)}$ be distribution functions of a flow of measures according to Definition 10.3.1. Then for all $x \in [0, 1]$*

$$m_x(s, t) \subset m_x(s', t), \quad \forall s' < s \leq t \in I \qquad (10.98)$$

Proof On the one hand, by definition (10.95) and compatibility (10.94) we have the inequality

$$(S^{(s,t)})^{-1}(x) \leq S^{(s',s)} \circ (S^{(s',s)})^{-1} \circ (S^{(s,t)})^{-1}(x)$$
$$= S^{(s',s)} \circ (S^{(s',t)})^{-1}(x) \tag{10.99}$$

leading to the upper bound

$$S^{(s,t)} \circ (S^{(s,t)})^{-1}(x) \leq S^{(s,t)} \circ S^{(s',s)} \circ (S^{(s',t)})^{-1}(x)$$
$$= S^{(s',t)} \circ (S^{(s',t)})^{-1}(x) \tag{10.100}$$

On the other hand, for any $\epsilon > 0$, $x > \Theta(\Theta^{-1}(x) - \epsilon)$ by definition (10.95). Then

$$(S^{(s,t)})^{-1}(x) > S^{(s',s)} \circ \left((S^{(s',s)})^{-1} \circ (S^{(s,t)})^{-1}(x) - \epsilon\right)$$
$$= S^{(s',s)} \circ \left((S^{(s',t)})^{-1}(x) - \epsilon\right) \tag{10.101}$$

Then for any $\epsilon > 0$ one can find $\delta(\epsilon) > 0$ such that $\delta(\epsilon) \downarrow 0$ as $\epsilon \downarrow 0$ and

$$(S^{(s,t)})^{-1}(x) - \delta(\epsilon) > S^{(s',s)} \circ \left((S^{(s',t)})^{-1} - \epsilon\right) \tag{10.102}$$

Thus

$$S^{(s,t)}\left((S^{(s,t)})^{-1}(x) - \delta(\epsilon)\right) > S^{(s,t)} \circ S^{(s',s)}\left((S^{(s',t)})^{-1}(x) - \epsilon\right)$$
$$= S^{(s',t)}\left((S^{(s',t)})^{-1}(x) - \epsilon\right) \tag{10.103}$$

Letting $\epsilon \downarrow 0$ in the inequality (10.103) yields the necessary lower bound, which together with (10.100) proves the lemma. □

Whenever $t = T$ is fixed, the function $|m_x(\cdot, T)|$ is the family size of the individual x in generation T as a function of the degree of relatedness. By Lemma 10.3.2 it is a decreasing function on I. Finally, we define the associated empirical distribution of the functions $|m_x(\cdot, T)|$ ('genealogical function')

$$\mathcal{K}_T = \int_0^1 dx \delta_{|m_x(\cdot, T)|} \tag{10.104}$$

This construction (10.104) allows us to associate an empirical distribution $\mathcal{K}_T$ with any flow of probability measures in the sense of Definition 10.3.1.

If, in addition, we assume that $[0, T] \subset I$ and $|m_x(\cdot, T)|$ are right-continuous, then $1 - |m_x(\cdot, T)|$ are probability distribution functions. Then we will think of $\mathcal{K}_T$ as a map from flows of probability measures into $M_1(M_1([0, 1]))$, which we call the genealogical map.

10.3.2 Coalescent processes

Now, let us define the exact degree of relatedness between two individuals $x, y \in [0, 1]$ with respect to a flow of measures (10.94) as

$$\gamma_T(x, y) \equiv \sup(s \in I : y \in m_x(s, T)) \tag{10.105}$$

Lemma 10.3.3 $1 - \gamma_T$ *defines an ultrametric distance on the unit interval.*

Proof By Lemma 10.3.2, for all $x, y \in [0, 1]$, if $s = \gamma_T(x, y)$, then

$$y \in m_x(s', T), \quad \forall s' < s, \quad s' \in I \tag{10.106}$$

It follows from (10.106) that, for any $x, y, z \in [0, 1]$, if $\gamma_T(x, y) \neq \gamma_T(x, z)$, then $\gamma_T(y, z) = \min\{\gamma_T(x, z), \gamma_T(x, y)\}$. In fact, let, e.g., $\gamma_T(x, z) > \gamma_T(x, y)$. Then $z \in m_x(s, T)$ for all $s \in I$ such that $s \leq \gamma_T(x, y)$ and then $\gamma_T(y, z) \geq \gamma_T(x, y)$. From the other point of view, if $\gamma_T(y, z) > \gamma_T(x, y)$, then either $\gamma_T(x, z) \geq \gamma_T(y, z) > \gamma_T(x, y)$ or $\gamma_T(y, z) > \gamma_T(x, z) > \gamma(x, y)$. In the first case, by (10.106) $x \in m_z(s, T)$ for all $s \in I$ such that $s < \gamma_T(y, z)$ and then $\gamma_T(x, y) \geq \gamma_T(y, z)$, which is impossible. In the second, $y \in m_z(s, T)$ for all $s \in I$ such that $s \leq \gamma_T(x, z)$ from where $\gamma_T(x, y) \geq \gamma_T(x, z)$, which is again impossible. Thus $\gamma_T(y, z) = \gamma_T(x, y)$.

Note also that, if $\gamma_T(x, y) = \gamma_T(x, z)$, then $\gamma_T(y, z) \geq \gamma_T(x, y) = \gamma_T(x, z)$. These observations imply that $1 - \gamma_T$ is an ultrametric distance on $[0, 1]$. □

The function γ_T is trivial if the $S^{(s,t)}$ are all continuous, for then $\gamma_T(x, y) = T$ if $x = y$ and $\gamma_T(x, y) = -\infty$ if $x \neq y$, and in a strict sense nobody has any relatives. Of course we are not interested in this situation.

We will have to consider situations where the flow $S^{(s,t)}$ of Definition 10.3.1 is *random*. We will now describe a useful way of characterizing a random genealogical map K_T in this case.

Having defined a distance $1 - \gamma_T$ on $[0, 1]$, we can define in a very natural way the analogous distance on the integers. To do this, consider a family of i.i.d. random variables $\{U_i\}_{i \in \mathbb{N}}$ distributed according to the uniform law on $[0, 1]$. Given such a family, we set

$$\rho_T(i, j) = \gamma_T(U_i, U_j) \tag{10.107}$$

Due to the ultrametric property of the γ_T and the independence of the U_i, for fixed T, the sets $B_i(s) \equiv \{j : \rho_T(i, j) \geq s\}$ form an exchangeable random partition of the integers. Moreover, the family of these partitions as a function of $T - s$ is a stochastic process on the space of integer partitions with the property that, for any $s > s'$, the partition $B_i(s')$ is a coarsening of the partition $B_i(s)$. Such a process is called a *coalescent process* (see, e.g., [23, 24, 25, 26, 27, 33, 206]).

The key observation for our purposes is the possibility of expressing the moments of K_T in terms of this coalescent [23]. Namely, it is plain by the law of large numbers that

$$\lim_{n \uparrow \infty} n^{-1} \sum_{j=1}^{n} \mathbb{1}_{j \in B_k(s)} = |m_{U_i}(s, T)|, \quad \text{a.s.} \tag{10.108}$$

for any i such that $i \in B_k(s)$. This implies, for instance, as shown in [23] that

$$\mathbb{E} \int dx |m_x(s, T)| = \mathbb{P}[2 \in B_1(s)] \tag{10.109}$$

and more generally that

$$\mathbb{E} \int dx |m_x(s, T)|^k = \mathbb{P}[2, 3, \ldots, k+1 \in B_1(s)] \tag{10.110}$$

Here the expectation $\mathbb{E}$ is with respect to the randomness of the family of measures $S^{(s,t)}$, and $\mathbb{P}$ is the law with respect to the random genealogy (depending on both the random measures

10.3 Continuous state branching and coalescent processes

and the i.i.d. random variables U_i). We will need slightly more general expressions, namely a family of moments that determine the law of K_T. These can be written as follows: Take any positive integer p, a collection of positive real numbers $0 < t_1 < \cdots < t_p \leq T$, a positive integer ℓ, and non-negative integers $k_{11}, \ldots, k_{1p}, k_{21}, \ldots, k_{2p}, \ldots, k_{\ell 1}, \ldots, k_{\ell p}$. Then we need

$$M(p, \underline{t}, \underline{\underline{k}}) \equiv \mathbb{E}\left(\int dx\, |m_x(t_1, T)|^{k_{11}} \cdots |m_x(t_p, T)|^{k_{1p}}\right) \cdots$$
$$\cdots \left(\int dx\, |m_x(t_1, T)|^{k_{\ell 1}} \cdots |m_x(t_p, T)|^{k_{\ell p}}\right) \quad (10.111)$$

By (10.108) we have that

$$\int dx\, |m_x(t_1, T)|^{k_{11}} \cdots |m_x(t_p, T)|^{k_{1p}} = \lim_{n \uparrow \infty} n^{-1-k_{11}-\cdots-k_{1p}} \sum_{i=1}^{n}$$
$$\times \sum_{j_1^1, \ldots, j_{k_{11}}^1, \ldots, j_1^p, \ldots, j_{k_{1p}}^p} \mathbb{1}_{j_1^1, \ldots, j_{k_{11}}^1 \in B_{k(i,t_1)}(t_1)} \cdots \mathbb{1}_{j_1^p, \ldots, j_{k_{1p}}^p \in B_{k(i,t_p)}(t_p)} \quad (10.112)$$

where $k(i, t_p)$ is the smallest integer such that $k(i, t_p) \in B_i(t_p)$. Let us note first that in these expressions contributions from terms where two indices are equal can be neglected. Second, since

$$B_{k(i,t_p)}(t_p) \subset B_{k(i,t_{p-1})}(t_{p-1}) \subset \cdots \subset B_{k(i,t_1)}(t_1) \quad (10.113)$$

the summand in (10.112) is the same as

$$\mathbb{1}_{j_1^1, \ldots, j_{k_{11}}^1, \ldots, j_1^p, \ldots, j_{k_{1p}}^p \in B_{k(i,t_1)}(t_1)} \cdots \mathbb{1}_{j_1^p, \ldots, j_{k_{1p}}^p \in B_{k(i,t_p)}(t_p)} \quad (10.114)$$

Then

$$M(p, \underline{t}, \underline{\underline{k}})$$
$$= \lim_{n \uparrow \infty} n^{-\ell - k_{11} - \cdots - k_{\ell p}} \mathbb{E} \sum_{i_1, \ldots, i_\ell} \sum_{j_1^{1,1}, \ldots, j_{k_{11}}^{1,1}, \ldots, j_1^{p,1}, \ldots, j_{k_{1p}}^{p,1}} \cdots$$
$$\cdots \sum_{j_1^{1,\ell}, \ldots, j_{k_{\ell 1}}^{1,\ell}, \ldots, j_1^{p,\ell}, \ldots, j_{k_{\ell p}}^{p,\ell}} \mathbb{1}_{j_1^{1,1}, \ldots, j_{k_{11}}^{1,1}, \ldots, j_1^{p,1}, \ldots, j_{k_{1p}}^{p,1} \in B_{k(i_1,t_1)}(t_1)} \cdots$$
$$\cdots \mathbb{1}_{j_1^{1,\ell}, \ldots, j_{k_{\ell 1}}^{1,\ell}, \ldots, j_1^{p,\ell}, \ldots, j_{k_{\ell p}}^{p,\ell} \in B_{k(i_\ell,t_1)}(t_1)} \cdots$$
$$\cdots \mathbb{1}_{j_1^{p,1}, \ldots, j_{k_{1p}}^{p,1} \in B_{k(i_1,t_p)}(t_p)} \cdots \mathbb{1}_{j_1^{p,\ell}, \ldots, j_{k_{\ell p}}^{p,\ell} \in B_{k(i_\ell,t_p)}(t_p)}$$
$$= \mathbb{P}\left[J_{11} \in B_1(t_1), \ldots, J_{1p} \in B_1(t_p), \ldots J_{\ell 1} \in B_\ell(t_1), \ldots, J_{\ell p} \in B_\ell(t_p)\right] \quad (10.115)$$

where

(i) $J_{11}, \ldots, J_{\ell 1}$ is a disjoint partition of $\{\ell + 1, \ell + 2, \ldots, k_{11} + \cdots + k_{\ell p} + \ell\}$,
(ii) for all $j = 1, \ldots, \ell, i = 1, \ldots, p$, $J_{ji} \supset J_{ji+1}$, and
(iii) $|J_{ji}| = k_{ji} + k_{ji+1} + \cdots + k_{jp}$.

By exchangeability, the choice of the partition and the subsets is irrelevant. The probabilities (10.115) can be expressed alternatively in the form

$$\mathbb{P}(\rho_T(i,j) \geq t_{m(i,j)}, \ \forall i \in \{\ell+1,\ldots,k_{11}+\cdots+k_{\ell p}+\ell\}, j \in \{1,\ldots,\ell\}) \quad (10.116)$$

where $m(i,j) \in \{1,\ldots,p\}$. Thus the genealogical map K_T is completely determined by the probabilities (10.115) or (10.116) of the corresponding coalescent through the family of its moments.

10.3.3 Finite N setting for the CREM

We will now show that for finite N we can use the general construction to relate the geometric description of the Gibbs measure on $\mathcal{S}_N$ to the genealogical description of a family of embedded measures on $[0,1]$. We can map the Gibbs measure on $\mathcal{S}_N$ to a flow of probability measures as follows. Let, for $t \in (0,1]$,

$$\tilde{\mu}^t_{\beta,N} \equiv \sum_{\sigma \in \mathcal{S}_N} \delta_{r_{[tN]}(\sigma)} \mu_{\beta,N}(\sigma) \quad (10.117)$$

and let $\theta^t_{\beta,N}$ be the probability distribution function of $\tilde{\mu}^t_{\beta,N}$:

$$\theta_{\beta,N}(x) = \mu_{\beta,N}(\sigma : r_{[tN]}(\sigma) \leq x) \quad (10.118)$$

It will be useful to look at the jumps of these distribution functions as an image of the hypercube: e.g., for $t=1$, each σ is represented as a jump of size $\mu_{\beta,N}(\sigma)$. Varying t, we get a picture where all blocks, $\Sigma_\sigma(t)$, are attributed an interval of the size corresponding to their mass. We will want to think of these intervals as 'individuals' at time t, and all the σ that share the first $[tN]$ coordinates as the 'offspring' of this individual. We want to describe the genealogical structure of all these families.

For $s, t \in [0,1]$ we define a family of compatible distribution functions in the sense of Definition 10.3.1 by

$$S^{(s,t)}_{\beta,N}(a) \equiv \theta^t_{\beta,N} \circ \left(\theta^s_{\beta,N}\right)^{-1}(a) = \sum_\sigma \mu_{\beta,N}(\sigma) \mathbb{I}_{\{\theta^t(r_{[sN]}(\sigma)) \leq a\}} \quad (10.119)$$

Let us note that, for all $t \geq s$, it turns out that $S^{(s,t)}_{\beta,N} = S^{(s,s)}_{\beta,N}$. In fact, $S^{(s,t)}_{\beta,N}$ is simply the function that has jumps at the $2^{[sN]}$ points $\mu_{\beta,N}(r_{[sN]}(\sigma)), \sigma \in \mathcal{S}_N$, and that at each of these points jumps precisely to the value $\mu_{\beta,N}(r_{[sN]}(\sigma))$. From this we get very easily that:

Lemma 10.3.4 *The functions (10.119) satisfy the assumptions of Definition 10.3.1 with $I = [0,1]$.*

Proof We need to show that, for any $s' < s \leq t$,

$$S^{(s,t)}_{\beta,N} \circ S^{(s',s)}_{\beta,N} = S^{(s',t)}_{\beta,N} \quad (10.120)$$

But

$$S^{(s,t)}_{\beta,N} \circ S^{(s',s)}_{\beta,N} = \theta^s_{\beta,N} \circ \left(\theta^t_{\beta,N}\right)^{-1} \theta^{s'}_{\beta,N} \circ \left(\theta^s_{\beta,N}\right)^{-1} \quad (10.121)$$

$$= \theta^s_{\beta,N} \circ \left(\theta^s_{\beta,N}\right)^{-1} \theta^{s'}_{\beta,N} \circ \left(\theta^{s'}_{\beta,N}\right)^{-1} \quad (10.122)$$

$$= \theta^{s'}_{\beta,N} \circ \left(\theta^{s'}_{\beta,N}\right)^{-1} = S^{(s',t)} \quad (10.123)$$

as is easily checked (see Fig. 10.6), since $S^{(s',t)}_{\beta,N}$ does not depend on t. □

10.3 Continuous state branching and coalescent processes

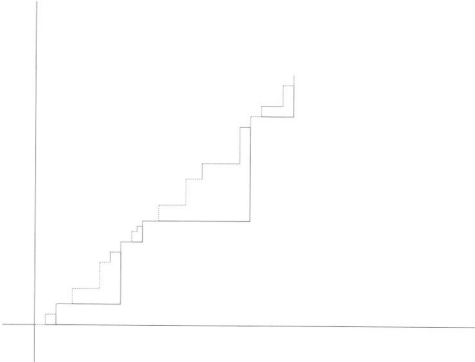

Figure 10.6 Two functions $S_{\beta,N}^{(s,t)}$, $S^{(s',s)}$ for $s' < s$.

Since the functions (10.119) satisfy Definition 10.3.1, we are entitled to apply to them the construction of the genealogy explained above. Indeed, if y_i^s, $i = 1, 2, \ldots$, is the ordered enumeration of the positions where $S_{\beta,N}^{(s,t)}$ jumps, then

$$m_x(s, t) = \left(y_{i-1}^s, y_i^s\right] \text{ with } |m_x(s, t)|$$
$$= |x_i^s|, \text{ if } x \in \left(y_{i-1}^s, y_i^s\right], \ i = 1, \ldots, 2^{[sN]} \qquad (10.124)$$

Consequently, we can also define the genealogical distance, $1 - \gamma_1$, as in (10.105). Note, however, that there is an explicit formula for this quantity in the finite N setting, namely

$$\gamma_1(x, y) = d_N\left(r_N^{-1} \circ \theta_{b,N}^{-1}(x), r_N^{-1} \circ \theta_{b,N}^{-1}(x)\right) \qquad (10.125)$$

Note that the inverse of $\theta_{\beta,N}$ maps $(0, 1]$ onto the sets of points where $\theta_{\beta,N}$ jumps, which is the range of r_N, so that the inverse of r_N is well defined in this equation.

We may associate to this genealogy the genealogical map (10.104) K_T and the coalescent process on the integers. The next lemma expresses the geometry of the Gibbs measure of the CREM contained in the empirical distance distribution function (10.45) $\mathcal{K}_{\beta,N}$ in terms of the genealogy induced by the functions (10.119).

Lemma 10.3.5 *We have*

$$\mathcal{K}_{\beta,N} = K_1^{\beta,N}, \qquad (10.126)$$

where the empirical distance distribution function $\mathcal{K}_{\beta,N}$ is defined in (10.45) and $K_1^{\beta,N}$ is the genealogical map (10.104) with $T = 1$ of the flow of measures with probability distribution functions (10.119).

Proof Clearly we have that

$$K_1^{\beta,N} = \int_0^1 dx \delta_{|m_x(\cdot,1)|} = \sum_{\sigma \in \sigma_N} \mu_{\beta,N}(\sigma) \delta_{|m_{r_N(\sigma)}(\cdot,1)|} \qquad (10.127)$$

On the other hand, $m_{r_N(\sigma)}(s, 1)$ is precisely the interval that corresponds to the jump of $S_{\beta,N}^{(s,1)}$ that contains the point $r_N(\sigma)$, and the the size of the jump is $\mu_{\beta,N}\left(\sigma' : d_N(\sigma, \sigma') > s\right)$. Therefore, $K_1^{\beta,N} = \mathcal{K}_{\beta,N}$. □

10.3.4 Neveu's continuous state branching process

Another example of flows of probability measures satisfying Definition 10.3.1 arises in the context of continuous state branching processes [25]. The basic object here is a continuous-state branching process $X(t)$, i.e. a continuous-time Markov process on $\mathbb{R}^+$ that is killed when it reaches zero, and that has the following branching property: Denote by $X(\cdot, a)$ the process started with initial condition $X(0, a) = a$. For any $a, b > 0$, if $X'(\cdot, b)$ and $X(\cdot, a)$ are independent copies, then $X(\cdot, a + b)$ has the same law as $X'(\cdot, b) + X(\cdot, a)$. This relation allows us to interpret $X(t, a)$ as the mass at time t of a population with initial mass a. Moreover, this property allows us to construct a genuine two-parameter process $(X(t, a), t, a \geq 0)$. The process $X(t, a)$ is characterized by the property that, for any $a, b \geq 0$, $X(\cdot, a + b) - X(\cdot, a)$ is independent of the processes $X(\cdot, c)$, for all $c \leq a$, and its law is the same as that of $X(\cdot, b)$. The right-continuous version of $X(t, \cdot)$ is a subordinator. Bertoin and Le Gall [25] prove the following proposition, based on the Markov property of this process.

Proposition 10.3.6 *On some probability space there exists a process* $(\tilde{S}^{(s,t)}(a), 0 \leq s \leq t, a \geq 0)$ *such that:*

(i) *For any* $0 \leq s \leq t$, $\tilde{S}^{(s,t)}$ *is a subordinator with Laplace exponent* $u_{t-s}(\lambda)$.
(ii) *For any integer* $p \geq 3$ *and* $0 \leq t_1 \leq t_2 \leq \cdots \leq t_p$, *the subordinators* $\tilde{S}^{(t_1,t_2)}$, $\tilde{S}^{(t_2,t_3)}$, $\ldots, \tilde{S}^{(t_{p-1},t_p)}$ *are independent, and*

$$\tilde{S}^{(t_1,t_p)}(a) = \tilde{S}^{(t_{p-1},t_p)} \circ \tilde{S}^{(t_{p-1},t_p)} \circ \cdots \circ \tilde{S}^{(t_2,t_3)} \circ \tilde{S}^{(t_1,t_2)}(a), \quad \forall a \geq 0, \text{ a.s.} \quad (10.128)$$

(iii) *The processes* $\tilde{S}^{(0,t)}(a)$ *and* $X(t, a)$ *have the same finite-dimensional marginals.*

The process $\tilde{S}^{(s,t)}$ allows us to construct a flow of probability distribution functions by setting

$$S^{(s,t)}(x) \equiv \frac{1}{X(t,1)} \tilde{S}^{(s,t)}(X(s,1)x), \quad 0 \leq s \leq t \leq 1 \quad (10.129)$$

Given I any countable subset of $\mathbb{R}^+$, they satisfy the assumptions of Definition 10.3.1 almost surely.

We are interested in a particular case of Neveu's continuous state branching process X_t with

$$E(e^{-\lambda X_t} \mid X_0 = a) = e^{-u_t(\lambda)a}, \quad u_t(\lambda) = \lambda^{e^{-t}} \quad (10.130)$$

In this case the $\tilde{S}^{(s,t)}$ are stable subordinators with index e^{s-t}. Then the normalized stable subordinators $S^{(s,t)}$ of (10.129) form a family of random probability distribution functions verifying Definition 10.3.1. Thus the genealogical construction of Section 10.3.1 applies to them.

Finally, note that, if we take an increasing function $t(y) \geq 0$ for $y \in [0, 1]$, then we may consider the time-changed flow $\bar{S}^{(y,z)} = S^{(t(y),t(z))}$, $0 \leq y \leq z$, satisfying again Definition 10.3.1 and therefore allowing the genealogical construction of Section 10.2.

Bertoin and Le Gall [25] showed that the coalescent process on the integers induced by $S^{(s,t)}$ of (10.129) associated to Neveu's process (10.130) coincides with the coalescent process constructed by Bolthausen and Sznitman [33]. They also proved the following remarkable result connecting the collection of subordinators to Ruelle's generalized random energy model. Take the parameters $0 < x_1 < \cdots < x_p < 1$ and $0 < t_1 < \cdots < t_p$ linked by the identities

$$t_k = \ln x_{k+1} - \ln x_1 \tag{10.131}$$

for $k = 0, \ldots, p-1$, and $t_p = -\ln x_1$. Then the law of the family of jumps of the normalized subordinators $S^{(t_k,t_p)}$, for $k = 0, \ldots, p-1$, is the same as the law of Ruelle's probability cascades with parameters x_i, $i = 1, \ldots, p$.

Now consider a GREM with finitely many hierarchies and parameters such that the points $y_0 = 0$ and $0 < y_1 < \ldots < y_p \leq 1$ are the extremal points of the concave hull of A. Let us recall that $\lim_{N \uparrow \infty} \mathbb{E}\psi_{\beta,N}(y) = \mathbb{E}\psi_\beta(y)$ can be computed by (10.43) for any $y \in [0, 1]$. Now set

$$\mathbb{E}\psi_\beta(y_{i-1}) = x_i, \quad i = 1, \ldots, p \tag{10.132}$$

where all of the $x_i < 1$. In Theorem 1.9 of [52] we proved that the point process

$$\sum_\sigma \delta_{\{\mu_{\beta,N}(\sigma': d_N(\sigma,\sigma') > y_1), \ldots, \mu_{\beta,N}(\sigma': d_N(\sigma,\sigma') > y_p)\}} \tag{10.133}$$

in $[0, 1]^p$ converges to Ruelle's probability cascades with parameters x_i, $i = 1, \ldots, p$. (The convergence of the marginals of the process (10.133) for the GREM, under the assumption that for any given hierarchy $i = 1, \ldots, p$ and $N > 0$ the number of configurations $\{\sigma' : d_N(\sigma, \sigma') > y_i\}$ is the same for all $\sigma \in \Sigma_N$, has been also established in Proposition 9.6 of [34].)

Combining these two results yields:

Proposition 10.3.7 *Let $\mu_{\beta,N}$ be the Gibbs measure associated to a GREM with finitely many hierarchies satisfying (10.132) at the extremal points y_i, $i = 1, \ldots, p$ of the concave hull of the function A. Then the family of distribution functions $S_{\beta,N}^{(y_k,y_p)}$, $k = 1, 2, \ldots, p$ defined according to (10.119) converges in law, and the limit has the same distribution as the family of normalized stable subordinators (10.129) $S^{(t_k,t_p)}$, $k = 0, 1, \ldots, p-1$ in the sense that the joint distribution of their jumps has the same law, provided t_k are chosen according to (10.131) and (10.132).*

From the preceding proposition we expect that Neveu's process will provide the universal limit for all of our CREMs. The dependence on the particular model (i.e. the function A) and on the temperature must come from a rescaling of time. Set

$$x(y) \equiv \mathbb{E}\psi_\beta(y) = \begin{cases} \frac{\sqrt{2\ln 2}}{\beta\sqrt{\bar{a}(t)}}, & \text{if } t < t_\beta \\ 1, & \text{if } t \geq t_\beta \end{cases} \tag{10.134}$$

where $\bar{a}$ is the right-derivative of the convex hull of the function A, $t_\beta = \sup(t : \frac{\sqrt{2\ln 2}}{\beta\sqrt{\bar{a}(t)}} < 1)$ (here $\mathbb{E}\psi_\beta(y)$ is defined by the function A through (10.43)). Set also

$$T = -\ln x(0), \quad t(y) = T + \ln x(y) \tag{10.135}$$

Define the flow of probability distribution functions

$$\bar{S}^{(y,z)}(x) \equiv S^{(t(y),t(z))}(x) \tag{10.136}$$

where $S^{(s,t)}$ is the flow of functions (10.129) associated with Neveu's process (10.130). Let $\bar{\mathcal{K}}_T^{t(y)}$ be the genealogical map (10.104) associated with this flow.

Theorem 10.3.8 *Consider a continuous random energy model with general function A such that A does not touch its convex hull $\bar{A}$ in the interior of any interval where $\bar{A}$ is linear. Then*

$$\mathcal{K}_{\beta,N} = K_1^{\beta,N} \xrightarrow{\mathcal{D}} \bar{\mathcal{K}}_1^{t(y)}. \tag{10.137}$$

Here $\mathcal{K}_{\beta,N}$ is the empirical distance distribution function (10.45), $K_1^{\beta,N}$ is the genealogical map (10.104) of the flow of probability distribution functions (10.119) and the equality $\mathcal{K}_{\beta,N} = K_1^{\beta,N}$ holds by Lemma 10.3.5. Theorem 10.3.8 is the main result of this section. It expresses the geometry of the limiting Gibbs measure contained in $\mathcal{K}_{\beta,N}$ in terms of the genealogy of Neveu's branching process via the deterministic time change (10.135).

10.3.5 Coalescence and Ghirlanda–Guerra identities

We now prove Theorem 10.3.8. As was remarked above, K_T associated with a flow of measures is completely determined by its moments (10.111), which can be expressed via genealogical distance distributions of the corresponding coalescent (10.116). So, we will prove that the moments of $\mathcal{K}_{\beta,N}$, which are the n-replica distance distributions in our spin-glass model (10.89), converge to the genealogical distance distributions on the integers (10.116) constructed from the flow of compatible measures with distribution functions $\bar{S}^{(y,z)}$ (10.136). But the flow $\bar{S}^{(y,z)}$ is the time-changed flow (10.129) of Neveu's branching process (10.130) that by [25] corresponds to the coalescent of Bolthausen–Sznitman. Therefore, its genealogical distance distributions on the integers are those of Bolthausen–Sznitman coalescent under this time change (10.135). Then the proof of Theorem 10.3.8 is reduced to the following Theorem 10.3.9 that gives in addition the connection between the n-replica distance distribution function of the CREM and the genealogical distance distribution function of the Bolthausen–Sznitman coalescent.

Theorem 10.3.9 *Under the same assumptions as in Theorem 10.3.8, for any $n \in \mathbb{N}$,*

$$\lim_{N\uparrow\infty} \mathbb{E}\mu_{\beta,N}^{\otimes n}\left(d_N(\sigma^1, \sigma^2) \leq y_1, \ldots, d_N(\sigma^{n-1}, \sigma^n) \leq y_{n(n-1)/2}\right)$$
$$= \mathbb{P}\left(\rho_T(1,2) \leq t(y_1), \ldots, \rho_T(n-1, n) \leq t(y_{n(n-1)/2})\right) \tag{10.138}$$

where $t(y)$ is defined in (10.135) via (10.134). The distance ρ_T is the distance on integers for the Bolthausen–Sznitman coalescent, induced through (10.107) by the genealogical distance γ_T of the flow of measures $S^{(s,t)}$ (10.129) of Neveu's branching process (10.130).

Proof The fact that in the Bolthausen–Sznitman coalescent $\mathbb{P}(\rho_T(1,2) \leq t) = e^{t-T}$ and the convergence (10.43) imply the statement of the theorem for $n = 2$:

$$\mathbb{E}\mu_{\beta,N}^{\otimes 2}(d_N(\sigma, \sigma') \leq y) \to x(y) = e^{t(y)-T} = \mathbb{P}(\rho_T(1,2) \leq t(y)) \tag{10.139}$$

The proof of the theorem for $n > 2$, and in fact the entire identification of the limiting processes with objects constructed from Neveu's branching process, relies on the Ghirlanda–Guerra identities [127] that were derived for the models considered here in [53]. We restate this result in a slightly modified form. Let us recall that the family of measures (10.89) $\mathbb{Q}_{\beta,N}^{(n)}$ is determined on the space $[0, 1]^{n(n-1)/2}$ as $\mathbb{E}\mu_{N,\beta}^{\otimes n}(\underline{d_N} \in \cdot)$ where $\underline{d_N}$ denotes the vector of replica distances $d_N(\sigma^k, \sigma^l), 1 \leq k < l \leq n$. Denote by $\mathcal{B}_k$ the vector of the first $k(k-1)/2$ coordinates.

Theorem 10.3.10 [51] *The family of measures $\mathbb{Q}_{\beta,N}^{(n)}$ converge to limiting measures $\mathbb{Q}_{\beta}^{(n)}$ for all finite n, as $N \uparrow \infty$. Moreover, these measures are uniquely determined by the distance distribution functions $\mathbb{E}\psi_\beta(y) = x(y)$ (10.43). They satisfy, for any $y \in [0, 1]$, $n \in \mathbb{N}$ and $k \leq n$,*

$$\mathbb{Q}_\beta^{(n+1)}\left(d(k, n+1) \leq y | \mathcal{B}_n\right) = \frac{x(y)}{n} + \frac{1}{n}\sum_{l \neq k}^{n} \mathbb{Q}_\beta^{(n)}\left(d(k, l) \leq y | \mathcal{B}_n\right) \quad (10.140)$$

Let us recall that due to the ultrametric property of d_N, these identities determine the measures $\mathbb{Q}_\beta^{(n)}$ uniquely. Thus, we must show that the right-hand side of (10.138) satisfies, for $t < T$,

$$\mathbb{P}\left(\rho_T(k, n+1) \leq t \mid \mathcal{B}_n\right) = \frac{1}{n}e^{t-T} + \frac{1}{n}\sum_{l \leq n, l \neq k} \mathbb{P}\left(\rho_T(k, l) \leq t \mid \mathcal{B}_n\right) \quad (10.141)$$

that can be equivalently written as

$$\mathbb{P}\left(\rho_T(k, n+1) > t \mid \mathcal{B}_n\right) = \frac{|l \in \{1, \ldots, n\} : \rho_T(k, l) > t| - e^{t-T}}{n}. \quad (10.142)$$

There are *two* ways to verify that (10.141) holds for the Bolthausen–Sznitman coalescent.

The first one is to observe that relation (10.141) involves only the marginals of the coalescent at a finite set of times. By Theorem 5 of Bertoin-Le Gall [25], these can be expressed in terms of Ruelle's probability cascades modulo the appropriate time change. Thus, by Theorem 1.9 of [52] these probabilities can be expressed as limits of a suitably constructed GREM (with finitely many hierarchies) for which the Ghirlanda–Guerra relations do hold by Proposition 1.12 of [52]. Thus (10.141) is satisfied.

The second way is to verify the Ghirlanda–Guerra relations (10.142) directly for the Bolthausen–Sznitman coalescent. As this leads us a bit too far afield into coalescence processes, we will not give the proof (which can be found in [51] and essentially is already contained in [33]). □

11

The SK models and the Parisi solution

> Deviner avant de démontrer! Ai-je besoin de rappeler que c'est ainsi que se sont faites toutes les découvertes importantes?[1]
>
> Henri Poincaré. *La valeur de la science.*

We now return to the class of Gaussian mean-field models whose covariance is a function of the Hamming distance, respectively the overlap R_N. They were the original mean-field models for spin-glasses, introduced by D. Sherrington and S. Kirkpatrick in [221], and are considered to be the most natural, or physically relevant ones. Certainly, a large part of the excitement and interest, both in physics and in mathematics, has come from the fact that Parisi suggested a very complex solution, more precisely a formula for the free energy of these models, that has in many respects been seen as rather mysterious and unexpected. Moreover, this formula for the free energy suggested a structure of the Gibbs measures, which is to some extent similar to what we have encountered already in the CREMs. Much of the mystery of Parisi's solution certainly resulted from the original method of its derivation, which involved objects such as zero-dimensional matrices, and that appeared to transcend the realm of standard mathematics.

Fortunately, these mysteries have been greatly clarified, mainly due to an ingenious discovery of F. Guerra [133] in 2002, and subsequent work by Aizenman, Sims, and Starr [7], and M. Talagrand [240], that allows a clear and mathematically rigorous formulation of the Parisi solution.

11.1 The existence of the free energy

A key observation that has led to the recent breakthroughs in spin-glass theory, made by F. Guerra and F.-L. Toninelli, was that Lemma 10.2.1, or rather the interpolation idea used in its proof, is a powerful tool in this context. This observation allowed them to solve a long-standing and embarrassing problem: to prove the existence of the mean free energy in the SK models (and for that matter most other mean-field spin-glasses).

Theorem 11.1.1 [134] *Assume that X_σ is a normalized Gaussian process on S_N with covariance*

$$\mathbb{E} X_\sigma X_\tau = \xi(R_N(\sigma, \tau)) \tag{11.1}$$

---

[1] Approximately: Guessing before proving! Need I remind you that it is so that all important discoveries have been made?

where $\xi : [-1, 1] \to [0, 1]$ is convex and even. Then

$$\lim_{N\uparrow\infty} \frac{-1}{\beta N} \mathbb{E} \ln \mathbb{E}_\sigma e^{\beta\sqrt{N}X_\sigma} \equiv f_\beta \qquad (11.2)$$

exists.

Proof The proof of this fact is frightfully easy, once you think about using Lemma 10.2.1. Choose any $1 < M < N$. Let $\sigma = (\hat\sigma, \check\sigma)$ where $\hat\sigma = (\sigma_1, \sigma_2, \ldots, \sigma_M)$, and $\check\sigma = (\sigma_{M+1}, \ldots, \sigma_N)$. Define independent Gaussian processes $\hat X$ and $\check X$ on $\mathcal{S}_M$ and $\mathcal{S}_{N-M}$, respectively, such that

$$\mathbb{E}\check X_{\hat\sigma}\hat X_{\hat\tau} = \xi(R_M(\hat\sigma, \hat\tau)) \qquad (11.3)$$

and

$$\mathbb{E}\check X_{\check\sigma}\hat X_{\check\tau} = \xi(R_{N-M}(\check\sigma, \check\tau)) \qquad (11.4)$$

Set

$$Y_\sigma \equiv \sqrt{\frac{M}{N}}\hat X_{\hat\sigma} + \sqrt{\frac{N-M}{N}}\check X_{\check\sigma} \qquad (11.5)$$

Clearly,

$$\begin{aligned}\mathbb{E}Y_\sigma Y_\tau &= \tfrac{M}{N}\xi(R_M(\hat\sigma, \hat\tau)) + \tfrac{N-M}{N}\xi(R_{N-M}(\check\sigma, \check\tau))\\ &\geq \xi\left(\tfrac{M}{N}R_M(\hat\sigma, \hat\tau) + \tfrac{N-M}{N}(R_{N-M}(\check\sigma, \check\tau))\right) = \xi(R_N(\sigma, \tau))\end{aligned} \qquad (11.6)$$

Define real-valued functions $F_N(x) \equiv \ln \mathbb{E}_\sigma e^{\beta\sqrt{N}x_\sigma}$ on $\mathbb{R}^{2^N}$. It is straightforward that

$$\mathbb{E}F_N(Y) = \mathbb{E}F_M(X) + \mathbb{E}F_{N-M}(X) \qquad (11.7)$$

A simple computation shows that, for $\sigma \neq \tau$,

$$\frac{\partial^2}{\partial x_\sigma \partial x_\tau} F_N(x) = -\frac{2^{-2N}\beta^2 N e^{\beta\sqrt{N}(x_\sigma + x_\tau)}}{Z_{\beta,N}^2} \leq 0 \qquad (11.8)$$

Thus, Lemma 10.2.1 tells us that

$$\mathbb{E}F_N(X) \geq \mathbb{E}F_N(Y) = \mathbb{E}F_M(X) + \mathbb{E}F_{N-M}(X) \qquad (11.9)$$

This implies that the sequence $-\mathbb{E}F_N(X)$ is subadditive, and this in turn implies (see Section 3.3) that the free energy exists, provided it is bounded, which is easy to verify (see, e.g., the discussion on the correct normalization in the SK model). $\square$

The same ideas can be used for other types of Gaussian processes, e.g. the GREM-type models discussed above (see [82]).

Convergence of the free energy in mean implies readily almost sure convergence. This follows from a general *concentration of measure* principle for functions of Gaussian random variables, analogous to Theorem 7.1.3. The following result can be found, e.g., in [167], page 23:

Theorem 11.1.2 *Let $X_1, \ldots, X_M$ be independent standard normal random variables, and let $f : R^M \to R$ be Lipschitz continuous with Lipschitz constant $\|f\|_{\text{Lip}}$. Set $g \equiv f(X_1, \ldots, X_M)$. Then*

$$\mathbb{P}[|g - \mathbb{E}g| > x] \leq 2\exp\left(-\frac{x^2}{2\|f\|_{\text{Lip}}^2}\right) \qquad (11.10)$$

Corollary 11.1.3 *Assume that the function ξ is analytic with positive Taylor coefficients. Then*

$$\mathbb{P}[|f_{\beta,N} - \mathbb{E}f_{\beta,N}| > x] \leq 2\exp\left(-\frac{Nx^2}{2\beta^2}\right) \quad (11.11)$$

In particular, $\lim_{N\uparrow\infty} f_{\beta,N} = f_\beta$, *almost surely.*

Proof If $\xi(x) = \sum_{p=1}^\infty a_p^2 x^p$, we can construct X_σ as

$$X_\sigma = \sum_{p=1}^\infty a_p N^{-p/2} \sum_{1 \leq i_1,\ldots,i_p \leq N} J^{(p)}_{i_1,\ldots,i_p} \sigma_{i_1} \cdots \sigma_{i_p} \quad (11.12)$$

with standard i.i.d. Gaussians $J^{(p)}_{i_1\cdots i_p}$. Check that, as a function of these variables, the free energy is Lipschitz with Lipschitz constant $\beta N^{-1/2}$. □

Remark 11.1.4 The first assertion of Corollary 11.1.3 had been noticed well before Theorem 11.1.1 was proven (see [45, 231]).

11.2 2nd moment methods in the SK model

In the GREM, the computation of the free energy can be achieved essentially by a clever improvement of the method of truncated second moments that we have explained in detail in the REM. Before we turn to more powerful tools in the next subsection, it is worthwhile to discuss the use and limitations of this method in the *p*-spin SK model.

11.2.1 Classical estimates on extremes

We have seen in the REM and GREM that estimates on the extremes of our processes are crucial. In this subsection we prove upper and lower bounds on extremes for the *p*-spin SK models, using classical techniques from extreme value theory. We will see that the results are not terribly satisfying, in particular if p is not large.

We set

$$X_\sigma \equiv -\frac{1}{\sqrt{N}} H_N^{I-SK}(\sigma) \quad (11.13)$$

Classical tools for estimating maxima of Gaussian processes are comparisons with simpler processes for which such estimates are more readily available. The most basic of these comparison methods rely on *Slepian's lemma*, which is just the specialization of Lemma 10.2.1 to the case where the function $F(x_1,\ldots,x_n) \equiv \max_i x_i$.

Lemma 11.2.1 [227] *Let X_σ, Y_σ be standardized Gaussian processes on some set $\mathcal{S}_N$. Assume that for any $\sigma, \tau \in \mathcal{S}_N$,*

$$\mathbb{E}X_\sigma X_\tau \leq \mathbb{E}Y_\sigma Y_\tau \quad (11.14)$$

Then, for any $x \in \mathbb{R}$,

$$\mathbb{P}\left[\sup_{\sigma \in \mathcal{S}_N} X_\sigma > x\right] \geq \mathbb{P}\left[\sup_{\sigma \in \mathcal{S}_N} Y_\sigma > x\right] \quad (11.15)$$

In particular,

$$\mathbb{E}\sup_{\sigma\in\mathcal{S}_N} X_\sigma \geq \mathbb{E}\sup_{\sigma\in\mathcal{S}_N} Y_\sigma \qquad (11.16)$$

An immediate consequence of Slepian's lemma is the fact that the ground-state energies of the p-spin models are monotone in p and dominated by that of the REM. Naturally one would like to get a lower bound.

Proposition 11.2.2 *With probability one, for all but finitely many N,*

$$\limsup_{N\uparrow\infty}\sup_\sigma \frac{|H_N^{p-SK}(\sigma)|}{N} \geq \sqrt{2\ln 2}(1-c_p) \qquad (11.17)$$

where for large p, $c_p \sim p2^{-p}$.

Proof We will now give a proof based on a standard Gaussian comparison lemma, originally due to Slepian [227] (see, e.g., [162], Theorem 4.2.1.). The idea is to see to what extent the independent random variables X_σ of the REM are reasonable approximations to the dependent X_σ of the SK models, in the sense that their extremes are comparable. To this end one would like to compare the probabilities $\mathbb{P}[\forall\sigma, X_\sigma \leq u_N]$ to the corresponding value in the i.i.d. case, $\Phi(u_N)^{2^N}$. The normal comparison lemma states the following:

Lemma 11.2.3 [162] *Let Z_i be a family of standard normal variables with covariance matrix R. Then, for any real u,*

$$\left|\mathbb{P}[Z_1 \leq u, \ldots, Z_n \leq u] - \Phi(u)^n\right|$$
$$\leq \frac{1}{2\pi}\sum_{1\leq i<j\leq n} \frac{|R_{ij}|}{1-R_{ij}^2}\exp\left(-\frac{u^2}{1+|R_{ij}|}\right) \qquad (11.18)$$

Applied to our problem, a first attempt would yield (with $n = 2^N$) the bound (due to the symmetry, we identify σ and $-\sigma$)

$$\left|\mathbb{P}[\forall\sigma, X_\sigma \leq u] - \Phi(u)^{2^N}\right| \qquad (11.19)$$
$$\leq \frac{1}{2\pi}\sum_{\sigma\neq\sigma'}\frac{|R(\sigma,\sigma')^p|}{1-R(\sigma,\sigma')^{2p}}\exp\left(-\frac{u^2}{1+|R(\sigma,\sigma')|^p}\right)$$
$$= \frac{1}{2\pi}\sum_{m\in\mathcal{M}_N\setminus\{-1,1\}}\sum_{\sigma\neq\sigma'}\mathbb{I}_{\{R(\sigma,\sigma')=m\}}\frac{|m|^p}{1-m^{2p}}e^{-\frac{u^2}{1+|m|^p}}$$
$$= \frac{1}{2\pi}\sum_{m\in\mathcal{M}_N\setminus\{-1,1\}}\frac{|m|^p}{1-m^{2p}}e^{-\frac{u^2}{1+|m|^p}}2^{2N}\frac{2e^{-NI(m)}}{\sqrt{2\pi N(1-m^2)}}$$

where

$$\mathcal{M}_N \equiv \{-1, -1+2/N, \ldots, 1-2/N, 1\} \qquad (11.20)$$

is the set of possible values the 'overlap' $R(\sigma,\sigma')$ can take on. We would need to have the right-hand side tend to zero for values of u at which the maximum of 2^N standard Gaussians is taken, i.e. where $\Phi(u)^{2^N}$ is between zero and one, i.e. when u is very close to

$u_N \equiv \sqrt{N 2 \ln 2}$. But (11.20) gives

$$\left| \mathbb{P}\left[\forall \sigma, X_\sigma \leq u_N \right] - \Phi(u_N)^{2^N} \right| \tag{11.21}$$

$$\leq \frac{1}{2\pi} \sum_{m \in \mathcal{M}_N \setminus \{-1,1\}} \frac{2}{\sqrt{2\pi N(1-m^2)}} \frac{|m|^p}{1-m^{2p}} e^{-N\left(I(m) - \frac{m^p 2 \ln 2}{1+|m|^p}\right)}$$

One now sees immediately the crucial difference between the cases $p=2$ and $p>2$. In the former, due to the fact that $I(m) \sim m^2/2$ for small m, spin configurations with overlap close to zero give a large contribution to this sum. On the other hand, if $p>3$, the contribution from a neighbourhood of $m=0$ is of the order $N^{-(p-1)/2}$ only. One might at first hope that the larger values of m also give no large contribution, but closer inspection shows that this is not the case. Indeed, for all p, there is a (shrinking with N) region near $|m|=1$ that gives a non-vanishing contribution to the sum, and thus it will not be true that the left-hand side of (11.21) will be small for any p. This means that there will be some effect coming from the strong correlation of spin configurations that are very similar.

In the sequel we need the following properties of the functions $I(m) - \frac{m^p 2 \ln 2}{1+|m|^p}$:

Lemma 11.2.4 *For any $p \geq 3$, there exist $m_p > 0$ such that, for all $m < m_p$,*

$$\psi_p(m) \equiv I(m) - \frac{m^p 2 \ln 2}{1+|m|^p} > 0 \tag{11.22}$$

Moreover, m_p can be chosen in such a way that $m_p \uparrow 1$, as $p \uparrow \infty$. For p large, $m_p \sim 1 - 2^{-p}$.

Based on this lemma, our strategy is simple: Find a set $\mathcal{K}_p \subset \{-1,1\}^N$ of cardinality K_p as large as possible such that, for any $\sigma, \sigma' \in \mathcal{K}_p$, $|R(\sigma, \sigma')| \leq m_p$. Suppose we can find such a set, with $K_p = 2^{N(1-\gamma_p)}$. Then, on the one hand,

$$\sup_{\sigma \in \{-1,1\}^N} X_\sigma \geq \sup_{\sigma \in \mathcal{K}_p} X_\sigma \tag{11.23}$$

while, on the other hand, if the X_σ with $\sigma \in \mathcal{K}_p$ behaved like independent Gaussians, their maximum would be around $\sqrt{N 2(1-\gamma_p) \ln 2}$. The normal comparison lemma applied to these variables then yields

$$\left| \mathbb{P}\left[\forall \sigma \in \mathcal{K}_p, X_\sigma \leq \sqrt{N 2(1-\gamma_p) \ln 2} \right] - \Phi\left(\sqrt{N 2(1-\gamma_p) \ln 2}\right)^{2^{N(1-\gamma_p)}} \right| \tag{11.24}$$

$$\leq \frac{1}{2\pi} \sum_{\sigma \neq \sigma' \in \mathcal{K}_p} \frac{|R(\sigma,\sigma')^p|}{1 - R(\sigma,\sigma')^{2p}} \exp\left(-\frac{N(1-\gamma_p) 2 \ln 2}{1+|R(\sigma,\sigma')|^p} \right)$$

$$= \sum_{m=-m_p}^{m_p} \frac{1}{2\pi} \sum_{\sigma \neq \sigma' \in \mathcal{K}_p} \mathbb{I}_{\{R(\sigma,\sigma')=m\}} \frac{|m|^p}{1-m^{2p}} e^{-\frac{N(1-\gamma_p) 2 \ln 2}{1+|m|^p}}$$

Assuming a homogeneous distribution of the set $\mathcal{K}_p$ would imply

$$\sum_{\sigma \neq \sigma' \in \mathcal{K}_p} \mathbb{I}_{\{R(\sigma,\sigma')=m\}} \approx 2^{2N(1-\gamma_p)} \frac{2}{2\pi N(1-m^2)} e^{-NI(m)} \tag{11.25}$$

in which case the right-hand side of (11.24) would simplify to

$$\frac{1}{2\pi}\sum_{m=-m_p}^{m_p}\frac{m^p}{1-m^{2p}}e^{-N\left(I(m)-\frac{(1-\gamma_p)2\ln 2|m|^p}{1+|m|^p}\right)} \leq C_p N^{-p/2} \tag{11.26}$$

and this would prove that the maximum of the X_σ is indeed of the order of $\sqrt{N2(1-\gamma_p)\ln 2}$.

It remains to establish that we can find sets $\mathcal{K}_p$ with the desired properties. The following construction may look funny, but it works.

Consider K i.i.d. random elements, $\sigma_1, \sigma_2, \ldots, \sigma_K$, in $\{-1, 1\}^N$ distributed according to the uniform distribution on $\{-1, 1\}^N$. We want to estimate the size of the largest subset, $\tilde{\mathcal{K}} \subset \{1, \ldots, K\}$, with the property that, for all $\sigma, \sigma' \in \tilde{\mathcal{K}}$, $R(\sigma, \sigma') < m_p$.

Lemma 11.2.5 *Let $\sigma_1, \sigma_2, \ldots, \sigma_K$ be random variables as described above. Then*

$$\mathbb{E}|\tilde{\mathcal{K}}| \geq K(1 - K\exp(-NI(m_p))) \tag{11.27}$$

Moreover,

$$\mathbb{P}[|\tilde{\mathcal{K}}| \geq K(1 - 2K\exp(-NI(m_p)))] \geq \frac{1}{4} \tag{11.28}$$

$\square$

Proof Note that

$$K - |\tilde{\mathcal{K}}| \leq |\{i \in 1, \ldots, K | \exists j \neq i \text{ s.t.} |R(\sigma_i, \sigma_j)| \geq m_p\}| \leq \sum_{i=1}^{K}\sum_{j\neq i} \mathbb{I}_{\{|R(\sigma_i, \sigma_j)|\geq m_p\}} \tag{11.29}$$

Now

$$\mathbb{E}_\sigma \sum_{j\neq i}^{K} \mathbb{I}_{\{|R(\sigma_i,\sigma_j)|\geq m_p\}} = K^2 \mathbb{P}_\sigma[|R(\sigma_i, \sigma_j)| \geq m_p] \leq K^2 e^{-NI(m_p)} \tag{11.30}$$

and from this (11.27) follows. Similarly, one verifies easily that

$$\mathbb{E}_\sigma \left(\sum_{j\neq 1}^{K}\mathbb{I}_{\{|R(\sigma_i,\sigma_j)|\geq m_p\}}\right)^2 \leq (K^4 + 6K^3)e^{-2NI(m_p)} + 2K^2 e^{-NI(m_p)} \tag{11.31}$$

From this one deduces (11.28) via the *Paley–Zygmund inequality*, which states that, for any positive random variable X and $0 \leq q \leq 1$,

$$\mathbb{P}[X \geq q\mathbb{E}X] \geq (1-q)^2 \frac{\mathbb{E}X^2}{(\mathbb{E}X)^2} \tag{11.32}$$

$\square$

Lemma 11.2.4 tells us that if we choose K such that $K\exp(-NI(m_p))$ is small, then almost all σ drawn will be isolated from the others as desired. Since $I(m)$ converges to $\ln 2$, as m tends to one, it is clear that we can find γ_p tending to zero, as p tends to infinity, such that this holds with $K = 2^{N(1-\gamma_p)}$. An asymptotic analysis of the function $\psi(x)$ for large p yields the estimates claimed in the proposition.

Having seen how we can prove a lower bound on the ground-state energy and thus the free energy of our model, it is tempting to try also to get an upper bound that improves on the trivial REM bound.

Proposition 11.2.6 *In the p-spin SK model, there exists $c_p > 0$ such that almost surely for all but finitely many values of N,*

$$\sup_{\sigma \in \mathcal{S}_N} \frac{H_N^{p-SK}(\sigma)}{N} \leq \sqrt{2 \ln 2}(1 - c_p) \tag{11.33}$$

Moreover, for p large,

$$c_p \geq 2^{-\alpha_p^2 p} \frac{p}{(2-\alpha)^2} \tag{11.34}$$

with $\alpha_p = 1 + \frac{1}{p \ln 2} + \sqrt{1 + \frac{1}{(p \ln 2)^2}}$ tending to 2 as $p \uparrow \infty$.

Since this result is not very impressive, and the proof is cumbersome and somewhat arbitrary, we will skip it. It can be found in [39].

11.2.2 The truncated partition function

The following result is the analogue to Theorem 9.1.2 in the REM.

Theorem 11.2.7 *In the p-spin SK model, (9.10) holds for all $\beta \leq \beta_p$, where*

$$\beta_p^2 \geq \hat{\beta}_p^2 \equiv \inf_{t \in [0,1]} \left(I(t) \frac{1+t^p}{t^p} \right) \tag{11.35}$$

Moreover,

$$\sqrt{2 \ln 2} \left(1 - \frac{2^{-p-1}}{\ln 2}\right) \leq \beta_p \leq \sqrt{2 \ln 2}(1 - c_p) \tag{11.36}$$

where c_p is the constant from Proposition 11.2.6.

Proof We use the same definitions as in Section 11.2.1. Note that the estimate (9.18) carries over unaltered. A difference occurs only when computing the mean square of the partition function. Namely, while in the REM only the two cases $\sigma = \sigma'$ and $\sigma \neq \sigma'$ had to be distinguished, here we have to distinguish all different values for the overlap $R_N(\sigma, \sigma')$. We can use the convenient fact that $X_\sigma + X_{\sigma'}$ has the same distribution as $\sqrt{2 + 2R_N(\sigma, \sigma')} X_\sigma$. This gives

$$\mathbb{E}\tilde{Z}_{\beta,p,N}(c)^2 = \mathbb{E}_\sigma \mathbb{E}_{\sigma'} \sum_{t \in \mathcal{M}_N} \mathbb{1}_{R(\sigma,\sigma')=t} \mathbb{E} e^{\beta\sqrt{N}(X_\sigma + X_{\sigma'})} \mathbb{1}_{X_\sigma < c\sqrt{N}, X'_\sigma < c\sqrt{N}}$$

$$\leq \sum_{t \in \mathcal{M}_N} \mathbb{E} e^{\beta\sqrt{N}\sqrt{2+2t^p} X_\sigma} \mathbb{1}_{X_\sigma < 2c\sqrt{N}/\sqrt{2+2t^p}} \mathbb{E}_\sigma \mathbb{E}_{\sigma'} \mathbb{1}_{R(\sigma,\sigma')=t} \tag{11.37}$$

□

Up to sub-leading corrections,

$$\mathbb{E}e^{\beta\sqrt{N}\sqrt{2+2t^p}X_\sigma}\mathbb{I}_{X_\sigma<2c\sqrt{N/(2+2t^p)}} \qquad (11.38)$$

$$= \begin{cases} \frac{\sqrt{1+t^p}}{\sqrt{2\pi N}(\beta(1+t^p)-c)}e^{N\left(2\beta c - \frac{c^2}{1+t^p}\right)}, & \text{if } 1+t^p > c/\beta \\ e^{\beta^2 N(1+t^p)}, & \text{if } 1+t^p \leq c/\beta \end{cases}$$

Therefore, using Stirling's approximation (3.37) for the binomial coefficients, we get:

(i) If $\beta < c$:

$$\mathbb{E}\tilde{Z}^2_{\beta,p,N}(c) \leq \frac{e^{\beta^2 N}}{\sqrt{2\pi N}}\left(\sum_{t:1+t^p<c/\beta}\frac{2e^{-NI(t)+N\beta^2 t^p}}{\sqrt{1-t^2}}\right. \qquad (11.39)$$

$$\left. + \sum_{t:1+t^p>c/\beta}\frac{2\sqrt{1+t^p}e^{-NI(t)+N\frac{c^2 t^p}{1+t^p}-N(c-\beta)^2}}{\sqrt{(1-t^2)}(\beta(1+t^p)-c)}\right)$$

(ii) If $\beta > c$:

$$\mathbb{E}\tilde{Z}^2_{\beta,p,N}(c) \leq \frac{e^{N(2\beta c - c^2)}}{2\pi N}\sum_{t\in\mathcal{M}_N}\frac{2\sqrt{1+t^p}}{\sqrt{1-t^2}(\beta(1+t^p)-c)}e^{-NI(t)+N\frac{c^2 t^p}{1+t^p}} \qquad (11.40)$$

We see that the expectation of the square of the truncated partition function is essentially equal to the square of the expectation, provided that c is chosen such that

$$\inf_t\left[-I(t)+\frac{c^2 t^p}{1+t^p}\right]\leq 0 \qquad (11.41)$$

Since we must choose $c > \beta$ if we want $\mathbb{E}Z_{\beta,N} = \mathbb{E}\tilde{Z}_{\beta,N}(c)$, we see that we can meet both conditions only if $\beta \leq \hat{\beta}_p$, defined in (11.35). More precisely, analyzing the function in the exponent near $t=0$, one finds:

Lemma 11.2.8 *For all $p \geq 3$, we have:*

(i) If $\beta < c$, and c satisfies (11.41), then

$$\frac{\mathbb{E}\tilde{Z}^2_{\beta,N}(c) - [\mathbb{E}\tilde{Z}_{\beta,N}(c)]^2}{[\mathbb{E}\tilde{Z}_{\beta,N}(c)]^2} \leq \begin{cases} C_p N^{-p/2+1}, & \text{for } p \text{ even} \\ C_p N^{-p+2}, & \text{for } p \text{ odd} \end{cases} \qquad (11.42)$$

(ii) If $\beta \geq c$, and c satisfies (11.41), then

$$\frac{\mathbb{E}\tilde{Z}^2_{\beta,N}(c)}{[\mathbb{E}\tilde{Z}_{\beta,N}(c)]^2} \leq C_p N^{+1/2} \qquad (11.43)$$

Proof Part (ii) is obvious and we leave the details as an exercise. To prove the more subtle result (i), note that, up to an exponentially small correction,

$$[\mathbb{E}\tilde{Z}_{\beta,N}(c)]^2 = \sum_{t\in\mathcal{M}_N}e^{\beta^2 N}\mathbb{E}_\sigma\mathbb{E}_{\sigma'}\mathbb{I}_{R(\sigma,\sigma')=t} \qquad (11.44)$$

Therefore,

$$\mathbb{E}\tilde{Z}_{\beta,N}^2(c) - [\mathbb{E}\tilde{Z}_{\beta,N}(c)]^2$$
$$\leq \sum_{t \in \mathcal{M}_N} \left(\mathbb{E}e^{\beta\sqrt{N}\sqrt{2+2t^p}X_\sigma}\mathbb{1}_{X_\sigma < 2c\sqrt{N/(2+2t^p)}} - e^{\beta^2 N}\right)\mathbb{E}_\sigma \mathbb{E}_{\sigma'} \mathbb{1}_{R(\sigma,\sigma')=t}$$
$$\leq \frac{e^{\beta^2 N}}{\sqrt{2\pi N}}\left(\sum_{t:1+t^p<c/\beta} \frac{2e^{-NI(t)}\left(e^{+N\beta^2 t^p} - 1\right)}{\sqrt{1-t^2}} \right.$$
$$\left. + \sum_{t:1+t^p>c/\beta} \frac{2e^{-NI(t)+N\frac{c^2 t^p}{1+t^p} - N(c-\beta)^2}}{\sqrt{1-t^2}}\right) \qquad (11.45)$$

By (11.41), the second sum is exponentially small in N; for the first case, if p is even, use that on the domain of summation,

$$\left|\frac{2e^{-NI(t)}(e^{+N\beta^2 t^p} - 1)}{\sqrt{1-t^2}}\right| \leq \frac{2N\beta^2 t^p \exp\left(-N\left[I(t) - \beta^2 t^2(c/\beta - 1)^{(p-2)/p}\right]\right)}{\sqrt{1 - (c/\beta - 1)^{2/p}}} \qquad (11.46)$$

The bound (11.42) is now obtained using that $I(t) \approx t^2/2$, for small t, comparing the sum in (11.45) to an integral, and performing a simple change of variables. In the odd case, we expand the exponential to second order, and use that the first-order term gives no contribution, by symmetry. This yields the sharper result in that case. $\square$

Remark 11.2.9 The bound (11.42) is the analogue of the exponential bound (9.22) valid in the REM. In [81] (for $p = 2$) and [56] (for $p \geq 3$) it has been shown that this variance estimate goes together with a central limit theorem for the free energy, analogous to parts (i) and (ii) of Theorem 9.2.1.

While (11.42) is weaker than its analogue in the REM, it is still sufficient to prove (9.24). From here the first assertion of the theorem follows just as in the REM.

The lower bound on β_p follows from the analysis of the right-hand side of (11.35). The upper bound in (11.36) is a consequence of Proposition 11.2.6. Namely, if $\lim_{N\uparrow\infty}\mathbb{E}\Phi_{\beta,N} = \beta^2/2$ for all $\beta \leq \beta_p$, then, for such β,

$$\beta = \frac{d}{d\beta}\lim_{N\uparrow\infty}\mathbb{E}\Phi_{\beta,N} = \lim_{N\uparrow\infty}\frac{d}{d\beta}\mathbb{E}\Phi_{\beta,N} = \lim_{N\uparrow\infty} N^{-1/2}\mu_{\beta,N}(X_\sigma)$$
$$\leq \limsup_{N\uparrow\infty} N^{-1/2} \sup_{\sigma \in \mathcal{S}_N} H_\sigma \leq \sqrt{2\ln 2}(1 - c_p) \qquad (11.47)$$

This concludes the proof of the theorem in the case $p \geq 3$.

We will not resist the temptation to present an alternative proof that does not need the *sharp* estimate of Lemma 11.2.8, and that also works in the case $p = 2$ (for $\beta < 1$). This clever trick was introduced by Talagrand in [232]. The argument uses the self-averaging property in Corollary 11.1.3. The crucial observation is that (11.11) and (11.42) are in contradiction, *unless* (9.10) holds. To see this, use the Paley–Zygmund inequality (11.32), which implies that, under our assumption,

$$\frac{C}{4} \leq \mathbb{P}[\tilde{Z}_{\beta,N}(c) \geq q\mathbb{E}\tilde{Z}_{\beta,N}(c)] \leq \mathbb{P}\left[Z_{\beta,N} \geq \frac{1}{2}\mathbb{E}\tilde{Z}_{\beta,N}(c)\right] \quad (11.48)$$

$$= \mathbb{P}\left[\Phi_{\beta,N} - \mathbb{E}\Phi_{\beta,N} \geq \frac{1}{N}\ln\mathbb{E}\tilde{Z}_{\beta,N}(c) - \mathbb{E}\Phi_{\beta,N} - N^{-1}\ln 2\right]$$

$$\leq 2\exp\left(-\frac{N}{2}\left(\frac{1}{N}\ln\mathbb{E}\tilde{Z}_{\beta,N}(c) + \mathbb{E}\Phi_{\beta,N} - N^{-1}\ln 2\right)^2\right)$$

provided $\frac{1}{N}\ln\mathbb{E}\tilde{Z}_{\beta,N}(c) - \mathbb{E}\Phi_{\beta,N} - N^{-1}\ln 2 \geq 0$; if this condition fails, by Jensen's inequality, $-N^{-1}\ln 2 \leq \mathbb{E}\Phi_{\beta,N} - \beta^2/2 \leq 0$, and we are done. If not, (11.48) can only hold if $|\mathbb{E}\Phi_{\beta,N} - \beta^2/2| \leq CN^{-1/2}$. Thus the first assertion of the theorem is proven. □

Remark 11.2.10 The argument yielding the upper bound on β_p can be inverted to show that $\hat{\beta}_p$ is a upper bound for the ground-state energy density, $-\liminf_{N\uparrow\infty} N^{-1}\sup_{\sigma\in\mathcal{S}_N}(-H_N^{p-SK}(\sigma))$. Namely, since

$$\mathbb{E}\frac{d}{d\beta}\Phi_{\beta,N} \leq N^{-1/2}\mathbb{E}\sup_{\sigma\in\mathcal{S}_N} H_\sigma \quad (11.49)$$

for all β, and for all $\beta < \beta_c$ it is true that $\mathbb{E}\frac{d}{d\beta}\Phi_{\beta,N} = -\beta$, it follows that

$$N^{-1}\mathbb{E}\sup_{\sigma\in\mathcal{S}_N} -H_N^{p-SK}(\sigma) \geq \beta_c \quad (11.50)$$

This gives a much simpler proof of Proposition 11.2.2, with an improved estimate. Thus, we have already proven the existence of a phase transition, and we have estimates for the critical temperature, which become sharp in the limit $p \uparrow \infty$, and, not surprisingly, the limiting value is the same as in the REM. In fact, a simple argument, which we leave as an exercise, shows that the free energy converges, as $p \uparrow \infty$, to that of the REM! However, it is far more interesting to understand what happens to the Gibbs measure at the phase transition for fixed p. We return to this question later.

11.3 The Parisi solution and Guerra's bounds

The original approach to the computation of the free energy in the SK models in the theoretical physics literature is quite remarkable. Morally, it constitutes an attempt to compute n moments, where n seems strangely to go to infinity and to zero simultaneously. This idea is the basis of the so-called *replica method*, or *replica trick*, which is a widely used tool in the heuristic analysis of disordered systems. While we must leave a detailed exposition of the heuristic approaches to the standard textbooks [96, 179, 193], it maybe worthwhile to discuss this approach briefly. The basic idea is the observation that $\lim_{n\downarrow 0}(x^n - 1) = \ln x$, and that, for integer n, $\mathbb{E}Z_{\beta,N}^n$ can be computed, at least in the sense that it is possible to perform the average over the disorder and to obtain a deterministic expression (as we have seen in the case $n = 2$). The obvious problem is that the computation of integer moments of $Z_{\beta,N}$ does not immediately allow us to infer information on the limit $n \downarrow 0$, where n is to be considered real valued. Parisi proposed to write down an expression for the n-th moment when n is considered large (and in fact is taken to infinity at the end). In the spirit of our 2nd moment computations, we would decompose the summation over the spin variables over fixed values of their overlaps, to get an expression of the form

$$\mathbb{E}Z_{\beta,N}^n = \sum_{q_{1,2},\ldots,q_{n-1,n}} \mathbb{E}_{\sigma^1} \cdots \qquad (11.51)$$

$$\cdots \mathbb{E}_{\sigma^n} \mathbb{1}_{R_N(\sigma^1,\sigma^2)=q_{1,2},\ldots,R_N(\sigma^{n-1},\sigma^n)=q_{n-1,n}} e^{\frac{1}{2}\beta^2 N(\xi(q_{1,2})+\ldots\xi(q_{n-1,n}))}$$

The remaining problem is then to compute the entropies. Parisi observed that this can be done for special choices of the *matrix q*, namely if q defines an ultrametric on the set $\{1,\ldots,n\}$. The idea that there should appear a non-trivial matrix is the basis of Parisi's *replica symmetry breaking* scheme. The computations then involve the representation of the indicator functions in terms of Fourier integrals, and will not be discussed further. The key point, however, is that while to compute the integrals for fixed integer n we should certainly choose the value of q that yields the maximal term, Parisi suggests that the correct answer in the limit $n \downarrow 0$ is obtained by performing these calculations as if n was a positive integer going to infinity, but then choosing the minimum over the possible values of q. This strange ad hoc assumption has proven very successful, but the entire procedure has surrounded the physics of spin-glasses with an aura of mysticism. We shall see that there is a simple approach that will lead to the same conclusion in a mathematically perfectly transparent way. It will again be based on a (very clever) use of Lemma 10.2.1. In the process we will also learn more about Parisi's solution.

Having seen that the comparison theorem yields the existence of the free energy almost for free, it is a bit more surprising that it provides a variational principle that allows us to compare the free energy to Parisi's solution. As we do not yet know what Parisi's solution is, we will learn about it in the process.

11.3.1 Application of a comparison lemma

Lemma 10.2.1 and its specializations such as Slepian's lemma have been devised to study properties of stochastic processes such as extremal values by comparing them to simpler processes where explicit computations are possible. A very natural idea in the context of the SK models would be to try comparison with the Derrida models, where we have seen that we can compute everything explicitly. However, this does not work excessively well: the only simple bound is obtained by comparing to the REM, and this is quite far from giving a good value.

The idea is to use comparison on a much richer class of processes. Basically, rather than comparing one process to another, we construct an extended process on a product space and use comparison on this richer space. Let us first explain this in an abstract setting. We have a process X on a space $\mathcal{S}$ equipped with a probability measure $\mathbb{E}_\sigma$. We want to compute as usual the average of the logarithm of the partition function $F(X) = \ln \mathbb{E}_\sigma e^{\beta X_\sigma}$. Now consider a second space $\mathcal{T}$ equipped with a probability law $\mathbb{E}_\alpha$. Choose a Gaussian process, Y, independent of X, on this space, and define a further independent process, Z, on the product space $\mathcal{S} \times \mathcal{T}$. Define real-valued functions, G, H, on the space of real-valued functions on

$\mathcal{T}$ and $\mathcal{S} \times \mathcal{T}$, respectively, via $G(y) \equiv \ln \mathbb{E}_\alpha e^{\beta \sqrt{N} y_\alpha}$ and $H(z) = \ln \mathbb{E}_\sigma \mathbb{E}_\alpha e^{\beta \sqrt{N} z_{\sigma,\alpha}}$. Note that $H(X + Y) = F(X) + G(Y)$. Assume that the covariances are chosen such that

$$\mathrm{cov}(X_\sigma, X_{\sigma'}) + \mathrm{cov}(Y_\alpha, Y_{\alpha'}) \geq \mathrm{cov}(Z_{\sigma,\alpha}, Z_{\sigma',\alpha'}) \quad (11.52)$$

Since we know that the second derivatives of H are negative, we get from Lemma 10.2.1 that

$$\mathbb{E} F(X) + \mathbb{E} G(Y) = \mathbb{E} H(X + Y) \leq \mathbb{E} H(Z) \quad (11.53)$$

This is a useful relation if we know how to compute $\mathbb{E} G(Y)$ and $\mathbb{E} H(Z)$. This idea may look a bit crazy at first, but we must remember that we have a lot of freedom in choosing the auxiliary spaces and processes. Before turning to the issue of whether we can find useful computable processes Y and Z, let us see why we could hope to find in this way *sharp* bounds.

To do so, we will show that, in principle, we can represent the free energy in the thermodynamic limit in the form $\mathbb{E} H(Z) - \mathbb{E} G(Y)$. To this end let $\mathcal{S} = \mathcal{S}_M$ and $\mathcal{T} = \mathcal{S}_N$, both equipped with their natural probability measure $\mathbb{E}_\sigma$. We will think of $N \gg M$, and both tending to infinity eventually. We write again $\mathcal{S} \times \mathcal{T} \ni \sigma = (\hat{\sigma}, \check{\sigma})$. Consider the process X_σ on $\mathcal{S}_{N+M}$ with covariance $\xi(R_{N+M}(\sigma, \sigma'))$. We would like to write this as

$$X_\sigma = \hat{X}_{\hat{\sigma}} + \check{X}_{\check{\sigma}} + Z_\sigma \quad (11.54)$$

where all three processes are independent. Note that here and in the sequel equalities between random variables are understood to hold in distribution. Moreover, we demand that

$$\mathrm{cov}(\hat{X}_{\hat{\sigma}}, \hat{X}_{\hat{\sigma}'}) = \xi\left(\tfrac{M}{N+M} R_M(\hat{\sigma}, \hat{\sigma}')\right) \quad (11.55)$$

and

$$\mathrm{cov}(\check{X}_{\check{\sigma}}, \check{X}_{\check{\sigma}'}) = \xi\left(\tfrac{N}{N+M} R_N(\check{\sigma}, \check{\sigma}')\right) \quad (11.56)$$

Obviously, this implies that

$$\mathrm{cov}(Z_\sigma, Z_{\sigma'}) = \xi\left(\tfrac{M}{N+M} R_M(\hat{\sigma}, \hat{\sigma}') + \tfrac{N}{N+M} R_N(\check{\sigma}, \check{\sigma}')\right) \quad (11.57)$$
$$- \xi\left(\tfrac{M}{N+M} R_M(\hat{\sigma}, \hat{\sigma}')\right) - \xi\left(\tfrac{N}{N+M} R_N(\check{\sigma}, \check{\sigma}')\right)$$

(we will not worry about the existence of such a decomposition; if $\xi(x) = x^p$, we can use the explicit representation in terms of p-spin interactions to construct them). Now we first note that, by super-additivity [7]

$$\lim_{M \uparrow \infty} \frac{1}{\beta M} \liminf_{N \uparrow \infty} \mathbb{E} \log \frac{Z_{\beta, N+M}}{Z_{\beta, N}} = -f_\beta \quad (11.58)$$

Thus we need a suitable representation for $\frac{Z_{\beta, N+M}}{Z_{\beta, N}}$. But

$$\frac{Z_{\beta, N+M}}{Z_{\beta, N}} = \frac{\mathbb{E}_\sigma e^{\beta \sqrt{N+M}(\check{X}_{\check{\sigma}} + Z_\sigma + \hat{X}_{\hat{\sigma}})}}{\mathbb{E}_{\check{\sigma}} e^{\beta \sqrt{N+M}(\sqrt{(1 - M/(N+M))} X_{\check{\sigma}})}} \quad (11.59)$$

Now we want to express the random variables in the denominator in the form

$$\sqrt{(1 - M/(N+M))} X_{\check{\sigma}} = \check{X}_{\check{\sigma}} + Y_{\check{\sigma}} \quad (11.60)$$

where Y is independent of $\check{X}$. Comparing covariances, this implies that

$$\text{cov}(Y_{\check{\sigma}}, Y_{\check{\sigma}'}) = (1 - M/(N+M))\xi(R_N(\check{\sigma},\check{\sigma}')) - \xi\left(\tfrac{N}{N+M} R_N(\check{\sigma},\check{\sigma}')\right) \quad (11.61)$$

As we will be interested in taking the limit $N \uparrow \infty$ before $M \uparrow \infty$, we may expand in $M/(N+M)$ to see that to leading order in $M/(N+M)$,

$$\text{cov}(Y_{\check{\sigma}}, Y_{\check{\sigma}'}) \sim \tfrac{M}{N+M} R_N(\check{\sigma},\check{\sigma}') \xi'\left(\tfrac{N}{N+M} R_N(\check{\sigma},\check{\sigma}')\right) - \tfrac{M}{N+M}\xi\left(\tfrac{N}{N+M} R_N(\check{\sigma},\check{\sigma}')\right) \quad (11.62)$$

Finally, we note that the random variables $\hat{X}_{\hat{\sigma}}$ are negligible in the limit $N \uparrow \infty$, since their variance is smaller than $\xi(M/(N+M))$ and hence their maximum is bounded by $\sqrt{\xi(M/(N+M))M\ln 2}$, which even after multiplication with $\sqrt{N+M}$ gives no contribution in the limit if ξ tends to zero faster than linearly at zero, which we can safely assume. Thus we see that we can indeed express the free energy as

$$f_\beta = -\lim_{M\uparrow\infty}\liminf_{N\uparrow\infty} \frac{1}{\beta M} \mathbb{E}\ln\frac{\mathbb{E}_{\hat{\sigma}}\tilde{\mathbb{E}}_{\check{\sigma}} e^{\beta\sqrt{N+M} Z_{\hat{\sigma},\check{\sigma}}}}{\tilde{\mathbb{E}}_{\check{\sigma}} e^{\beta\sqrt{N+M} Y_{\check{\sigma}}}} \quad (11.63)$$

where the measure $\tilde{\mathbb{E}}_{\check{\sigma}}$ can be chosen as a probability measure defined by $\tilde{\mathbb{E}}_{\check{\sigma}}(\cdot) = \mathbb{E}_{\check{\sigma}} e^{\beta\sqrt{N+M}\check{X}_{\check{\sigma}}}(\cdot)/\check{Z}_{\beta,N,M}$ where $\check{Z}_{\beta,N,M} \equiv \mathbb{E}_{\check{\sigma}} e^{\beta\sqrt{N+M}\check{X}_{\check{\sigma}}}$. Of course this representation is quite pointless, because it is certainly uncomputable, since $\tilde{\mathbb{E}}$ is effectively the limiting Gibbs measure that we are looking for. However, at this point there occurs a certain miracle: the (asymptotic) covariances of the processes X, Y, Z satisfy

$$\xi(x) + y\xi'(y) - \xi(y) \geq x\xi'(y) \quad (11.64)$$

for all $x, y \in [-1, 1]$ if ξ is convex and even. This comes as a surprise, since we did not do anything to impose such a relation! But it has the remarkable consequence that asymptotically, by virtue of Lemma 10.2.1, it implies the bound

$$\mathbb{E}\ln\mathbb{E}_{\hat{\sigma}} e^{\beta\sqrt{M}X_{\hat{\sigma}}} \leq \mathbb{E}\ln\frac{\mathbb{E}_{\hat{\sigma}}\tilde{\mathbb{E}}_{\check{\sigma}} e^{\beta\sqrt{N+M}Z_{\hat{\sigma},\check{\sigma}}}}{\tilde{\mathbb{E}}_{\check{\sigma}} e^{\beta\sqrt{N+M}Y_{\check{\sigma}}}} \quad (11.65)$$

(if the processes are taken to have the asymptotic form of the covariances). Moreover, this bound will hold *even* if we replace the measure $\tilde{\mathbb{E}}$ by some other probability measure, and even if we replace the overlap R_N on the space $\mathcal{S}_N$ by some other function, e.g. the ultrametric d_N. Seen the other way around, we can conclude that a lower bound of the form (11.53) can actually be made as good as we want, provided we choose the right measure $\tilde{\mathbb{E}}$. This observation is due to Aizenman, Sims, and Starr [7]. They call the auxiliary structure made from a space $\mathcal{T}$, a probability measure $\tilde{\mathbb{E}}$ on $\mathcal{T}$, a normalized distance q on $\mathcal{T}$, and the corresponding processes, Y and Z, a *random overlap structure* (random, because they actually think of the measure $\tilde{\mathbb{E}}$ and the distance q as random variables).

The key hypothesis in the derivation of Parisi's solution is now that the limiting Gibbs measure for any SK model, for given temperature, looks very much like that of some CREM, at least to the point that this approximation will produce the right value of the free energy. Fortunately, we know very well the Gibbs measures of the CREMs from the previous chapter. To derive the corresponding bounds, we could now proceed in two ways: (i) start with formula (11.63) and replace the distribution $\tilde{\mathbb{E}}_{\check{\sigma}}$ with the Gibbs measure of some CREM, and then pass to the limit; or (ii) ignore this step and immediately place on the second set $\mathcal{T}$ the limiting measure for the CREM. This second way gives a very appealing form of the bound which highlights the rôle of the genealogical structures associated to Neveu's branching process.

We have seen that the limiting Gibbs measure of the CREMs can be understood as random genealogical structures on the unit interval. This suggests that we choose as the auxiliary space $\mathcal{T}$ the unit interval, equipped with a distance q (which later will be the asymptotic genealogical distance of some CREM). Moreover, we endow the unit interval with a probability measure $\mathbb{E}_\alpha$. Next we define two Gaussian random processes: the process Y_α on the unit interval with covariance

$$\mathrm{cov}(Y_\alpha, Y_{\alpha'}) = q(\alpha, \alpha')\xi'(q(\alpha, \alpha')) - \xi(q(\alpha, \alpha')) \tag{11.66}$$

and the process $Z_{\sigma,\alpha}$ on $\mathcal{S}_N \times [0, 1]$ with covariance

$$\mathrm{cov}(Z_{\sigma,\alpha}, Z_{\sigma',\alpha'}) \equiv R_N(\sigma, \sigma')\xi'(q(\alpha, \alpha')) \tag{11.67}$$

With these choices, and naturally X_σ, our original process with covariance $\xi(R_N)$, the equation (11.52) is satisfied, and hence the inequality (11.53) holds, no matter what choice of q and $\mathbb{E}_\alpha$ we make. Restricting these choices to the random genealogies obtained from Neveu's process by a time change with some probability distribution function m, and $\mathbb{E}_\alpha$ the Lebesgue measure on $[0, 1]$, gives the bound we want. This is the form derived in Aizenman, Sims, and Starr [7].

This bound would be quite useless if we could not compute the right-hand side. Fortunately, one can get rather explicit expressions. We need to compute two objects:

$$\mathbb{E}_\alpha \mathbb{E}_\sigma e^{\beta\sqrt{N} Z_{\sigma,\alpha}} \tag{11.68}$$

and

$$\mathbb{E}_\alpha e^{\beta\sqrt{N} Y_\alpha} \tag{11.69}$$

In the former we use that Z has the representation

$$Z_{\sigma,\alpha} = N^{-1/2} \sum_{i=1}^{N} \sigma_i z_{\alpha,i} \tag{11.70}$$

where the processes $z_{\alpha,i}$ are independent for different i and have covariance

$$\mathrm{cov}(z_{\alpha,i}, z_{\alpha',i}) = \xi'(q(\alpha, \alpha')) \tag{11.71}$$

Thus at least the σ-average is trivial:

$$\mathbb{E}_\alpha \mathbb{E}_\sigma e^{\beta\sqrt{N} Z_{\sigma,\alpha}} = \mathbb{E}_\alpha \prod_{i=1}^{N} e^{\ln \cosh(\beta z_{\alpha,i})} \tag{11.72}$$

Thus we see that, in any case, we obtain bounds that only involve objects that we introduced ourselves and that thus can be manipulated to be computable. In fact, such computations have been done in the context of the Parisi solution [179]. A useful mathematical reference is [33].

11.3.2 Computations with the GREM

We will now see that what we have learned in the GREM can be put to good use to get more or less explicit expressions for our bounds. To this end, we find it convenient to use a finite N GREM with covariance function A as our auxiliary measure. Moreover, it will be enough to consider a GREM with finitely many hierarchies, i.e. choose A as a step function with n jumps at positions $x_1 \ldots, x_n$. Our auxiliary Gaussian processes are then also defined on $\{-1, 1\}^N$, and their covariance is a function of the ultrametric distance, i.e. of d_N. The measure $\mathbb{E}_\alpha$ is then taken as $\frac{\mathbb{E}_{\check\sigma} e^{\beta\sqrt{N}X_{\check\sigma}}(\cdot)}{\mathbb{E}_{\check\sigma} e^{\beta\sqrt{N}X_{\check\sigma}}}$, with $X_{\check\sigma}$ a GREM process. We will chose to label the distance distribution function, i.e. fix parameters such that $\lim_{N\uparrow\infty} \mathbb{E}\psi_N(x) = m(x)$, where $m(x)$ takes on the value m_ℓ in the interval $[x_{\ell-1}, x_\ell)$.

In this context, the quantities (11.67) and (11.68) take the form

$$\frac{\mathbb{E}_{\check\sigma} e^{\beta\sqrt{N}X_{\check\sigma} + \sum_{i=1}^{M} \ln \cosh(\beta z_{\check\sigma,i})}}{\mathbb{E}_{\check\sigma} e^{\beta\sqrt{N}X_{\check\sigma}}} \tag{11.73}$$

and

$$\frac{\mathbb{E}_{\check\sigma} e^{\beta\sqrt{N}X_{\check\sigma} + \beta\sqrt{M}Y_{\check\sigma}}}{\mathbb{E}_{\check\sigma} e^{\beta\sqrt{N}X_{\check\sigma}}} \tag{11.74}$$

where we are interested only in the limit $N \uparrow \infty$ for fixed M. Recall that the covariance of $Y_{\check\sigma}$ is given by

$$\mathrm{cov}\,(Y_{\check\sigma}, Y_{\check\sigma'}) = h(d_N(\check\sigma, \check\sigma')) \tag{11.75}$$

with $h(x) = x\xi'(x) - \xi(x)$, and that of $Z_{\sigma,\check\sigma}$ by

$$\mathrm{cov}\,(Z_{\sigma,i}, Z_{j,\check\sigma'}) = \delta_{i,j}\xi'(d_N(\check\sigma, \check\sigma')) \tag{11.76}$$

To compute these quantities, we want to benefit from our experience with the computations of partition functions in the GREM. That is, it will be natural to compute the expectations over the spin variables iteratively over the n hierarchies of the GREM. Each step then involves only computation in a REM and can be carried out with the help of the following lemma.

Lemma 11.3.1 *Assume that X_σ are i.i.d. standard normal random variables on $\mathcal{S}_N$. Let $Y_{\sigma,i}$, be centered Gaussian processes on $\{-1, 1\}^N \times \mathbb{N}$ with covariance*

$$\mathrm{cov}(Y_{\sigma,i}, Y_{\sigma',j}) = \delta_{ij}k(d_N(\sigma, \sigma')) \tag{11.77}$$

where $k(x) \downarrow 0$, as $x \downarrow 0$. Let Y be a random variable that has the same distribution as $Y_{\sigma,1}$. Let $g_i : \mathbb{R} \to \mathbb{R}$ be smooth functions such that, for all $|m| \leq 2$, there exist $C < \infty$, independent of N, such that

$$\mathbb{E}_Y e^{mg_i(Y)} \equiv e^{L_i(m)} < C \tag{11.78}$$

Then

$$\lim_{N\uparrow\infty} \mathbb{E} \ln \frac{\mathbb{E}_\sigma e^{\beta\sqrt{N}X_\sigma + \sum_{i=1}^{M} g_i(Y_{\sigma,i})}}{\mathbb{E}_\sigma e^{\beta\sqrt{N}X_\sigma}} = \sum_{i=1}^{M} \frac{L_i(m)}{m} \tag{11.79}$$

where $m = \lim_{N\uparrow\infty} \mathbb{E}\psi_{\beta,N}(x)$, for $x \in [0, 1)$.

Proof We give the proof only for the case $M = 1$; the general case is only notationally different. The idea of the proof is to consider the numerator in (11.79) as the partition function of a slightly perturbed REM, where the new random variables, $X_\sigma + \frac{1}{\beta\sqrt{N}}g(Y_\sigma)$ are slightly non-Gaussian and slightly dependent. If $m < 1$, the logarithm of the partition function in the numerator will behave, to leading order, like $\sqrt{N}\beta\max_\sigma(X_\sigma + \frac{1}{\beta\sqrt{N}}g(Y_\sigma))$. If we ignore the dependence, to compute the maximum we just need to find a such that

$$2^N \mathbb{P}\left[X_\sigma + \frac{1}{\beta\sqrt{N}}g(Y_\sigma) > \sqrt{N}a\right] = O(1) \quad (11.80)$$

But

$$\mathbb{P}\left[X_\sigma + \frac{1}{\beta\sqrt{N}}g(Y_\sigma) > \sqrt{N}a\right] \sim \mathbb{E}_Y \frac{e^{-\frac{1}{2}\left(\sqrt{N}a - \frac{1}{\beta\sqrt{N}}g(Y)\right)^2}}{\sqrt{2\pi}\left(\sqrt{N}a - \frac{1}{\beta\sqrt{N}}g(Y)\right)}$$

$$\sim \frac{e^{-\frac{1}{2}Na^2}}{\sqrt{2\pi Na}}\mathbb{E}_Y e^{\frac{a}{\beta}g(Y)}$$

where we used that we will have $a \geq \sqrt{2\ln 2}$, and $\beta > \sqrt{2\ln 2}$, so that

$$\mathbb{P}_Y\left[\frac{1}{\beta\sqrt{N}}g(Y_\sigma) > a\sqrt{N}/2\right] \leq e^{-Na\beta}\mathbb{E}_Y e^{2g(Y)} \leq C2^{-2N} \quad (11.81)$$

and the event $\{\frac{1}{\beta\sqrt{N}}g(Y_\sigma) > a\sqrt{N}/2\}$ is therefore negligible. To satisfy (11.78), a must be chosen such that $\frac{1}{2}a^2 - N^{-1}L(a/\beta) = \ln 2$. This implies $a = \sqrt{2\ln 2} + \frac{1}{N\sqrt{2\ln 2}}L(\sqrt{2\ln 2}/\beta)$, or $N\beta(a - \sqrt{2\ln 2}) = L(m)/m$. One can easily check that all sub-leading contributions to a cancel with the corresponding terms from the denominator, or vanish, as $N \uparrow \infty$. Thus, the assertion of the lemma follows if we can show that ignoring the dependence of the $g(Y_\sigma)$ can be justified. We leave this for the reader to check (a convenient tool is given by Proposition 13.3.1 in Chapter 13). Finally, in the case when $m = 1$, i.e. if the REM is in the high-temperature phase, the claimed result follows since now the law of large numbers holds. □

Remark 11.3.2 The same result can of course be proven (in the low-temperature phase) by using Ruelle's version of the REM. In that case we start with a Poisson process, $\mathcal{P} = \sum_j \delta_{z_j}$, and i.i.d. Gaussian random variables Y_j (for simplicity let $M = 1$). We must compute

$$\mathbb{E}\ln \frac{\sum_j e^{\alpha z_j + g(Y_j)}}{\sum_j e^{\alpha z_j}} \quad (11.82)$$

It is easy to see that $\sum_j \delta_{z_j + \alpha^{-1}g(Y_j)}$ is a Poisson point process with intensity measure the convolution of the measure $e^{-z}dz$ and the distribution of the random variable $\alpha^{-1}g(Y)$. A simple computation shows that this distribution is $\mathbb{E}_Y e^{g(Y)/\alpha}e^{-z}dz$. Thus the Poisson point process $\sum_j \delta_{e^{\alpha z_j + g(Y_j)}}$ has intensity measure $\mathbb{E}_Y e^{g(Y)/\alpha}\alpha^{-1}x^{-1/\alpha-1}dx$. Finally, one makes the elementary but surprising and remarkable observation that the Poisson point process $\sum_j \delta_{e^{\alpha z_j}[\mathbb{E}_Y e^{g(Y)/\alpha}]^\alpha}$ has the same intensity measure, and therefore $\sum_j e^{\alpha z_j + g(Y_j)}$ has the same law, as $\sum_j e^{\alpha z_j}[\mathbb{E}_Y e^{g(Y)/\alpha}]^\alpha$: multiplying each atom with an i.i.d. random variable leads to the same process as multiplying each atom by a suitable constant! The assertion of the lemma follows immediately. More material on the invariance properties of this Poisson process, and a slightly different proof can be found in [218].

We will now show how to put this lemma to use.

Let us look first at (11.74). We may write

$$Y_{\breve{\sigma}} = \sum_{\ell=1}^{n} \sqrt{h(x_\ell) - h(x_{\ell-1})} Y^{(\ell)}_{\breve{\sigma}_1 \cdots \breve{\sigma}_\ell} \tag{11.83}$$

where $\breve{\sigma}_\ell \in \{-1, 1\}^{[(x_\ell - x_{\ell-1})N]}$, and the standard normal processes $Y^{(\ell)}_{\breve{\sigma}_1 \cdots \breve{\sigma}_\ell}$ are independent for different indices ℓ and have covariance

$$\mathrm{cov}\left(Y^{(\ell)}_{\breve{\sigma}_1 \cdots \breve{\sigma}_\ell}, Y^{(\ell)}_{\breve{\sigma}'_1 \cdots \breve{\sigma}'_\ell}\right) = \mathbb{1}_{\breve{\sigma}_1 \cdots \breve{\sigma}_{\ell-1} = \breve{\sigma}'_1 \cdots \breve{\sigma}'_{\ell-1}} \frac{h(x_{\ell-1} + d_N(\breve{\sigma}_\ell, \breve{\sigma}'_\ell)) - h(x_{\ell-1})}{h(x_\ell) - h(x_{\ell-1})} \tag{11.84}$$

We will use the abbreviation

$$Y_{\breve{\sigma}_1 \cdots \breve{\sigma}_k} \equiv \sum_{\ell=1}^{k} \sqrt{h(x_\ell) - h(x_{\ell-1})} Y^{(\ell)}_{\breve{\sigma}_1 \cdots \breve{\sigma}_\ell} \tag{11.85}$$

We can then write

$$\frac{\mathbb{E}_{\breve{\sigma}} e^{\beta \sqrt{N} X_{\breve{\sigma}} + \beta \sqrt{M} Y_{\breve{\sigma}}}}{\mathbb{E}_{\breve{\sigma}} e^{\beta \sqrt{N} X_{\breve{\sigma}}}} = \frac{\mathbb{E}_{\breve{\sigma}} e^{\beta \sqrt{N} X_{\breve{\sigma}} + \beta \sqrt{M} Y_{\breve{\sigma}_1 \cdots \breve{\sigma}_{n-1}} + \sqrt{h(x_n) - h(x_{n-1})} Y^{(n)}_{\breve{\sigma}_1 \cdots \breve{\sigma}_n}}}{\mathbb{E}_{\breve{\sigma}} e^{\beta \sqrt{N} X_{\breve{\sigma}}}}$$

$$= \frac{\mathbb{E}_{\breve{\sigma}_1 \cdots \breve{\sigma}_{n-1}} e^{\beta \sqrt{N} \sum_{\ell=1}^{n-1} \sqrt{a_\ell} X_{\breve{\sigma}_1 \cdots \breve{\sigma}_{\ell-1}} + \beta \sqrt{M} Y_{\breve{\sigma}_1 \cdots \breve{\sigma}_{n-1}}} \mathbb{E}_{\breve{\sigma}_n} e^{\beta \sqrt{N a_n} X_{\breve{\sigma}_1 \cdots \breve{\sigma}_n}}}{\mathbb{E}_{\breve{\sigma}} e^{\beta \sqrt{N} X_{\breve{\sigma}}}}$$

$$\times \frac{\mathbb{E}_{\breve{\sigma}_n} e^{\beta \sqrt{N a_n} X_{\breve{\sigma}_1 \cdots \breve{\sigma}_n} + \beta \sqrt{M} \sqrt{h(x_n) - h(x_{n-1})} Y^{(n)}_{\breve{\sigma}_1 \cdots \breve{\sigma}_n}}}{\mathbb{E}_{\breve{\sigma}_n} e^{\beta \sqrt{N a_n} X_{\breve{\sigma}_1 \cdots \breve{\sigma}_n}}} \tag{11.86}$$

Using Lemma 11.3.1, the last factor can be replaced, in the limit $N \uparrow \infty$, by

$$\frac{\mathbb{E}_{\breve{\sigma}_n} e^{\beta \sqrt{N a_n} X_{\breve{\sigma}_1 \cdots \breve{\sigma}_n} + \beta \sqrt{M} \sqrt{h(x_n) - h(x_{n-1})} Y^{(n)}_{\breve{\sigma}_1 \cdots \breve{\sigma}_n}}}{\mathbb{E}_{\breve{\sigma}_n} e^{\beta \sqrt{N} \sqrt{a_n} X_{\breve{\sigma}_1 \cdots \breve{\sigma}_n}}} \tag{11.87}$$

$$\rightarrow \left[\int \frac{dz}{\sqrt{2\pi}} e^{-\frac{z^2}{2}} e^{z m_n \beta \sqrt{M} \sqrt{h(x_n) - h(x_{n-1})}}\right]^{1/m_n} \tag{11.88}$$

$$= e^{\frac{\beta^2 M}{2} m_n (h(x_n) - h(x_{n-1}))} \tag{11.89}$$

The remaining terms have almost the same form as if the n-th level were simply absent (in which case we could simply iterate the procedure) except that we should replace the random variables $X_{\breve{\sigma}_1 \cdots \breve{\sigma}_{n-1}}$ by

$$X_{\breve{\sigma}_1 \cdots \breve{\sigma}_n} + \frac{1}{\beta a_{n-1} N} \left(\ln \mathbb{E}_{\breve{\sigma}_n} e^{\beta \sqrt{N a_n} X_{\breve{\sigma}_1 \cdots \breve{\sigma}_n}} - \mathbb{E} \ln \mathbb{E}_{\breve{\sigma}_n} e^{\beta \sqrt{N a_n} X_{\breve{\sigma}_1 \cdots \breve{\sigma}_n}}\right) \tag{11.90}$$

From Theorem 9.2.1 we know that the quantity in brackets converges to a random variable. It is not hard to check that this small modification does not affect the next steps of the iteration, and that as a consequence, we get

$$\lim_{N \uparrow \infty} M^{-1} \mathbb{E} \ln \frac{\mathbb{E}_{\breve{\sigma}} e^{\beta \sqrt{N} X_{\breve{\sigma}} + \beta \sqrt{M} Y_{\breve{\sigma}}}}{\mathbb{E}_{\breve{\sigma}} e^{\beta \sqrt{N} X_{\breve{\sigma}}}} = \sum_{l=1}^{n} \frac{\beta^2}{2} m_l (h(x_l) - h(x_{l-1}))$$

$$= \frac{\beta^2}{2} \int_0^1 m(x) x \xi''(x) dx \tag{11.91}$$

Remark 11.3.3 An alternative derivation of (11.91) can be obtained by realizing that the numerator is the partition function of a CREM corresponding to the process $X_{\check\sigma} + \beta\sqrt{M/N}\,Y_{\check\sigma}$ with covariance

$$\bar A(d_N(\check\sigma,\check\sigma')) + \frac{M}{N}(d_N(\check\sigma,\check\sigma')\xi'(d_N(\check\sigma,\check\sigma')) - \xi(\delta_N(\check\sigma,\check\sigma'))) \qquad (11.92)$$

We leave the details as an exercise.

The computation of the expression (11.73) is now very very similar, but gives a more complicated result since the analogues of the expressions (11.87) cannot be computed explicitly. Thus, after the k-th step, we end up with a new function of the remaining random variables $Y_{\check\sigma_1\cdots\check\sigma_{n-k}}$. The result can be expressed in the form

$$\lim_{N\uparrow\infty}\frac{1}{M}\mathbb{E}\ln\frac{\mathbb{E}_{\check\sigma}\mathbb{E}_\sigma e^{\beta\sqrt{N}X_{\check\sigma}+\beta\sqrt{M}Z_{\sigma,\check\sigma}}}{\mathbb{E}_{\check\sigma}e^{\beta\sqrt{N}X_{\check\sigma}}} = \zeta(0,h,m,\beta) \qquad (11.93)$$

(here h is the magnetic field (which we have so far hidden in the notation) that can be taken as a parameter of the a-priori distribution on the σ such that $\mathbb{E}_{\sigma_i}(\cdot) \equiv \frac{1}{2\cosh(\beta h)}\sum_{\sigma_i=\pm 1} e^{\beta h\sigma_i}(\cdot))$, where $\zeta(1,h) = \ln\cosh(\beta h)$, and

$$\zeta(x_{a-1},h) = \frac{1}{m_a}\ln\int\frac{dz}{\sqrt{2\pi}}e^{-z^2/2}e^{m_a\zeta(x_a,h+z\sqrt{\xi'(x_a)-\xi'(x)})} \qquad (11.94)$$

We can now announce Guerra's bound in the following form:

Theorem 11.3.4 [133] *Let $\zeta(t,h,m,b)$ be the function defined in terms of the recursion (11.94). Then*

$$\lim_{N\uparrow\infty}N^{-1}\mathbb{E}\ln Z_{\beta,h,N} \leq \inf_m \zeta(0,h,m,\beta) - \frac{\beta^2}{2}\int_0^1 m(x)x\xi''(x)dx \qquad (11.95)$$

where the infimum is over all probability distribution functions m on the unit interval.

Remark 11.3.5 F. Guerra derived this theorem directly using the representation of $\zeta(0,h,m,\beta)$ in terms of the solution of (11.99) via (11.94), using the method of proof of Lemma 10.2.1, rather than the lemma itself. The proof following Aizenman *et al.* sheds more light on the nature of the bound.

The variational formula on the right-hand side of (11.94) is the celebrated formula for the free energy proposed by Parisi on the basis of heuristic computations. It is thanks to Guerra and Aizenman, Sims, and Starr that we have a mathematically clear interpretation of this formula.

Let us briefly point out the connection between the choice of the auxiliary structure in terms of GREMs and the choice of structures made at the end of the last subsection. Namely, consider the map $\theta^1_{\beta,N}\circ r_N : \{-1,+1\}^N \to (0,1]$, and define, for $\alpha\in(0,1]$,

$$\tilde Y_\alpha \equiv Y_{r_N^{-1}\circ(\theta^1_{\beta,N})^{-1}(\alpha)} \qquad (11.96)$$

Then, by the observation (10.125),

$$\mathrm{cov}(\tilde Y_\alpha,\tilde Y_{\alpha'}) = h\left(d_N\left(r_N^{-1}\circ(\theta^1_{\beta,N})^{-1}(\alpha), r_N^{-1}\circ(\theta^1_{\beta,N})^{-1}(\alpha')\right)\right)$$
$$= h(\gamma_1(\alpha,\alpha')) \qquad (11.97)$$

Thus, it is easily checked that

$$\frac{\mathbb{E}_{\tilde{\sigma}} e^{\beta N X_{\tilde{\sigma}}} e^{\beta \sqrt{M} Y_{\tilde{\sigma}}}}{\mathbb{E}_{\tilde{\sigma}} e^{\beta N X_{\tilde{\sigma}}}} = \int_0^1 d\alpha e^{\beta \sqrt{M} \tilde{Y}_\alpha} \quad (11.98)$$

passing to the limit $N \uparrow \infty$ in the last expression, we obtain the prescription given before.

Remark 11.3.6 It is also interesting to see that the recursive form of the function ζ above can also be represented in closed form as the solution of a partial differential equation. Consider the case $\xi(x) = x^2/2$. Then ζ is the solution of the differential equation

$$\frac{\partial}{\partial t}\zeta(t, h) + \frac{1}{2}\left(\frac{\partial^2}{\partial h^2}\zeta(t, h) + m(t)\left(\frac{\partial}{\partial h}\zeta(t, h)\right)^2\right) = 0 \quad (11.99)$$

with final condition

$$\zeta(1, h) = \ln \cosh(\beta h) \quad (11.100)$$

If m is a step function, it is easy to see that a solution is obtained by setting, for $x \in [x_{a-1}, x_a)$,

$$\zeta(x, h) = \frac{1}{m_a} \ln \mathbb{E}_z e^{m_a \zeta(x_a, h + z\sqrt{x_a - x})} \quad (11.101)$$

For general convex ξ, analogous expressions can be obtained through changes of variables [133].

11.3.3 Talagrand's theorem

In both approaches, it pays to write down the expression of the difference between the free energy and the lower bound, since this takes a very suggestive form.

To do this, we just have to use formula (10.54) with

$$X^t_{\sigma,\alpha} \equiv \sqrt{t}(X_\sigma + Y_\alpha) + \sqrt{1-t} Z_{\sigma,\alpha} \quad (11.102)$$

and $f(X^t)$ replaced by $H(X^t) = \ln \mathbb{E}_\sigma \mathbb{E}_\alpha e^{\beta \sqrt{N} Z^t_{\sigma,\alpha}}$. This gives the equality

$$H(X+Y) - H(Z) = \frac{1}{2}\mathbb{E}\int_0^1 dt \tilde{\mu}^{\otimes 2}_{\beta,t,N}(d\sigma, d\alpha)(\xi(R_N(\sigma, \sigma')) + q(\alpha, \alpha')\xi'(q(\alpha, \alpha'))$$
$$- \xi(q(\alpha, \alpha')) - R_N(\sigma, \sigma')\xi'(q(\alpha, \alpha'))) \quad (11.103)$$

where the measure $\tilde{\mu}_{\beta,t,N}$ is defined as

$$\tilde{\mu}_{\beta,t,N}(\cdot) \equiv \frac{\mathbb{E}_\sigma \mathbb{E}_\alpha e^{\beta\sqrt{N} X^t_{\sigma,\alpha}}(\cdot)}{\mathbb{E}_\sigma \mathbb{E}_\alpha e^{\beta\sqrt{N} X^t_{\sigma,\alpha}}} \quad (11.104)$$

where we interpret the measure $\tilde{\mu}_{\beta,t,N}$ as a joint distribution on $\mathcal{S}_N \times [0, 1]$. Note that for convex and even ξ, the function $\xi(R_N(\sigma, \sigma')) + q(\alpha, \alpha')\xi'(q(\alpha, \alpha')) - \xi(q(\alpha, \alpha'))$ vanishes if and only if $R_N(\sigma, \sigma') = q(\alpha, \alpha')$. Thus for the left-hand side of (11.103) to vanish, the replicated interpolating product measure should (for almost all t), concentrate on configurations where the overlaps in the σ-variables coincide with the genealogical distances of the α-variables. Thus we see that the inequality in Theorem 11.3.4 will turn into an equality if it is possible to choose the parameters of the reservoir system in such a way that the overlap distribution on $\mathcal{S}_N$ aligns with the genealogical distance distribution in the reservoir once the systems are coupled by the interpolation.

This latter fact was proven very recently, and not long after the discovery of Guerra's bound, by M. Talagrand [240].

Theorem 11.3.7 [240] *Let $\zeta(t, h, m, b)$ be the function defined in terms of (11.99) and (11.100). Then*

$$\lim_{N\uparrow\infty} N^{-1}\mathbb{E}\ln Z_{\beta,h,N} = \inf_m \left(\zeta(0, h, m, \beta) - \frac{\beta^2}{2}\int_0^1 m(x)x\xi''(x)\mathrm{d}x\right) \quad (11.105)$$

where the infimum is over all probability distribution functions m on the unit interval.

I will not give the complex proof, which the interested reader should study in the original paper [240], but I will make some comments on the key ideas. First, Talagrand proves more than the assertion (11.105). What he actually proves is the following. For any $\epsilon > 0$, there exists a positive integer $n(\epsilon) < \infty$, and a probability distribution function m_n that is a step function with n steps, such that for all $t > \epsilon$,

$$\lim_{N\uparrow\infty} \mathbb{E}\tilde{\mu}_{\beta,t,N}^{\otimes 2}(\mathrm{d}\sigma, \mathrm{d}\alpha)(\xi(R_N(\sigma, \sigma')) + q(\alpha, \alpha')\xi'(q(\alpha, \alpha'))$$
$$- \xi(q(\alpha, \alpha')) - R_N(\sigma, \sigma')\xi'(q(\alpha, \alpha'))) = 0 \quad (11.106)$$

if the measure $\tilde{\mu}_{b,t,N}$ corresponds to the genealogical distance obtained from this function m. That is to say, if the coupling parameter t is large enough, the SK model can be aligned to a GREM with any desired number of hierarchies.

Second, the proof naturally proceeds by showing that the measure $\mathbb{E}\tilde{\mu}_{\beta,t,N}^{\otimes 2}$ seen as a distribution of the 'overlaps' concentrates on the set where the R_N and q's are the same. Such a fact is usually proven by looking at a suitable Laplace transform, whose calculation amounts again to the estimate of a free energy, this time in a replicated, coupled system. Since the main effort goes into an upper bound, Guerra's techniques can again be used to provide help, even though the details of the computations now get very involved.

Remark 11.3.8 It should have become clear that Guerra's ideas, in particular in the transparent form discovered by Aizenman, Sims, and Starr, provide an extremely efficient and simple tool for proving good bounds on free energies. Their applications go well beyond the Gaussian SK models. Soon after they appeared, they were applied, e.g., to the site dilute version of the SK model, the Viana–Bray model, by Franz and Toninelli [104] and De Sanctis [83]. The general interpolation idea behind Lemma 10.2.1 can also be used very efficiently to compare different models. One of the nice results obtained in this way is the proof by Franz, Leone, and Toninelli [105] of the fact that the free energy of the Kac version of the SK model converges to that of the SK model itself, in the limit when the range of interaction tends to infinity. This result, analogous to the same result in the ferromagnetic setting, shows that there is after all some connection between mean-field and lattice models, even in the case of spin-glasses.

Another important result that can be proven with the help of the same method of proof is that the value of the free energy does not depend strongly on the distribution of the couplings. In particular, it remains the same if in the p-spin SK models the Gaussian couplings $J_{i_1,\ldots,i_p}$ are replaced by non-Gaussian couplings [72, 237].

Theorem 11.3.7 suggests that the Gibbs measure of the SK models should be described by the genealogy of Neveu's branching process with time change corresponding to the choice of m that realizes the infimum. This clearly appears to be the belief of the physics community (when translated into mathematics). However, nothing of what we have seen so far would prove such a claim. For the time being, we are stuck with the formula for the free energy. Let us recall that, in Derrida's models, we have been able to derive the structure of the limiting Gibbs measures essentially from the free energy. The key to success, however, was the Ghirlanda–Guerra identities, together with the fact that the lexicographic distance used for the description of the Gibbs measure was an ultrametric distance, which implied that the GG identities determined the empirical distance distribution up to the function ψ_β. In the SK models, there are some serious problems with both issues, which for the time being prevent us from reaching a full understanding of the nature the Gibbs measures in this case. It will be worthwhile to discuss this point in some more detail.

11.4 The Ghirlanda–Guerra relations in the SK models

Recall that the Ghirlanda–Guerra relations in Derrida's model follow from the Gaussian partial integration formula used with respect to the increments of the processes $X_\sigma(t)$ defined through (10.57). In their original paper, Ghirlanda and Guerra did not consider Derrida's models, but rather the p-spin SK models, using the representation of (8.8) and integration by parts with respect to the independent random variables $J_{i_1,\ldots,i_p}$.

The analogue of Proposition 9.5.1 is:

Lemma 11.4.1 *For any value of β, in the p-spin model,*

$$\mathbb{E}\frac{d}{d\beta}\Phi_{\beta,N} = \beta\bigl(1 - \mathbb{E}\mu_{\beta,N}^{\otimes 2}\bigl(R_N^p(\sigma,\sigma')\bigr)\bigr) \tag{11.107}$$

Proof We just repeat the steps of the proof of Proposition 9.5.1 with the obvious modifications. Namely

$$\mathbb{E}\frac{d}{d\beta}\Phi_{\beta,N} = N^{-1/2}\mathbb{E}\frac{\mathbb{E}_\sigma X_\sigma e^{\beta\sqrt{N}X_\sigma}}{\mathbb{E}_\sigma e^{\beta\sqrt{N}X_\sigma}}$$

$$= N^{-(p+1)/2}\sum_{1\le i_1,\ldots,i_p\le N}\mathbb{E}J_{i_1,\ldots,i_p}\frac{\mathbb{E}_\sigma \sigma_{i_1},\ldots,\sigma_{i_p} e^{\beta\sqrt{N}X_\sigma}}{\mathbb{E}_\sigma e^{\beta\sqrt{N}X_\sigma}}$$

$$= N^{-(p+1)/2}\sum_{1\le i_1,\ldots,i_p\le N}\mathbb{E}\Biggl(\beta N^{-(p-1)/2}\frac{\mathbb{E}_\sigma \sigma_{i_1}^2,\ldots,\sigma_{i_p}^2 e^{\beta\sqrt{N}X_\sigma}}{\mathbb{E}_\sigma e^{\beta\sqrt{N}H_\sigma}}$$

$$-\beta N^{-(p-1)/2}\frac{\mathbb{E}_\sigma\mathbb{E}_{\sigma'}\sigma_{i_1},\ldots,\sigma_{i_p}\sigma_{i_1}',\ldots,\sigma_{i_p}' e^{\beta\sqrt{N}(X_\sigma+X_{\sigma'})}}{\mathbb{E}_\sigma\mathbb{E}_{\sigma'}e^{\beta\sqrt{N}(X_\sigma+X_{\sigma'})}}\Biggr)$$

$$= \beta - \beta\mathbb{E}\frac{\mathbb{E}_\sigma\mathbb{E}_{\sigma'}\bigl(N^{-1}\sum_{i=1}^N \sigma_i\sigma_i'\bigr)^p e^{\beta\sqrt{N}(X_\sigma+X_{\sigma'})}}{\mathbb{E}_\sigma\mathbb{E}_{\sigma'}e^{\beta\sqrt{N}(X_\sigma+X_{\sigma'})}} \tag{11.108}$$

This is (11.107). □

Proposition 11.4.2 *In the p-spin model, for any bounded function* $h : \mathcal{S}_N^n \to \mathbb{R}$,

$$\left| \mathbb{E}\mu_{\beta,N}^{\otimes n+1}\left(h(\sigma^1, \ldots, \sigma^n) R_N^p(\sigma^k, \sigma^{n+1})\right) \right. \tag{11.109}$$

$$\left. - \frac{1}{n}\mathbb{E}\mu_{\beta,N}^{\otimes n+1}\left(h(\sigma^1, \ldots, \sigma^n)\left(\sum_{l \neq k}^n R_N^p(\sigma^l, \sigma^k) + \mathbb{E}\mu_{\beta,N}^{\otimes 2}\left(R_N^p(\sigma_1, \sigma_2)\right)\right)\right) \right| = \delta_N(\beta)$$

with $\delta_N(\beta)$ as in Lemma 9.5.4.

Proof We leave the proof, which follows that of Proposition 9.5.7 in detail, as an exercise. See also [166] for a good and general exposition. □

Choosing h to be the indicator function

$$h(\sigma^1, \ldots, \sigma^n) = \mathbb{I}_{\forall_{k \neq l} R_N(\sigma^k, \sigma^l) = q_{kl}} \tag{11.110}$$

we get that (for almost all β)

$$\mathbb{E}\mu_{N,\beta}^{\otimes n+1}\left[R_N^p(\sigma^k, \sigma^{n+1}) \mid \forall_{k \neq l} R_N(\sigma^k, \sigma^l) = q_{kl}\right]$$
$$= \frac{1}{n}\sum_{l \neq k}^n q_{kl}^p + \frac{1}{n}\mathbb{E}\mu_{N,\beta}^{\otimes 2}\left(R_N^p(\sigma^1, \sigma^2)\right) + \delta_N(\beta) \tag{11.111}$$

which is the relation (17) of [127]. This relation provides a recursion for the p-th moments of the measures $\mathbb{Q}_\beta^{(n)}$, but does not determine fully the distributions recursively. Ghirlanda and Guerra remarked that the situation can be improved by adding a small perturbation to the Hamiltonian, which provides enough random variables to produce the relations (11.111) with p replaced by any integer ℓ. This will happen if we take

$$\tilde{H}_N(\sigma) = H_N(\sigma) + N^{1/2-\alpha}\sum_{\ell=1}^\infty a_\ell X_\sigma^\ell \tag{11.112}$$

where $\text{cov}(X_\sigma^\ell, X_{\sigma'}^{\ell'}) = \delta_{\ell\ell'} R_N^\ell(\sigma, \sigma')$, for some square-summable sequence a_ℓ, and use partial integration with respect to the i.i.d. random variables $J_{i_1,\ldots,i_\ell}$ through which the processes X_σ^ℓ can be constructed. The resulting set of identities then provides recursive identities for all moments of the laws $\mathbb{Q}_\beta^{(n)}$, from which we recover the form (10.90) of the identities. It should be noted, however, that even these *extended Ghirlanda–Guerra identities* do not fully determine the measures $\mathbb{Q}_\beta^{(n)}$ in terms of the $\mathbb{Q}_\beta^{(2)}$. To see this, consider the case $n = 3$ and consider the distribution $\mathbb{Q}_\beta^{(3)}(d_{31}, d_{32}, d_{12})$. While the GG identities allow us to express $\mathbb{Q}_\beta^{(3)}(d_{31}, d_{12})$ in terms of $\mathbb{Q}_\beta^{(2)}(d_{12})$, namely

$$\mathbb{Q}_\beta^{(3)}(d_{31} \in \mathcal{A}, d_{12} \in \mathcal{B}) = \frac{1}{2}\mathbb{Q}_\beta^{(2)}(d_{12} \in \mathcal{A})\mathbb{Q}_\beta^{(2)}(d_{12} \in \mathcal{B}) + \frac{1}{2}\mathbb{Q}_\beta^{(2)}(d_{12} \in \mathcal{A}, d_{12} \in \mathcal{B}) \tag{11.113}$$

and likewise for $\mathbb{Q}_\beta^{(3)}(d_{32} \in \mathcal{A}, d_{12} \in \mathcal{B})$, we have no expression for the general probability $\mathbb{Q}_\beta^{(3)}(d_{31} \in \mathcal{A}, d_{32} \in \mathcal{C}, d_{12} \in \mathcal{B})$. Recall that, in the case of Derrida's models, this shortcoming was overcome by the fact that the distance used was ultrametric. It is believed that, in the SK models, ultrametricity will hold asymptotically with probability one on the support of the Gibbs measures, similar to the situation in the GREM, where by Theorem 10.1.15 the distribution of d_N and R_N coincide in the thermodynamic limit. In the SK models, this question appears now to be the main obstacle to a full understanding of the structure of the Gibbs states.

Let us make some comments on the problem of adding perturbations to the Hamiltonian as in (11.112). First, it is clear that such additions do not change the value of the free energy, as follows from a simple estimate on the supremum of this perturbation, which will be non-extensive. However, this does not at all imply that the structure of the Gibbs states remains unchanged. On the other hand, if the covariance function ξ is a generic analytic function whose Taylor coefficients are all non-vanishing, the extended GG identities do hold. One could of course go further and add to the Hamiltonian another Gaussian process of the GREM type, possibly with a small coefficient that ensures that the free energy is unaltered. In that case we would immediately recover the extended GG inequalities for the distribution of the ultrametric overlap distance, and that would imply that the the genealogy of this measure is derived from Neveu's process, just as in the case of Derrida's model. Again we could argue that the presence of such terms is 'generic' in the space of all Gaussian processes. From this we can deduce two lessons: first, it is much more difficult to study a specific model than a 'generic' one, and second, we cannot exclude the possibility that there are models that have the same free energy but different Gibbs states.[2]

11.5 Applications in the p-spin SK model

To conclude our discussion of the SK models, we want to point out that the Ghirlanda–Guerra relations, even in their limited form (11.111), can give important information about the properties of the Gibbs measures. To do so, we have to get some a-priori estimates.

As first observed by Talagrand, the equality (11.107) implies that the replica overlap cannot remain concentrated on the value zero for all values of β. Namely, assume that, for all values of $\beta \in [0, b]$, $\lim_{N \uparrow \infty} \mathbb{E}\mu_{\beta,N}^{\otimes 2}(R_N^p(\sigma, \sigma')) = 0$. Then it is plain from (11.107) that, in this interval, $\lim_{N \uparrow \infty} \mathbb{E}\Phi_{\beta,N} = -\beta^2/2$. Thus the replica overlap cannot remain zero beyond the critical value, β_p (defined to be the largest value for which this behaviour holds). This actually suggests that for all values $\beta > \beta_p$, $\liminf_{N \uparrow \infty} \mathbb{E}\mu_{\beta,N}^{\otimes 2}(R_N^p(\sigma, \sigma')) > 0$, but there is unfortunately no monotonicity argument available to prove this.[3] However, an upper bound on the ground-state energy implies strict positivity at least soon after the critical point. In fact:

Proposition 11.5.1 *There exists $c_p > 0$ such that, for all $\beta > \sqrt{2 \ln 2}(1 - c_p)$,*

$$\liminf_{N \uparrow \infty} \mathbb{E}\mu_{\beta,N}^{\otimes 2}\left(R_N^p(\sigma, \sigma')\right) > 0 \tag{11.114}$$

Proof Just note that by the first lines of (11.108) and (11.107),

$$\beta\left(1 - \mathbb{E}\mu_{\beta,N}^{\otimes 2}\left(R_N^p(\sigma, \sigma')\right)\right) \leq N^{-1/2}\mathbb{E} \sup_{\sigma \in \mathcal{S}_N} X_\sigma \tag{11.115}$$

---

[2] For example, in the GREM the free energy depends only on the concave hull of the function A. It should be possible to fine-tune models with different A but the same $\bar{A}$ to have different Gibbs states. The same should be possible in degenerate cases of the models proposed by Bolthausen and Kistler in [32].

[3] That is to say, it is in principle possible that the overlap would exhibit a very erratic oscillating behaviour just above the critical value of β!

Now already comparison with the i.i.d. case shows that $\mathbb{E} \sup_{\sigma \in \mathcal{S}_N} X_\sigma \leq N^{1/2}\sqrt{2\ln 2}$. This estimate can be improved to $\sqrt{2\ln 2}(1 - c_p)$, either by elementary methods, or using the Guerra bounds (for an exposition of classical methods, see, e.g., [39]). Thus we get that

$$\left(1 - \mathbb{E}\mu_{\beta,N}^{\otimes 2}\left(R_N^p(\sigma,\sigma')\right)\right) \leq \beta^{-1}\sqrt{2\ln 2}(1 - c_p) \tag{11.116}$$

from which the assertion of the proposition follows. □

Proposition 11.5.1 is not quite enough information on the distribution of the overlap. We need to complement this by an estimate that asserts that there exists an interval of forbidden values.

Theorem 11.5.2 *For any $\epsilon > 0$ there exists $p_0 < \infty$ such that for all $p \geq p_0$, and for all $0 \leq \beta < \infty$,*

$$\lim_{N\uparrow\infty} \mathbb{E}\mu_{N,\beta}^{\otimes 2}(|R_N(\sigma,\sigma')| \in [\epsilon, 1-\epsilon]) = 0 \tag{11.117}$$

If, moreover, $\beta < \hat{\beta}_p$, then for any $\epsilon > 0$ there exists $p_0 < \infty$ such that for all $p \geq p_0$, such that for some $\delta > 0$, for all large enough N,

$$\mathbb{E}\mu_{N,\beta}^{\otimes 2}(|R_N(\sigma,\sigma')| \in [\epsilon, 1]) \leq e^{-\delta N} \tag{11.118}$$

Remark 11.5.3 Note that we prove this result without any restriction on the temperature, while Talagrand requires some upper bound on β both in [238] and (largely improved) in [235].

Proof Let us first consider the high-temperature region. In view of the fact that we know how to compute the mean of the square of the *truncated* partition function and that this would give the desired result, we only need to (i) justify the truncation and (ii) show that the partition functions in the denominator can be replaced by a constant without harm. But (ii) follows from Lemma 11.2.8: by Chebyshev's inequality, it implies that, for $\beta < c < \hat{\beta}_p$,

$$\begin{aligned}
\mathbb{P}[Z_{\beta,N} < (1-\delta)\mathbb{E}\tilde{Z}_{\beta,N}(c)] &\leq \mathbb{P}[\tilde{Z}_{\beta,N}(c) < (1-\delta)\mathbb{E}\tilde{Z}_{\beta,N}(c)] \\
&\leq \mathbb{P}[|\tilde{Z}_{\beta,N}(c) - \mathbb{E}\tilde{Z}_{\beta,N}(c)| \geq \delta\mathbb{E}\tilde{Z}_{\beta,N}(c)] \\
&\leq \delta^{-2} N^{1-p/2}
\end{aligned} \tag{11.119}$$

To justify point (i), we use (11.119) to show that

$$\begin{aligned}
\mathbb{E}\mu_{\beta,N}^{\otimes 2}(\{X_\sigma > c\sqrt{N}\} \cup \{X_{\sigma'} > c\sqrt{N}\}) &\leq 2\mathbb{E}\mu_{\beta,N}(\{X_\sigma > c\sqrt{N}\}) \\
&\leq 2\frac{\mathbb{E}\mathbb{E}_\sigma e^{\beta X_\sigma} \mathbb{1}_{X_\sigma > c\sqrt{N}}}{(1-\delta)\mathbb{E}\tilde{Z}_{\beta,N}(c)} + \mathbb{P}[Z_{\beta,N} < (1-\delta)\mathbb{E}\tilde{Z}_{\beta,N}(c)] \\
&\leq 2\frac{e^{-N(c-\beta)^2/2}}{(1-\delta)\sqrt{N}(c-\beta)} + \delta^{-2} N^{1-p/2}
\end{aligned}$$

Now we can conclude readily that

$$\mathbb{E}\mu_N^{\otimes 2}(|R_N(\sigma,\sigma')| \in [\epsilon, 1]) \leq \frac{\mathbb{E}\mathbb{E}_{\sigma,\sigma'} \mathbb{I}_{X_\sigma < c\sqrt{N}, X_{\sigma'} < c\sqrt{N}} \mathbb{I}_{R_N(\sigma,\sigma') \in [\epsilon, 1]} e^{\beta\sqrt{N}(X_\sigma + X_{\sigma'})}}{(1-\delta)[\mathbb{E}\tilde{Z}_{\beta,N}(c)]^2}$$

$$+ 2\frac{e^{-N(c-\beta)^2/2}}{(1-\delta)\sqrt{N}(c-\beta)} + 2\delta^{-2}N^{1-p/2} \quad (11.120)$$

Using the representation (11.39), we see that the principal term in (11.120) satisfies

$$\frac{\mathbb{E}\mathbb{E}_{\sigma,\sigma'} \mathbb{I}_{X_\sigma < c\sqrt{N}, X_{\sigma'} < c\sqrt{N}} \mathbb{I}_{R_N(\sigma,\sigma') \in [\epsilon, 1]} e^{\beta\sqrt{N}(X_\sigma + X_{\sigma'})}}{(1-\delta)[\mathbb{E}\tilde{Z}_{\beta,N}(c)]^2} \leq C \exp(-Nc\epsilon^2) \quad (11.121)$$

with constants of order unity. This means that ϵ can even be chosen as $\epsilon \sim N^{-1/2}\ln N$, showing that the measure at high temperatures concentrates as sharply on zero overlap as the uniform measure.

Let us now turn to the more intricate and more delicate low-temperature case. Again we have to justify truncation, and we need a uniform lower bound on the partition functions in the denominator. But first we must decide how to truncate. Note that so far we used truncation only for high temperatures and in such a way that the truncations did not alter the mean partition function. At low temperatures this will not do, and we must dare truncation with $c < \beta$. In the REM we have actually seen a much finer truncation at work that isolated the main contributions to the Gibbs measures coming from the extremal order statistics. This suggest that we should try to truncate at the ground-state energy, which we could extract from Guerra's bound. The important fact is that its random fluctuations are quite small, which we will prove using concentration of measure.

Lemma 11.5.4 *For any $\epsilon > 0$, and for all N large enough,*

$$\mathbb{P}\left[\left|\sup_\sigma X_\sigma - \mathbb{E}\sup_\sigma X_\sigma\right| > x\sqrt{N}\right] \leq \exp\left(-N\frac{x^2}{2}\right) \quad (11.122)$$

Proof This is in fact a simple corollary of Theorem 11.1.2. All we have to do is compute the Lipschitz norm of $\sup_\sigma X_\sigma$. But

$$\left|\sup_\sigma X_\sigma[\omega] - \sup_\sigma X_\sigma[\omega']\right| \leq \sup_\sigma |X\sigma[\omega] - X_\sigma[\omega']|$$

$$= N^{-p/2} \sup_\sigma \left|\sum_{i_1,\ldots,i_p}(J_{i_1,\ldots,i_p}[\omega] - J_{i_1,\ldots,i_p}[\omega'])\sigma_{i_1}\cdots\sigma_{i_p}\right|$$

$$\leq N^{-p/2}\|J[\omega] - J[\omega']\|_2 N^{p/2} = \|J[\omega] - J[\omega']\|_2 \quad (11.123)$$

which means that the Lipschitz norm of $\sup_\sigma(X_\sigma)$ is equal to one. □

As a consequence, we can introduce without harm the indicator function of the event that $\sup_\sigma X_\sigma \leq \mathbb{E}\sup_\sigma X_\sigma + c_N$, where $c_N \sim C \ln N$. On the other hand, we can bound the partition functions in the denominator by $2^{-N} \exp(\beta \sqrt{N}(\mathbb{E} \sup_\sigma(X_\sigma) - c_N))$, since clearly

$$\mathbb{P}\left[Z_{\beta,N} < 2^{-N} \exp\left(\beta \sqrt{N} \left(\mathbb{E} \sup_\sigma X_\sigma\right) - c_N\right)\right]$$
$$\leq \mathbb{P}\left[2^{-N} \sup_\sigma e^{\beta\sqrt{N}X_\sigma} < 2^{-N} \exp\left(\beta\sqrt{N}(\mathbb{E}\sup_\sigma X_\sigma) - c_N\right)\right]$$
$$= \mathbb{P}\left[\sup_\sigma X_N(\sigma) < \mathbb{E} \sup_\sigma(X_\sigma) - c_N\right] \leq \exp\left(-\frac{c_N^2}{2}\right) \quad (11.124)$$

Now let $\Delta \subset (-1, 1)$. Set $E_N \equiv N^{-1/2} \mathbb{E} \sup_\sigma X_\sigma$. Then

$$\mathbb{E}\mu_N^{\otimes 2}(R_N(\sigma, \sigma') \in \Delta) \mathbb{E}\,\mathbb{I}_{\sup_\sigma X_\sigma \leq \sqrt{N}E_N + c_N}$$
$$\times \frac{\mathbb{E}_{\sigma,\sigma'} e^{\beta\sqrt{N}(X_\sigma + X_{\sigma'})} \mathbb{I}_{R_N(\sigma,\sigma') \in \Delta}}{2^{-2N} e^{2\beta N(E_N - c_N/\sqrt{N})}} + 2\exp\left(-\frac{c_N^2}{2}\right) \quad (11.125)$$
$$\leq \frac{\mathbb{E}\mathbb{E}_{\sigma,\sigma'} e^{\beta\sqrt{N}(X_\sigma + X_{\sigma'})} \mathbb{I}_{X_\sigma + X_{\sigma'} \leq 2\sqrt{N}E_N + 2c_N} \mathbb{I}_{R_N(\sigma,\sigma') \in \Delta}}{2^{-2N} e^{2\beta N(E_N - c_N/\sqrt{N})}} + 2\exp\left(-\frac{c_N^2}{2}\right)$$

Now assume that $\beta > E_N + c_N/\sqrt{N}$. Then, as in (11.40),

$$\frac{\mathbb{E}\mathbb{E}_{\sigma,\sigma'} e^{\beta\sqrt{N}(X_\sigma + X_{\sigma'})} \mathbb{I}_{X_\sigma + X_{\sigma'} \leq 2(\sqrt{N}E_N + c_N)} \mathbb{I}_{R_N(\sigma,\sigma') \in \Delta}}{2^{-2N} e^{2\beta N(E_N - c_N/\sqrt{N})}}$$
$$\leq \frac{\exp(N(2\ln 2 - (E_N + c_N/\sqrt{N})^2 + 4\beta N^{-1/2} c_N))}{2\pi N}$$
$$\times \sum_{t \in \Delta} \frac{2\sqrt{1+t^p} \exp\left(-NI(t) + N\frac{(E_N + c_N N^{-1/2})^2 t^p}{1+t^p}\right)}{(1-t^2)(\beta(1+t^p) - E_N - N^{-1/2}c_N)} \quad (11.126)$$

The pleasant aspect of this expression is that it is essentially independent of β, since c_N goes to zero with N. Moreover, we know that $\sqrt{2\ln 2} > E_N \geq \sqrt{2\ln 2}(1 - c_p)$, with $c_p = 2^{-p}/\ln 2$. Thus, any interval Δ on which

$$I(t) - N\frac{2\ln 2\, t^p}{1+t^p} > c_p \sim 2^{-p} \quad (11.127)$$

gives only an exponentially small contribution. This applies to all intervals except $(-2^{-p/2}, 2^{-p/2})$ and $[1 - 2^{-p}, 1]$. This proves the theorem. $\square$

Of course we would like to know more precisely how the the overlap is distributed on the remaining two little intervals. In particular it seems more than reasonable that the phase transition is accompanied, as in the REM, by a charging of mass to the neighbourhood of the value one, which would imply a 'lumping phenomenon' of the Gibbs measure. But it looks hopeless at the moment to get such information from a computation like the preceding one, in particular, as long as we do not know the precise value of E_N (remember how sharply we had to estimate in the REM to get such information out!). Thus to get more information, one has to use some more subtle tricks.

If we combine Proposition 11.5.1 with Theorem 11.5.2, we see that, above the critical value of β, the overlap has a non-trivial distribution supported on the neighbourhoods of

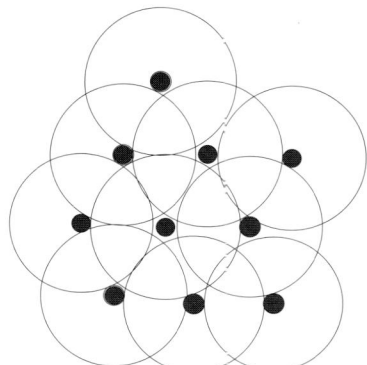

Figure 11.1 Lumps surrounded by empty regions.

zero and ± 1, while intermediate values are excluded. As Talagrand observed in [236], this fact alone allows us to draw rather stringent conclusions about the Gibbs measures. Roughly speaking, and maybe not too surprisingly, they amount to saying that the Gibbs measure looks like some softened-up version of that of the REM. More precisely, one can conclude that asymptotically, as $N \uparrow \infty$, the Gibbs measure will be concentrated on a small subset of the hypercube that consists of even smaller disjoint components that Talagrand called 'lumps', as depicted in Fig. 11.1. Each lump is of size no bigger than $O(2^{-p})$, while the distance between any two lumps is at least $1 - O(2^{-p/2})$. Naturally, one may want to think of the lumps as candidates for pure states. More precisely, is possible to decompose the state space $\mathcal{S}_N$ into a collection of disjoint subsets $\mathcal{C}_k$ such that

(i) $\lim_{N \uparrow \infty} \mathbb{E} \mu_{\beta,N}^{\otimes 2}(\{(\sigma, \sigma') | |R_N(\sigma, \sigma')| > \epsilon\} \setminus \cup_k \mathcal{C}_k \times \mathcal{C}_k) = 0$ (11.128)

(where the $\mathcal{C}_k$ depend both on N and on the random parameter!), and

(ii) if $\sigma, \sigma' \in \mathcal{C}_k$, then $R_N(\sigma, \sigma') \geq 1 - \epsilon$.

These facts follow from quite simple geometric considerations that we invite the reader to work out or to look up in [236].

The most important contribution of Proposition 11.5.1 to this picture is that it implies that at least one of these lumps has positive mass. Of course one would expect that the distribution of the masses of the individual lumps should be similar to the situation in the REM as well, i.e. the total mass should be distributed according to some law on a countable set of lumps. In the REM this was the consequence of a fundamental theorem on the Poisson convergence of the extreme order statistics of an i.i.d. family of random variables. There seems no immediate way to obtain a similar proff in the p-spin case. On the other hand, we have seen that substantial information on the lump distribution could also be obtained via the Ghirlanda–Guerra identities. Talagrand has shown that this path is feasible in the p-spin case, and we will follow him at least some way along this road.

Theorem 11.5.5 *Assume that $\beta > \beta_p$. Let $\mathcal{C}_k$ be ordered such that, for all k, $\mu_{N,\beta}(\mathcal{C}_k) \geq \mu_{N,\beta}(\mathcal{C}_{k+1})$. Then for all $k \in \mathbb{N}$, there exists $p_k < \infty$ such that, for all $p \geq p_k$,*

$$\mathbb{E}\mu_{N,\beta}\left(\cup_{l=1}^k \mathcal{C}_l\right) < 1 + \delta_N(\beta) \quad (11.129)$$

Moreover, for k large, $p_k \sim \frac{2}{3} \frac{\ln k}{\ln 2}$.

11.5 Applications in the p-spin SK model

Proof In fact, just as we could prove an analogue of Proposition 9.5.1 for the p-spin model, we can also generalize, virtually without changing the proof, Proposition 9.5.7.

Assume that the assertion of Theorem 11.5.5 fails. Then there exists a first instance k^* such that

$$\mathbb{E}\mu_{N,\beta}\left(\cup_{l=1}^{k^*}\mathcal{C}_l\right) = 1 + \delta_N(\beta) \tag{11.130}$$

Now define events $Q_{\epsilon_0}^{(n)} \in \mathcal{B}_n$ by

$$Q_{\epsilon_0}^{(n)} \equiv \left\{\underline{R} \in [-1,1]^{n(n-1)/2} | \forall_{1 \leq l < k \leq n} |R_{lk}| \leq \epsilon_0\right\} \tag{11.131}$$

The important observation is that, if $\{R_N(\sigma_l, \sigma_k)\}_{1 \leq l < k \leq k^*} \in Q_{\epsilon_0}^{(k^*)}$, then there exists some permutation $\pi \in S_{k^*}$ such that, with probability one, $\sigma^k \in \mathcal{C}_{\pi(k)}$ for all $k \leq k^*$. In particular

$$\mathbb{E}\mu_{N,\beta}^{\otimes k^*+1}\left[R_N^p\left(\sigma^k, \sigma^{k^*+1}\right)\mathbb{1}_{\{R_N(\sigma^l,\sigma_m)\}_{1 \leq l < \leq k^*} \in Q_{\epsilon_0}^{(k^*)}}\right]$$
$$= \mathbb{E}\mu_{N,\beta}^{\otimes k^*+1}\left[R_N^p\left(\sigma^k, \sigma^{k^*+1}\right)\mathbb{1}_{\exists_\pi \forall_{l=1}^{k^*} \sigma^l \in \mathcal{C}_{\pi(l)}}\right] + \delta_N(\beta) \tag{11.132}$$

But

$$\mathbb{E}\mu_{N,\beta}^{\otimes k^*+1}\left[R_N^p\left(\sigma^k, \sigma^{k^*+1}\right)\mathbb{1}_{\exists_\pi \forall_{l=1}^{k^*} \sigma^l \in \mathcal{C}_{\pi(l)}}\right]$$
$$= \sum_{\pi \in S_{k^*}} \mathbb{E}\mu_{N,\beta}^{\otimes k^*+1}\left[R_N^p\left(\sigma^k, \sigma^{k^*+1}\right)\mathbb{1}_{\forall_{l=1}^{k^*} \sigma^l \in \mathcal{C}_{\pi(l)}}\right]$$
$$= \sum_{\pi \in S_{k^*}}\sum_{j=1}^{k^*} \mathbb{E}\mu_{N,\beta}^{\otimes k^*+1}\left[R_N^p\left(\sigma^k, \sigma^{k^*+1}\right)\mathbb{1}_{\sigma^{k^*+1} \in \mathcal{C}_{\pi(j)}}\mathbb{1}_{\forall_{l=1}^{k^*} \sigma^l \in \mathcal{C}_{\pi(l)}}\right]$$
$$= \sum_{\pi \in S_{k^*}}\sum_{\substack{j=1 \\ j \neq k}}^{k^*} \mathbb{E}\mu_{N,\beta}^{\otimes k^*+1}\left[R_N^p\left(\sigma^k, \sigma^j\right)\mathbb{1}_{\sigma^{k^*+1} \in \mathcal{C}_{\pi(j)}}\mathbb{1}_{\forall_{l=1}^{k^*} \sigma^l \in \mathcal{C}_{\pi(l)}}\right]$$
$$+ \sum_{\pi \in S_{k^*}} \mathbb{E}\mu_{N,\beta}^{\otimes k^*+1}\left[R_N^p\left(\sigma^k, \sigma^{k^*+1}\right)\mathbb{1}_{\sigma^{k^*+1} \in \mathcal{C}_{\pi(k)}}\mathbb{1}_{\forall_{l=1}^{k^*} \sigma^l \in \mathcal{C}_{\pi(l)}}\right] \tag{11.133}$$

where we used the symmetry between replicas in the terms $j \neq k$ to exchange σ^{k^*+1} with σ^j. Note that for the first term we have the obvious (though not very good) bound

$$0 \leq \sum_{\pi \in S_{k^*}}\sum_{j \neq k}^{k^*} \mathbb{E}\mu_{N,\beta}^{\otimes k^*+1}\left[R_N^p(\sigma^k, \sigma^j)\mathbb{1}_{\sigma^{k^*+1} \in \mathcal{C}_{\pi(j)}}\mathbb{1}_{\forall_{l=1}^{k^*} \sigma^l \in \mathcal{C}_{\pi(l)}}\right]$$
$$\leq \epsilon_0^p \mathbb{E}\mu_{N,\beta}^{\otimes k^*}\left[\mathbb{1}_{\forall_{l=1}^{k^*} \sigma^l \in \mathcal{C}_{\pi(l)}}\right] = \epsilon_0^p \mathbb{E}\mu_{N,\beta}^{\otimes k^*}\left[Q_{\text{e}}^{k^*}\right]$$

while the second term satisfies

$$\sum_{\pi \in S_{k^*}} \mathbb{E}\mu_{N,\beta}^{\otimes k^*+1}\left[R_N^p\left(\sigma^k, \sigma^{k^*+1}\right)\mathbb{1}_{\sigma^{k^*+1} \in \mathcal{C}_{\pi(k)}}\mathbb{1}_{\forall_{l=1}^{k^*} \sigma^l \in \mathcal{C}_{\pi(l)}}\right]$$
$$\geq (1-\epsilon_1)^p \sum_{\pi \in S_{k^*}} \mathbb{E}\mu_{N,\beta}^{\otimes k^*+1}\left[\mathbb{1}_{\sigma^{k^*+1} \in \mathcal{C}_{\pi(k)}}\mathbb{1}_{\forall_{l=1}^{k^*} \sigma^l \in \mathcal{C}_{\pi(l)}}\right]$$
$$= \frac{1}{k^*}(1-\epsilon_1)^p \mathbb{E}\mu_{N,\beta}^{\otimes k^*+1}\left[\mathbb{1}_{\forall_{l=1}^{k^*} \sigma^l \in \mathcal{C}_{\pi(l)}}\right]$$
$$= \frac{1}{k^*}(1-\epsilon_1)^p \mathbb{E}\mu_{N,\beta}^{\otimes k^*}\left[Q_{\epsilon_0}^{k^*}\right] \tag{11.134}$$

where we used the obvious permutation symmetry among the first k^* replicas. Let us now use (11.109) with f chosen as the indicator function of the event $Q_{\epsilon_0}^{(k^*)}$. Clearly we get

$$\mathbb{E}\mu_{N,\beta}^{\otimes k^*+1}\left[R_N^p(\sigma^k,\sigma^{k^*+1})\mathbb{1}_{\{R_N(\sigma^l,\sigma_m)\}_{1\leq l<\leq k^*}\in Q_{\epsilon_0}^{(k^*)}}\right]$$
$$\leq \frac{1}{k^*}\mathbb{E}\mu_{N,\beta}^{\otimes k^*+1}\left[\mathbb{1}_{\{R_N(\sigma^l,\sigma_{l_2})\}_{1\leq l<\leq k^*}\in Q_{\epsilon_0}^{(k^*)}}\right]$$
$$\times \left((k^*-1)\epsilon_0^p + \mathbb{E}\mu_{N,\beta}^{\otimes 2}R^p(\sigma,\sigma')\right) + \delta_N(\beta) \qquad (11.135)$$

Comparing (11.134) to (11.135) we see that

$$(1-\epsilon_1)^p \leq (k^*-1)\epsilon_0^p + \lim_{N\uparrow\infty}\mathbb{E}\mu_{N,\beta}^{\otimes 2}R^p(\sigma,\sigma') \leq (k^*-1+p_0)\epsilon^p + p_1 \qquad (11.136)$$

This implies the lower bound

$$k^* \geq \frac{(1-\epsilon_1)^p - p_1}{\epsilon_0^p} \qquad (11.137)$$

Quantitatively, this estimate can be refined to

$$k^* \geq C^{-1}2^{3p/2}((1-C2^{-p})^p - p_1) = 2^p p_0(1 - O(2^{-2p})) \qquad (11.138)$$

This proves the theorem. □

Remark 11.5.6 Talagrand has actually improved this result dramatically in two directions: First, he has shown that if the temperature is not too low, the overlap is sharply concentrated on two values, and that the distribution of the lump masses has the same statistics as in the REM, a fact that would follow easily from the extended Ghirlanda–Guerra relations, but which are proven in [238] even without this assumption. Moreover, he proved that in this case the Gibbs measure conditioned on the lumps is a product measure, i.e. a pure state. The interested reader can find this material in Talagrand's textbook [239] and in [236, 238].

12

Hopfield models

> I have hardly ever known a mathematician who was capable of reasoning.
> *Plato, The Republic.*

In this chapter we turn our attention to the second line of generalizations of the Curie–Weiss model that we mentioned at the beginning of Chapter 3. That is we will study models that share with the Curie–Weiss model the feature that their Hamiltonian is expressed as a function of so-called 'overlap parameters' or 'macroscopic functions'. The model we will focus on here is the so-called 'Hopfield model' [139]. In this case, the macroscopic functions are just the 'overlaps' with a random set of a-priori chosen spin configurations, typically denoted by $\xi^1, \ldots, \xi^M \in \mathcal{S}_N$ and called 'patterns'. We set

$$m_\mu(\sigma) \equiv R_N(\sigma, \xi^\mu) \tag{12.1}$$

If the ξ^μ are chosen at random, these quantities become random functions on the space of spin configurations, and a random Hamiltonian is defined, e.g., via

$$H_N(\sigma) \equiv -N \sum_{\mu=1}^{M} m_\mu^2(\sigma) \tag{12.2}$$

This choice yields the famous Hopfield Hamiltonian.

12.1 Origins of the model

The story of the Hopfield model is quite interesting and worth telling. The name goes back to John Hopfield, a physicist at Caltech interested in modelling the behaviour of networks of neurons (for a more general survey, see [1], and from a more mathematical perspective [202]), such as the human brain. Now the brain is a really messy system, composed of a giant mesh of roughly 10^{10} cells, called neurons. These neurons are all linked up via dendrites, essentially long organic wires that are capable of transmitting electric impulses from one neuron to another. What these neurons basically do is send out and receive sequences of such electric pulses at various frequencies. The important thing is that the frequency (or 'firing rate' in the jargon) at which a neuron sends out its pulses depends (among other things) in a complicated manner on the signals coming in from all the other neurons it is connected with. That is to say, each neuron is a small device that processes incoming information and transmits the result to other neurons. Clearly, the way these things are connected produces a device that can perform amazing computational tasks (like reading these pages and possibly making some sense out of them ...). How can one possibly understand how such a system works? Clearly, a single neuron is already a complicated system whose dynamics is far from easy to analyze; trying to analyze the joint behaviour of billions of them thus looks hopeless. In such a situation physicists like to simplify the models, and to abstract from details while keeping what are believed to be the essential features. The first step in this simplification goes the same way as our old friend Ising: simplify the 'state space' of a single neuron to the simplest possible one, $\{-1, 1\}$, suggesting that the neuron fires 'rapidly' or 'slowly'. This idea goes back at least to McCulloch and Pitts [176] in 1943. Now it was known that a neuron changes its state at a rate depending on the compound, but weighted, effect of all its input signals, which are functions of the states of those neurons that are connected to it.[1] We can think of this effective input signal as some field

$$h_i = f_i\left(\{\sigma_j\}_{j \in N(i)}\right) \tag{12.3}$$

where $N(i)$ denotes the set of all neurons 'firing' into the neuron i. The way the network processes information then depends on the structure of these neighbourhoods (the 'graph' of connections), and the properties of the functions f_i. Setting these up corresponds in a way to programming the system. Clearly the simplest choice for a function f_i is a linear one, that is

$$f_i\left(\{\sigma_j\}_{j \in N(i)}\right) = \sum_{j \in N(i)} J_{ij} \sigma_j \tag{12.4}$$

It is known that the effects of a signal on a neuron can go both ways, inducing it to fire or to stop firing. Thus the coefficients J_{ij} should have the possibility of taking both signs, and different strengths. Thus we see that this field looks like the local field acting on site i in a *spin-glass model*. Of course, in contrast to a spin-glass, the coupling should not be arbitrary, but rather be programmed to ensure a particular task. But how can such programming happen? As early as 1949, D. Hebb [137] suggested a sort of progressive self-programming of such a network that should give rise to the network functioning as a *memory*, more precisely an *autoassociative memory*. The idea is that the connection between two neurons should be altered in a direction to 'favour' the current states present, i.e. one should add a term proportional to $\sigma_i \sigma_j$. In this way, if the network over time has passed through a number M of different 'states', denoted by $\xi^1, \ldots, \xi^M$, the couplings would take the form

$$J_{ij} = \sum_{\mu=1}^{M} \xi_i^\mu \xi_j^\mu \tag{12.5}$$

This form of the coupling is called *Hebb's rule*. While it is a bit difficult to believe that things should be that simple, this rule has been a widely accepted one, and in any case is interesting enough for us to start investigating the ensuing model. Of course, if this idea was to be taken seriously, the couplings should evolve in time; we will, however, assume that the couplings have reached a state of saturation where they do not change anymore, and what we are interested in should be the evolution of the state of the network from some initial state $\sigma(0)$ with fixed couplings of the form (12.5). We still have to define the dynamics: here Hopfield, in his seminal 1982 paper [139], proposed a *Markov chain*, where at independent

---

[1] Connections in real neural networks are directed and not reciprocal, a fact that we shall ignore here.

exponentially distributed random times a clock rings at site i and the neuron changes its state $\sigma_i(t)$ to a new value ± 1 with probabilities proportional to $\exp(\pm\beta h_i)$. Hopfield's key observation was that (in the case of symmetric connections) such a Markov chain would have as its invariant measure a *Gibbs measure* corresponding to the Hamiltonian

$$H_N(\sigma) = -\sum_{i,j:j\in N(i)} J_{ij}\sigma_i\sigma_j \tag{12.6}$$

with J given by (12.5). Finally, simplifying the model further by assuming that any neuron is connected to any other, and normalizing properly, one arrives at the Hamiltonian

$$H_N(\sigma) = -\frac{1}{N}\sum_{\mu=1}^{M}\sum_{i,j=1}^{N} \xi_i^\mu \xi_j^\mu \sigma_i\sigma_j \tag{12.7}$$

which one sees easily to be equal to the expression given in (12.2). This is, in brief, the reasoning that led Hopfield to derive what could be called the 'Ising model' of neural networks. The formal resemblance to spin-glass models then of course sparked immediate interest among physicists who saw a chance to bring their expertise to bear in an entirely new context.

Interestingly enough, though, the Hamiltonian (12.7) had been introduced five years before Hopfield, in three papers by Figotin and Pastur [197, 198, 199] that had apparently received very little attention.[2] Not surprisingly, they were advocated as simple, exactly solvable models of *spin-glasses!* Figotin and Pastur more or less gave a complete solution, which, however, failed to exhibit the key features expected of spin-glasses but revealed that the model behaves very much like the ordinary Curie–Weiss ferromagnet, except that the number of stable magnetized states was equal to $2M$ instead of 2. In a way, from this point of view the model was about as disappointing as the model introduced in 1976 by Mattis [174] that corresponded to the case $M = 1$ and was seen to be totally equivalent to the Curie–Weiss model.

Fortunately, Hopfield did not repeat the analysis of Figotin and Pastur (this was done a few years later by various people), but performed numerical experiments.[3] Moreover, in these experiments he had an objective that was motivated by the interpretation of the model as a memory. This meant that, starting from an initial configuration close to one of the 'patterns' ξ^μ, the system should approach ξ^μ and stay close to it for a long time, if not forever. Now Hopfield observed that this was indeed true (in a sense), but only if M was not too big: in fact the allowed value of M depended on the size N of the network and was roughly $M^* = 0.14N$. This was interpreted as a limited *memory capacity* of the network. So something interesting happened, but only if M was taken as a function of N! Naturally, this fact had eluded Figotin and Pastur who studied the thermodynamic limit $N \uparrow \infty$ with M fixed!

The seemingly small modification brought to the model by considering M as a function $M(N)$ of the system size N thus turns what otherwise would be a simple mean-field model into something much more interesting and also much more complicated to analyze. In [42]

---

[2] This fact was brought to my attention only in 1993 by L. Pastur; it seems that these papers have almost never been cited in the entire literature on Hopfield's model.
[3] This is not to discourage people from doing analytic work. However, the example shows that interesting facts are often found by experimenting with things one does not understand.

we have called such models *generalized random mean-field models*. I will not discuss here the more general setting introduced there but stay with the single example of the standard Hopfield model.

While Hopfield's results suggested that this model is much more interesting than was first thought, the choice of the function $M(N)$ provides a parameter that could make a rigorous analysis possible at least under certain conditions on the growth rate of this function. Indeed, at least the rigorous study of the model can be seen as a constant struggle to push our understanding to larger and larger growth rates, ranging from Figotin and Pastur's constant M through logarithmic [117, 153, 252] via sub-linear growth [44] to what we can control today, linear growth $M(N) = \alpha N$ with sufficiently small α [41, 42, 45, 233, 235]. I will in the sequel explain the most significant steps in this development.

12.2 Basic ideas: finite M

To get a feeling for the model, it will be worthwhile to discuss first the case when M is finite. In this case, everything is similar to the Curie–Weiss model from Section 3.5, and we can use in principle the same two tools introduced there: large deviations and combinatorial calculations, or the Hubbard–Stratonovich transformation.

The large deviation approach looks a-priori more rational and has the advantage that it can be applied to a very general class of (generalized) mean-field models. Its strategy is to study first the distribution of the macroscopic order parameters $m_N(\sigma)$ under the Gibbs measure, i.e. the measures $\mathbb{Q}_N$ such that

$$\mathbb{Q}_{\beta,N}(m \in \mathcal{A}) \equiv \mu_{\beta,N}(m_N(\sigma) \in \mathcal{A}) \tag{12.8}$$

for any Borel set in $\mathbb{R}^M$. Now it is fairly straightforward to see that the family of measures $\mathbb{Q}_{\beta,N}$ satisfies a *large deviation principle*, i.e.

$$\sup_{m \in \mathcal{A}^o} \Psi_\beta(m) = \liminf_{N \uparrow \infty} -\tfrac{1}{N} \ln \mathbb{Q}_{\beta,N}(\mathcal{A}) \tag{12.9}$$
$$\leq \limsup -\tfrac{1}{N} \ln \mathbb{Q}_{\beta,N}(\mathcal{A}) = \sup_{m \in \bar{\mathcal{A}}} \Psi_\beta(m)$$

with probability one, and for a *rate function*, Ψ_β, that is independent of the realization of the random variables ξ. This observation goes back to van Hemmen and co-workers [252, 253, 254, 255, 256] and, in greater generality, Comets [79]. The computation of the rate function is greatly simplified by the fact that H_N is just a function of the m_N. In fact, finding the rate function is reduced to the *combinatorial problem* of counting the number of spin configurations σ that give rise to the same value m_N. This problem is in principle elementary, even though the final expression is fairly involved.

The second approach, based on what is frequently called the *Hubbard–Stratonovich (HS) transformation*, works well only in cases where the Hamiltonian is a quadratic function of the order parameters. When it works, however, it is much simpler, and some results have only been obtained with this method. Since the large deviation approach has been explained extensively in [43], I will stick with the Hubbard–Stratonovich approach here. This approach is incidentally also the one used in [197, 198]. One way to view the HS transformation is to say that it consists in constructing the convolution of the induced measure $\mathbb{Q}_{\beta,N}$ with a

Gaussian measure of mean zero and variance $1/\beta N$,

$$\mathcal{Q}_{\beta,N} \equiv \mathbb{Q}_{\beta,N} \star \mathcal{N}(0, 1/\beta N) \tag{12.10}$$

This measure can be written down in a very explicit form (much more explicit than the measure $\mathbb{Q}_{\beta,N}$), due to the simple identity (3.59) from Section 3.5 applied to the Boltzmann factor:

$$\exp\left(-\frac{1}{2}N\sum_{\mu=1}^{M}(m_N(\sigma)^{\mu})^2\right) \tag{12.11}$$

$$= \int_{-\infty}^{\infty} \frac{dz_1 \cdots dz_M}{(2\pi)^{M/2}} \exp\left(-\frac{1}{2}\sum_{\mu} z_{\mu}^2 + N^{-1/2}\sum_{i=1}^{N}\sigma_i\left(\sum_{\mu}\xi_i^{\mu} z_{\mu}\right)\right)$$

The exponent in this expression is now linear in the variables σ_i. It follows immediately, after a convenient change of variables, that

$$Z_{\beta,N} = \int_{-\infty}^{\infty} \frac{dz_1 \cdots dz_M}{(2\pi)^{M/2}} \exp(-N\beta \Phi_{\beta,N}(z)) \tag{12.12}$$

and

$$\mathcal{Q}_{\beta,N}(dz) = \frac{\exp(-N\beta \Phi_{\beta,N}(z))}{Z_{\beta,N}} dz \tag{12.13}$$

with

$$\Phi_{\beta,N}(z) \equiv \frac{1}{2}\sum_{\mu} z_{\mu}^2 - \frac{1}{\beta N}\sum_{i=1}^{N} \ln \cosh\left(\beta \sum_{\mu}\xi_i^{\mu} z_{\mu}\right) \tag{12.14}$$

This function will play a central rôle in the remainder of this section. In the context of the HS approach, it replaces the rate function in the large deviation approach; the big advantage of the HS method is that it can be written down explicitly for any values of N and M and looks thus much more suitable in the cases when M depends on N. Moreover, the rate function Ψ and the function Φ are quite intimately connected, as is explained at length in [42].

Looking at the function $\Phi_{\beta,N}$, we see that the second part of it is an empirical mean over the sample of N random vectors ξ_i, as shown in Fig. 12.1. Thus we may expect that, in the limit $N \uparrow \infty$, this function will converge to a deterministic one, namely

$$\Phi_{\beta}(z) \equiv \frac{1}{2}\|z\|_2^2 - \beta^{-1}\mathbb{E}\ln\cosh(\beta(\xi_1, z)) \tag{12.15}$$

This follows e.g. in the topology of uniform convergence on compact sets from the *law of large numbers* in Banach spaces (see, e.g., [167]). Since I will later show a related result in the case when M depends on N, I will not discuss this any further at this point. In any case, it is clear that the first thing to do is to understand what this function looks like if we want to understand the properties of the measure $\mathcal{Q}_{\beta,N}$ for large N.

The first observation, due to Figotin and Pastur, is that:

Lemma 12.2.1 *For any $M \in \mathbb{N}$, if ξ_1^{μ} are i.i.d. Rademacher random variables taking the values ± 1 with equal probability, then the function Φ_{β} takes its minimal values $\phi_{\beta}(m^*)$,*

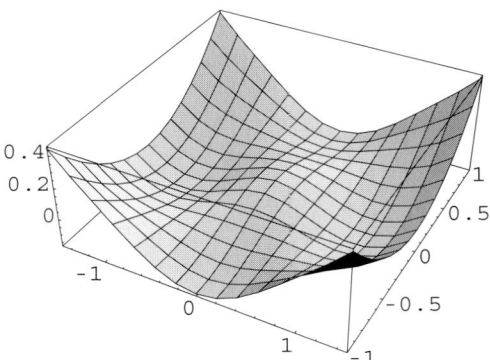

Figure 12.1 The function $\Phi_2(z)$ for $M = 2$.

where m^* is the largest minimizer of the function $\phi_\beta(m) \equiv \frac{1}{2}m^2 - \beta^{-1}\ln\cosh(\beta m)$, i.e. the largest solution of the equation

$$x = \tanh(\beta x) \qquad (12.16)$$

on the set $\mathcal{M}_{\beta,M}$ given by

$$\mathcal{M}_{\beta,M} \equiv \bigcup_{\pm,\mu \in \{1,\ldots,M\}} \{\pm m^* e^\mu\}, \qquad (12.17)$$

where e^μ denotes the μ-th unit vector in $\mathbb{R}^M$. Note that $m^* = 0$, whence $|\mathcal{M}_{\beta,M}| = 1$ if and only if $\beta \leq 1$.

We even have more:

Lemma 12.2.2 *Under the assumptions of the preceding lemma, for all $\beta \neq 1$ there exists $c(\beta) > 0$ such that, for all $M \in \mathbb{N}$,*

$$\Phi_\beta(z) - \phi_\beta(m^*) \geq c(\beta) \min_{y \in \mathcal{M}_{\beta,M}} \|z - y\|_2^2 \qquad (12.18)$$

Proof I will only give a short proof of Lemma 12.2.1. Note that

$$\|z\|_2^2 = \mathbb{E}\left((\xi_1, z)\right)^2 \qquad (12.19)$$

so that

$$\Phi_\beta = \mathbb{E}\phi_\beta((\xi_1, z)) \qquad (12.20)$$

where ϕ_β is the Curie–Weiss function defined above. This function attains its minima at the points $\pm m^*$. Thus it is clear that, if z is such that the random variable (ξ_1, z) is supported on the set $\{-m^*, m^*\}$, then Φ_β attains its absolute minimum at this point; moreover, if such a value exists, then the absolute minimum is attained precisely on the set of values z for which this is true. Now, if ξ_i^μ are Rademacher, then any z of the form $z = \pm m^* e^\mu$ has this property. Moreover, it is very easy to see that these are the only possible candidates. Namely, our condition is

$$\sum_{\mu=1}^M \xi_1^\mu z_\mu = \pm m^*, \; \forall \xi_i \in \{-1, 1\}^M \qquad (12.21)$$

Without loss of generality we can assume $\xi_1^1 = 1$. Then (12.21) implies that

$$z_1 + b = \pm m^*$$
$$z_1 - b = \pm m^* \tag{12.22}$$

Quite obviously this can only be true if either $z_1 = \pm m^*$ and $b = 0$, or $z_1 = 0$ and $b = \pm m^*$. In the second case we are done if $M = 2$, and we can proceed inductively otherwise. In the first case we argue that $b = 0$ implies that all z_μ, $\mu > 1$ must be zero. This is trivial in the case $M = 2$, while for $M > 2$ we can split b again into two pieces that would need to satisfy $z_2 + b_2 = 0$ and $z_2 - b_2 = 0$, which is obviously only possible if both $z_2 = 0$ and $b_2 = 0$. This proves Lemma 12.2.1. □

Noting the elementary fact that $\phi_\beta(x) - \phi_\beta(m^*) \geq a(\beta)(|x| - m^*)^2$, it is quite obvious that Lemma 12.2.2 holds with some constant $c(\beta)$. The proof given in [43] yields a numerical estimate of the constant, but I am not terribly happy with either the proof (which is cumbersome) or the estimate (which is poor). Thus I encourage the reader to try to improve this estimate!

In the case of finite M, it follows readily from these observations that any ϵ-neighbourhood of the set $\mathcal{M}_{\beta,M}$ carries all but an exponentially small fraction of the total mass of the measures $\mathcal{Q}_{\beta,N}$, with probability that tends to one very rapidly. In particular, it is very easy to see that the measure conditioned on, say, any ball[4] of radius $r < m^*$ centered at $m^* e^\mu$ converges, as $N \uparrow \infty$, to the Dirac measure on the point $m^* e^\mu$, almost surely. The same is then true, naturally, for the measures $\mathbb{Q}_{\beta,N}$.

We see that the vicinities of the points $\pm m^* e^\mu$ play here the same rôle as the 'lumps' in the REM or the p-spin SK model, with the difference that they are not randomly placed but deterministically put in by the construction of the model. A natural question is then whether we can control in this model the respective 'lump-masses', that is whether we can control the behaviour of the unconditioned Gibbs measures. This problem was considered only lately as an illustration of the concept of 'metastates' in two papers by Külske [157, 158] in 1997. It is clear that this requires a much more precise analysis of the function $\Phi_{\beta,N}$ than what we have given so far. How can this be obtained?

The first idea should be to use the functional central limit theorem (see, e.g., [167]) to extract the sub-leading corrections. Indeed the following holds:

$$\sqrt{N}(\Phi_{\beta,N}(z) - \Phi_\beta(z)) \xrightarrow{\mathcal{D}} g_\beta(z) \tag{12.23}$$

where g_β is a Gaussian process on $\mathbb{R}^M$ with covariance

$$C_\beta(z, z') = \mathbb{E} \ln \cosh(\beta(\xi_1, z)) \ln \cosh(\beta(\xi_1, z')) \tag{12.24}$$
$$- \mathbb{E} \ln \cosh(\beta(\xi_1, z)) \mathbb{E} \ln \cosh(\beta(\xi_1, z'))$$

At first glance this would suggest that the fluctuations of $\Phi_{\beta,N}$ are of order $1/\sqrt{N}$, and thus the relative weights of the different 'lumps' should differ by factors of order $\exp(\beta\sqrt{N})$. However, a closer inspection shows that this is not true. Namely, note that we are interested in the process g_β essentially only very near the points $z = \pm m^* e^\mu$. But at these points, the variance turns out to be zero, as should be the case, since at these points $\Phi_{\beta,N}$ is non-random!

---
[4] Or, for that matter, any closed set containing the single point $m^* e^\mu$ from $\mathcal{M}_{\beta,M}$.

Note that this relies crucially on the fact that the random variables ξ_i^μ take only the values ± 1, and as soon other distributions are considered, this will change dramatically.[5] In any case, we see that the precision of the CLT is not enough to solve our problem, and we have to look for the next order corrections. In fact, given that the fluctuations are strictly zero at the points $\pm m^* e^\mu$, one might first suspect that maybe the weights could all be equal. However, the random effects will induce small shifts of the position of the minima of $\Phi_{\beta,N}$ away from these points, and we will have to control these shifts and the values of the function at these real minima to solve our problem. Since we expect these shifts to be very small (tending to zero with N), a natural approach is to use Taylor expansions. Let us consider the minimum near the point $m^* e^\mu$. Its location $z^{(\mu)}$ must satisfy the equations

$$z_\nu^{(\mu)} = \frac{1}{N} \sum_i \xi_i^\nu \tanh\left(\beta(\xi_i, z^{(\mu)})\right) \tag{12.25}$$

Now write $z^{(\mu)} = m^* e^\mu + \delta$. Then δ satisfies

$$\delta_\mu = \frac{1}{N} \sum_i \xi_i^\mu \tanh(\beta(m^* \xi_i^\mu + (\delta, \xi_i))) - m^*, \tag{12.26}$$

$$\delta_\nu = \frac{1}{N} \sum_i \xi_i^\nu \tanh(\beta(m^* \xi_i^\mu + (\delta, \xi_i))), \quad \nu \neq \mu$$

Taylor expanding, and using that $m^* = \tanh(\beta m^*)$, we get

$$\delta_\mu = \beta \cosh^{-2}(m^*\beta) \frac{1}{N} \sum_i \xi_i^\mu (\delta, \xi_i) + O(\|\delta\|_2^2) \tag{12.27}$$

$$= \beta \cosh^{-2}(m^*\beta) \delta_\mu + \beta \cosh^{-2}(m^*\beta) \frac{1}{N} \sum_{\nu \neq \mu} \sum_i \xi_i^\mu \xi_i^\nu \delta_\nu + O(\|\delta\|_2^2),$$

$$\delta_\nu = \frac{1}{N} \sum_i \xi_i^\nu \xi_i^\mu \tanh(\beta m^*) \tag{12.28}$$

$$+ \frac{\beta}{N} \sum_i \xi_i^\nu \cosh^{-2}(\beta m^*)(\delta, \xi_i) + O(\|\delta\|_2^2),$$

$$= \beta \cosh^{-2}(m^*\beta) \delta_\nu + \tanh(\beta m^*) \frac{1}{N} \sum_i \xi_i^\nu \xi_i^\mu$$

$$+ \beta \cosh^{-2}(m^*\beta) \frac{1}{N} \sum_{\nu' \neq \nu} \sum_i \xi_i^\nu \xi_i^{\nu'} \delta_{\nu'} + O(\|\delta\|_2^2), \quad \nu \neq \mu$$

Since $\frac{1}{N} \sum_i \xi_i^\nu \xi_i^\mu = O(N^{-1/2})$, one checks readily that to leading order the solution of these equations is

$$\delta_\mu = 0 + O(1/N), \tag{12.29}$$

$$\delta_\nu = \frac{\tanh(\beta m^*)}{1 - \beta \cosh^{-2}(\beta m^*)} \frac{1}{N} \sum_i \xi_i^\nu \xi_i^\mu + O(1/N)$$

---

[5] A particularly interesting situation arises if the distribution of the ξ is taken to be Gaussian (see [60, 248]).

It follows that

$$\Phi_{\beta,N}(z^{(\mu)}) - \phi_\beta(m^*) = \sum_\nu \delta_\nu^2 (1 - \beta \cosh^{-2}(\beta m^*)) + o(1/N) \qquad (12.30)$$

$$= \frac{1}{N} \frac{(m^*)^2}{1 - \beta(1-(m^*)^2)} \sum_{\nu \neq \mu} \left(N^{-1/2} \sum_i \xi_i^\nu \xi_i^\mu \right)^2 + o(1/N)$$

where we used that $\cosh^{-2}(x) = 1 - \tanh^2(x)$. As a consequence, the measure $\mathcal{Q}_{\beta,N}$ has a decomposition

$$\mathcal{Q}_{\beta,N} \approx \sum_{\mu=1}^M p_{\beta,N}(\mu)(\mathcal{Q}_{\beta,N}(\cdot|B_\epsilon(m^* e^\mu)) + \mathcal{Q}_{\beta,N}(\cdot|B_\epsilon(-m^* e^\mu)) \qquad (12.31)$$

where the conditional measures converge almost surely to Dirac measures, and

$$p_{\beta,N}(\mu) = \frac{\exp\left(\beta \frac{(m^*)^2}{1-\beta(1-(m^*)^2)} \sum_{\nu \neq \mu} \left(N^{-1/2} \sum_i \xi_i^\nu \xi_i^\mu\right)^2\right)}{\sum_{\mu=1}^M \exp\left(\beta \frac{(m^*)^2}{1-\beta(1-(m^*)^2)} \sum_{\nu \neq \mu} \left(N^{-1/2} \sum_i \xi_i^\nu \xi_i^\mu\right)^2\right)} \qquad (12.32)$$

We see that these weights are, as random variables, functions of the $M(M-1)$ sums of i.i.d. random variables

$$b_N^{\mu\nu} \equiv N^{-1/2} \sum_i \xi_i^\nu \xi_i^\mu \qquad (12.33)$$

Since for $\mu < \nu$, these variables are uncorrelated (check!), it follows again by the central limit theorem that this family of variables converges weakly to independent normal variables. This permits us to formulate a convergence result in the spirit of the metastate formalism.

Theorem 12.2.3 [157] *Assume that $\beta > 1$ and $M < \infty$. Then*

$$\mathcal{Q}_{\beta,N} \xrightarrow{\mathcal{D}} \mathcal{Q}_\beta \qquad (12.34)$$

where $\mathcal{Q}_\beta$ is the random measure on $\mathbb{R}^M$ given by

$$\mathcal{Q}_\beta \equiv \sum_{\mu=1}^M p_\beta^\mu (\delta_{m^* e^\mu} + \delta_{-m^* e^\mu}) \qquad (12.35)$$

where

$$p_\beta^\mu \equiv \frac{\exp\left(\beta \frac{(m^*)^2}{1-\beta(1-(m^*)^2)} \sum_{\nu \neq \mu} g_{\mu\nu}^2\right)}{\sum_{\mu=1}^M \exp\left(\beta \frac{(m^*)^2}{1-\beta(1-(m^*)^2)} \sum_{\nu \neq \mu} g_{\mu\nu}^2\right)} \qquad (12.36)$$

and, for $\mu < \nu$, the family $\{g_{\mu\nu}\}$ are independent standard Gaussian random variables.

Remark 12.2.4 Of course, the same result also holds with $\mathcal{Q}$ replaced by $\mathbb{Q}$.

It is clear that we could fix (condition on) a finite number of the components of ξ^μ without at all affecting the result. Thus the rôle of conditioning on the disorder that was emphasized in the construction of the metastates does not really come to bear in this setting. In that respect it will be more instructive to look at the Gibbs measures as measures on the original spin space. Since as usual we are interested in the convergence of finite-volume measures in the product topology, it will be enough to consider probabilities

$$\mu_{\beta,N}(\sigma_I = s_I) \tag{12.37}$$

for any finite $I \subset \mathbb{N}$ and $s_I \in \{-1, 1\}^I$, and to prove joint convergence of arbitrary finite collections of such probabilities. A simple computation shows that these can be represented as follows:

$$\mu_{\beta,N}(\sigma_I = s_I) = \frac{\int dz e^{-\beta N \Phi_{\beta',N'}(z)} \prod_{i \in I} e^{\beta s_i(\xi_i, z)}}{\int dz e^{-\beta N \Phi_{\beta',N'}(z)} \prod_{i \in I} 2 \cosh(\beta(\xi_i, z))} \tag{12.38}$$

where $N' = N - |I|$, $\beta' = \beta N/N'$. Note that the difference between N and N' and β and β' is negligible in the limit $N \uparrow \infty$. It is clear that Theorem 12.2.3 implies that $\mu_{\beta,N}$ converges in distribution to a random measure, μ_β, whose finite-dimensional marginals are given by

$$\mu_{\beta,N}(\sigma_I = s_I) = \int \mathcal{Q}_\beta(dz) \prod_{i \in I} \frac{e^{\beta s_i(\xi_i, z)}}{2 \cosh(\beta(\xi_i, z))} \tag{12.39}$$

where ξ_i^μ are i.i.d. symmetric Rademacher random variables, and $\mathcal{Q}_\beta$ is the random measure from Theorem 12.2.3. We see that, when looking at convergence in distribution, we lose the information on our patterns: the ξ_i^μ appearing in (12.39) have the same distribution as the patterns that we used to construct the Hamiltonian, but they are not the same random variables. But clearly we can do better. Namely, conditioning on the ξ_i^μ, for all $i \in J$, for an arbitrarily large finite set $J \subset \mathbb{N}$, does not change anything concerning the convergence of the measures $\mathcal{Q}_{\beta,N}$ (since these finitely many variables give no contribution to the limits of the variables $b_N^{\mu\nu}$!); however, they do appear in (12.39). Therefore, the Aizenman–Wehr metastate will be the same random measure as μ_β, except that now the ξ_i^μ appearing on the right are precisely the original patterns from the definition of the Hamiltonian. Combining these observations, we can state the following theorem:

Theorem 12.2.5 [157] Assume that $\beta > 1$ and $M < \infty$. Then, given ξ,

$$\mu_{\beta,N}[\xi] \xrightarrow{\mathcal{D}} \sum_{\nu=1}^{M} p_\beta^\nu \left(\mu_\beta^{+,\nu}[\xi] + \mu_\beta^{-,\nu}[\xi] \right) \tag{12.40}$$

where p_β^ν are defined in (12.36) and $\mu_\beta^{\pm,\nu}[\xi]$ are product measures on $\{-1, 1\}^\mathbb{N}$ with marginals

$$\mu_\beta^{\pm,\nu}[\xi](\sigma_i = s_i) \equiv \frac{e^{\pm \beta m^* \xi_i^\nu s_i}}{2 \cosh(\beta m^*)} \tag{12.41}$$

Remark 12.2.6 The product measures $\mu_\beta^{\pm,\nu}[\xi]$ can be viewed naturally as the *pure states* for this model. Theorem 12.2.5 then says that the *metastate* is a random convex combination of these extremal states, in accordance with Theorem 6.2.8. One can easily construct sequences of measures converging almost surely to one of these extremal measures by adding an *external field term*

$$\pm h \sum_{i=1}^{N} \xi_i^\nu \sigma_i \qquad (12.42)$$

to the Hamiltonian, and taking the limit $N \uparrow \infty$ first and $h \downarrow 0$ after this (Exercise!). This is in fact much simpler than proving Theorems 12.2.3 and 12.2.5 and requires much less information about the fluctuations of the process $\Phi_{\beta,N}(z)$. It is also much more robust and can be proven even when M grows with N, as long as $\lim_{N \uparrow \infty} M/N = 0$ [44].

Since all of these results follow from the explicit representation of the lump-weights as functions of sums of i.i.d. random variables, one can obtain further results. In particular, from the *invariance principle* one can construct the so-called *superstate* (proposed in [43]), i.e. the measure-valued stochastic process constructed as the (conditioned) distributional limit of the process $\{\mu_{\beta,[tN]}\}_{t \in (0,1]}$ and which is obtained from the expression for μ_β by just replacing the Gaussian random variables $g_{\mu\nu}$ by independent standard Brownian motions $b_{\mu\nu}(t)$. This and further results relating the different notions of metastates can be found in [157, 158].

12.3 Growing M

The finite M model is, as we have seen, readily solvable using just standard results from probability theory: the law of large numbers, central limit theorems, and the Laplace method. As soon as M turns into a function of N, these results are no longer immediately applicable and require substantial modifications. Before discussing the results obtained in this direction, let me identify the main steps in the analysis of the finite M case that need to be reconsidered.

The first difficulty is the fact that the law of large numbers no longer provides the convergence of the function $\Phi_{\beta,N}$ to its deterministic limit. Indeed, since these functions are now defined on $\mathbb{R}^{M(N)}$, with $M(N)$ growing to infinity, we would not even know what we should mean by such convergence.

Even if we control $\Phi_{\beta,N}$, the Laplace method will have to be adapted to a situation when the integral is over a space whose dimension grows together with the large parameter. This is, however, a minor difficulty.

A precise analysis of the local properties of the minima of $\Phi_{\beta,N}$ can no longer rely on simple Taylor expansions as used in the derivation of Theorem 12.2.3. The main point we have used is that, by the Schwartz inequality, $|(\xi_i, \delta)| \leq \sqrt{M} \|\delta\|_2$. If M is finite, this implies that if δ is small in norm, then all the shifts in the arguments of the functions appearing are also small; if M grows with N, this is no longer the case.

The faster M grows as a function of N, the more serious all these difficulties get. As a consequence, the history of the mathematical analysis of the Hopfield model is marked by a sequence of steps reaching larger and larger rates of growth: first results (using combinatorial large deviation methods) by Koch and Piasko [153] in 1989 and van Enter and van Hemmen [252] in 1986 reached logarithmic growth $M(N) \ll \ln N$. The next stage reached sublinear growth ($M(N)/N \downarrow 0$). After the first computation of the free energy in 1993 (Koch [152] and Shcherbina and Tirozzi [220]), the extremal Gibbs measures were constructed in 1995 in a collaboration with Gayrard and Picco [44], a large deviation principle was proven with Gayrard [40] (see also [80] for an interesting variant), and finally a central limit theorem was proven in the same collaboration in 1997 [40] (first results on the CLT, under stronger growth conditions, are due to Gentz [119, 120]). An interesting result on a non-central limit theorem *at the critical temperature* was found more recently by Gentz and Löwe [121, 122]). The only result missing in this regime is the analogue of Theorem 12.2.3, which we have obtained only under more stringent growth conditions (namely $M(N) \ll \sqrt{N}$) in a collaboration with D. Mason [57] in 2001. I will not go into the details of these results, but pass to the next step, the case when $M(N) = \alpha N$ with $\alpha > 0$, but small. Here I will distinguish two steps of progress: first, a-priori estimates of the support of the Gibbs measures, and exponential estimates of the respective weights [41, 45], and second, analysis of the conditional measures corresponding to one pattern, and justification of the replica symmetric solution [42, 43, 233, 235]. Let me mention that there is another line of research that I will not discuss in this book that relates solely to the analysis of local minima of the Hamiltonian of the model and that principally investigates the question of how large α can be chosen to be if one wants to guarantee the existence of local minima of H_N 'near' the stored patterns (the so-called problem of the *storage capacity*). Key references are [154, 170, 184, 233].

12.3.1 Fluctuations of Φ

In a way one can say that the key to analyzing the Hopfield model with growing M is to understand how to use $\alpha = M/N$ as a small parameter instead of $1/N$. This means in particular that we would like to say that $\Phi_{\beta,N}$ is still close to its mean as long as α is small.

Our aim is to show that for small α, the minima of the function $\Phi_{\beta,N}$ are reasonably close to $\pm m^* e^\mu$, and that beyond a small neighbourhood of these points, the function grows like $\mathbb{E}\Phi_{\beta,N}$. Clearly, this requires an estimate of $\Phi_{\beta,N}(z) - \mathbb{E}\Phi_{\beta,N}(z)$, which may get worse as z is farther away from the minima of $\mathbb{E}\Phi_{\beta,N}$. On the other hand, we need estimates that are uniform in z. That is to say, a desirable estimate will be of the form:

Proposition 12.3.1 *Let $M(N) = \alpha N$. Then there exists a constant $C < \infty$ such that, for all $\beta > 1$,*

$$\mathbb{P}\left[\forall_{z:d_N(z,\mathcal{M}_{\beta,M(N)}) > C\frac{\sqrt{\alpha}}{m^*}} : \Phi_{\beta,N}(z) \geq \frac{1}{2}\mathbb{E}\Phi_{\beta,N}(z)\right] \geq 1 - e^{-M(N)/C} \quad (12.43)$$

Proof Our task is to control the fluctuations of a stochastic process defined on a space of dimension $M(N)$. In principle this is a classical problem in the theory of stochastic processes and there are well-developed tools available that we will not fail to employ: exponential estimates and *chaining* (see in particular [167]). Let me briefly explain the ideas behind this. Probably the most elementary estimate used in probability is that $\mathbb{P}[\max_{i \in I} X_i > x_i] \leq \sum_{i \in I} \mathbb{P}[X_i > x_i]$. This estimate tends to be good if the variables X_i are independent and the probabilities are small. In our situation, this is not directly applicable, since we are considering suprema over uncountable sets. The standard remedy is to consider a grid and to group the points close to a grid-point together, hoping that they will not vary too much from the value on the grid-point, while bounding the maximum over the grid by the sum. Note that we have already used a similar procedure in the proofs of Propositions 7.1.8 and 11.2.6.

12.3 Growing M

Let us first look at an exponential estimate for the deviation at a fixed point:

$$\Phi_{\beta,N}(z) - \mathbb{E}\Phi_{\beta,N}(z) = \frac{1}{N\beta}\sum_{i=1}^{N}(\ln\cosh(\beta(\xi_i, z)) - \mathbb{E}\ln\cosh(\beta(\xi_i, z))) \quad (12.44)$$

and so, as we have already seen above, this difference vanishes strictly whenever z has a single non-vanishing component. It will be crucial to exploit this and to get a bound that shows that the fluctuations decrease, as we approach one of the minima of $\mathbb{E}\Phi_{\beta,N}$. Thus, we exploit that, with z^* being any of the values $\pm m^* e^\mu$, we have

$$\ln\cosh(\beta(\xi_i, z)) - \mathbb{E}\ln\cosh(\beta(\xi_i, z)) \quad (12.45)$$
$$= \ln\cosh(\beta(\xi_i, z)) - \ln\cosh(\beta(\xi_i, z^*))$$
$$\quad - \mathbb{E}(\ln\cosh(\beta(\xi_i, z)) - \ln\cosh(\beta(\xi_i, z^*)))$$
$$\equiv \beta[f_i(z, z^*) - \mathbb{E}f_i(z, z^*)]$$

Next we use Taylor's formula to bound

$$|f_i(x, y)| \leq |(\xi_i, (x-y))||\tanh(\beta(\xi_i, \bar{z}))| \leq |(\xi_i, (x-y))| \quad (12.46)$$

We will want to use Chebyshev's inequality to estimate

$$\mathbb{P}\left[\frac{1}{N}\sum_{i=1}^{N}(f_i(x,y) - \mathbb{E}f_i(x,y)) \geq \delta\right] \leq \inf_{t \geq 0} e^{-t\delta N}\prod_{i=1}^{N}\mathbb{E}e^{t(f_i(x,y)-\mathbb{E}f_i(x,y))} \quad (12.47)$$

We must estimate the Laplace transforms of f_i. Using the standard second-order bound on the exponential function, $e^x \leq 1 + x + \frac{1}{2}x^2 e^{|x|}$, together with (12.46), we get

$$\mathbb{E}e^{t(f_i(x,y)-\mathbb{E}f_i(x,y))} \leq 1 + \frac{t^2}{2}\mathbb{E}(f_i(x,y) - \mathbb{E}f_i(x,y))^2 e^{t|f_i(x,y)-\mathbb{E}f_i(x,y)|} \quad (12.48)$$
$$\leq 1 + \frac{t^2}{2}\left[\mathbb{E}(f_i(x,y) - \mathbb{E}f_i(x,y))^4 \mathbb{E}e^{2t|f_i(x,y)-\mathbb{E}f_i(x,y)|}\right]^{1/2}$$

where we have used the Cauchy–Schwarz inequality to separate the expectation of the polynomial and exponential terms. Clearly,

$$\mathbb{E}e^{2t|f_i(x,y)-\mathbb{E}f_i(x,y)|} \leq \mathbb{E}e^{4t|f_i(x,y)|} \leq \mathbb{E}e^{4t|(\xi_i,(x-y))|} \quad (12.49)$$

and

$$\mathbb{E}(f_i(x,y) - \mathbb{E}f_i(x,y))^4 \leq 7\mathbb{E}\mathbb{E}(f_i(x,y))^4 \leq 7\mathbb{E}(\xi_i, (x-y))^4 \quad (12.50)$$

Now we can use the Marcinkiewicz–Zygmund inequalities (see, e.g., [78], pp. 366ff.), in particular $\mathbb{E}(z, \xi_i)^k \leq k^k \|z\|_2^{2k}$ and $\mathbb{E}e^{s|(z,\xi_i)|} \leq 2e^{s^2\|z\|_2^2/2}$. This yields

$$\mathbb{E}e^{t(f_i(x,y)-\mathbb{E}f_i(x,y))} \leq 1 + t^2 32\|x-y\|_2^2 e^{4t^2\|x-y\|_2^2} \quad (12.51)$$
$$\leq \exp\left(32t^2\|x-y\|_2^2 e^{4t^2\|x-y\|_2^2}\right)$$

We now get:

Lemma 12.3.2 *Let z^* be any of the points $\pm m^* e^\mu$, and set $R \equiv R(z) \equiv \|z - z^*\|_2$. Then, for all $\delta \leq 1$,*

$$\mathbb{P}[|\Phi_{\beta,N}(z) - \mathbb{E}\Phi_{\beta,N}(z)| \geq \delta R] \leq 2e^{-\frac{N}{150}\delta^2} \tag{12.52}$$

Proof Insert (12.51) into (12.46) and choose $t = \delta/(64R)$. This gives the bound for the upper deviation. For the lower deviation, the analogous procedure gives the same bound, and this implies (12.52). □

Lemma 12.3.2 shows that typical deviations at a given point z are of order $R/\sqrt{N}$, where R is the distance of z from the nearest coordinate axis. But this does not yet tell us anything about maximal fluctuations. The first idea would be to introduce a suitable lattice, $\mathcal{W}$ in $\mathbb{R}^M$, to use Lemma 12.3.2 to bound the maximal fluctuations on the lattice (as a function of R), and to prove some uniform bound on $\Phi_{\beta,N}(x) - \Phi_{\beta,N}(y)$ to control the deviations from the nearest lattice point. Using again (12.46), and the Cauchy–Schwarz inequality, we easily get

$$|\Phi_{\beta,N}(x) - \Phi_{\beta,N}(y)|^2 \leq \frac{1}{N}\sum_{i=1}^{N}(x-y,\xi_i)^2 \frac{1}{N}\sum_{i=1}^{N}\tanh^2(\beta(\bar{z},\xi_i))$$

$$\leq \frac{1}{N}\sum_{i=1}^{N}(x-y,\xi_i)^2 \tag{12.53}$$

Even though this bound will not be sufficient to get optimal results (bounding tanh by 1 everywhere is exaggerated when β is close to 1), it presents us with the occasion to consider an important object, namely the $M \times M$ random matrix A, with elements

$$A_{\mu\nu} \equiv \frac{1}{N}\sum_{i=1}^{N}\xi_i^\mu \xi_i^\nu \tag{12.54}$$

In terms of this matrix we can of course write

$$\frac{1}{N}\sum_{i=1}^{N}(z,\xi_i)^2 = (z, Az) \tag{12.55}$$

and thus obtain from (12.53) the bound

$$|\Phi_{\beta,N}(x) - \Phi_{\beta,N}(y)| \leq \|x-y\|_2 \|A\|^{1/2} \tag{12.56}$$

where $\|A\|$ is the operator norm of the matrix A in $\ell_2(\mathbb{R}^M)$.

Random matrices of this form belong to one of the classical ensembles of random matrix theory, the so-called *Marchenko–Pastur matrices* [173]. They are also well known in statistics where they appear as *sample covariance matrices*. As a result, their spectral properties, and in particular their norm (coinciding with the maximal eigenvalue), have been widely investigated (a certainly incomplete selection of references is [118, 128, 223, 260]; some results have been rediscovered or even improved in the course of the investigation of the Hopfield model in [43, 44, 58, 152, 220]). In particular it is known that:

Theorem 12.3.3 [260] *Let A be the $M \times M$ random matrix defined in (12.54), with ξ_i^μ i.i.d. random variables with mean zero, variance one, and $\mathbb{E}(\xi_i^\mu)^4 < \infty$. Assume that*

$\lim_{N \uparrow \infty} \frac{M}{N} = \alpha < \infty$. Then

$$\lim_{N \uparrow \infty} \|A\| = (1 + \sqrt{\alpha})^2, \text{ a.s.} \tag{12.57}$$

In fact, much more precise results are available, but they will not be relevant to us for the moment. In fact, in the remainder of this book we will simply pretend that A always has norm bounded by $(1 + \sqrt{\alpha})^2$, effectively placing ourselves on a subspace of full measure where this is true (for large enough N). It should be noted that the observation that this matrix and the bound on its norm are important goes back to two papers by Shcherbina and Tirozzi [220] and Koch [152] and triggered much of the later progress.

Exercise: Use Lemma 12.3.2 and (12.56) with $\|A\| \leq (1 + \sqrt{\alpha})^2$ to show that the assertion of Proposition 12.3.1 holds with the supremum taken over $z : d_N(z, \mathcal{M}_{\beta, M(N)}) > C\sqrt{\alpha}|\ln \alpha|$ for any fixed $\beta \geq \beta_0 > 1$.

Since this one-step approach does not yield a result that we deem sharp enough, we must use a refined approach known as chaining that consists of introducing a hierarchy of lattices. Let us denote by $\mathcal{W}_{M,r} \equiv (rM^{-1/2}\mathbb{Z})^M$ the hypercubic lattice of spacing $rM^{-1/2}$ in $\mathbb{R}^M$. Note that no point in $\mathbb{R}^M$ is farther away from $\mathcal{W}_{M,r}$ than $r/2$. It is not difficult to see that, if $B_R(0)$ denotes the ball of radius R centered at the origin, then the number of lattice points in this ball satisfies the bound, for $R > r$,

$$|\mathcal{W}_{\beta,r} \cap B_R(0)| \leq e^{M[\ln(R/r)+2]} \tag{12.58}$$

Now choose a sequence of spacings, $r_n = e^{-n}R$, and set $\mathcal{W}(n) \equiv \mathcal{W}_{M,r_n} \cap B_{r_{n-1}}$. For $x \in \mathbb{R}^M$, let $k_n(x) \in \mathcal{W}_{M,r_n}$ be the (in case of non-uniqueness, one of the) closest point(s) to x in $\mathcal{W}_{M,r_n}$. Note that, by construction, $\|k_n(x) - x\|_2 \leq r_n/2$ and $k_n(x) - k_{n-1}(x) \in \mathcal{W}(n)$. Clearly we have the telescopic expansion

$$\begin{aligned}
\Phi_{\beta,N}(z) - \mathbb{E}\Phi_{\beta,N}(z) &= \Phi_{\beta,N}(z) - \mathbb{E}\Phi_{\beta,N}(z) - (\Phi_{\beta,N}(z^*) - \mathbb{E}\Phi_{\beta,N}(z^*)) \\
&= \Phi_{\beta,N}(k_0(z)) - \mathbb{E}\Phi_{\beta,N}(k_0(z)) - (\Phi_{\beta,N}(z^*) - \mathbb{E}\Phi_{\beta,N}(z^*)) \\
&\quad + \Phi_{\beta,N}(k_1(z)) - \mathbb{E}\Phi_{\beta,N}(k_1(z)) - (\Phi_{\beta,N}(k_0(z)) - \mathbb{E}\Phi_{\beta,N}(k_0(z))) \\
&\quad + \Phi_{\beta,N}(k_2(z)) - \mathbb{E}\Phi_{\beta,N}(k_2(z)) - (\Phi_{\beta,N}(k_1(z)) - \mathbb{E}\Phi_{\beta,N}(k_1(z))) \\
&\quad + \cdots \\
&\quad + \cdots \\
&\quad + \Phi_{\beta,N}(k_n(z)) - \mathbb{E}\Phi_{\beta,N}(k_n(z)) \\
&\quad - (\Phi_{\beta,N}(k_{n-1}(z)) - \mathbb{E}\Phi_{\beta,N}(k_{n-1}(z))) \\
&\quad + \Phi_{\beta,N}(z) - \mathbb{E}\Phi_{\beta,N}(z) - (\Phi_{\beta,N}(k_n(z)) - \mathbb{E}\Phi_{\beta,N}(k_n(z)))
\end{aligned} \tag{12.59}$$

Now let $\delta_\ell > 0$, $\ell = 0, \ldots, n$, be a sequence of numbers such that $\sum_{\ell=0}^n \delta_\ell = \delta$. Then, the event that $\Phi_{\beta,N}(z) - \mathbb{E}\Phi_{\beta,N}(z) \geq \delta$ occurs, only if, for at least one $1 \leq \ell \leq n-1$,

$$\Phi_{\beta,N}(k_\ell(z)) - \mathbb{E}\Phi_{\beta,N}(k_\ell(z)) - (\Phi_{\beta,N}(k_{\ell-1}(z)) - \mathbb{E}\Phi_{\beta,N}(k_{\ell-1}(z))) \geq \delta_\ell, \tag{12.60}$$

$$\Phi_{\beta,N}(k_0(z)) - \mathbb{E}\Phi_{\beta,N}(k_0(z)) - (\Phi_{\beta,N}(z^*) - \mathbb{E}\Phi_{\beta,N}(z^*)) \geq \delta_0, \tag{12.61}$$

or

$$\Phi_{\beta,N}(z) - \mathbb{E}\Phi_{\beta,N}(z) - (\Phi_{\beta,N}(k_n(z)) - \mathbb{E}\Phi_{\beta,N}(k_n(z))) \geq \delta_n \tag{12.62}$$

and, consequently, the probability of the event in question is smaller than the sum of the probabilities of these $n+1$ events. The probability of the event (12.60) is bounded, by Lemma 12.3.2. Using the uniform bound (12.56), if we choose $\delta_n > e^{-n}R(1+\sqrt{\alpha})$, the event (12.62) is excluded. By (12.44), the probabilities of the events (12.60) can be estimated using (12.47) and the arguments leading to Lemma 12.3.2. This gives

$$\mathbb{P}[|\Phi_{\beta,N}(k_\ell(z)) - \Phi_{\beta,N}(k_{\ell-1}(z)) - \mathbb{E}(\Phi_{\beta,N}(k_\ell(z)) - \Phi_{\beta,N}(k_{\ell-1}(z)))| \geq \delta_\ell]$$
$$\leq 2\exp\left(-\frac{N}{150}\frac{\delta_\ell^2}{r_{\ell-1}^2}\right) = 2\exp\left(-\frac{N}{150}\frac{e^{2(\ell-1)}\delta_\ell^2}{R^2}\right) \qquad (12.63)$$

On the other hand, when z varies over $B_R(z^*)$, the pairs $(k_\ell(z), k_{\ell-1}(z))$ take only a small number of values, since their difference lies in $\mathcal{W}(\ell)$. This yields that

$$\text{Card}\,\{(k_\ell(z), k_{\ell-1}(z))|x \in B_R(z^*)\} \leq |\mathcal{W}_{M,r_{\ell-1}} \cap B_R(z^*)||\mathcal{W}(\ell)|$$
$$\leq e^{M[\ln(R/r_{\ell-1})+5]} \qquad (12.64)$$

Putting these observations together, we get that

$$\mathbb{P}\left[\sup_{z \in B_R(z^*)} |\Phi_{\beta,N}(z) - \mathbb{E}\Phi_{\beta,N}(z)| \geq \sum_{\ell=0}^{n-1} e^{-n}R(1+\sqrt{\alpha}) + Re^{-n}\right]$$
$$\leq 2\exp\left(M[\ln R/r_0 + 2] - \frac{N}{150}\frac{\delta_0^2}{R^2}\right) \qquad (12.65)$$
$$+ 2\sum_{\ell=1}^{n-1} \exp\left(M[\ln(R/r_{\ell-1})+5] - \frac{N}{150}\frac{e^{2(\ell-1)}\delta_\ell^2}{R^2}\right)$$

If we chose $\delta_\ell = C\sqrt{\alpha}e^{-\ell}\sqrt{\ell}$, for some C large enough, then there is a constant $c > 0$, depending only on the choice of C but not on α or R, such that

$$\mathbb{P}\left[\sup_{z \in B_R(z^*)} |\Phi_{\beta,N}(z) - \mathbb{E}\Phi_{\beta,N}(z)|\right. \qquad (12.66)$$
$$\left. \geq \sqrt{\alpha}(1+\sqrt{\alpha})RC\sum_{\ell=0}^{n-1} e^{-\ell}\ell^{1/2} + Re^{-n}\right] \leq 2ne^{-cM}$$

We see that it suffices to chose $n = -\frac{1}{2}\ln\alpha$ to achieve that

$$\sqrt{\alpha}(1+\sqrt{\alpha})RC\sum_{\ell=0}^{n-1} e^{-\ell}\ell^{1/2} + Re^{-n} \leq C'R\sqrt{\alpha} \qquad (12.67)$$

with $C' \sim C$ independent of α and R. One can easily see that (12.67) can be improved to show that, for C large enough, there exists $c > 0$ such that for, say all $R \leq 1$,

$$\mathbb{P}\left[\sup_{z \in B_R(z^*)} |\Phi_{\beta,N}(z) - \mathbb{E}\Phi_{\beta,N}(z)| \geq \sqrt{\alpha}CR(z)\right] \leq 2ne^{-cM} \qquad (12.68)$$

and, using very similar arguments,

$$\mathbb{P}\left[\sup_{z:d_N(z,\mathcal{M}_{\beta,M(N)})\geq 1}|\Phi_{\beta,N}(z) - \mathbb{E}\Phi_{\beta,N}(z)| \geq \sqrt{\alpha}CR(z)\right] \leq 2ne^{-cM} \quad (12.69)$$

We leave it to the reader to check that this implies the assertion of the proposition for all β, such that $m^*(\beta) \geq m > 0$; however, so far, we have no uniform control on the constants when $m^*(\beta)$ tends to 0 (which happens when $\beta \downarrow 1$). Inspection of our proof shows that the only place where we have exaggerated is when in (12.53) we estimated $\tanh^2(\beta(\bar{z}, \xi_i)) \leq 1$. This estimate becomes poor when $\|\bar{z}\|_2$ tends to zero, which becomes relevant precisely when $\beta \downarrow 1$. In this case, we have to use $\tanh^2 x \leq x^2$ and to replace (12.53) by

$$|\Phi_{\beta,N}(x) - \Phi_{\beta,N}(y)| \leq \sqrt{\frac{1}{N}\sum_{i=1}^{N}(x-y,\xi_i)^2}\sqrt{\frac{\beta^2}{N}\sum_{i=1}^{N}(\bar{z},\xi_i)^2}$$
$$\leq \beta\|A\|\|x-y\|_2 \max(\|x\|_2, \|y\|_2) \quad (12.70)$$

We leave it to the reader to fill in the details showing that from here we obtain the assertion of the proposition in the case $1 < \beta < \beta_0$. □

Proposition 12.3.1 is a key result that allows us immediately to conclude that the measure $\mathcal{Q}_{\beta,N}$ is concentrated on the union of $2M(N)$ disjoint balls of radius $c\sqrt{\alpha}/m^*$, provided $\alpha \leq \gamma(m^*)^2$, for finite positive constants γ and c. This is in perfect agreement with the prediction of the replica method [9], and even the scaling of the upper bound on α, as $\beta \downarrow 0$, is in accordance with these predictions. Of course, our constants are pretty lousy, as one might expect. We state this result for easy reference as:

Theorem 12.3.4 *There exist $0 < c_0, C, \gamma_a < \infty$, such that, for all $\beta > 1$, $\sqrt{\alpha} < \gamma_a(m^*)^2$, and all ρ satisfying $c_0(\frac{\sqrt{\alpha}}{m^*} \wedge N^{-1/4}) < \rho < m^*/\sqrt{2}$, we have, with probability one, for all but a finite number of indices N,*

$$\mathcal{Q}_{N,\beta}\left(\cup_{\mu=1}^{M} \cup_{s=\pm 1} B_{\rho}(sm^*e^{\mu})\right) \geq 1 - e^{-C(M \wedge N^{1/2})}. \quad (12.71)$$

The same result holds for the measures $\mathbb{Q}_{\beta,N}$.

Remark 12.3.5 A version of Theorem 12.3.4, with worse bounds on the radii of the balls and on the maximal value of α, was first proven in [45]. The correct asymptotics near $\beta = 1$ was proven in [41]. We have been following closely the version of the arguments given in [42]. An alternative proof was also given by Talagrand [233].

As we have already explained in the analysis of the p-spin SK model, having established a result like Theorem 12.3.4 two questions remain open: How is the mass distributed over individual 'lumps' (here 'balls'), and what are the properties of the measure conditioned on one lump (ball)? In the case of finite M we could completely answer both questions. When M grows, both become much more subtle. As we will soon see, amazingly enough, the second question can be answered in full under additional conditions on α and β. Concerning the first question, a full answer has been given only under very strong conditions on the growth rate of M, namely $M^2 \ll N$, in [57]. The approach used there consisted essentially of pushing the analysis of the finite M case to its limits, employing in the process some very strong Gaussian approximation results. Since these appear to be special methods that work in a non-canonical regime, I will not include a discussion in this book. On the other hand, there are some weaker, but very general results concerning these weights based on concentration of measure estimates that I will discuss in the next subsection.

12.3.2 Logarithmic equivalence of lump-weights

Theorem 12.3.4 suggests quite naturally that as in the finite M case, there should be (at least) a pair of pure states corresponding to each pattern and its mirror image. However, on closer inspection one sees that this conclusion is premature. The point is that the theorem says nothing about the mass of any given pattern: it could well be that the mass of some of the balls is exponentially small (in N) and thus there would be no reason to give it preference over any other region in the state space. In particular, if one adopts the *external field* construction of extremal infinite-volume limits of Gibbs states, in such a situation we would not recover a limit state corresponding to such a pattern. This problem has been an obstacle for quite some time. Namely, a straightforward estimation (see [44]) of the relative weights of these balls would only show that they differ by no more than a factor of $\exp(O(M))$; this suffices in the case $M(N)/N \downarrow 0$. This allowed us to construct the extremal measures under these hypotheses [44], but it remained unclear what would happen if α is strictly positive. This problem was solved in [45], where it was realized that the right tools to address this situation are *concentration inequalities*. By today's standards, the approach used in [45] was rather clumsy, and due to some new concentration inequalities proven subsequently by Talagrand [232] (cited as Theorem 7.1.3 in Chapter 7), this is now a very simple and standard routine.

Theorem 12.3.6 *Let ρ be as in Theorem 12.3.4. Set*

$$I_N^\mu \equiv \frac{1}{\beta N} \ln \int_{B_\rho(e^\mu m^*)} dz e^{-\beta N \Phi_{\beta,N}(z)} \tag{12.72}$$

Then, for any $\mu, \nu \leq M(N)$,

$$\mathbb{P}\left[\left|I_N^\mu - I_N^\nu\right| \geq x\right] \leq 4 \exp\left(-N \frac{x^2}{128(m^*)^2}\right) \tag{12.73}$$

The same result holds if I_N^μ is replaced by $J_N^\mu \equiv (\beta N)^{-1} \ln \mathcal{Q}_{\beta,N}(B_\rho(e^\mu m^*))$.

Proof Note that by symmetry $\mathbb{E}I_N^\mu = \mathbb{E}I_N^\nu$, and so

$$\left|I_N^\mu - I_N^\nu\right| \leq \left|I_N^\mu - \mathbb{E}I_N^\mu\right| + \left|I_N^\nu - \mathbb{E}I_N^\nu\right| \tag{12.74}$$

Thus

$$\mathbb{P}\left[\left|I_N^\mu - I_N^\nu\right| \geq x\right] \leq 2\mathbb{P}\left[\left|I_N^\mu - \mathbb{E}I_N^\mu\right| \geq x/2\right] \tag{12.75}$$

We want to use Theorem 7.1.3 to bound this probability. To do so, we must prove a Lipschitz bound on I_N^μ. Note first that, using Cauchy–Schwarz in a very similar way as in (12.53), we get that

$$\left|\Phi_{\beta,N}[\xi](z) - \Phi_{\beta,N}[\xi'](z)\right| \leq \|\xi - \xi'\|_2 \|z\|_2 \tag{12.76}$$

while

$$\begin{aligned}
|I_N^\mu[\xi] - I_N^\mu[\xi']| &= \frac{1}{\beta N} \left| \ln \left(\frac{\int_{B_\rho(m^* e^\mu)} dz\, e^{-\beta N \Phi_{\beta,N}[\xi](z)}}{\int_{B_\rho(m^* e^\mu)} dz\, e^{-\beta N \Phi_{\beta,N}[\xi'](z)}} \right) \right| \\
&= \frac{1}{\beta N} \left| \ln \left(\frac{\int_{B_\rho(m^* e^\mu)} dz\, e^{-\beta N \Phi_{\beta,N}[\xi'](z) + \beta N [\Phi_{\beta,N}[\xi'](z) - \Phi_{\beta,N}[\xi](z)]}}{\int_{B_\rho(m^* e^\mu)} dz\, e^{-\beta N \Phi_{\beta,N}[\xi'](z)}} \right) \right| \\
&\leq \sup_{z \in B_\rho(m^* e^\mu)} |\Phi_{\beta,N}[\xi](z) - \Phi_{\beta,N}[\xi]'(z)| \\
&\leq \|\xi - \xi'\|_2 (m^* + \rho) \leq 2m^* \|\xi - \xi'\|_2
\end{aligned} \tag{12.77}$$

(12.73) is now straightforward.

It remains to show that this result for the measure $\mathcal{Q}_{N,\beta}$ extends also to the measure $\mathbb{Q}_{\beta,N}$, and therefore to the Gibbs measure itself. But this is quite simple, using the fact that $\mathcal{Q}_{\beta,N}$ is a convolution of $\mathbb{Q}_{\beta,N}$ with an M-dimensional Gaussian measure with mean zero and covariance $\beta N \mathbb{I}$. This allows us to bound

$$\mathbb{Q}_{\beta,N}(B_\rho(m^* e^\mu)) \leq \mathcal{Q}_{\beta,N}(B_{\rho+\delta}(m^* e^\mu)) + 2^M e^{-\beta N \delta^2/4} \tag{12.78}$$

that is, up to an exponentially small correction, $\mathbb{Q}_{\beta,N}(B_\rho(m^* e^\mu))$ and $\mathcal{Q}_{\beta,N}(B_\rho(m^* e^\mu))$ differ by at most the $\mathcal{Q}_{\beta,N}$ mass of the shell between the radii $\rho - \delta$ and $\rho + \delta$. Choosing δ sufficiently small, and ρ not too small, this has exponentially small mass, and this implies the result for $\mathbb{Q}$. □

At this stage a reasonably satisfactory qualitative picture is reached, that confirms the heuristic and numerical findings that, for small α and not too small β, Gibbs measures corresponding to each of the patterns (and their mirror images) exist, implying in some sense that the model, in this regime, does what it was conceived for, namely to 'store a number of preselected random patterns'.

12.4 The replica symmetric solution

From the point of view of our general philosophy, having established certain localization properties of the Gibbs measures, we should now ask what the measure conditioned on one ball looks like. One natural approach (although, as it turned out, not the only one) appeared to be to analyze more carefully the properties of the function $\Phi_{\beta,N}$ in the vicinity of its minima. The aim of such an analysis should should clearly be to:

(i) Localize more precisely the position of the minima (which so far are only localized in a ball of radius $\sim \sqrt{\alpha}$).
(ii) Determine the value of Φ at the minimum.
(iii) Determine whether or not there is a *unique minimum* in a sufficiently small ball around $m^* e^\mu$.

This analysis was started in 1997 with [41]. Our idea there was to extend the use of the Taylor expansion, which had been very useful for finite M beyond its natural realm of applicability. The idea behind this was quite simple: in the finite M case, we used Taylor expansions in the arguments of functions like $\frac{1}{N} \sum_i \ln \cosh(\beta(z, \xi_i))$, and we took advantage of the uniform bound $|(z, \xi_i)| \leq \|z\|_1 \leq \sqrt{M} \|z\|_2$. This bound is realized essentially when $\xi_i^\mu = \text{sign}(z_\mu)$, for all μ. But of course, for given z, unless the ξ_i are very atypical, it is impossible that this holds true for a large number of indices i. Rather, for most values of i, it should be true that $|(z, \xi_i)| \sim \|z\|_2$, leaving room for a Taylor expansion to work even when $M = \alpha N$.

While the main thrust of the paper [41] was directed towards answering question (i), and to determine good bounds on the numerical constants allowing for the existence of local minima near $m^* e^\mu$, the most consequential result proved to be the answer to question (iii), which could accidentally also be given in a limited domain of α and β values. I will therefore concentrate on reviewing this issue.

12.4.1 Local convexity

How does one prove that a function has a unique minimum in a region where the existence of a minimum is already established? In the absence of a better idea, prove that the function is convex in this region. This was done in [41], where the following result was proved:

Theorem 12.4.1 *There exist finite positive constants c_1, c_2, γ_a such that, if (i) $\alpha \leq \gamma_a^2 m^*(\beta)^4$ and (ii) $\beta > c\alpha^{-1}$, then, for ρ as in Theorem 12.3.4, with probability one, for all but a finite number of values of N, the function $\Phi_{\beta,N}(z)$ is strictly convex on any of the balls $B_\rho(\pm m^* e^\mu)$, and there exists $\epsilon > 0$ such that the Hessian matrix $\nabla^2 \Phi_{\beta,N}(z)$ has a smallest eigenvalue larger than ϵ, for all $z \in B_\rho(\pm m^* e^\mu)$.*

Remark 12.4.2 The lower bound (ii) on β may come as a surprise, but we will explain that it is qualitatively optimal, which we will see is a pity.

Proof Let us consider without loss of generality the neighbourhood of the point $m^* e^1$. It will be convenient to set $z = m^* e^1 + v$. We are interested in $\|v\|_2 \leq \rho$. Then we have

$$\nabla^2 \Phi_{\beta,N}(z) = \mathbb{I} - \frac{1}{N} \sum_{i=1}^N \frac{\beta}{\cosh^2\left(\beta(m^*\xi_i^1 + (\xi_i, v))\right)} \xi_i \xi_i \quad (12.79)$$

It is instructive to first consider the point $v = 0$. Here

$$\nabla^2 \Phi_{\beta,N}(m^* e^1) = \mathbb{I} - \frac{\beta}{\cosh^2(\beta(m^*))} A = \mathbb{I} - \beta(1 - (m^*)^2) A \quad (12.80)$$

with A the matrix introduced earlier (12.54). Here we used that $\cosh^{-2} x = 1 - \tanh^2 x$ and $m^* = \tanh \beta m^*$. This matrix is positive if and only if

$$(1 + \sqrt{\alpha})^2 \beta(1 - (m^*(\beta))^2) < 1 \quad (12.81)$$

Note that, with $\alpha = 0$, this is just the condition for the positivity of the second derivative at m^* in the Curie–Weiss model, and thus we know that, for all $\beta > 1$, there exists $\alpha_0(\beta)$ such that (12.82) holds for $\alpha < \alpha_0(\beta)$. Moreover, as $\beta \uparrow \infty$, α_0 tends to infinity. Thus, so far, we have not seen any sign of condition (ii).

To understand this point, it is best to think of β being large. Then positivity requires that the $\cosh^2$ in the denominators be large, to compensate for the β in the numerator. This requires the argument to be roughly of order $\ln \beta$. Now assume that, even for a single term in the sum in (12.79), $m^*\xi_i^1 + (\xi_i, v) \sim 0$. Then[6]

$$\nabla^2 \Phi_{\beta,N}(z) \leq \mathbb{I} - \frac{\beta}{N}\xi_i\xi_i \qquad (12.82)$$

and, since the matrix $\xi_i\xi_i$ has norm M, this cannot be positive definite if $\alpha\beta > 1$. But such a point will exist in $B_\rho(e^1 m^*)$: just take v with components $v_\mu = -\xi_i^\mu \xi_i^1 m^*/M$. Then $m^*\xi_i^1 + (\xi_i, v) = 0$, while $\|v\|_2 = m^*/\sqrt{M}$, so that $m^* e^\mu + v$ lies within the ball $B_\rho(e^1 m^*)$. Thus, condition (ii) is surely necessary. To prove that it is also sufficient, we must show that the condition $m^*\xi_i^1 + (\xi_i, v) \sim 0$ cannot be realized for too many indices i at the same time. To make this precise, fix $0 < \tau < 1$ and write

$$\nabla^2 \Phi_{\beta,N}(z) = \mathbb{I} - \frac{\beta}{N}\sum_{i=1}^N \xi_i\xi_i \qquad (12.83)$$
$$+ \frac{\beta}{N}\sum_{i=1}^N \tanh^2(\beta(m^*\xi_i^1 + (\xi_i, v)))\xi_i\xi_i \mathbb{I}_{|(\xi_i,v)|\leq \tau}$$
$$+ \frac{\beta}{N}\sum_{i=1}^N \tanh^2(\beta(m^*\xi_i^1 + (\xi_i, v)))\xi_i\xi_i \mathbb{I}_{|(\xi_i,v)|> \tau}$$

Using positivity of $\xi_i\xi_i$, this can be bounded by

$$\nabla^2 \Phi_{\beta,N}(z) \geq \mathbb{I} - \frac{\beta}{N}\sum_{i=1}^N \xi_i\xi_i + \tanh^2(\beta m^*(1-\tau))\frac{\beta}{N}\sum_{i=1}^N \xi_i\xi_i$$
$$+ \frac{\beta}{N}\sum_{i=1}^N \Big(\tanh^2\big(\beta(m^*\xi_i^1 + (\xi_i, v))\big)$$
$$- \tanh^2(\beta m^*(1-\tau))\Big)\xi_i\xi_i \mathbb{I}_{|(\xi_i,v)|>\tau}$$
$$\geq \mathbb{I} - \beta[1 - \tanh^2(\beta m^*(1-\tau))]A$$
$$- \beta \tanh^2(\beta m^*(1-\tau))\frac{1}{N}\sum_{i=1}^N \xi_i\xi_i \mathbb{I}_{|(\xi_i,v)|>\tau} \qquad (12.84)$$

Clearly the only dangerous and difficult term is the last one. We see that, as β grows, it behaves like β times a certain matrix, whose norm we therefore need to control.

We start from the observation that, for any symmetric matrix B, $\|B\| = \sup_{w:\|w\|_2=1}(w, Bw)$. Thus

$$\sup_{v\in B_\rho}\left\|\frac{1}{N}\sum_{i=1}^N \mathbb{I}_{\{|(\xi_i,v)|>\tau m^*\}}\xi_i^T\xi_i\right\| = \sup_{v\in B_\rho}\sup_{w:\|w\|_2=\rho}\frac{1}{\rho^2}\frac{1}{N}\sum_{i=1}^N \mathbb{I}_{\{|(\xi_i,v)|>\tau m^*\}}(\xi_i, w)^2 \qquad (12.85)$$
$$\leq \frac{1}{\rho^2}\sup_{v\in B_\rho}\sup_{w\in B_\rho}\frac{1}{N}\sum_{i=1}^N \mathbb{I}_{\{|(\xi_i,v)|>\tau m^*\}}(\xi_i, w)^2$$

It will be convenient to use that

$$\mathbb{I}_{\{|(\xi_i,v)|>\tau m^*\}}(\xi_i, w)^2 = \mathbb{I}_{\{|(\xi_i,v)|>\tau m^*\}}(\xi_i, w)^2\big(\mathbb{I}_{\{|(\xi_i,w)|<|(\xi_i,v)|\}} + \mathbb{I}_{\{|(\xi_i,w)|\geq|(\xi_i,v)|\}}\big) \qquad (12.86)$$
$$\leq \mathbb{I}_{\{|(\xi_i,v)|>\tau m^*\}}(\xi_i, v)^2 + \mathbb{I}_{\{|(\xi_i,w)|>\tau m^*\}}(\xi_i, w)^2$$

---
[6] We use the notation $A > B$ for matrices to mean that $A - B$ is positive definite.

which allows us to bound (12.85). Thus

$$2\rho^{-2} \sup_{v \in B_\rho} \frac{1}{N} \sum_{i=1}^{N} \mathbb{I}_{\{|(\xi_i, v)| > \tau m^*\}} (\xi_i, v)^2 \equiv 2 \sup_{v \in B_\rho} X_{\tau m^*}(v) \qquad (12.87)$$

Thus our task is to bound the supremum of the quantities X_a, which actually can be seen as the partial second moments of the empirical measure of the family of random variables (ξ_i, v). This was done in [41], and I give here only a sketch of the arguments. As in the analysis of the function $\Phi_{\beta,N}$, we are faced with the problem of controlling the supremum over a continuous family of random variables indexed by a high-dimensional set. Thus we may expect to have to use the chaining technology again. As we already know, this requires exponential estimates on the $X_a(v)$, as well as on differences $X_a(v) - X_a(v')$. Actually, this is a tricky business, and things will be more complicated, and we will be forced to study simultaneously a second object, the empirical distribution function of the same variables (ξ_i, v),

$$Y_a(v) \equiv \frac{1}{N} \sum_{i=1}^{N} \mathbb{I}_{\{|(\xi_i, v)| > \tau m^*\}} \qquad (12.88)$$

Instead of considering simply the differences $X_a(v) - X_a(v')$, we will use the following lemma:

Lemma 12.4.3 *Let $a_1, b_1 > 0$ and $v, \epsilon \in B_\rho$. Then*

$$X_{a_1+b_1}(v + \epsilon) \leq X_{a_1}(v) + 2\sqrt{X_{a_1}(v)(\epsilon, A\epsilon)} + 2a_1^2 Y_{b_1}(\epsilon) + 3(\epsilon, A\epsilon) \qquad (12.89)$$

and

$$Y_{a_1+b_1}(v + \epsilon) \leq Y_{a_1}(v) + Y_{b_1}(\epsilon) \qquad (12.90)$$

Proof The basic idea behind this lemma is the simple observation that $|(\xi_i, v + \epsilon)| > a_1 + b_1$ can only be true if $|(\xi_i, v)| > a_1$, or if $|(\xi_i, v)| \leq a_1$ and $|(\xi_i, \epsilon)| > b_1$. Therefore,

$$\mathbb{I}_{|(\xi_i, v+\epsilon)| > a_1 + b_1} \leq \mathbb{I}_{|(\xi_i, v)| > a_1} + \mathbb{I}_{|(\xi_i, v)| \leq a_1} \mathbb{I}_{|(\xi_i, \epsilon)| > b_1} \qquad (12.91)$$

(12.90) is already obvious from this. To get (12.89), we still have to work with $(\xi_i, v + \epsilon)^2$. Squaring out the sum and using (12.91) again, together with the Schwarz inequality, gives the result easily. □

The second basic ingredient is the exponential bounds on both $X_a(v)$ and $Y_a(v)$.

Lemma 12.4.4 *Set $p_a \equiv 2 \exp(-a^2/2)$. Then, for all v with $\|v\| = 1$,*

$$\mathbb{P}[X_a(v) \geq x] \leq \exp(N[2p_a^{1/2} - x/4]) \qquad (12.92)$$

and, for $x \geq p_a$,

$$\mathbb{P}[Y_a(v) \geq x] \leq \exp(N[(2p_a)^{1/2} - xa^2/4]) \qquad (12.93)$$

The proof of this lemma can be found in [41] (Lemma 4.2). It is, as usual, an involved application of the exponential Chebyshev inequality. Note that p_a is roughly equal to the mean of $Y_a(v)$. Note also that the corresponding estimates for other values of $\|v\|_2$ follow by scaling, since $Y_a(Cv) = Y_{a/C}(v)$ and $X_a(Cv) = C^2 X_{a/C}(v)$.

We now have all the ingredients for the analysis of the supremum together: Choosing a lattice $\mathcal{W}_{M,r_1}$, we can control $X_a(v)$ everywhere in terms of $X_{a_1}(v_1)$, with v_1 on the lattice and the supremum over $Y_{a-a_a}(\epsilon)$ with $\|\epsilon\|_2 \leq r_1$. This latter supremum is then controlled via the usual chaining, using (12.90).

This allows us to prove the following estimate:

Proposition 12.4.5 *There exist finite positive constants C, c such that, if*

$$\Gamma(\alpha, a) = Ce^{-ca^2} + C\alpha|\ln \alpha| \tag{12.94}$$

then, for all $\rho > 0$,

$$\mathbb{P}\left[\sup_{v \in B_\rho} X_a(v) \geq \rho^2 \Gamma(\alpha, a/\rho)\right] \leq Ce^{\alpha N} \tag{12.95}$$

Remark 12.4.6 Proposition 4.8 in [41] is a quantitatively more precise version of this statement.

We can now combine this proposition with (12.87) and (12.84) to get immediately a uniform lower bound on the Hessian of Φ:

Lemma 12.4.7 *Let $\Gamma(\alpha, a)$ be as in Proposition 12.4.5. Then, with probability large than $1 - Ce^{-\alpha N}$,*

$$\inf_{z \in B_\rho(m^*e^1)} \nabla^2 \Phi_{\beta, N}(z) \geq 1 - \beta[1 - \tanh^2(\beta(m^*(1-\tau)))](1 + \sqrt{\alpha})^2$$
$$- \beta \tanh^2(\beta(m^*(1-\tau)))\Gamma(\alpha, \tau m^*/\rho) \tag{12.96}$$

If we choose $\rho = c\gamma m^*$, the lower bound is

$$1 - \beta[1 - \tanh^2(\beta(m^*(1-\tau)))](1 + \sqrt{\alpha})^2$$
$$- \gamma^2(m^*)^2 \beta \tanh^2(\beta(m^*(1-\tau)))\Gamma(\alpha, \tau/\gamma) \tag{12.97}$$
$$\sim 1 - \beta[1 - \tanh^2(\beta(m^*(1-\tau)))](1 + \sqrt{\alpha})^2$$
$$- C\beta \tanh^2(\beta(m^*(1-\tau)))(e^{-c/\gamma^2} + \alpha|\ln \alpha|)$$

Thus we see that this bound is strictly positive on a non-empty domain of parameters α and β, which has the shape claimed in the theorem. $\square$

Heuristics Before investigating the consequences of this convexity result, it may be instructive to go through some very heuristic considerations that, however, will illustrate our goal.

To this end we go back to the equations (12.25) determining the location of a minimum of $\Phi_{\beta, N}$ near a point $e^\mu m^*$. To simplify the notation, we consider the case $\mu = 1$, and we may take, without loss of generality, $\xi_i^1 \equiv 1$. This time let us write $z^{(1)} = m_1 + x$ where $x_1 = 0$. Then we get, instead of (12.26), the system of equations

$$m_1 = \frac{1}{N} \sum_{i=1}^{N} \tanh(\beta(m_1 + (x, \xi_i))),$$

$$x_\nu = \frac{1}{N} \sum_{i=1}^{N} \xi_i^\nu \tanh(\beta(m_1 + (x, \xi_i))) \qquad (12.98)$$

Let us denote by R_N the empirical measure

$$R_N \equiv \frac{1}{N} \sum_{i=1}^{N} \delta_{(x,\xi_i)} \qquad (12.99)$$

Then m_1 is only a function of this empirical measure, through the equation

$$m_1 = \int R_N(dg) \tanh(\beta(m_1 + g)) \qquad (12.100)$$

It is not difficult to see that, provided all x_ν tend to zero sufficiently fast, as $N \uparrow \infty$, R_N converges to a Gaussian distribution with mean zero and variance $\|x\|_2^2$. Thus, if we knew that this convergence held for the (random!) solution of these equations, all we would need to determine was this variance. This should be the rôle of the remaining equations. Naively, one might want simply to square both sides and sum over ν, but a bit more care must be taken to disentangle the dependence between the argument of the tanh and the coefficient ξ_i^ν. Thus, it will be more useful to write

$$(x, \xi_i) = x_\nu \xi_i^\nu + \left(x^{(\nu)}, \xi_i\right) \qquad (12.101)$$

and to Taylor expand

$$x_\nu = \frac{1}{N} \sum_{i=1}^{N} \xi_i^\nu \tanh(\beta(m_1 + (x^{(\nu)}, \xi_i)))$$

$$+ \frac{1}{N} \sum_{i=1}^{N} \xi_i^\nu \beta x_\nu \xi_i^\nu \cosh^{-2}(\beta(m_1 + (x^{(\nu)}, \xi_i)))$$

$$= x_\nu \beta \frac{1}{N} \sum_{i=1}^{N} \cosh^{-2}(\beta(m_1 + (x^{(\nu)}, \xi_i)))$$

$$+ \frac{1}{N} \sum_{i=1}^{N} \xi_i^\nu \tanh(\beta(m_1 + (x^{(\nu)}, \xi_i))) \qquad (12.102)$$

which can be written, using that $1 - \cosh^{-2}(y) = \tanh^2(y)$, as

$$x_\nu \left(1 - \beta + \beta \frac{1}{N} \sum_{i=1}^{N} \tanh^2(\beta(m_1 + (x^{(\nu)}, \xi_i)))\right)$$

$$= \frac{1}{N} \sum_{i=1}^{N} \xi_i^\nu \tanh(\beta(m_1 + (x^{(\nu)}, \xi_i))) \qquad (12.103)$$

Now, if we ignore the small difference between (ξ_i, x) and $(\xi_i, x^{(\nu)})$, the coefficient of x_ν is just $1 - \beta + \beta \int R_N(\mathrm{d}g) \tanh^2(\beta(m_1 + g))$. Then squaring and summing over ν gives

$$\|x\|_2^2 \left(1 - \beta + \beta \int R_N(\mathrm{d}g) \tanh^2(\beta(m_1 + g))\right)^2 = \alpha \int R_N(\mathrm{d}g) \tanh^2(\beta(m_1 + g))$$
$$+ \frac{1}{N^2} \sum_{i \neq j} \sum_{\nu=2}^{M} \xi_i^\nu \xi_j^\nu \tanh(\beta(m_1 + (x^{(\nu)}, \xi_i))) \tanh(\beta(m_1 + (x^{(\nu)}, \xi_j))) \quad (12.104)$$

The important point is that the second term in the numerator on the right has mean zero, by construction. Thus, if it were true that, in the limit $N \uparrow \infty$, $\|x\|_2^2$ is almost surely constant and equal to its mean value, then this limit must satisfy

$$\|x\|_2^2 = \mathbb{E}\|x\|_2^2 = \frac{\alpha \int R_\infty(\mathrm{d}g) \tanh^2(\beta(m_1 + g))}{\left(1 - \beta + \beta \int R_\infty(\mathrm{d}g) \tanh^2(\beta(m_1 + g))\right)^2} \quad (12.105)$$

Assuming that R_∞ is a Gaussian distribution, with mean zero and variance $\|x\|_2^2 \equiv \alpha r$, we see that we arrive at a closed system of equations for the two parameters m_1 and r, namely

$$m_1 = \frac{1}{\sqrt{2\pi}} \int \mathrm{d}g\, e^{-g^2/2} \tanh(\beta(m_1 + \sqrt{\alpha r})),$$
$$r = \frac{q}{(1 - \beta + \beta q)^2}, \quad \text{with} \quad (12.106)$$
$$q \equiv \frac{1}{\sqrt{2\pi}} \int \mathrm{d}g\, e^{-g^2/2} \tanh^2(\beta(m_1 + \sqrt{\alpha r}))$$

The solution of this system of equations is known as the 'replica symmetric' solution, first derived by Amit, Gutfreund, and Sompolinsky [9] in 1987, using what is called the 'replica trick'. Although the derivation of the equations given above may look less striking than the replica method, it is hardly more rigorous, given the numerous ad hoc assumptions we had to make. It is surprising that the equations can indeed be derived rigorously, as we shall explain in the next section.

12.4.2 The cavity method 1

To see how our convexity results are related to the question of the replica symmetric solution, we have to take a step back and look at an approach to the analysis of disordered mean-field models that was originally introduced by Parisi et al. (see [179]) as an alternative to the replica method, called the *cavity method*. This method is in principle nothing other than induction, more precisely induction over the volume, N, of the system. The basic idea is simple. Let f_N be any thermodynamic quantity of interest. Suppose we could derive a relation of the form $f_{N+1} = F(f_N) + o(1)$. Then, if f_N converges to a limit, this limit must be a fixed point of the map F. Moreover, under certain hypothesis, we may even be able to show that f_N will converge by virtue of this recursion relation. Of course, the difficulty will be that one will not be able in general to find such relations; in particular, it will not be true that f_{N+1} will be a function of f_N only. However, at least heuristically, i.e. ignoring the problem of proving that the error terms are really $o(1)$, it is indeed possible to obtain such recursions for certain, sufficiently large sets of thermodynamic quantities. The first endeavours to use the cavity method to obtain rigorous results were made by Pastur and Shcherbina [204] in the SK model, and in 1994 (together with Tirozzi) [200] in the Hopfield model. A major breakthrough was achieved in 1998 when Talagrand [233] began to systematically develop this method.

Let us look at this problem in our model. The way to proceed is in fact not quite obvious, but we will to some extent be guided by the preceding heuristic discussion. Note that we are now interested in local properties of the Gibbs measure, i.e. we want to consider the measure restricted to, say, the ball $B_\rho(m^* e^1)$. We denote by $\mu_{\beta,N}^{(1,1)}$, $\mathcal{Q}_{\beta,N}^{(1,1)}$, etc., the conditioned measures

$$\mu_{\beta,N}(\cdot | m_N(\sigma) \in B_\rho(m^* e^1)), \text{ resp.} \mathcal{Q}_{\beta,N}(\cdot | z \in B_\rho(m^* e^1)) \tag{12.107}$$

Consider the Hamiltonian in a volume $N + 1$. We may write it as

$$H_{N+1}(\sigma) = -\frac{1}{2(N+1)} \sum_{\mu=1}^{M} \sum_{i,j=1}^{N+1} \xi_i^\mu \xi_j^\mu \sigma_i \sigma_j \tag{12.108}$$

$$= -\frac{1}{2(N+1)} \sum_{\mu=1}^{M} \sum_{i,j=1}^{N} \xi_i^\mu \xi_j^\mu \sigma_i \sigma_j$$

$$-\frac{1}{N+1} \sigma_{N+1} \sum_{\mu=1}^{M} \sum_{j=1}^{N} \xi_{N-1}^\mu \xi_j^\mu \sigma_j - \frac{M}{2(N+1)}$$

$$= \frac{N+1}{N} H_N(\sigma) - \frac{N}{N+1} \sigma_{N+1}(\xi_{N+1}, m_N(\sigma)) - \frac{M}{2(N+1)}$$

It will also be important to note that

$$\|m_{N+1}(\sigma) - m_N(\sigma)\|_2 \leq \frac{1}{N} \|m_N(\sigma)\|_2 + \frac{\sqrt{M}}{N} \tag{12.109}$$

Let us now consider

$$\frac{Z_{\beta,N+1}}{Z_{\beta,N}} = \sum_{\sigma_{N+1}=\pm 1} \mathbb{E}_\sigma e^{\beta' \sigma_{N+1}(\xi_{N+1}, m_N(\sigma))} e^{-\beta' H_N(\sigma)} \frac{1}{Z_{\beta,N}} e^{\frac{M}{2(N+1)}}$$

$$= \sum_{\sigma_{N+1}=\pm 1} \mu_{\beta',N} \left(e^{\beta' \sigma_{N+1}(\xi_{N+1}, m_N(\sigma))} \right) \frac{Z_{\beta',N}}{Z_{\beta,N}} e^{\frac{M}{2(N+1)}}$$

$$\tag{12.110}$$

where $\beta' \equiv \frac{N}{N+1}\beta$. Similarly, we get

$$\mu_{\beta,N+1}^{(1,1)}(\sigma_{N+1}) = \frac{\sum_{s=\pm 1} s \mu_{\beta',N}^{(1,1)} \left(e^{\beta' s(\xi_{N+1}, m_N(\sigma))} \right)}{\sum_{s=\pm 1} \mu_{\beta',N}^{(1,1)} \left(e^{\beta' s(\xi_{N+1}, m_N(\sigma))} \right)} \tag{12.111}$$

up to a small error (of order $\exp(-\alpha N)$), coming from the fact that the conditioning on the left is on the vector $m_{N+1}(\sigma)$, while on the right it is on $m_N(\sigma)$. But, due to (12.109), this difference gives only contributions of size of the order of the mass of the shell $\rho - \frac{\sqrt{M}}{N} \leq \|\mu_N(\sigma)\|_2 \leq \rho + \frac{\sqrt{M}}{N}$, which is exponentially small. We will ignore all errors of that order in the sequel.

Equation (12.111) can easily be extended to a formula representing all finite-dimensional marginal distributions. It shows that a central rôle is played by the Laplace transform

$$\widetilde{\mathcal{L}}_{\beta,N}(t) \equiv \mu^{(1,1)}_{\beta',N}\bigl(e^{\beta' t(\xi_{N+1},m_N(\sigma))}\bigr) = \mathcal{Q}^{(1,1)}_{\beta',N}\bigl(e^{\beta' t(\xi_{N+1},m)}\bigr)$$
$$= e^{\alpha/2}\mathcal{Q}^{(1,1)}_{\beta',N}\bigl(e^{\beta' t(\xi_{N+1},z)}\bigr) \equiv e^{\alpha/2}\mathcal{L}_{\beta,N}(t) \quad (12.112)$$

Thus, we can rewrite (12.111) as

$$\mu^{(1,1)}_{\beta,N+1}(\sigma_{N+1}) = \frac{\sum_{s=\pm 1} s\, \mathcal{Q}^{(1,1)}_{\beta',N}\bigl(e^{\beta' s(\xi_{N+1},z)}\bigr)}{\sum_{s=\pm 1} \mathcal{Q}^{(1,1)}_{\beta',N}\bigl(e^{\beta' s(\xi_{N+1},z)}\bigr)} \quad (12.113)$$

Being able to replace the measure $\mathbb{Q}$ by $\mathcal{Q}$ will actually be very useful (although it shouldn't be). Equation (12.111) appeared in the work of Pastur et al. [200, 204], where it was realized that being able to control the Laplace transform $\mathcal{L}_{\beta,N}$ was a key step to getting the replica symmetric solution. This was later pushed towards full rigour by Talagrand. Let us see why this is the case.

We compute first the mean of $\mathcal{L}_{\beta,N}(t)$ with respect to the variables ξ_{N+1}. This is possible, since $\mathcal{Q}_{\beta,N}$ is independent of them. Of course, we simply get

$$\mathbb{E}_{\xi_{N+1}}\mathcal{L}_{\beta,N}(t) = \mathcal{Q}_{\beta,N}\bigl(e^{\sum_\mu \ln\cosh(\beta t z_\mu)}\bigr) \quad (12.114)$$

Now this would make sense if $\mathcal{L}$ could then be shown to be self-averaging. But this is not the case. In a way, we can see that taking this average throws out too much information. One thing we can try then is to extract first a random part, and try to show that what is left is self-averaging. A natural possibility is to center the variable z in the exponent. Let Z denote the random variable whose distribution is $\mathcal{Q}^{(1,1)}_{\beta,N}$, and write $\bar{Z} \equiv Z - \mathcal{Q}^{(1,1)}_{\beta,N}(Z)$; actually, at least at this point, we start to feel that our notation is getting too heavy. Let us therefore denote henceforth by $\mathbb{E}_\mathcal{Q}$ the expectation with respect to the random measure $\mathcal{Q}^{(1,1)}_{\beta,N}$. Then we can of course write

$$\mathcal{L}_{\beta,N}(t) = e^{t(\xi_{N+1},\mathbb{E}_\mathcal{Q} Z)} \mathbb{E}_\mathcal{Q} e^{t(\xi_{N+1},\bar{Z})} \quad (12.115)$$

As in (12.114), we get

$$\mathbb{E}\mathbb{E}_\mathcal{Q} e^{t(\xi_{N+1},\bar{Z})} = \mathbb{E}_\mathcal{Q}\bigl(e^{\sum_\mu \ln\cosh(\beta t \bar{Z}_\mu)}\bigr) \quad (12.116)$$

Now, if all $\bar{Z}_\mu$ are small, $\ln\cosh(\beta t \bar{Z}_\mu) \sim \frac{1}{2}\beta^2 t^2 \|\bar{Z}\|_2^2$. Assume for a moment that the distribution of $\bar{Z}$ was an M-dimensional Gaussian distribution with variance of order $1/N$. Then the length of $\bar{Z}$ would be sharply concentrated about its mean, since its distribution would have a density proportional to $\exp(-N(r^2/2 - \ln r))$. Thus in such a case we would get that

$$\mathbb{E}\mathbb{E}_\mathcal{Q} e^{t(\xi_{N+1},\bar{Z})} = e^{\frac{1}{2}\beta^2 t^2 \mathbb{E}_\mathcal{Q} \|\bar{Z}\|_2^2 + o(1)} \quad (12.117)$$

which would be quite desirable.

12.4.3 Brascamp–Lieb inequalities

Of course, $\mathcal{Q}^{(1,1)}_{\beta,N}$ is not a Gaussian distribution, but maybe it is sufficiently similar to one that we still get (12.117)? Indeed, the local convexity proven in Theorem 12.4.1 does imply (12.117), as well as a number of similar results that we will need in the sequel. Interestingly, the proof of this fact uses a sophisticated result from functional analysis, the so-called Brascamp–Lieb inequalities [62]. Unfortunately, while convexity gives a very elegant way of advancing here, it is known not to be necessary, neither for the Brascamp–Lieb inequalities to hold, nor for the replica symmetric solution to be correct. Therefore, the proof I will present here does not give the best available conditions. In principle, there are two ways to improve the results: (i) One is to give a proof that does not use the Brascamp–Lieb inequalities (and (12.117)). This was Talagrand's original approach [233]; in its original version, however, this gave conditions that were comparable to those under which convexity holds; improved conditions, closer to those expected by physicists, required substantially more work [235]. (ii) There has been considerable work done to establish the Brascamp–Lieb inequalities without the assumption of convexity [13]. In fact, the real conditions needed for the Brascamp–Lieb inequalities to hold concern the minimal eigenvalue of a certain matrix-differential operator. Bach, Jecko, and Sjöstrand [13] established criteria that allow us to bound this operator from below, even when convexity fails, but so far they have not been shown to hold in our situation. This remains an interesting problem to study.

Let us now state the Brascamp–Lieb inequalities in their original form.

Lemma 12.4.8 *Let $V : \mathbb{R}^M \to \mathbb{R}$ be nonnegative and strictly convex with $\mathrm{Hess}(V(x)) \geq K > 0$. Denote by $\mathbb{E}_V$ the expectation with respect to the probability measure $e^{-V(x)} dx / \int e^{-V(y)} dy$. Let $f : \mathbb{R}^M \to \mathbb{R}$ be any continuously differentiable function that is square integrable with respect to $\mathbb{E}_V$. Then*

$$\mathbb{E}_V (f - \mathbb{E}_f)^2 \leq \frac{1}{K} \mathbb{E}_V \|\nabla f\|_2^2 \tag{12.118}$$

Remark 12.4.9 It is not difficult to see that the result also holds up to an exponentially small error term if V is an extended convex function, i.e. it is a convex function on its domain D and equal to $+\infty$ outside of D. See, e.g., [43] for details. In our situation this is what we will actually have to use.

Remark 12.4.10 The condition on $\mathrm{Hess}(V(x))$ can be replaced by the condition that the smallest eigenvalue of the matrix-differential operator

$$(-\Delta + \nabla V \cdot \nabla) \otimes \mathbb{I} + \mathrm{Hess}(V(x)) \tag{12.119}$$

is larger than K.

It is easy to see that the first two terms are a positive operator, so that, if the Hessian matrix is strictly positive, this gives an immediate bound. However, as pointed out above, this condition is not necessary.

The following simple applications of Lemma 12.4.8 show how we will actually use these inequalities (see [42]).

Corollary 12.4.11 *Let $\mathbb{E}_V$ be as in Lemma 12.4.8. Then:*

(i) $\mathbb{E}_V \|x - \mathbb{E}_V x\|_2^2 \leq \frac{M}{K}$
(ii) $\mathbb{E}_V \|x - \mathbb{E}_V x\|_4^4 \leq 4 \frac{M}{K^2}$

(iii) For any function, f such that $V_t(x) \equiv V(x) - tf(x)$, for $t \in [0, 1]$, is still strictly convex and $\mathrm{Hess}(\nabla^2 V_t) \geq K' > 0$, then

$$0 \leq \ln \mathbb{E}_V e^f - \mathbb{E}_V f \leq \frac{1}{2K'} \sup_{t \in [0,1]} \mathbb{E}_{V_t} \|\nabla f\|_2^2 \quad (12.120)$$

In particular:
(iv) $\ln \mathbb{E}_V e^{(t,(x-\mathbb{E}_V x))} \leq \frac{\|t\|_2^2}{2K}$
(v) $\ln \mathbb{E}_V e^{\|x - \mathbb{E}_V x\|_2^2} - \mathbb{E}_V \|x - \mathbb{E}_V x\|_2^2 \leq \frac{M}{K^2}$

The point of these relations is that the measure $\mathbb{E}_V$ behaves with regard to its covariance structure essentially like a Gaussian measure. A first important consequence of Corollary 12.4.11 is equation (12.117). In fact, this follows from the estimates above and the bound $x^2/2 - x^4/4 \leq \ln \cosh(x) \leq x^2/2$ (Exercise!). Moreover, the same tool allows us to estimate the variance of $\mathbb{E}_\mathbb{Q} e^{t(\xi_{N+1}, \bar{Z})}$, namely

$$\mathbb{E}\big(\mathbb{E}_\mathbb{Q} e^{t(\xi_{N+1}, \bar{Z})} - \mathbb{E}\mathbb{E}_\mathbb{Q} e^{t(\xi_{N+1}, \bar{Z})}\big)^2 \leq \frac{C}{N} \quad (12.121)$$

12.4.4 The local mean field

These results combine to create the following important observation:

Lemma 12.4.12 *Whenever the conclusions of Theorem 12.4.1 hold, there exists a constant $C < \infty$ such that*

$$\ln \mathcal{L}_{\beta,N}(t) = \beta t(\xi_{N+1}, \mathbb{E}_\mathbb{Q}(z)) + \frac{t^2 \beta^2}{2} \mathbb{E}_\mathbb{Q} \|\bar{Z}\|_2^2 + R_N \quad (12.122)$$

where

$$\mathbb{E} R_N^2 \leq \frac{C}{N} \quad (12.123)$$

Remark 12.4.13 Note that this lemma can be seen as a statement about the distribution of the random field (ξ_{N+1}, Z) under the Gibbs measure, stating that it is asymptotically Gaussian with random mean $(\xi_{N+1}, \bar{\mathbb{E}}_\mathbb{Q}(Z))$ and variance $\mathbb{E}_\mathbb{Q} \|\bar{Z}\|_2^2$.

From our previous results, we certainly expect that the vector $\mathbb{E}_\mathbb{Q} z$ has one component (the first one) of order m^*, while all other components should be 'microscopic', i.e. tend to zero as $N \uparrow \infty$. Thus we write

$$(\xi_{N+1}, \mathbb{E}_\mathbb{Q}(Z)) = \xi_{N+1}^1 \mathbb{E}_\mathbb{Q}(Z_1) + (\hat{\xi}_{N+1}, \mathbb{E}_\mathbb{Q}(\hat{Z})) \quad (12.124)$$

where $\hat{Z}_1 = 0$ and $\hat{Z}_\nu = Z_\nu$, for $\nu \neq 1$. It is now very natural to expect that the second term in (12.124), being a sum of αN independent random variables (under the conditional distribution given $\xi_1, \ldots, \xi_N$), will converge in distribution to a *Gaussian random variable* with mean zero and variance $\|\mathbb{E}_\mathbb{Q} \hat{Z}\|_2^2$. If, moreover, as one might also expect, the quantity $\|\mathbb{E}_\mathbb{Q} \hat{Z}\|_2^2$ converges to a constant, almost surely, as $N \uparrow \infty$, this second term would converge in distribution to the same Gaussian, also unconditionally. In that case, the entire Laplace transform $\mathcal{L}_{\beta,N}(t)$ would be fully characterized in terms of the three constants $m_1(N) \equiv \mathbb{E}_\mathbb{Q}(Z_1)$, $U_N \equiv \|\mathbb{E}_\mathbb{Q} \hat{Z}\|_2^2$, and $T_N \equiv \mathbb{E}_\mathbb{Q} \|\bar{Z}\|_2^2$.

Thus, we are left with three problems to solve: (1) Show that the central limit theorem alluded to holds.[7] (2) Show that the three quantities mentioned above are *self-averaging*. (3) Prove that these converge, and characterize their limits. Technically, both (1) and (2) will rely essentially on concentration of measure estimates. Problem (3) will then be solved by the cavity method, i.e. we will derive a system of recursive equations that can be proven to have a unique stable fixed point in the domain where these quantities are a-priori located.[8]

We will immediately formulate a more general version of this central limit theorem (which we will actually need to construct the metastate).

Proposition 12.4.14 *Let $I \subset \mathbb{N}\backslash\{1,\ldots,N\}$ be finite, independent of N. For $i \in I$, set $X_i(N) \equiv \frac{1}{\sqrt{T_N}} \sum_{\mu=2} \xi_i^\mu \mathbb{E}_Q Z_\mu$. Whenever the conclusions of Theorem 12.4.1 hold, either this family converges to a family of i.i.d. standard normal random variables, or $\sqrt{T_N} X_i(N)$ converges to zero in probability.*

Proof To prove this result requires us essentially to show that the $\mathbb{E}_Q Z_\mu$, for all $\mu \geq 2$, tend to zero, as $N \uparrow \infty$. We note first that, by symmetry, for all $\mu \geq 2$, $\mathbb{E}\mathbb{E}_Q Z_\mu = \mathbb{E}\mathbb{E}_Q Z_2$. On the other hand,

$$\sum_{\mu=2}^M [\mathbb{E}\mathbb{E}_Q Z_\mu]^2 \leq \mathbb{E} \sum_{\mu=2}^M [\mathbb{E}_Q Z_\mu]^2 \leq \rho^2 \qquad (12.125)$$

so that $|\mathbb{E}\mathbb{E}_Q Z_\mu| \leq \rho M^{-1/2}$.

To use information about the mean values, we will need a concentration estimate for derivatives of self-averaging quantities (since all expectations with respect to $\mathbb{E}_Q$ can be represented as derivatives of some log-Laplace transforms).

Lemma 12.4.15 *Assume that $f(x)$ is a random function defined on some open neighbourhood, $U \subset \mathbb{R}$. Assume that f satisfies, for all $x \in U$ and for all $0 \leq r \leq 1$,*

$$\mathbb{P}[|f(x) - \mathbb{E}f(x)| > r] \leq c \exp\left(-\frac{Nr^2}{c}\right) \qquad (12.126)$$

and that, at least with probability $1 - p$, $|f'(x)| \leq C$ and $|f''(x)| \leq C < \infty$, uniformly in $x \in U$. Then, for any $0 < \zeta \leq 1/2$ and for any $C < \delta < N^{\zeta/2}$,

$$\mathbb{P}\left[|f'(x) - \mathbb{E}f'(x)| > \delta N^{-\zeta/2}\right] \leq \frac{32C^2}{\delta^2} N^\zeta \exp\left(-\frac{\delta^4 N^{1-2\zeta}}{256c}\right) + p \qquad (12.127)$$

The proof of this lemma can be found in [42] and [43].

We will now use Lemma 12.4.15 to control $\mathbb{E}_Q Z_\mu$. We define

$$f(x) = \frac{1}{\beta N} \ln \int_{B_\rho^{(1,1)}} d^M z e^{\beta N x z_\mu} e^{-\beta N \Phi_{\beta,N,M}(z)} \qquad (12.128)$$

and denote by $\mathbb{E}_{Q,x}$ the corresponding modified expectation. By exactly the same arguments

---

[7] This fact is assumed in [200] without proof.
[8] This approach is in principle contained in [200]; Talagrand gave the first fully rigorous version, without using the a-priori estimates furnished by the Brascamp–Lieb inequalities, making the entire inductive scheme even more complicated.

12.4 The replica symmetric solution

as in the proof of Theorem 12.3.6, $f(x)$ satisfies (12.126). Moreover, $f'(x) = \mathbb{E}_{Q,x} Z_\mu$ and

$$f''(x) = \beta N \mathbb{E}_{Q,x}(Z_\mu - \mathbb{E}_{Q,x} Z_\mu)^2 \tag{12.129}$$

The Brascamp–Lieb inequalities again allow us to bound this second derivative:

$$\mathbb{E}_{Q,x}(Z_\mu - \mathbb{E}_{Q,x} Z_\mu)^2 \leq \frac{1}{\epsilon N \beta} \tag{12.130}$$

and so $f''(x) \leq c = \frac{1}{\epsilon}$.

Thus we arrive at:

Corollary 12.4.16 *There are finite, positive constants c, C such that, for any $0 < \zeta \leq \frac{1}{2}$ and for any μ,*

$$\mathbb{P}\left[|\mathbb{E}_Q Z_\mu - \mathbb{E}\mathbb{E}_Q Z_\mu| \geq N^{-\zeta/2}\right] \leq C N^\zeta \exp\left(-\frac{N^{1-2\zeta}}{c}\right) \tag{12.131}$$

We are now ready to conclude the proof of our proposition. We may choose $\zeta = 1/4$, and denote by Ω_N the subset of Ω where, for all μ, $|\mathbb{E}_Q Z_\mu - \mathbb{E}\mathbb{E}_Q Z_\mu| \leq N^{-1/8}$. Then $\mathbb{P}[\Omega_N^c] \leq O\left(e^{-N^{1/2}}\right)$.

We will now show that the characteristic function converges to that of a product of standard normal distributions, i.e. we show that for any $t \in \mathbb{R}^I$, $\mathbb{E}\prod_{j \in I} e^{it_j X_j(N)}$ converges to $\prod_{j \in I} e^{-\frac{1}{2} t_j^2}$. We have

$$\mathbb{E}\prod_{j \in I} e^{it_j X_j(N)} = \mathbb{E}_{\xi_{I^c}}\left[\mathbb{1}_{\Omega_N} \mathbb{E}_{\xi_I} e^{i\sum_{j \in I} t_j X_j(N)} + \mathbb{1}_{\Omega_N^c} \mathbb{E}_{\xi_I} e^{i \sum_{j \in I} t_j X_j(N)}\right]$$

$$= \mathbb{E}_{\xi_{I^c}}\left[\mathbb{1}_{\Omega_N} \prod_{\mu \geq 2} \prod_{j \in I} \cos\left(\frac{t_j}{\sqrt{T_N}} \mathbb{E}_Q Z_\mu\right)\right] + O\left(e^{-N^{1/2}}\right) \tag{12.132}$$

Thus the second term tends to zero rapidly and can be forgotten. On the other hand, on Ω_N,

$$\sum_{\mu=2}^{M} (\mathbb{E}_Q Z_\mu)^4 \leq N^{-1/4} \sum_{\mu=2}^{M} (\mathbb{E}_Q Z_\mu)^2 \leq N^{-1/4} T_N \tag{12.133}$$

Moreover, for any finite t_j, for N large enough, $\left|\frac{t_j}{\sqrt{T_N}} \mathbb{E}_Q Z_\mu\right| \leq 1$. Thus, using that $|\ln \cos x + x^2/2| \leq cx^4$ for $|x| \leq 1$, and that

$$\mathbb{E}_{\xi_{I^c}} \mathbb{1}_{\Omega_N} \mathbb{E}_\eta e^{i \sum_{j \in I} t_j X_j(N)} \tag{12.134}$$

$$\leq e^{-\sum_{j \in I} t_j^2 / 2} \sup_{\Omega_N}\left[\prod_{j \in I} \exp\left(c \frac{t_j^4 N^{-1/4}}{T_N}\right)\right] \mathbb{P}_\xi(\Omega_N)$$

Clearly, the right-hand side converges to $e^{-\sum_{j \in I} t_j^2/2}$, provided only that $N^{1/4} T_N \uparrow \infty$. Otherwise, $\mathbb{E} T_N X_i(N)^2 \downarrow 0$. Thus the lemma is proven. □

12.4.5 Gibbs measures and metastates

We now control the convergence of our Laplace transform, except for the two parameters $m_1(N) \equiv \mathbb{E}_Q Z_1$ and $T_N \equiv \sum_{\mu=2}^M [\mathbb{E}_Q Z_\mu]^2$. What we have to show is that these quantities converge almost surely, and that the limits satisfy the equations of the replica symmetric solution of Amit, Gutfreund, and Sompolinsky [9].

While the issue of convergence is crucial, the technical intricacies of its proof are largely disconnected to the question of the convergence of the Gibbs measures. We will therefore assume for the moment that these quantities do converge to some limits and draw the conclusions for the Gibbs measures from the results of this section under this assumption (which will later be proven to hold).

To this end we first note that all of the preceding discussion may be carried out without substantial changes for the Laplace transforms of the local mean fields acting on a finite family of singled out spins, σ_i, $i \in I \subset \mathbb{N}$ (the details of the computations can be found in [43]). As a result, one obtains the following expression for the Gibbs mass of cylinder events:

$$\mu_{\Lambda,\beta,\rho}^{(1,1)}[\omega](\{\sigma_I = s_I\}) \tag{12.135}$$
$$= \frac{\exp\left(\beta_N' \sum_{i \in I} s_i \left[m_1(N)\xi_i^1 + X_i(N)\sqrt{T_N}\right] + R_N(s_I)\right)}{2^I \mathbb{E}_{\sigma_I} \exp\left(\beta_N' \sum_{i \in I} \sigma_i \left[m_1(N)\xi_i^1 + X_i(N)\sqrt{T_N}\right] + R_N(\sigma_I)\right)}$$

where

$$\beta_N' \to \beta$$
$$R_N(s_I) \to 0 \quad \text{in probability}$$
$$X_i(N) \to g_i \quad \text{in law}$$
$$T_N \to \alpha r \quad \text{a.s.}$$
$$m_1(N) \to m_1 \quad \text{a.s.}$$

for some numbers, r, m_1, and where $\{g_i\}_{i \in \mathbb{N}}$ is a family of i.i.d. standard Gaussian random variables. Note that the first three of these assertions are proven, while the last two are for the moment *assumed*. From this representation we obtain:

Proposition 12.4.17 *In addition to our general assumptions, assume that $T_N \to \alpha r$, a.s., and that $m_1(N) \to m_1$, a.s. Then, for any finite $I \subset \mathbb{N}$*

$$\mu_{\Lambda,\beta,\rho}^{(1,1)}(\{\sigma_I = s_I\}) \to \prod_{i \in I} \frac{\exp(\beta s_i [m_1 \bar{\xi}_i^1 + g_i \sqrt{\alpha r}])}{2 \cosh(\beta \sigma_i [m_1 \bar{\xi}_i^1 + g_i \sqrt{\alpha r}])} \tag{12.136}$$

where the convergence holds in law with respect to the measure $\mathbb{P}$, $\{g_i\}_{i \in \mathbb{N}}$ is a family of i.i.d. standard normal random variables, and $\{\bar{\xi}_i^1\}_{i \in \mathbb{N}}$ are independent Rademacher random variables, independent of the g_i, having the same distribution as the variables ξ_i^1.

To arrive at the convergence in law of the random Gibbs measures, it is enough to show that (12.136) holds jointly for any finite family of cylinder sets, $\{\sigma_i = s_i, \forall_{i \in I_k}\}$, $I_k \subset \mathbb{N}$, $k = 1, \ldots, \ell$ (cf. [148], Theorem 4.2). But this is easily seen to hold from the same arguments.

12.4 The replica symmetric solution

Therefore, denoting by $\mu_{\infty,\beta}^{(1,1)}$ the random measure

$$\mu_{\infty,\beta}^{(1,1)}[\omega](\sigma) \equiv \prod_{i\in\mathbb{N}} \frac{\exp(\beta\sigma_i[m_1\xi_i^1[\omega] + \sqrt{\alpha r}g_i[\omega]])}{2\cosh(\beta[m_1\xi_i^1[\omega] + \sqrt{\alpha r}g_i[\omega]])} \tag{12.137}$$

we have:

Theorem 12.4.18 *Under the assumptions of Proposition 12.4.17, and with the same notation,*

$$\mu_{\Lambda,\beta,\rho}^{(1,1)} \to \mu_{\infty,\beta}^{(1,1)}, \text{ in law, as } \Lambda \uparrow \infty, \tag{12.138}$$

This result can easily be extended to the language of metastates. The following theorem gives an explicit representation of the Aizenman–Wehr metastate in our situation:

Theorem 12.4.19 *Let $\kappa_\beta(\cdot)[\omega]$ denote the Aizenman–Wehr metastate. Under the hypothesis of Theorem 12.4.18, for almost all ω, for any continuous function $F : \mathbb{R}^k \to \mathbb{R}$, and cylinder functions f_i on $\{-1,1\}^{I_i}$, $i = 1, \ldots, k$, one has*

$$\int_{\mathcal{M}_1(\mathcal{S}_\infty)} \kappa_\beta(d\mu)[\omega] F(\mu(f_1), \ldots, \mu(f_k))$$

$$= \int \prod_{i\in I} d\mathcal{N}(g_i) F\left(\mathbb{E}_{s_{I_1}} f_1(s_{I_1}) \prod_{i\in I_1} \frac{\exp(\beta[\sqrt{\alpha r}g_i + m_1\xi_i^1[\omega]])}{2\cosh(\sqrt{\alpha r}g_i + m_1\xi_i^1[\omega])}, \ldots \right.$$

$$\left. \ldots, \mathbb{E}_{s_{I_k}} f_k(s_{I_k}) \prod_{i\in I_k} \frac{\exp(\beta[\sqrt{\alpha r}g_i + m_1\xi_i^1[\omega]])}{2\cosh(\sqrt{\alpha r}g_i + m_1\xi_i^1[\omega])}\right) \tag{12.139}$$

where $\mathcal{N}$ denotes the standard normal distribution.

Proof This theorem is proven just as Theorem 12.4.18, except that the 'almost sure version' of the central limit theorem, Proposition 12.4.14, is used. The details are left to the reader. □

Remark 12.4.20 Our conditions on the parameters α and β place us in the regime where, according to [9] the 'replica symmetry' is expected to hold.

Some remarks concerning the implications of this proposition are in place. First, it shows (modulo a small argument that can be found in [43]) that, if the standard definition of limiting Gibbs measures as weak limit points is adopted, then we have discovered that in the Hopfield model all product measures on $\{-1,1\}^\mathbb{N}$ are extremal Gibbs states. Such a statement contains some information, but it is clearly not useful as information on the approximate nature of a finite-volume state. This confirms our discussion in Section 6.2 on the necessity of using a metastate formalism.

Second, one may ask whether conditioning on the application of external fields of vanishing strength can improve the convergence behaviour of our measures. The answer appears obviously to be no. Contrary to a situation where a symmetry is present whose breaking biases the system to choose one of the possible states, the application of an arbitrarily weak field cannot alter anything.

Third, we note that the total set of limiting Gibbs measures does not depend on the conditioning on the ball $B_\rho^{(1,1)}$, while the metastate obtained does depend on it. Thus the conditioning allows us to construct two metastates corresponding to each of the stored patterns. These metastates are in a sense extremal, since they are concentrated on the set of extremal (i.e. product) measures of our system. Without conditioning, one can construct other metastates (which, however, we cannot control explicitly in our situation).

12.4.6 The cavity method 2

We now conclude our analysis by showing that the quantities $U_N \equiv \mathbb{E}_Q \|\bar{Z}\|_2^2$, $m_1(N) \equiv \mathbb{E}_Q Z_1$ and $T_N \equiv \sum_{\mu=2}^M [\mathbb{E}_Q Z_\mu]^2$ actually do converge almost surely under our general assumptions. The proof consist of two steps: first we show that these quantities are self-averaging, and then the convergence of their mean values is proven by induction. We will assume throughout this section that the parameters α and β are such that local convexity holds. The basic ideas of this section are otherwise due to Pastur, Shcherbina, and Tirozzi [200], and Talagrand [233].

We first need some more concentration of measure results.

Proposition 12.4.21 *Let A_N denote any of the three quantities U_N, $m_1(N)$, or T_N. Then there are finite positive constants c, C such that, for any $0 < \zeta \leq \frac{1}{2}$,*

$$\mathbb{P}\left[|A_N - \mathbb{E}A_N| \geq N^{-\zeta/2}\right] \leq CN^\zeta \exp\left(-\frac{N^{1-2\zeta}}{c}\right) \tag{12.140}$$

Proof The proofs of these three statements, for U_N, $m_1(N)$ and T_N, are all very similar to that of Corollary 12.4.16. Indeed, for $m_1(N)$, (12.140) is a special case of that corollary. In the two other cases, we just need to define the appropriate analogues of the 'generating function' f from (12.128). They are

$$g(x) \equiv \frac{1}{\beta N} \ln \mathbb{E}_Q^{\otimes 2} \exp(\beta N x(\bar{Z}, \bar{Z}')) \tag{12.141}$$

in the case of T_N, and

$$\tilde{g}(x) \equiv \frac{1}{\beta N} \ln \mathbb{E}_Q^{\otimes 2} \exp(\beta N x \|\bar{Z}\|_2^2) \tag{12.142}$$

The proof then proceeds as in that of Corollary 12.4.16. □

We now turn to the induction part of the proof and derive a recursion relation for the three quantities above. To simplify notation we will henceforth set $\eta \equiv \xi_{N+1}$. Let us define

$$u_N(\tau) \equiv \ln \mathbb{E}_Q \exp(\beta \tau(\eta, Z)) \tag{12.143}$$

We also set $v_N(\tau) \equiv \tau\beta(\eta, \mathbb{E}_Q Z)$ and $w_N(\tau) \equiv u_N(\tau) - v_N(\tau)$. In the sequel we will need the following auxiliary result:

Lemma 12.4.22 *Under our general assumptions:*

(i) $\frac{1}{\beta\sqrt{T_N}} \frac{d}{d\tau}(v_N(\tau) - \beta\tau\eta^1 \mathbb{E}_Q Z_1)$ *converges weakly to a standard Gaussian random variable.*

(ii) $\left|\frac{d}{d\tau} w_N(\tau) - \tau\beta^2 \mathbb{E}\mathbb{E}_Q \|\bar{Z}\|_2^2\right|$ *converges to zero in probability.*

Proof (i) is obvious from Proposition 12.4.14 and the definition of $v_N(\tau)$. To prove (ii), note that $w_N(\tau)$ is convex and $\frac{d^2}{d\tau^2}w_N(\tau) \leq \frac{\beta\alpha}{\epsilon}$. Thus, *if* $\mathrm{var}(w_N(\tau)) \leq \frac{C}{\sqrt{N}}$, then $\mathrm{var}\left(\frac{d}{d\tau}w_N(\tau)\right) \leq \frac{C'}{N^{1/4}}$, by a standard result similar in spirit to Lemma 12.4.15 (see, e.g., [234], Proposition 5.4). On the other hand, $|\mathbb{E}w_N(\tau) - \frac{\tau^2\beta^2}{2}\mathbb{E}\mathbb{E}_\mathcal{Q}\|\bar{Z}\|_2^2| \leq \frac{K}{\sqrt{N}}$, by Lemma 12.4.12, which, together with the boundedness of the second derivative of $w_N(\tau)$, implies that $|\frac{d}{d\tau}\mathbb{E}w_N(\tau) - \tau\beta^2\mathbb{E}\mathbb{E}_\mathcal{Q}\|\bar{Z}\|_2^2| \downarrow 0$. This means that $\mathrm{var}(w_N(\tau)) \leq \frac{C}{\sqrt{N}}$ implies the lemma. Since we already know from (12.123) that $\mathbb{E}R_N^2 \leq \frac{C}{N}$, it is enough to prove $\mathrm{var}(\mathbb{E}_\mathcal{Q}\|\bar{Z}\|_2^2) \leq \frac{C}{\sqrt{N}}$. This follows just as the corresponding concentration estimate for U_N. □

We are now ready to start the induction procedure. We will place ourselves on a subspace, $\tilde{\Omega} \subset \Omega$, where, for all but finitely many N, it is true that $|U_N - \mathbb{E}U_N| \leq N^{-1/4}$, $|T_N - \mathbb{E}T_N| \leq N^{-1/4}$, etc. This subspace has probability one by our estimates.

Let us note that $\mathbb{E}_\mathcal{Q} Z_\mu$ and $\int d\mathcal{Q}_{N,\beta,\rho}^{(1,1)}(m)m_\mu$ differ only by an exponentially small term. Thus

$$\mathbb{E}_\mathcal{Q} Z_\mu = \frac{1}{N}\sum_{i=1}^N \xi_i^\mu \int \mu_{N,\beta,\rho}^{(1,1)}(d\sigma)\sigma_i + O\left(e^{-cM}\right) \qquad (12.144)$$

Since we want to perform induction over N, we will have to add an index referring to the volume to the measures $\mathcal{Q}$. Note that, by symmetry, from (12.144) we get

$$\mathbb{E}\mathbb{E}_{\mathcal{Q}_{N+1}}(Z_\mu) = \mathbb{E}\eta^\mu \int \mu_{N+1,\beta,\rho}^{(1,1)}(d\sigma)\sigma_{N+1} + O\left(e^{-cM}\right) \qquad (12.145)$$

Using (12.113) and the definition of u_N, this gives

$$\mathbb{E}\mathbb{E}_{\mathcal{Q}_{N+1}}(Z_\mu) = \mathbb{E}\eta^\mu \frac{e^{u_N(1)} - e^{u_N(-1)}}{e^{u_N(1)} + e^{u_N(-1)}} + O\left(e^{-cM}\right) \qquad (12.146)$$

where, to be precise, one should note that the left- and right-hand sides are computed at temperatures β and $\beta' = \frac{N}{N}\beta$, respectively, and that the value of M is equal to $M(N+1)$ on both sides; that is, both sides correspond to slightly different values of α and β, but we will see that this causes no problems.

Using our concentration results and Lemma 12.4.12, this gives

$$\mathbb{E}\mathbb{E}_{\mathcal{Q}_{N+1}}(Z_\mu) = \mathbb{E}\eta^\mu \tanh\left(\beta(\eta^1 \mathbb{E}m_1(N) + \sqrt{\mathbb{E}T_N}X_{N+1}(N))\right) + O(N^{-1/4}) \qquad (12.147)$$

Using further Proposition 12.4.14, we get a first recursion for $m_1(N)$:

$$m_1(N+1) = \int d\mathcal{N}(g)\tanh\left(\beta(\mathbb{E}m_1(N) + \sqrt{\mathbb{E}T_N}g)\right) + o(1) \qquad (12.148)$$

Of course, we need a recursion for T_N as well. From here on there is no great difference from the procedure in [200], except that the N-dependences have to be kept track of carefully. To simplify the notation, we ignore all the $o(1)$ error terms and only restore them at the end. Also, the remarks concerning β and α made above apply throughout.

Note that $T_N = \|\mathbb{E}_Q Z\|_2^2 - (\mathbb{E}_Q Z_1)^2$ and

$$\mathbb{E}\|\mathbb{E}_{Q_{N+1}} Z\|_2^2 = \sum_{\mu=1}^{M} \mathbb{E}\left(\frac{1}{N+1}\sum_{i=1}^{N+1}\xi_i^\mu \mu_{\beta,N+1,M}(\sigma_i)\right)^2$$
$$= \frac{M}{N+1}\mathbb{E}\left(\mu_{\beta,N+1,M}^{(1,1)}(\sigma_{N+1})\right)^2 \qquad (12.149)$$
$$+ \sum_{\mu=1}^{M} \mathbb{E}\xi_{N+1}^\mu \mu_{\beta,N+1,M}^{(1,1)}(\sigma_{N+1})\left(\frac{1}{N+1}\sum_{i=1}^{N}\xi_i^\mu \mu_{\beta,N+1,M}(\sigma_i)\right)$$

Using (12.113), as in the step leading to (12.146), we get for the first term in (12.149)

$$\mathbb{E}\left(\mu_{\beta,N+1,M}^{(1,1)}(\sigma_{N+1})\right)^2 = \mathbb{E}\tanh^2\left(\beta(\eta_1 \mathbb{E}_Q Z_1 + \sqrt{\mathbb{E}T_N} X_{N+1}(N))\right)$$
$$\equiv \mathbb{E} Q_N \qquad (12.150)$$

For the second term, we use the identity from [200]:

$$\sum_{\mu=1}^{M} \xi_{N+1}^\mu \left(\frac{1}{N}\sum_{i=1}^{N}\xi_i^\mu \mu_{\beta,N+1,M}(\sigma_i)\right) \qquad (12.151)$$
$$= \frac{\sum_{\sigma_{N+1}} \mathbb{E}_Q(\xi_{N+1}, Z) e^{\beta \sigma_{N+1}(\xi_{N+1}, Z)}}{\sum_{\sigma_{N+1}} \mathbb{E}_Q e^{\beta \sigma_{N+1}(\xi_{N+1}, Z)}}$$
$$= \beta^{-1} \frac{\sum_{\tau=\pm 1} u_N'(\tau) e^{u_N(\tau)}}{\sum_{\tau=\pm 1} e^{u_N(\tau)}}$$

Together with Lemma 12.4.22 one concludes that, up to small errors,

$$\sum_{\mu=1}^{M} \xi_{N+1}^\mu \left(\frac{1}{N+1}\sum_{i=1}^{N}\xi_i^\mu \mu_{\beta,N+1,M}(\sigma_i)\right)$$
$$= \xi_{N+1}^1 \mathbb{E}_Q Z_1 + \sqrt{\mathbb{E}T_N} X_N$$
$$+ \beta \mathbb{E}_Q \|\bar{Z}\|_2^2 \tanh\beta\left(\xi_{N+1}^1 \mathbb{E}_Q Z_1 + \sqrt{\mathbb{E}T_N} X_N\right) \qquad (12.152)$$

and so

$$\mathbb{E}\|\mathbb{E}_{Q_{N+1}} Z\|_2^2 = \alpha \mathbb{E} Q_N$$
$$+ \mathbb{E}\left[\tanh\beta\left(\xi_{N+1}^1 \mathbb{E}_Q Z_1 + \sqrt{\mathbb{E}T_N} X_N\right)\left[\xi_{N+1}^1 \mathbb{E}_Q Z_1 + \sqrt{\mathbb{E}T_N} X_N\right]\right]$$
$$+ \beta \mathbb{E}\mathbb{E}_Q \|\bar{Z}\|_2^2 \tanh^2\beta\left(\xi_{N+1}^1 \mathbb{E}_Q Z_1 + \sqrt{\mathbb{E}T_N} X_N\right) \qquad (12.153)$$

Using the self-averaging properties of $\mathbb{E}_Q \|\bar{Z}\|_2^2$, the last term is of course essentially equal to

$$\beta \mathbb{E}\mathbb{E}_Q \|\bar{Z}\|_2^2 \mathbb{E} Q_N \qquad (12.154)$$

The reappearance of $\mathbb{E}_Q \|\bar{Z}\|_2^2$ (remember that this was the variance of the local mean field!) may seem disturbing, as it introduces a new quantity into the system. Fortunately, it is the

12.4 The replica symmetric solution

last one. Namely, proceeding as above, we can show that

$$\mathbb{E}\mathbb{E}_{\mathcal{Q}_{N+1}}\|Z\|_2^2 = \alpha + \mathbb{E}\left[\tanh\beta\left(\xi_{N+1}^1\mathbb{E}_{\mathcal{Q}}Z_1 + \sqrt{\mathbb{E}T_N}X_N\right)\right. \tag{12.155}$$
$$\left. \times \left[\xi_{N+1}^1\mathbb{E}_{\mathcal{Q}}Z_1 + \sqrt{\mathbb{E}T_N}X_N\right]\right] + \beta\mathbb{E}\mathbb{E}_{\mathcal{Q}}\|\bar{Z}\|_2^2\mathbb{E}Q_N$$

so that by setting $U_N \equiv \mathbb{E}_{\mathcal{Q}}\|\bar{Z}\|_2^2$ we get, subtracting (12.153) from (12.155), the simple recursion

$$\mathbb{E}U_{N+1} = \alpha(1 - \mathbb{E}Q_N) + \beta(1 - \mathbb{E}Q_N)\mathbb{E}U_N \tag{12.156}$$

From this we get (since all quantities considered are self-averaging, we drop the $\mathbb{E}$ to simplify the notation), setting $m_1(N) \equiv \mathbb{E}_{\mathcal{Q}}Z_1$,

$$\begin{aligned} T_{N+1} &= -(m_1(N+1))^2 + \alpha Q_N + \beta U_N Q_N \\ &\quad + \int d\mathcal{N}(g)[m_1(N) + \sqrt{T_N}g]\tanh\beta(m_1(N) + \sqrt{T_N}g) \\ &= m_1(N+1)(m_1(N) - m_1(N+1)) + \beta U_N Q_N \\ &\quad + \beta T_N(1 - Q_N) + \alpha Q_N \end{aligned} \tag{12.157}$$

where we used integration by parts. The complete system of recursion relations can be written as

$$\begin{aligned} m_1(N+1) &= \int d\mathcal{N}(g)\tanh\beta\big(m_1(N) + \sqrt{T_N}g\big) + O(N^{-1/4}), \\ T_{N+1} &= m_1(N-1)(m_1(N) - m_1(N+1)) + \beta U_N Q_N \\ &\quad + \beta T_N(1 - Q_N) + \alpha Q_N + O(N^{-1/4}), \\ U_{N+1} &= \alpha(1 - Q_N) + \beta(1 - Q_N)U_N + O(N^{-1/4}), \\ Q_{N+1} &= \int d\mathcal{N}(g)\tanh^2\beta\big(m_1(N) + \sqrt{T_N}g\big) + O(N^{-1/4}) \end{aligned} \tag{12.158}$$

If the solutions to this system of equations converge, then the limits $r = \lim_{N\uparrow\infty} T_N/\alpha$, $q = \lim_{N\uparrow\infty} Q_N$ and $m_1 = \lim_{N\uparrow\infty} m_1(N)$ ($u \equiv \lim_{N\uparrow\infty} U_N$ can be eliminated) must satisfy the replica symmetric equations (12.106).

In principle, one might think that to prove convergence it is enough to study the stability of the dynamical system above without the error terms. However, this is not quite true. Note that the parameters β and α of the quantities on the two sides of the equation differ slightly (although this is suppressed in the notation). In particular, if we iterate too often, α will tend to zero. The way out of this difficulty was proposed by Talagrand [233]. With simplified notation, we are in the following situation: We have a sequence, $X_n(p)$, of functions depending on a parameter p. There is an explicit sequence, p_n, satisfying $|p_{n+1} - p_n| \leq c/n$, and functions, F_p, such that

$$X_{n+1}(p_{n+1}) = F_{p_n}(X_n(p_n)) + O(n^{-1/4}) \tag{12.159}$$

In this setting, we have the following lemma:

Lemma 12.4.23 *Assume that there exists a domain D containing a single fixed point $X^*(p)$ of F_p. Assume that $F_p(X)$ is Lipschitz continuous as a function of X, Lipschitz continuous* as a function of p uniformly for $X \in D$, and that, for all $X \in D$, $F_p^n(X) \to X^*(p)$. Assume we know that for all n large enough, $X_n(p) \in D$. Then

$$\lim_{n \uparrow \infty} X_n(p) = X^*(p) \tag{12.160}$$

Proof Let us choose an integer-valued monotone increasing function $k(n)$ such that $k(n) \uparrow \infty$ as n goes to infinity. Assume, e.g., $k(n) \leq \ln n$. We will show that

$$\lim_{n \uparrow \infty} X_{n+k(n)}(p) = X^*(p) \tag{12.161}$$

To see this, note first that $|p_{n+k(n)} - p_n| \leq \frac{k(n)}{n}$. By (12.159), we have that, using the Lipschitz properties of F,

$$X_{n+k(n)}(p) = F_p^{k(n)}(X_n(p_n)) + O(n^{-1/4}) \tag{12.162}$$

where we choose p_n such that $p_{n+k(n)} = p$. Now, since $X_n(p_n) \in D$, $\left|F_p^{k(n)}(X_n(p_n)) - X^*(p)\right| \downarrow 0$, as n, and thus $k(n)$ goes to infinity, so that (12.162) implies (12.161). But (12.161), for any slowly diverging function $k(n)$, implies the convergence of $X_n(p)$, as claimed. □

This lemma can be applied to the recurrence (12.157). The main point to check is whether the corresponding F_β attracts a domain in which the parameters $m_1(N), T_N, U_N, Q_N$ are a-priori located, due to the support properties of the measure $\widetilde{\mathcal{Q}}_{N,\beta,\rho}^{(1,1)}$. This stability analysis was carried out (for an equivalent system) by Talagrand and answered in the affirmative. We do not want to repeat this tedious, but in principle elementary, computation here.

We would like to make some remarks, however. It is clear that, if we consider conditional measures, then we can always force the parameters $m_1(N), R_N, U_N, Q_N$ to be in some domain. Thus, in principle, we could first study the fixed points of (12.157), determine their domains of attraction, and then define corresponding conditional Gibbs measures. However, these measures may then be metastable. Also, of course, at least in our derivation, we need to verify local convexity in the corresponding domains since this was used in the derivation of the equations (12.157).

13

The number partitioning problem

> La pensée n'est qu'un éclair au millieu d'une longue nuit. Mais c'est cet éclair qui est tout.[1]
>
> *Henri Poincaré, La valeur de la science.*

With the Hopfield model we have seen that the range of applications of statistical mechanics goes beyond standard physics into biology and neuroscience. In recent years, many such extensions have been noted. A particularly lively area is that of combinatorial optimization, notably with applications to computer science. While I cannot cover this subject adequately, I will in this last chapter concentrate on possibly the simplest example, the *number partitioning problem*. The connection to statistical mechanics has been pointed out by S. Mertens [18, 177, 178], from whom I learned about this interesting problem.

13.1 Number partitioning as a spin-glass problem

The number partitioning problem is a classical optimization problem: Given N numbers $n_1, n_2, \ldots, n_N$, find a way of distributing them into two groups such that their sums in each group are as similar as possible. One can easily imagine that this problem occurs all the time in real life, albeit with additional complication: Imagine you want to stuff two moving boxes with books of different weights. You clearly have an interest in making both boxes more or less the same weight, so that neither of them is too heavy. In computing, you want to distribute a certain number of jobs on, say, two processors, in such a way that all of your processes are executed in the fastest way, etc. Clearly, the two examples indicate that the problem has a natural generalization to partitionings into more than two groups. What is needed in practice is an algorithm that, when presented with the N numbers, finds the optimal partitioning as quickly as possible. Since the number of possible partitions is 2^N, simply trying them all out is not going to be a very clever way of doing this. Surprisingly, however, it is quite hard to do very much better, as this problem is (believed) to be NP-hard, i.e. no algorithm can be found that is sure to solve the problem in a time that is polynomial in the size N!

As pointed out by Mertens, this fact can to some extent be understood by realizing that the problem is closely related to mean-field spin-glasses, and in particular the random energy model. Of course, the occurrence of the number 2^N should already have made us suspect that.

---

[1] Approximately: Thought is just a flash in the middle of a long night. But it is this flash that is everything.

Indeed, any partition of the set $\{1, \ldots, N\}$ into two disjoint subsets, Λ_1, Λ_2, is equivalent to a spin configuration $\sigma \in \mathcal{S}_N$ via $\Lambda_1 \equiv \{i : \sigma_i = +1\}$ and $\Lambda_2 == \equiv \{i : \sigma_i = -1\}$. Moreover, the quantity to be minimized is

$$\left| \sum_{i \in \Lambda_1} n_i - \sum_{i \in \Lambda_2} n_i \right| = \left| \sum_{i=1}^{N} n_i \sigma_i \right| \equiv H_N(\sigma) \tag{13.1}$$

This is a spin-system Hamiltonian depending on the parameters n_i. It even looks rather simple, since it is linear in σ. The problem, however, is the sign: if the issue were to *maximize* $H_N(\sigma)$, we would soon be done, choosing $\sigma_i = \text{sign}(n_i)$ (should the n_i be allowed to be negative). But we want to *minimize* H_N, which is quite a different task! Note that our problem has an obvious symmetry: $H_N(\sigma) = H_N(-\sigma)$. It will be reasonable to factor out this symmetry and consider σ to be an element of the set $\Sigma \equiv \{\sigma \in \{-1, 1\}N : \sigma_1 = +1\}$.

Statistical mechanics will not undertake to find algorithms that solve this problem, but will say something about the solutions, i.e. the minima, and more generally the energy landscape. Knowledge about the energy landscape may then be useful to people who want to find algorithms. The first step we have to undertake, in order to bring the problem within the realm of our field, is to turn the numbers n_i into random variables. This is again in the same spirit as in the Hopfield model: we will not claim to say something about all possible assignments of the numbers n_i, but we want to give results of typical assignments if they are drawn from a probability distribution. It is by now well accepted that answering such questions is reasonable in practice.

We will reformulate our problem slightly in the following way. Let $X_i, i \in \mathbb{N}$, be a family of i.i.d. positive random variables with common distribution $\mathbb{P}$. Let $Y_N(\sigma)$ be the random process indexed by Σ_N, given by

$$Y_N(\sigma) \equiv N^{-1/2} \sum_{i=1}^{N} \sigma_i X_i \tag{13.2}$$

and let

$$H_N(\sigma) = \sqrt{N} |Y_N(\sigma)| \tag{13.3}$$

The first result will concern the distribution of the smallest values of $H_N(\sigma)$. It was conjectured by Mertens [177] and proven by Borgs, Chayes, and Pittel [35], and by Borgs, Chayes, Mertens, and Pittel [36] in the so-called restricted case, where the size of the sets is constrained to be equal.

Theorem 13.1.1 *Assume that the random variables X_i are uniformly distributed on the interval $[0, 1]$, and let $C_N = 2^{-N+1} \sqrt{\frac{2\pi N}{12}}$, and define a point process on $\mathbb{R}_+$ by*

$$\mathcal{P}_N \equiv \sum_{\sigma \in \Sigma_N} \delta_{C_N H_N(\sigma)} \tag{13.4}$$

Then, as $N \uparrow \infty$, $\mathcal{P}_N$ converges to the Poisson point process on $\mathbb{R}_+$ with intensity measure equal to the Lebesgue measure. Moreover, for any $\epsilon > 0$, and any $x \in \mathbb{R}_+$,

$$\mathbb{P}\left[\forall_{N_0} \exists_{N \geq N_0} : \exists_{\sigma, \sigma' : |R_N(\sigma, \sigma')| < \epsilon} : H_N(\sigma) \leq H_N(\sigma') \leq C_N(x)\right] = 0 \tag{13.5}$$

Clearly this result is very reminiscent of the REM and says that the minima of H_N behave just as if the values, $H_N(\sigma)$, were i.i.d. random variables. This looks strange, because we have a feeling that there are considerable correlations in the process H_N! Interestingly, they do not seem to play a rôle, in spite of the fact that the correlations decay more slowly even than in the SK model, since

$$\text{cov}(Y_N(\sigma), Y_N(\sigma')) = (\mathbb{E}X^2 - (\mathbb{E}X)^2)R_N(\sigma, \sigma') \tag{13.6}$$

In fact, if we were to look for the maximum of Y_N, we would encounter an extremal structure quite different from the REM. The point is that here we are looking for the local distribution of energy values near a specific value (in our case 0), in the center of the distribution of the random variable $Y_N(\sigma)$. It turns out, and we will see why in the course of the proof, that correlations play a much lesser rôle in such a situation than in the tails of the distributions. Bauke and Mertens [19] conjectured that the REM-like behaviour in Theorem 13.1.1 should be *universal* in the sense that it holds for 'any' random model if one considers the set of energy levels 'closest' to any specific fixed value of the energy. In [55], this conjecture is proven in a large class of models, with some caveats with regard to the presence of symmetries.

13.2 An extreme value theorem

In this section we give a useful criterion that implies convergence to a Poisson process for correlated random variables. Recall how we obtained the distribution of $\max_{\sigma \in S_N} X_\sigma$ in the random energy model. The main point had been that the probability that the maximum does not exceed a very large number was equal to the 2^N-th power of 1 minus a number of order 2^{-N}, which converges to an exponential. We would like to get such a result without having to use independence. To do this, we use a general theorem that requires only asymptotic factorization of finitely many of the random variables X_σ.

Notation: We write $\sum_{\alpha(\ell)}$ when the sum is taken over all possible *ordered* sequences of *different* indices $\{i_1, \ldots, i_\ell\} \subset \{1, \ldots, M\}$. We also write $\sum_{\alpha(r_1), \ldots, \alpha(r_R)}(\cdot)$ when the sum is taken over all possible *ordered* sequences of disjoint ordered subsets $\alpha(r_1) = (i_1, \ldots, i_{r_1})$, $\alpha(r_2) = (i_{r_1+1}, \ldots, i_{r_2}), \ldots, \alpha(r_R) = (i_{r_1+\cdots+r_{R-1}+1}, \ldots, i_{r_1+\cdots+r_R})$ of $\{1, \ldots, M\}$.

Theorem 13.2.1 *Let $V_{i,M}$, $i \in \mathbb{N}$, be a family of positive random variables satisfying the following assumptions: for any $\ell \in \mathbb{N}$, and all sets of constants $c_j > 0$, $j = 1, \ldots, \ell$,*

$$\lim_{M \uparrow \infty} \sum_{\alpha(\ell)=(i_1,\ldots,i_\ell)} \mathbb{P}\left[V_{i_j,M} < c_j, \forall_{j=1}^\ell\right] \to \prod_{j=1,\ldots,\ell} c_j \tag{13.7}$$

Then the point process

$$\mathcal{P}_M = \sum_{i=1}^M \delta_{V_{i,M}} \tag{13.8}$$

on $\mathbb{R}_+$, converges weakly in distribution, as $M \uparrow \infty$, to the Poisson point process $\mathcal{P}$, on $\mathbb{R}_+$, whose intensity measure is the Lebesgue measure.

Remark 13.2.2 Theorem 13.2.1 was proven (in a more general form, involving vector-valued random variables) in [54]; it is very similar in spirit to an analogous theorem for the case of exchangeable variables proven in [57] in an application to the Hopfield model.

Proof We use Kallenberg's theorem [148] on the weak convergence of a point process $\mathcal{P}_M$ to the Poisson process $\mathcal{P}$. Applying his theorem in our situation, weak convergence holds whenever:

(i) For all intervals, $A = [a, b)$,
$$\mathbb{E}\mathcal{P}_M(A) \to |A|, \quad M \to \infty \tag{13.9}$$

(ii) For any finite union, $A = \bigcup_{\ell=1}^{L}[a_\ell, b_\ell)$, of disjoint intervals,
$$\mathbb{P}[\mathcal{P}_M^p(A) = 0] \to e^{-|A|}, \quad M \to \infty \tag{13.10}$$

The tool for checking (i) and (ii) is the inclusion–exclusion principle: for any $\ell \in \mathbb{N}$ and any events, $O_1, \ldots, O_\ell$,

$$\mathbb{P}\left[\bigcap_{i=1,\ldots,\ell} O_i\right] = \sum_{k=0}^{\ell} \sum_{\substack{A_k=(i_1,\ldots,i_k)\subset\{1,\ldots,\ell\} \\ i_1<i_2<\cdots<i_k}} (-1)^k \mathbb{P}\left[\bigcap_{j=1}^{k} \bar{O}_{i_j}\right] \tag{13.11}$$

where $\bar{O}_{i_j}$ denotes the complements of O_{i_j}. We use (13.11) to 'invert' the inequalities of type $\{V_{i,M} \geq a\}$, i.e. to represent their probability as the sum of probabilities of complementary events, which can be estimated by (13.7). The power of the inclusion–exclusion principle comes from the fact that the partial sums of the right-hand side provide upper and lower bounds. (These are known as Bonferroni inequalities, see [99].) Namely, for any $n \leq [\ell/2]$,

$$\sum_{k=0}^{2n} \sum_{\substack{A_k=(i_1,\ldots,i_k) \\ \subset\{1,\ldots,\ell\} \\ i_1<i_2<\cdots<i_k}} (-1)^k \mathbb{P}\left[\bigcap_{j=1}^{k} \bar{O}_{i_j}\right] \geq \mathbb{P}\left[\bigcap_{i=1,\ldots,\ell} O_i\right] \tag{13.12}$$

$$\geq \sum_{k=0}^{2n+1} \sum_{\substack{A_k=(i_1,\ldots,i_k) \\ \subset\{1,\ldots,\ell\} \\ i_1<i_2<\cdots<i_k}} (-1)^k \mathbb{P}\left[\bigcap_{j=1}^{k} \bar{O}_{i_j}\right]$$

Using (13.11), we derive from the assumption of the theorem the following statement: Let $A_1, \ldots, A_\ell \in \mathbb{R}_+$ be disjoint intervals. Then, for any $m_1, \ldots, m_\ell$,

$$\sum_{\alpha(m_1),\alpha(m_2),\ldots,\alpha(m_\ell)} \mathbb{P}[V_{i,M} \in A_r \,\forall i \in \alpha(m_r), \forall r = 1, \ldots, \ell] \to \prod_{r=1}^{\ell} |A_r|^{m_r} \tag{13.13}$$

Let us first concentrate on the proof of this statement. Consider first the case $\ell = 1$, i.e. $A = [a, b)$. If also $m = 1$, we clearly have

$$\sum_{i=1}^{M} \mathbb{P}[V_{i,M} \in [a, b)] = \sum_{i=1}^{M} (\mathbb{P}[V_{i,M} < b] - \mathbb{P}[V_{i,M} < a]) \tag{13.14}$$

By assumption, the left-hand side of (13.14) converges to $(b - a)$ by the assumption (13.7). To treat the general case, one writes

$$\{V_{i,M} \in A_r, \forall i \in \alpha_r\} = \{V_{i,M} \geq a_r, \forall i \in \alpha_r\} \bigcup \{V_{i,M} < b_r, \forall i \in \alpha_r\} \tag{13.15}$$

and applies the inclusion-exclusion principle (13.11) to the first event. The result then follows easily. This proves (i).

To verify (ii), note that

$$\mathbb{P}[\mathcal{P}_M(A) = 0] = \mathbb{P}\left[\bigcap_{i=1}^{M} V_{i,M} \notin A\right] \tag{13.16}$$

Applying the inclusion-exclusion principle (13.12), we get

$$\sum_{k=0}^{2n} \frac{(-1)^k}{k!} \sum_{\alpha(k)} \mathbb{P}[V_{i,M} \in A, \forall i \in \alpha(k)] \geq \mathbb{P}[\mathcal{P}_M(A) = 0] \tag{13.17}$$

$$\geq \sum_{k=0}^{2n+1} \frac{(-1)^k}{k!} \sum_{\alpha(r)} \mathbb{P}[V_{i,M} \in A, \forall i \in \alpha(k)]$$

Using (13.13), this yields

$$\sum_{k=0}^{2n} \frac{(-1)^k |A|^k}{k!} \geq \lim_{M \to \infty} \mathbb{P}[\mathcal{P}_M(A) = 0] \geq \sum_{k=0}^{2n+1} \frac{(-1)^k |B|^k}{k!} \tag{13.18}$$

Since n can be fixed arbitrarily large, it follows that

$$\lim_{M \to \infty} \mathbb{P}[\mathcal{P}_M(A) = 0] = e^{-|A|} \tag{13.19}$$

This concludes the proof of the theorem. $\square$

13.3 Application to number partitioning

We will now prove Theorem 13.1.1. It will follow directly from Theorem 13.2.1 and the following proposition:

Proposition 13.3.1 *Let* $K_N = 2^{N-1}(2\pi)^{-1/2}$. *We write* $\sum_{\sigma^1,\ldots,\sigma^\ell \in \Sigma_N}(\cdot)$ *for the sum over all possible ordered sequences of different elements of* Σ_N. *Then, for any* $l \in \mathbb{N}$ *and any constants* $c_j > 0$, $j = 1, \ldots, \ell$, *we have*

$$\sum_{\sigma^1,\ldots,\sigma^\ell \in \Sigma_N} \mathbb{P}\left[K_N \frac{|Y_N(\sigma^j)|}{\text{var}X} < c_j, \forall_{j=1}^{\ell}\right] \to \prod_{j=1,\ldots,\ell} c_j \tag{13.20}$$

Heuristics Let us first outline the main steps of the proof. The random variables $Y_N(\sigma)$ are sums of i.i.d. random variables with mean $\mathbb{E}Y_N(\sigma) = \frac{1}{2\sqrt{N}} \sum_{i=1}^{N} \sigma_i$ and covariance matrix $B_N(\sigma^1, \ldots, \sigma^l)$, whose elements are

$$b_{m,n} = \text{cov}(Y_N(\sigma^m), Y_N(\sigma^n)) = \frac{\text{var}X}{N} \sum_{n=1}^{N} \sigma_i^m \sigma_i^n \tag{13.21}$$

In particular, $b_{m,m} = \text{var}(X) = \frac{1}{12}$. Moreover, for the vast majority of choices, $\sigma^1, \ldots, \sigma^\ell$, $b_{i,j} = o(1)$, for all $i \neq j$; in fact, this fails only for an exponentially small fraction of configurations. Thus in the typical choices, the central limit theorem suggests that the random variables $\frac{Y(\sigma^j)}{\sqrt{\text{var}X}}$ should behave, asymptotically, like independent Gaussian random variables with mean $\sqrt{\frac{3}{N}} \sum_{i=1}^{N} \sigma_i$ and variance 1. The probability defined in (13.20) is then the probability that these Gaussians belong to the exponentially small intervals $[-c_j 2^{-N} 2\sqrt{2\pi}, c_j 2^{-N} 2\sqrt{2\pi}]$, and is thus of the order

$$\prod_{j=1,\dots,\ell} 4c_j 2^{-N} \exp\left(-\frac{(\mathbb{E}Y_N)^2}{2}\right) \tag{13.22}$$

Moreover, $N^{-1/2} \sum_{i=1}^{N} \sigma_i$ converge, as $N \uparrow \infty$, to standard normal Gaussian random variables, which, by the same arguments, are independent for almost all configurations σ. Thus,

$$\lim_{N\uparrow\infty} \sum_{\sigma^1,\dots,\sigma^\ell \in R(N,\ell)} \mathbb{P}\left[\forall_{j=1}^\ell, K_N |H_N(\sigma^j)| < c_j\right] = \prod_{j=1,\dots,\ell} 2c_j \mathbb{E}_Z e^{-\frac{3Z^2}{2}} = \prod_{j=1}^\ell c_j \tag{13.23}$$

This estimate would yield the assertion of the proposition, if all remaining terms could be ignored.

Let us turn to the remaining tiny part of $\Sigma_N^{\otimes l}$ where $\sigma^1, \dots, \sigma^l$ are such that $b_{i,j} \not\to 0$ for some $i \neq j$ as $N \to \infty$. A priori, we would be inclined to believe that there should be no problem, since the number of terms in the sum is by an exponential factor smaller than the total number of terms. In fact, we only need to worry if the corresponding probability is also going to be exponentially larger than for the bulk of terms. As it turns out, the latter situation can only arise when the covariance matrix is degenerate.

Namely, if the covariance matrix, $B_N(\sigma^1, \dots, \sigma^l)$, is non-degenerate, the central limit theorem suggests that the probability $\mathbb{P}[\cdot]$ is of the order

$$(\det B_N(\sigma^1, \dots, \sigma^l))^{-1/2} \prod_{j=1,\dots,\ell} 2(2\pi)^{-1/2} c_j K_N^{-1} \tag{13.24}$$

But, from the definition of $b_{i,j}$, $(\det B_N(\sigma^1, \dots, \sigma^\ell))^{-1/2}$ may grow at most polynomially. Thus, the probability $\mathbb{P}[\cdot]$ is $K_N^{-\ell}$ up to a polynomial factor, while the number of sets $\sigma^1, \dots, \sigma^\ell$ in this part is exponentially smaller than K_N^ℓ. Hence, the contribution of all such $\sigma^1, \dots, \sigma^\ell$ in (13.20) is exponentially small.

The case when $\sigma^1, \dots, \sigma^\ell$ give rise to a degenerate $B(\sigma^1, \dots, \sigma^\ell)$ is more delicate. Degeneracy of the covariance implies that there are linear relations between the random variables $\{Y(\sigma^i)\}_{i=1,\dots,\ell}$, and hence the probabilities $\mathbb{P}[\cdot]$ can be exponentially bigger than $K_N^{-\ell}$. A detailed analysis shows, however, that the total contribution from such terms is still negligible.

Proof (of Proposition 13.3.1) Let

$$f_N^{\sigma^1,\dots,\sigma^\ell}(\vec{t}) = \mathbb{E} \exp\left(\frac{i}{\sqrt{N \operatorname{var} X}} \sum_{j=1,\dots,\ell} t_j Y(\sigma^j)\right) \tag{13.25}$$

be the characteristic function of the random vector $\frac{\{Y(\sigma^j)\}_{j=1,\ldots,\ell}}{\sqrt{\mathrm{var}\,X}}$. It is easy to see that $f_N^{\sigma^1,\ldots,\sigma^l}(\vec{t})$ can be written as

$$f_N^{\sigma^1,\ldots,\sigma^l}(\vec{t}) = \prod_{n=1}^N \mathbb{E}\exp\left(\frac{iX_n}{\sqrt{N\mathrm{var}\,X}}(\vec{t},\vec{\sigma}_n)\right) \quad (13.26)$$

$$= \prod_{n=1}^N \exp\left(\frac{i\mathbb{E}X}{\sqrt{N\mathrm{var}\,X}}(\vec{t},\vec{\sigma}_n)\right) \frac{\sin\left(\frac{1}{2\sqrt{N\mathrm{var}\,X}}(\vec{t},\vec{\sigma}_n)\right)}{\frac{1}{\sqrt{N\mathrm{var}\,X}}(\vec{t},\vec{\sigma}_n)}$$

Here $\vec{t}$ denotes the vector with components t_j; we will use the notation $\vec{\sigma}$ and $\vec{\sigma}_n$ analogously. We will write $(\vec{t},\vec{\sigma}_n) \equiv \sum_{j=1}^\ell \tau_j \sigma_n^j$, and $\langle \vec{t},\vec{\sigma} \rangle \equiv N^{-1/2} \sum_{j=1}^\ell \sum_{n=1}^N t_j \sigma_n^j$. Then

$$\mathbb{P}\left[\forall_{j=1}^\ell |Y(\sigma^j)| < \frac{c_j}{K_N}\right] \quad (13.27)$$

$$= \lim_{D\uparrow\infty} \int_{[-D,D]^\ell} f_N^{\sigma^1,\ldots,\sigma^\ell}(\vec{t}) \prod_{j=1,\ldots,\ell} \frac{e^{it_j c_j K_N^{-1}} - e^{-it_j c_j K_N^{-1}}}{2\pi i t_j} dt_j$$

Let us denote by $C(\vec{\sigma})$ the $\ell \times N$ matrix with elements σ_n^j.

We will split the sum of (13.20) into two terms

$$\sum_{\sigma^1,\ldots,\sigma^l \in \Sigma_N} \mathbb{P}[\cdot] = \sum_{\substack{\sigma^1,\ldots,\sigma^\ell \in \Sigma_N \\ \mathrm{rank}\,C(\vec{\sigma})=\ell}} \mathbb{P}[\cdot] + \sum_{\substack{\sigma^1,\ldots,\sigma^l \in \Sigma_N \\ \mathrm{rank}\,C(\vec{\sigma})<\ell}} \mathbb{P}[\cdot] \quad (13.28)$$

and show that the first term converges to the right-hand side of (13.20), while the second term converges to zero.

Lemma 13.3.2 *Assume that the matrix $C(\vec{\sigma})$ contains all ℓ possible different rows. Assume that a configuration $\tilde{\sigma}$ is such that it is a linear combination of the columns of the matrix $C(\vec{\sigma})$. Then there exists $1 \leq j \leq \ell$ such that either $\tilde{\sigma} = \sigma^{(j)}$ or $\tilde{\sigma} = -\sigma^{(j)}$.*

Remark 13.3.3 Lemma 13.3.2 has been an important ingredient in the analysis of the Hopfield model; see, e.g., the proof of Lemma 12.2.2. It possibly appeared first in a paper by Koch and Piasko [153].

Lemma 13.3.2 implies the following: Assume that there are $r < \ell$ linearly independent vectors, $\sigma^{i_1},\ldots,\sigma^{i_r}$, among the ℓ vectors $\sigma^1,\ldots,\sigma^\ell$. The number of such vectors is at most $(2^r - 1)^N$. In fact, if the matrix $C(\sigma^{i_1},\ldots,\sigma^{i_r})$ contains all 2^r different rows, then by Lemma 13.3.2 the remaining configurations, σ^j with $j \in \{1,\ldots,l\}\setminus\{i_1,\ldots,i_r\}$, would be equal to one of $\sigma^{i_1},\ldots,\sigma^{i_r}$ as elements of Σ_N, which is impossible since we sum over *different* elements of Σ_N. Thus there can be at most $O((2^r - 1)^N)$ ways to construct these r columns. Furthermore, there is only an N-independent number of possibilities for completing the set of vectors by $\ell - r$ linear configurations of these columns to get $C(\sigma^1,\ldots,\sigma^l)$.

The next lemma gives an a-priori estimate of the probability corresponding to each of these terms.

Lemma 13.3.4 *There exists a constant $C = C(\ell) > 0$ such that, for any distinct $\sigma^1, \ldots, \sigma^\ell \in \Sigma_N$, any $r = \mathrm{rank}\, C(\sigma^1, \ldots \sigma^\ell) \leq \ell$, and all $N > 1$,*

$$\mathbb{P}\left[\forall_{j=1}^{\ell}\, |Y(\sigma^j)| < \frac{c_j}{K_N}\right] \leq C K_N^{-r} N^{3r/2} \tag{13.29}$$

Lemma 13.3.4 implies that each term in the second sum in (13.28) is smaller than $C K_N^{-r} N^{3r/2} \sim 2^{-Nr}$. It follows that the sum over these terms is of order $O\big([(2^r - 1)2^{-r}]^N\big) \to 0$ as $N \to \infty$.

We now turn to the first sum in (13.28), where the covariance matrix is non-degenerate. Let us fix $\alpha \in (0, 1/2)$ and introduce a subset, $\mathcal{R}_{l,N}^\alpha \subset \Sigma_N^{\otimes l}$, through

$$\mathcal{R}_{N,\ell}^\alpha = \left\{\sigma^1, \ldots, \sigma^\ell \in \Sigma_N : \forall_{1 \leq i < r \leq 2}, \left|\sum_{n=1}^{N} \sigma_n^i \sigma_n^r\right| < N^{\alpha+1/2}\right\} \tag{13.30}$$

It is easy to estimate

$$\left|\Sigma_N^{\otimes l} \setminus \mathcal{R}_{l,N}^\alpha\right| \leq 2^{Nl} \exp(-h N^{2\alpha}) \tag{13.31}$$

By definition, for any $(\sigma^1, \ldots, \sigma^\ell) \mathcal{R}_{N,\ell}^\alpha$, the elements $b_{k,m}$ of the covariance matrix (see (13.21)) satisfy, for all $k \neq m$,

$$|b_{k,m}| = \left|\frac{\sum_{i=1}^{N} \sigma_i^k \sigma_i^m}{N \,\mathrm{var}\, X}\right| \leq N^{\alpha-1/2} \tag{13.32}$$

Therefore, for any $\sigma^1, \ldots, \sigma^l \in \mathcal{R}_{l,n}^\alpha$, $\det B_N(\vec{\sigma}) = 1 + o(1)$ and, in particular, the rank of $C(\vec{\sigma})$ equals ℓ.

By Lemma 13.3.4 and the estimate (13.31),

$$\sum_{\substack{\sigma^1,\ldots,\sigma^\ell \notin \mathcal{R}_{\ell,N}^\alpha \\ \mathrm{rank}_{C(\sigma^1,\ldots,\sigma^\ell)=\ell}}} \mathbb{P}[\cdot] \leq 2^{N\ell} e^{-h N^{2\alpha}} C N^{3\ell/2} K_N^{-\ell} \to 0 \tag{13.33}$$

To complete the study of the first term of (13.28), let us show that

$$\sum_{\sigma^1,\ldots,\sigma^\ell \in \mathcal{R}_{\ell,N}^\alpha} \mathbb{P}[\cdot] \to \prod_{j=1,\ldots,\ell} c_j \tag{13.34}$$

with $\mathbb{P}[\cdot]$ defined by (13.27). Using the representation (13.27), we will divide the normalized probability $\mathbb{P}[\cdot]$ of (13.27) into five parts

$$\prod_{j=1,\ldots,\ell} c_j^{-1} \mathbb{P}[\cdot] = \sum_{i=1}^{5} I_N^i(\sigma^1, \ldots, \sigma^\ell) \tag{13.35}$$

where

$$I_N^1 \equiv (2\pi)^{-\ell} \int_{\|\vec{t}\|<\epsilon N^{1/6}} \exp\left(\frac{i}{2\sqrt{\text{var }X}}\langle \vec{t}, \vec{\sigma}\rangle - \vec{t}B_N\vec{t}/2\right) \prod_{j=1,\ldots,\ell} dt_j, \qquad (13.36)$$

$$I_N^2 \equiv (2\pi)^{-\ell} \int_{\|\vec{t}\|<\epsilon N^{1/6}} \left(f_N^{\sigma^1,\ldots,\sigma^\ell}(\vec{t}) - \exp\left(\frac{i}{2\sqrt{\text{var }X}}\langle \vec{t}, \vec{\sigma}\rangle - \vec{t}B_N\vec{t}/2\right)\right) \prod_{j=1,\ldots,\ell} dt_j,$$

$$(13.37)$$

$$I_N^3 \equiv (2\pi)^{-\ell} \int_{\epsilon N^{1/6}<\|\vec{t}\|<\delta\sqrt{N}} f_N^{\sigma^1,\ldots,\sigma^\ell}(\vec{t}) \prod_{j=1,\ldots,\ell} dt, \qquad (13.38)$$

$$I_N^4 \equiv (2\pi)^{-\ell} \int_{\|\vec{t}\|\leq\delta\sqrt{N}} f_N^{\sigma^1,\ldots,\sigma^\ell}(\vec{t}) \left[\prod_{j=1,\ldots,\ell} \frac{\sin(t_j c_j/K_N)}{t_j c_j/K_N} - 1\right] \prod_{j=1,\ldots,\ell} dt_j, \qquad (13.39)$$

and

$$I_N^5 \equiv (2\pi)^{-\ell} \lim_{D\to\infty} \int_{[-D,D]^\ell \cap \|\vec{t}\|>\delta\sqrt{N}} f_N^{\sigma^1,\ldots,\sigma^\ell}(\vec{t}) \prod_{j=1,\ldots,\ell} \frac{\sin(t_j c_j/K_N)}{t_j c_j/K_N} dt_j \qquad (13.40)$$

for values $\delta, \epsilon > 0$ to be chosen appropriately later. We will show that there is a choice such that $I_N^i(\sigma^1,\ldots,\sigma^\ell) \to 0$, for $i = 2, 3, 4, 5$, while $I_N^1(\sigma^1,\ldots,\sigma^\ell)$ gives the desired contribution. These facts combined with (13.33) imply the assertion (13.34) and complete the proof of the proposition.

The following lemma gives control over some of the terms appearing above.

Lemma 13.3.5 *There exist constants $C > 0$, $\epsilon > 0$, $\delta > 0$, and $\zeta > 0$ such that, for all $\sigma^1,\ldots,\sigma^\ell \in \mathcal{R}_{l,N}^\alpha$, the following estimates hold:*

(i) *For $\|\vec{t}\| < \epsilon N^{1/6}$,*

$$\left| f_N^{\sigma^1,\ldots,\sigma^\ell}(\vec{t}) - \exp\left(\frac{i}{2\sqrt{N\text{var }X}}\langle \vec{t}, \vec{\sigma}\rangle - \vec{t}B_N\vec{t}/2\right) \right| \leq \frac{C\|\vec{t}\|^3}{\sqrt{N}} e^{-\vec{t}B_N\vec{t}/2} \qquad (13.41)$$

(ii) *For $\|\vec{t}\| < \delta\sqrt{N}$,*

$$\left| f_N^{\sigma^1,\ldots,\sigma^\ell}(\vec{t}) \right| \leq \exp(-\vec{t}B_N\vec{t}/2 + C|t|^3 N^{-1/2}) \qquad (13.42)$$

(iii) *For $\|\vec{t}\| < \delta\sqrt{N}$,*

$$\left| f_N^{\sigma^1,\ldots,\sigma^\ell}(\vec{t}) \right| \leq \exp(-\zeta\|\vec{t}\|^2) \qquad (13.43)$$

Assuming this lemma, we can now estimate the terms I_N^i. First, by a standard estimate on Gaussian integrals, we note that we can extend the domain of integration in the definition of I_N^1 to all of $\mathbb{R}^\ell$, and then use the standard formula for the Fourier transform of the Gaussian distribution to find that

$$I_N^1(\vec{\sigma}) = ((2\pi)^\ell \det B_N(\vec{\sigma}))^{-1/2} \exp\left(-\frac{\langle \vec{s}_N, B_N^{-1} \vec{s}_N \rangle}{8 \operatorname{var} X}\right) + o(1)$$

$$= (2\pi)^{-\ell/2} \exp\left(-\frac{3}{2}\|\vec{s}_N\|_2^2\right) + o(1) \qquad (13.44)$$

where $s_N^j \equiv N^{-1/2} \sum_{n=1}^N \sigma_n^j$, and $o(1)$ is uniform for $\sigma^1, \ldots, \sigma^\ell \in \mathcal{R}_{\ell,N}^\alpha$ by (13.32). Moreover, by the central limit theorem,

$$2^{-(N-1)\ell} \sum_{\sigma^1, \ldots, \sigma^\ell} (2\pi)^{-\ell/2} e^{-\frac{1}{2}\|\vec{s}_N\|_2^2} \to \left[(2\pi)^{-1/2} \int e^{-\frac{1}{2}z^2} e^{-\frac{3}{2}z^2} dz\right]^\ell = 2^{-\ell} \qquad (13.45)$$

Thus I_N^1 gives the desired main contribution.

As for the second term, $I_N^2(\vec{\sigma}) = O(N^{-1/2})$ uniformly for $\vec{\sigma} \in \mathcal{R}_{\ell,N}^\alpha$ by the estimates (13.41) and (13.32). The third part I_N^3 is found to be exponentially small by (13.43). To treat I_N^4, we note that for any $\epsilon > 0$ one can find N_0 such that for all $N \geq N_0$ and all $\vec{t}$ with $\|\vec{t}\| \leq \delta\sqrt{N}$ the quantity in square brackets is smaller than ϵ in absolute value, and apply again (13.43). Finally, we estimate

$$|I_N^5(\vec{\sigma})| \leq (2\pi)^{-\ell} \int_{\|\vec{t}\| > \delta\sqrt{N}} |f_N^{\vec{\sigma}}(\vec{t})| \prod_{j=1,\ldots,\ell} dt_j \qquad (13.46)$$

For any $\vec{\sigma} \in \mathcal{R}_{\ell,N}^\alpha$ the matrix $C(\vec{\sigma})$ contains all 2^ℓ possible different rows and by (13.26) $f_N^{\vec{\sigma}}(\vec{t})$ is the product of 2^ℓ different characteristic functions, where each is taken to the power $N2^{-\ell}(1 + o(1))$. Let us fix from the 2^ℓ different rows of $C(\vec{\sigma})$ ℓ linearly independent ones, and denote by $\bar{C}$ the $\ell \times \ell$ matrix composed of them. There exists $\eta(\delta) > 0$ such that $\sqrt{\vec{t} \bar{C}^T \bar{C} \vec{t}/(\operatorname{var} X)} \geq \eta$, for all $\vec{t}$ with $\|\vec{t}\| > \delta$. Changing variables $\vec{s} = \vec{t} \bar{C}^T / \sqrt{N \operatorname{var} X}$, one gets the bound

$$|I_N^5(\vec{\sigma})| \leq (2\pi)^{-\ell} (N \operatorname{var} X)^\ell (\det \bar{C})^{-1} \int_{\|\vec{s}\|>\eta} \prod_{j=1,\ldots,\ell} \left|\frac{e^{is_j}-1}{is_j}\right|^{N 2^{-\ell}(1+o(1))} ds_j \qquad (13.47)$$

$$\leq C N^{\ell/2} (1-h(\eta))^{N 2^{-\ell}(1+o(1))-2} \int_{\|\vec{s}\|>\eta} \prod_{j=1,\ldots,\ell} \left|\frac{e^{is_j}-1}{is_j}\right|^2 ds_j$$

where $h(\eta) > 0$ is chosen such that $|(e^{is}-1)/s| < 1 - h(\eta)$, for all s with $|s| > \eta/\ell$, and C is a constant, independent of $\vec{\sigma}$ and N. Thus, $I_N^5(\vec{\sigma}) \to 0$, uniformly in $\vec{\sigma} \in \mathcal{R}_{\ell,N}^\alpha$, and exponentially fast, as $N \to \infty$. This concludes the proof of (13.34) and of Proposition 13.3.1. $\square$

Proof (of Lemma 13.3.4) Let us remove from the matrix $C(\sigma^1, \ldots, \sigma^\ell)$ linearly dependent columns and leave only r linearly independent columns. These correspond to a certain subset of r configurations, σ^j, $j \in A_r \subset \{1, \ldots, \ell\}$, $|A_r| = r$. We denote by $\bar{C}^r(\vec{\sigma})$ the $N \times r$ matrix composed by them. Then the probability on the right-hand side of (13.29) is not greater than the probability of the same events for $j \in A_r$ only. Let $\bar{f}_N^{\sigma^1, \ldots, \sigma^\ell}(\vec{t})$, $j \in A_r$, be the characteristic function of the vector $\frac{\{Y(\sigma^j)\}_{j \in A_r}}{\sqrt{\operatorname{var} X}}$. Then

$$\mathbb{P}\left[\forall_{j=1}^{\ell} \frac{|Y(\sigma^j)|}{\sqrt{\operatorname{var} X}} < \frac{c_j}{K_N}\right] \qquad (13.48)$$

$$\leq \frac{1}{(2\pi)^r} \lim_{D \to \infty} \int_{[-D,D]^r} \bar{f}_N^{\sigma^1,\ldots,\sigma^\ell}(\vec{t}) \prod_{j \in A_r} \frac{e^{it_j c_j K_N^{-1}} - e^{-it_j c_j K_N^{-1}}}{it_j} dt_j$$

To bound the integrand in (13.48), we use that

$$\left|\frac{e^{it_j c_j^\beta K_N^{-1}} - e^{-it_j c_j K_N^{-1}}}{it_j}\right| \leq \min\left(2c_j K_N^{-1}, 2(t_j)^{-1}\right) \qquad (13.49)$$

Next, let us choose in the matrix $\bar{C}^r(\sigma^1, \ldots, \sigma^\ell)$ any r linearly independent rows and construct from them an $r \times r$ matrix, $\hat{C}$. Then

$$\left|\bar{f}_N^{\sigma^1,\ldots,\sigma^\ell}(\vec{t})\right| = \prod_{n=1}^{N} \left|\mathbb{E} \exp\left(\frac{iX_n}{\sqrt{N \operatorname{var} X}} \sum_{j \in A_r} t_j \sigma_n^j\right)\right| \qquad (13.50)$$

$$\leq \prod_{s=1}^{r} \left|\mathbb{E} \exp\left(\frac{iX_s}{\sqrt{N \operatorname{var} X}} \sum_{j \in A_r} \{\hat{C}\}_{s,j} t_j\right)\right|$$

$$\leq \prod_{s=1}^{r} \min\left(1, \frac{2\sqrt{N \operatorname{var} X}}{|\{\hat{C}\vec{t}\}_s|}\right)$$

where $\vec{t} = \{t_j\}_{j \in A_r}$. Hence, the absolute value of the integral (13.48) is bounded by the sum of two terms

$$\frac{K_N^{-r} \prod_{j \in A_r}(2c_j)}{(2\pi)^r} \int_{|t_j|<K_N} \prod_{s=1}^{r} \min\left(1, \frac{2\sqrt{N \operatorname{var} X}}{|\{\hat{C}\vec{t}\}_s|}\right) dt_j$$

$$+ \frac{1}{(2\pi)^r} \int_{|t_j|>K_N} \prod_{j \in A_r} \frac{2}{t_j} \prod_{s=1}^{r} \min\left(1, \frac{2\sqrt{N \operatorname{var} X}}{|\{\hat{C}\vec{t}\}_s|}\right) dt_j \qquad (13.51)$$

The change of variables $\vec{\eta} = \hat{C}\vec{t}$ in the first term shows that the integral over $\|\vec{t}\| < K_N$ is at most $O(N^{r/2}(\ln K_N)^r)$, where $\ln K_N = O(N)$ as $N \to \infty$. Thus the first term of (13.51) is bounded by $C_1(\hat{C}, \ell) N^{3r/2} K_N^{-r}$, with some constant $C_1(\hat{C}, \ell) > 0$, independent of N. Using the change of variables $\vec{\eta} = K_N^{-1}\vec{t}$ in the second term of (13.51), one finds that the integral over $\|\vec{t}\| > K_N$ is at most $O(K_N^{-r})$. Thus (13.51) is not greater than $C_2(\hat{C}, \ell) N^{3r/2} K_N^{-r}$, with some positive constant $C_2(\hat{C}, \ell)$, independent of N.

To conclude the proof, let us recall that there is a finite, N-independent number of possibilities to construct the matrix $\hat{C}$ from $C(\sigma^1, \ldots, \sigma^\ell)$, since each of its elements may take only three values ± 1 and 0. Thus, there exist fewer than 3^{r^2} different constants, $C_2(\hat{C}, \ell)$, corresponding to different matrices $\hat{C}$. It remains to take the maximal one over them to get (13.29). □

Proof (of Lemma 13.3.5)   The statement (13.43) is an immediate consequence of (13.42) and (13.32) if $\delta > 0$ is small enough.

The proof of (13.41) and (13.42) only requires the elementary estimates

$$\left|\sin x - x - \frac{x^3}{3!}\right| \leq \frac{x^4}{4!}e^{|x|} \quad \text{and} \quad \left|e^{-x^2/2}1 - \frac{x^2}{2}\right| \leq \frac{x^4}{8}e^{|x|^2/2}.$$

Therefore, recalling that $2\sqrt{\operatorname{var} X} = \sqrt{1/3}$,

$$\frac{\sin\left(\frac{(\vec{t},\vec{\sigma}_n)}{\sqrt{N/3}}\right)}{\frac{(\vec{t},\vec{\sigma}_n)}{\sqrt{N/3}}} = 1 - \frac{(\vec{t},\vec{\sigma}_n)^2}{2N} + O\left(\|\vec{t}\|^3 N^{-3/2} e^{\|\vec{t}\|^3 N^{-3/2}/4!}\right) \tag{13.52}$$

Note that, when $\|\vec{t}\| \leq N^{1/2}$, all annoying exponentials are bounded by constants; if $\|\vec{t}\| \leq N^{1/6}$, the error term is even $\leq O(1/N)$. Now write

$$\prod_{n=1}^{N} \frac{\sin\left(\frac{(\vec{t},\vec{\sigma}_n)}{\sqrt{N/3}}\right)}{\frac{(\vec{t},\vec{\sigma}_n)}{\sqrt{N/3}}} = \exp\left(-\sum_n \frac{(\vec{t},\vec{\sigma}_n)^2}{2N}\right) \prod_{n=1}^{N} \exp\left(\frac{(\vec{t},\vec{\sigma}_n)^2}{2N}\right) \frac{\sin\left(\frac{(\vec{t},\vec{\sigma}_n)}{\sqrt{N/3}}\right)}{\frac{(\vec{t},\vec{\sigma}_n)}{\sqrt{N/3}}} \tag{13.53}$$

But by the preceding observations,

$$\exp\left(\frac{(\vec{t},\vec{\sigma}_n)^2}{2N}\right) \frac{\sin\left(\frac{(\vec{t},\vec{\sigma}_n)}{\sqrt{N/3}}\right)}{\frac{(\vec{t},\vec{\sigma}_n)}{\sqrt{N/3}}} = 1 + O\left(\|\vec{t}\|^3 N^{-3/2}\right) \tag{13.54}$$

and so

$$\prod_{n=0}^{N} \exp\left(\frac{(\vec{t},\vec{\sigma}_n)^2}{2N}\right) \frac{\sin\left(\frac{(\vec{t},\vec{\sigma}_n)}{\sqrt{N/3}}\right)}{\frac{(\vec{t},\vec{\sigma}_n)}{\sqrt{N/3}}} = 1 + O\left(\|\vec{t}\|^3 N^{-1/2}\right) \tag{13.55}$$

From here the assertions (i) and (ii) of the lemma follow immediately. This concludes the proof of Lemma 13.3.5 □

References

[1] R. J. Adler. On excursion sets, tube formulas and maxima of random fields. *Ann. Appl. Probab.*, **10**(1):1–74, 2000.

[2] M. Aizenman. Translation invariance and instability of phase coexistence in the two-dimensional Ising system. *Comm. Math. Phys.*, **73**(1):83–94, 1980.

[3] M. Aizenman, J. T. Chayes, L. Chayes, and C. M. Newman. The phase boundary in dilute and random Ising and Potts ferromagnets. *J. Phys. A*, **20**(5):L313–18, 1987.

[4] M. Aizenman and P. Contucci. On the stability of the quenched state in mean-field spin-glass models. *J. Statist. Phys.*, **92**(5–6):765–83, 1998.

[5] M. Aizenman, J. L. Lebowitz, and D. Ruelle. Some rigorous results on the Sherrington-Kirkpatrick spin glass model. *Comm. Math. Phys.*, **112**(1):3–20, 1987.

[6] M. Aizenman and E. H. Lieb. The third law of thermodynamics and the degeneracy of the ground state for lattice systems. *J. Statist. Phys.*, **24**(1):279–97, 1981.

[7] M. Aizenman, R. Sims, and S. L. Starr. An extended variational principle for the SK spin-glass model. *Phys. Rev. B*, **6821**(21):4403, 2003.

[8] M. Aizenman and J. Wehr. Rounding effects of quenched randomness on first-order phase transitions. *Comm. Math. Phys.*, **130**(3):489–528, 1990.

[9] D. J. Amit, H. Gutfreund, and H. Sompolinsky. Statistical mechanics of neural networks near saturation. *Ann. Phys.*, **173**:30–67, 1987.

[10] Ph. W. Anderson and S. F. Edwards. Theory of spin glasses. *J. Phys. F.*, **5**:965–74, 1975.

[11] G. Ben Arous, L. V. Bogachev, and S. A. Molchanov. Limit theorems for random exponentials. *Probab. Theor. Rel. Fields*, **132**(4):579–612, 2005.

[12] J. E. Avron, G. Roepstorff, and L. S. Schulman. Ground state degeneracy and ferromagnetism in a spin glass. *J. Statist. Phys.*, **26**(1):25–36, 1981.

[13] V. Bach, T. Jecko, and J. Sjöstrand. Correlation asymptotics of classical lattice spin systems with nonconvex Hamilton function at low temperature. *Ann. Henri Poincaré*, **1**(1):59–100, 2000.

[14] L. A. Bassalygo and R. L. Dobrushin. Uniqueness of a Gibbs field with a random potential – an elementary approach. *Teor. Veroyatnost. i Primenen.*, **31**(4):651–70, 1986.

[15] G. A. Battle. A new combinatoric estimate for cluster expansions. *Comm. Math. Phys.*, **94**(1):133–9, 1984.

[16] G. A. Battle and P. Federbush. A note on cluster expansions, tree graph identities, extra $1/N!$ factors! *Lett. Math. Phys.*, **8**(1):55–7, 1984.

[17] G. A. Battle and L. Rosen. The FKG inequality for the Yukawa$_2$ quantum field theory. *J. Statist. Phys.*, **22**(2):123–92, 1980.

[18] H. Bauke, S. Franz, and St. Mertens. Number partitioning as random energy model. *J. Statist. Mech.: Theory and Experiment*, page P04003, 2004.

[19] H. Bauke and St. Mertens. Universality in the level statistics of disordered systems. *Phys. Rev. E*, 70:025102(R), 2004.

[20] R. J. Baxter. Eight-vertex model in lattice statistics. *Phys. Rev. Lett.*, **26**(14):832–3, 1971.

[21] C. Berge. *Graphs and Hypergraphs*. Amsterdam: North-Holland Publishing Co., 1973.

[22] A. Berretti. Some properties of random Ising models. *J. Statist. Phys.*, **38**(3–4):483–96, 1985.

[23] J. Bertoin. Homogeneous fragmentation processes. *Probab. Theory Related Fields*, **121**(3):301–18, 2001.

[24] J. Bertoin. Self-similar fragmentations. *Ann. Inst. H. Poincaré Probab. Statist.*, **38**(3):319–40, 2002.

[25] J. Bertoin and J.-F. Le Gall. The Bolthausen–Sznitman coalescent and the genealogy of continuous-state branching processes. *Probab. Theory Related Fields*, **117**(2):249–66, 2000.

[26] J. Bertoin and J. Pitman. Two coalescents derived from the ranges of stable subordinators. *Electron. J. Probab.*, 5:no. 7, 17 pp. (electronic), 2000.

[27] J. Bertoin and M. Yor. On subordinators, self-similar Markov processes and some factorizations of the exponential variable. *Electron. Comm. Probab.*, **6**:95–106 (electronic), 2001.

[28] P. Billingsley. *Probability and Measure*. Wiley Series in Probability and Mathematical Statistics. New York: John Wiley & Sons Inc., 1995.

[29] P. M. Bleher, J. Ruiz, and V. A. Zagrebnov. One-dimensional random-field Ising model: Gibbs states and structure of ground states. *J. Statist. Phys.*, **84**(5–6):1077–93, 1996.

[30] T. Bodineau. Translation invariant Gibbs states for the Ising model. *Probab. Theory Related Fields* (online) DOI 10.1007/s00440–005–0457–0, 2005.

[31] T. Bodineau, D. Ioffe, and Y. Velenik. Rigorous probabilistic analysis of equilibrium crystal shapes. *J. Math. Phys.*, **41**(3):1033–98, 2000.

[32] E. Bolthausen and N. Kistler. On a non-hierarchical version of the generalized random energy model. Preprint, Univ. Zürich, 2004.

[33] E. Bolthausen and A.-S. Sznitman. On Ruelle's probability cascades and an abstract cavity method. *Comm. Math. Phys.*, **197**(2):247–76, 1998.

[34] E. Bolthausen and A.-S. Sznitman. *Ten Lectures on Random Media*, vol. 32 of *DMV Seminar*. Basel: Birkhäuser Verlag, 2002.

[35] C. Borgs, J. Chayes, and B. Pittel. Phase transition and finite-size scaling for the integer partitioning problem. *Random Structures Algorithms*, **19**(3–4):247–88, 2001.

[36] C. Borgs, J. T. Chayes, S. Mertens, and B. Pittel. Phase diagram for the constrained integer partitioning problem. *Random Structures Algorithms*, **24**(3):315–80, 2004.

[37] C. Borgs and R. Kotecký. to appear.

[38] A. Bovier. The Kac version of the Sherrington–Kirkpatrick model at high temperatures. *J. Statist. Phys.*, **91**(1–2):459–74, 1998.

[39] A. Bovier. *Statistical Mechanics of Disordered Systems*, vol. 10 of *MaPhySto Lecture Notes*. Aarhus University, 2001.

[40] A. Bovier and V. Gayrard. An almost sure central limit theorem for the Hopfield model. *Markov Process. Related Fields*, **3**(2):151–73, 1997.

[41] A. Bovier and V. Gayrard. The retrieval phase of the Hopfield model: a rigorous analysis of the overlap distribution. *Probab. Theory Related Fields*, **107**(1):61–98, 1997.

[42] A. Bovier and V. Gayrard. Hopfield models as generalized random mean field models. In *Mathematical Aspects of Spin Glasses and Neural Networks*, vol. 41 of *Progr. Probab.*, pp. 3–89. Boston, MA: Birkhäuser Boston, 1998.

[43] A. Bovier and V. Gayrard. Metastates in the Hopfield model in the replica symmetric regime. *Math. Phys. Anal. Geom.*, **1**(2):107–44, 1998.

[44] A. Bovier, V. Gayrard, and P. Picco. Gibbs states of the Hopfield model in the regime of perfect memory. *Probab. Theory Related Fields*, **100**(3):329–63, 1994.

[45] A. Bovier, V. Gayrard, and P. Picco. Gibbs states of the Hopfield model with extensively many patterns. *J. Statist. Phys.*, **79**(1–2):395–414, 1995.

[46] A. Bovier, V. Gayrard, and P. Picco. Large deviation principles for the Hopfield model and the Kac-Hopfield model. *Probab. Theory Related Fields*, **101**(4):511–46, 1995.

[47] A. Bovier, V. Gayrard, and P. Picco. Distribution of overlap profiles in the one-dimensional Kac–Hopfield model. *Comm. Math. Phys.*, **186**(2):323–79, 1997.

[48] A. Bovier and C. Külske. Stability of hierarchical interfaces in a random field model. *J. Statist. Phys.*, **69**(1–2):79–110, 1992.

[49] A. Bovier and C. Külske. Hierarchical interfaces in random media. II. The Gibbs measures. *J. Statist. Phys.*, **73**(1–2):253–66, 1993.

[50] A. Bovier and C. Külske. A rigorous renormalization group method for interfaces in random media. *Rev. Math. Phys.*, **6**(3):413–96, 1994.

[51] A. Bovier and I. Kurkova. Gibbs measures of Derrida's generalized random energy models and genealogies of Neveu's continuous state branching processe. Preprint 854, WIAS, 2003.

[52] A. Bovier and I. Kurkova. Derrida's generalised random energy models. I. Models with finitely many hierarchies. *Ann. Inst. H. Poincaré Probab. Statist.*, **40**(4):439–80, 2004.

[53] A. Bovier and I. Kurkova. Derrida's generalized random energy models. II. Models with continuous hierarchies. *Ann. Inst. H. Poincaré Probab. Statist.*, **40**(4):481–95, 2004.

[54] A. Bovier and I. Kurkova. Poisson convergence in the restricted k-partitioning problem. Preprint 964, WIAS, 2004.

[55] A. Bovier and I. Kurkova. Level statistics of disordered systems: a proof of the local REM conjecture. Preprint 1023, WIAS, 2005.

[56] A. Bovier, I. Kurkova, and M. Löwe. Fluctuations of the free energy in the REM and the p-spin SK models. *Ann. Probab.*, **30**(2):605–51, 2002.

[57] A. Bovier and D. M. Mason. Extreme value behavior in the Hopfield model. *Ann. Appl. Probab.*, **11**(1):91–120, 2001.

[58] A. Bovier and B. Niederhauser. The spin-glass phase-transition in the Hopfield model with p-spin interactions. *Adv. Theor. Math. Phys.*, **5**(6):1001–46, 2001.

[59] A. Bovier and P. Picco. Stability of interfaces in a random environment. A rigorous renormalization group analysis of a hierarchical model. *J. Statist. Phys.*, **62**(1–2):177–99, 1991.

[60] A. Bovier, A. C. D. van Enter, and B. Niederhauser. Stochastic symmetry-breaking in a Gaussian Hopfield model. *J. Statist. Phys.*, **95**(1–2):181–213, 1999.

[61] A. Bovier and M. Zahradník. A simple inductive approach to the problem of convergence of cluster expansions of polymer models. *J. Statist. Phys.*, **100**(3–4):765–78, 2000.

[62] H. J. Brascamp and E. H. Lieb. On extensions of the Brunn–Minkowski and Prékopa–Leindler theorems, including inequalities for log concave functions, and with an application to the diffusion equation. *J. Functional Analysis*, **22**(4):366–89, 1976.

[63] J. Bricmont and A. Kupiainen. Phase transition in the 3d random field Ising model. *Comm. Math. Phys.*, **116**(4):539–72, 1988.

[64] J. Bricmont and J. Slawny. Phase transitions in systems with a finite number of dominant ground states. *J. Statist. Phys.*, **54**(1–2):89–161, 1989.

[65] D. Brydges and P. Federbush. A new form of the Mayer expansion in classical statistical mechanics. *J. Math. Phys.*, **19**(10):2064–7, 1978.

[66] D. C. Brydges and J. Z. Imbrie. Branched polymers and dimensional reduction. *Ann. of Math. (2)*, **158**(3):1019–39, 2003.

[67] D. C. Brydges. A short course on cluster expansions. In *Phénomènes Critiques, Systèmes Aléatoires, Théories de Jauge, Part I, II* (Les Houches, 1984), pp. 129–83. Amsterdam: North-Holland, 1986.

[68] L. A. Bunimovič and J. G. Sinaĭ. The fundamental theorem of the theory of scattering billiards. *Mat. Sb. (N.S.)*, **90**(132):415–31, 479, 1973.

[69] L. A. Bunimovich, S. G. Dani, R. L. Dobrushin *et al. Dynamical Systems, Ergodic Theory and Applications*, vol. 100 of *Encyclopaedia of Mathematical Sciences*. Berlin: Springer-Verlag, 2000.

[70] C. Cammarota. Decay of correlations for infinite range interactions in unbounded spin systems. *Comm. Math. Phys.*, **85**(4):517–28, 1982.

[71] D. Capocaccia, M. Cassandro, and P. Picco. On the existence of thermodynamics for the generalized random energy model. *J. Statist. Phys.*, **46**(3–4):493–505, 1987.

[72] P. Carmona and Y. Hu. Universality in Sherrington–Kirkpatrick's spin glass model. Preprint arXiv:math.PR/0403359 v2, May 2004.

[73] M. Cassandro, E. Orlandi, and P. Picco. Typical configurations for one-dimensional random field Kac model. *Ann. Probab.*, **27**(3):1414–1467, 1999.

[74] M. Cassandro, E. Orlandi, P. Picco, and M. E. Vares. One-dimensional random field Kac's model: localization of the phases. *Electronic J. Probab.*, **10**:786–864, 2005.

[75] R. Cerf. Large deviations of the finite cluster shape for two-dimensional percolation in the Hausdorff and L^1 metric. *J. Theoret. Probab.*, **13**(2):491–517, 2000.

[76] R. Cerf and Á. Pisztora. Phase coexistence in Ising, Potts and percolation models. *Ann. Inst. H. Poincaré Probab. Statist.*, **37**(6):643–724, 2001.

[77] J. T. Chalker. On the lower critical dimensionality of the Ising model in a random field. *J. Phys. C*, **16**:6615–22, 1983.

[78] Y. S. Chow and H. Teicher. *Probability Theory*. Springer Texts in Statistics. New York: Springer-Verlag, 1997.

[79] F. Comets. Large deviation estimates for a conditional probability distribution. Applications to random interaction Gibbs measures. *Probab. Theory Related Fields*, **80**(3):407–32, 1989.

[80] F. Comets and A. Dembo. Ordered overlaps in disordered mean-field models. *Probab. Theory Related Fields*, **121**(1):1–29, 2001.

[81] F. Comets and J. Neveu. The Sherrington–Kirkpatrick model of spin glasses and stochastic calculus: the high temperature case. *Comm. Math. Phys.*, **166**(3):549–64, 1995.

[82] P. Contucci, M. Degli Esposti, C. Giardinà, and S. Graffi. Thermodynamical limit for correlated Gaussian random energy models. *Comm. Math. Phys.*, **236**(1):55–63, 2003.

[83] L. De Sanctis. Random multi-overlap structures and cavity fields in diluted spin glasses. *J. Statist. Phys.*, **117**:785–99, 2005.

[84] W. Th. F. den Hollander and M. Keane. Inequalities of FKG type. *Physica*, **138A**:167–82, 1986.

[85] B. Derrida. Random-energy model: limit of a family of disordered models. *Phys. Rev. Lett.*, **45**(2):79–82, 1980.

[86] B. Derrida. Random-energy model: an exactly solvable model of disordered systems. *Phys. Rev. B (3)*, **24**(5):2613–26, 1981.

[87] B. Derrida. A generalisation of the random energy model that includes correlations between the energies. *J. Phys. Lett.*, **46**:401–7, 1985.

[88] R. Dobrushin, R. Kotecký, and S. Shlosman. *Wulff Construction*, vol. 104 of *Translations of Mathematical Monographs*. Providence, RI: American Mathematical Society, 1992.

[89] R. L. Dobrushin. Existence of a phase transition in the two-dimensional and three-dimensional Ising models. *Soviet Physics Dokl.*, **10**:111–13, 1965.

[90] R. L. Dobrushin. Gibbs states that describe coexistence of phases for a three-dimensional Ising model. *Theor. Probab. Appl.*, **17**:582–600, 1972.

[91] R. L. Dobrushin. Estimates of semi-invariants for the Ising model at low temperatures. In *Topics in Statistical and Theoretical Physics*, vol. 177 of *Amer. Math. Soc. Transl. Ser. 2*, pp. 59–81. Providence, RI: American Mathematical Society, 1996.

[92] R. L. Dobrushin and S. B. Shlosman. 'Non-Gibbsian' states and their Gibbs description. *Comm. Math. Phys.*, **200**(1):125–79, 1999.

[93] R. L. Dobrušin. Description of a random field by means of conditional probabilities and conditions for its regularity. *Teor. Verojatnost. i Primenen*, **13**:201–29, 1968.

[94] R. L. Dobrušin. Definition of a system of random variables by means of conditional distributions. *Teor. Verojatnost. i Primenen.*, **15**:469–97, 1970.

[95] T. C. Dorlas and J. R. Wedagedera. Large deviations and the random energy model. *Internat. J. Modern Phys. B*, **15**(1):1–15, 2001.

[96] V. Dotsenko. *Introduction to the Replica Theory of Disordered Statistical Systems*. Collection Aléa-Saclay: Monographs and Texts in Statistical Physics. Cambridge: Cambridge University Press, 2001.

[97] T. Eisele. On a third-order phase transition. *Comm. Math. Phys.*, **90**(1):125–59, 1983.

[98] R. S. Ellis. *Entropy, Large Deviations, and Statistical Mechanics*, vol. 271 of *Grundlehren der Mathematischen Wissenschaften [Fundamental Principles of Mathematical Sciences]*. New York: Springer-Verlag, 1985.

[99] P. Federbush. The semi-Euclidean approach in statistical mechanics. II. The cluster expansion, a special example. *J. Mathematical Phys.*, **17**(2):204–7, 1976.

[100] D. S. Fisher and D. A. Huse. Absence of many states in realistic spin glasses. *J. Phys. A*, **20**(15):L1005–10, 1987.

[101] D. S. Fisher, J. Fröhlich, and T. Spencer. The Ising model in a random magnetic field. *J. Statist. Phys.*, **34**(5–6):863–70, 1984.

[102] M. E. Fisher. The free energy of a macroscopic system. *Arch. Rational Mech. Anal.*, **17**:377–410, 1964.

[103] C. M. Fortuin, P. W. Kasteleyn, and J. Ginibre. Correlation inequalities on some partially ordered sets. *Comm. Math. Phys.*, **22**:89–103, 1971.

[104] S. Franz, M. Leone, and F. L. Toninelli. Replica bounds for diluted non-Poissonian spin systems. *J. Phys. A*, **36**(43):10967–85, 2003.

[105] S. Franz and F. L. Toninelli. Finite range spin glasses in the Kac limit: free energy and local observables. *J. Phys. A*, **37**(30):7433–46, 2004.

[106] J. Fröhlich and J. Z. Imbrie. Improved perturbation expansion for disordered systems: beating Griffiths singularities. *Comm. Math. Phys.*, **96**(2):145–80, 1984.

[107] J. Fröhlich and B. Zegarliński. The high-temperature phase of long-range spin glasses. *Comm. Math. Phys.*, **110**(1):121–55, 1987.

[108] J. Fröhlich and B. Zegarliński. Spin glasses and other lattice systems with long range interactions. *Comm. Math. Phys.*, **120**(4):665–88, 1989.

[109] G. Gallavotti. *Statistical mechanics*. Texts and Monographs in Physics. Berlin: Springer-Verlag, 1999.

[110] G. Gallavotti and S. Miracle-Solé. Correlation functions of a lattice system. *Comm. Math. Phys.*, **7**:274–88, 1968.

[111] G. Gallavotti and S. Miracle-Solé. Equilibrium states of the Ising model in the two-phase region. *Phys. Rev.*, **B5**:2555–9, 1973.

[112] G. Gallavotti, S. Miracle-Solé, and D. W. Robinson. Analyticity properties of a lattice gas. *Phys. Letters. A*, **25**:493–4, 1968.

[113] A. Galves, S. Martínez, and P. Picco. Fluctuations in Derrida's random energy and generalized random energy models. *J. Statist. Phys.*, **54**(1–2):515–29, 1989.

[114] E. Gardner and B. Derrida. Magnetic properties and function $q(x)$ of the generalised random energy model. *J. Phys. C*, **19**:5783–98, 1986.

[115] E. Gardner and B. Derrida. Solution of the generalised random energy model. *J. Phys. C*, **19**:2253–74, 1986.

[116] E. Gardner and B. Derrida. The probability distribution of the partition function of the random energy model. *J. Phys. A*, **22**(12):1975–81, 1989.

[117] V. Gayrard. Thermodynamic limit of the q-state Potts–Hopfield model with infinitely many patterns. *J. Statist. Phys.*, **68**(5–6):977–1011, 1992.

[118] S. Geman. A limit theorem for the norm of random matrices. *Ann. Probab.*, **8**(2):252–61, 1980.

[119] B. Gentz. A central limit theorem for the overlap in the Hopfield model. *Ann. Probab.*, **24**(4):1809–41, 1996.

[120] B. Gentz. On the central limit theorem for the overlap in the Hopfield model. In *Mathematical Aspects of Spin Glasses and Neural Networks*, vol. 41 of *Progr. Probab.*, pp. 115–49. Boston, MA: Birkhäuser Boston, 1998.

[121] B. Gentz and M. Löwe. Fluctuations in the Hopfield model at the critical temperature. *Markov Process. Related Fields*, **5**(4):423–49, 1999.

[122] B. Gentz and M. Löwe. The fluctuations of the overlap in the Hopfield model with finitely many patterns at the critical temperature. *Probab. Theory Related Fields*, **115**(3):357–81, 1999.

[123] H.-O. Georgii. Spontaneous magnetization of randomly dilute ferromagnets. *J. Statist. Phys.*, **25**(3):369–96, 1981.

[124] H.-O. Georgii. On the ferromagnetic and the percolative region of random spin systems. *Adv. in Appl. Probab.*, **16**(4):732–65, 1984.

[125] H.-O. Georgii. *Gibbs Measures and Phase Transitions*, vol. 9 of *de Gruyter Studies in Mathematics*. Berlin: Walter de Gruyter & Co., 1988.

[126] H.-O. Georgii and Y. Higuchi. Percolation and number of phases in the two-dimensional Ising model. *J. Math. Phys.*, **41**(3):1153–69, 2000.

[127] S. Ghirlanda and F. Guerra. General properties of overlap probability distributions in disordered spin systems. Towards Parisi ultrametricity. *J. Phys. A*, **31**(46):9149–55, 1998.

[128] V. L. Girko. Limit theorems for distributions of eigenvalues of random symmetric matrices. *Teor. Veroyatnost. i Mat. Statist.*, **41**:23–9, 129, 1989.

[129] J. Glimm and A. Jaffe. *Quantum Physics*. New York: Springer-Verlag, 1981.

[130] R. B. Griffiths. Peierls proof of spontaneous magnetization in a two-dimensional Ising ferromagnet. *Phys. Rev. (2)*, **136**:A437–9, 1964.

[131] D. H. E. Gross. The microcanonical entropy is multiply differentiable. No dinosaurs in microcanonical gravitation. No special 'microcanonical phase transitions'. Preprint cond-mat/0423582, 2004.

[132] C. Gruber and H. Kunz. General properties of polymer systems. *Comm. Math. Phys.*, **22**:133–61, 1971.

[133] F. Guerra. Broken replica symmetry bounds in the mean field spin glass model. *Comm. Math. Phys.*, **233**(1):1–12, 2003.

[134] F. Guerra and F. L. Toninelli. The thermodynamic limit in mean field spin glass models. *Comm. Math. Phys.*, **230**(1):71–9, 2002.

[135] Y. M. Guttmann. *The Concept of Probability in Statistical Physics*. Cambridge Studies in Probability, Induction, and Decision Theory. Cambridge: Cambridge University Press, 1999.

[136] P. Hall and C. C. Heyde. *Martingale Limit Theory and its Application*. New York: Academic Press Inc., 1980.

[137] D. Hebb. *The Organisation of Behaviour: a Neurophysiological Theory*. New York: Wiley, 1949.

[138] Y. Higuchi. On the absence of non-translation invariant Gibbs states for the two-dimensional Ising model. In *Random Fields, Vol. I, II* (Esztergom, 1979), vol. 27 of *Colloq. Math. Soc. János Bolyai*, pp. 517–34. Amsterdam: North-Holland, 1981.

[139] J. J. Hopfield. Neural networks and physical systems with emergent collective computational abilities. *Proc. Nat. Acad. Sci. USA*, **79**(8):2554–8, 1982.

[140] J. Hubbard. Calculation of partition functions. *Phys. Rev. Lett.*, **3**:77–8, 1959.

[141] D. A. Huse and D. S. Fisher. Pure states in spin glasses. *J. Phys. A*, **20**(15):L997–1003, 1987.

[142] Y. Imry and S. Ma. Random-field instability of the ordered state of continuous symmetry. *Phys. Rev. Lett.*, **35**:1399–401, 1975.

[143] D. Ioffe and R. H. Schonmann. Dobrushin–Kotecký–Shlosman theorem up to the critical temperature. *Comm. Math. Phys.*, **199**(1):117–67, 1998.

[144] E. Ising. Beitrag zur Theorie des Ferro- und Paramagnetismus. Ph.D. thesis, Univ. Hamburg, 1924.

[145] E. Ising. Beitrag zur Theories des Ferromagnetismus. *Zeitschrift. f. Physik*, **31**:253–8, 1925.

[146] R. B. Israel. *Convexity in the Theory of Lattice Gases*. Princeton, NJ: Princeton University Press, 1979.

[147] M. Kac, G. E. Uhlenbeck, and P. C. Hemmer. On the van Waals theory of the vapor-liquid equilibrium. I. Discussion of a one-dimensional model. *J. Math. Phys.*, **4**:216–28, 1963.

[148] O. Kallenberg. *Random Measures*. Berlin: Akademie-Verlag, 1983.

[149] J. G. Kirkwood and Z. W. Salsburg. The statistical mechanical theory of molecular distribution functions in liquids. *Discussions Farday Soc.*, **15**:28–34, 1953.

[150] A. Klein and S. Masooman. Taming Griffiths' singularities in long range random Ising models. *Comm. Math. Phys.*, **189**(2):497–512, 1997.

[151] S. Kobe. Ernst Ising, physicist and teacher. *J. Phys. Stud.*, **2**(1):1–2, 1998.

[152] H. Koch. A free energy bound for the Hopfield model. *J. Phys. A*, **26**(6):L353–5, 1993.

[153] H. Koch and J. Piasko. Some rigorous results on the Hopfield neural network model. *J. Statist. Phys.*, **55**(5–6):903–28, 1989.

[154] J. Komlos and R. Paturi. Convergence results in a autoassociative memory model. *Neural Networks*, **1**:239–50, 1988.

[155] R. Kotecký and D. Preiss. Cluster expansion for abstract polymer models. *Comm. Math. Phys.*, **103**(3):491–8, 1986.

[156] M. F. Kratz and P. Picco. A representation of Gibbs measure for the random energy model. *Ann. Appl. Probab.*, **14**(2):651–77, 2004.

[157] C. Külske. Metastates in disordered mean-field models: random field and Hopfield models. *J. Statist. Phys.*, **88**(5–6):1257–93, 1997.

[158] C. Külske. Metastates in disordered mean-field models. II. The superstates. *J. Statist. Phys.*, **91**(1–2):155–76, 1998.

[159] C. Külske. Weakly Gibbsian representations for joint measures of quenched lattice spin models. *Probab. Theory Related Fields*, **119**(1):1–30, 2001.

[160] C. Külske. Analogues of non-Gibbsianness in joint measures of disordered mean-field models. *J. Statist. Phys.*, **112**(5–6):1079–108, 2003.

[161] O. E. Lanford, III and D. Ruelle. Observables at infinity and states with short range correlations in statistical mechanics. *Comm. Math. Phys.*, **13**:194–215, 1969.

[162] M. R. Leadbetter, G. Lindgren, and H. Rootzén. *Extremes and Related Properties of Random Sequences and Processes*. Springer Series in Statistics. New York: Springer-Verlag, 1983.

[163] J. L. Lebowitz and A. Martin-Löf. On the uniqueness of the equilibrium state for Ising spin systems. *Comm. Math. Phys.*, **25**:276–82, 1972.

[164] J. L. Lebowitz, A. Mazel, and E. Presutti. Liquid–vapor phase transitions for systems with finite-range interactions. *J. Statist. Phys.*, **94**(5–6):955–1025, 1999.

[165] J. L. Lebowitz and A. E. Mazel. Improved Peierls argument for high-dimensional Ising models. *J. Statist. Phys.*, **90**(3–4):1051–9, 1998.

[166] M. Ledoux. On the distribution of overlaps in the Sherrington–Kirkpatrick spin glass model. *J. Statist. Phys.*, **100**(5–6):871–92, 2000.

[167] M. Ledoux and M. Talagrand. *Probability in Banach Spaces*, vol. 23 of *Ergebnisse der Mathematik und ihrer Grenzgebiete (3) [Results in Mathematics and Related Areas (3)]*. Berlin: Springer-Verlag, 1991.

[168] E. H. Lieb. Exact solution of the problem of the entropy of two-dimensional ice. *Phys. Rev. Lett.*, **18**(17):692–4, 1967.

[169] E. H. Lieb and J. Yngvason. The physics and mathematics of the second law of thermodynamics. *Phys. Rep.*, **310**(1):96, 1999.

[170] D. Loukianova. Lower bounds on the restitution error in the Hopfield model. *Probab. Theory Related Fields*, **107**(2):161–76, 1997.

[171] C. Maes, F. Redig, and A. Van Moffaert. Almost Gibbsian versus weakly Gibbsian measures. *Stochastic Process. Appl.*, **79**(1):1–15, 1999.

[172] V. A. Malyshev. Complete cluster expansions for weakly coupled Gibbs random fields. In *Multicomponent Random Systems*, vol. 6 of *Adv. Probab. Related Topics*, pp. 505–530. New York: Dekker, 1980.

[173] V. A. Marčenko and L. A. Pastur. Distribution of eigenvalues in certain sets of random matrices. *Mat. Sb. (N. S.)*, **72**(114):507–36, 1967.

[174] D. C. Mattis. Solvable spin system with random interactions. *Phys. Lett.*, **56**(A):421–2, 1976.

[175] J. E. Mayer. *Handbuch der Physik*. Berlin, Heidelberg: Springer-Verlag, 1958.

[176] W. S McCulloch and W. Pitts. A logical calculus of the ideas immanent in nervous activity. *Bull. Math. Biophysics*, **5**:115–33, 1943.

[177] St. Mertens. Phase transition in the number partitioning problem. *Phys. Rev. Lett.*, **81**(20):4281–4, 1998.

[178] St. Mertens. A physicist's approach to number partitioning. *Theoret. Comput. Sci.*, **265**(1–2):79–108, 2001.

[179] M. Mézard, G. Parisi, and M. A. Virasoro. *Spin Glass Theory and Beyond*, vol. 9 of *World Scientific Lecture Notes in Physics*. Teaneck, NJ: World Scientific Publishing Co. Inc., 1987.

[180] S. Miracle-Solé. On the convergence of cluster expansions. *Phys. A*, **279**(1–4):244–9, 2000.

[181] F. R. Nardi, E. Olivieri, and M. Zahradník. On the Ising model with strongly anisotropic external field. *J. Statist. Phys.*, **97**(1–2):87–144, 1999.

[182] J. Neveu. A continuous state branching process in relation with the GREM model of spin glass theory. Rapport interne 267, Ecole Polytechnique Paris, 1992.

[183] C. M. Newman and D. L. Stein. Simplicity of state and overlap structure in finite-volume realistic spin glasses. *Phys. Rev. E (3)*, **57**(2, part A):1356–66, 1998.

[184] C. M. Newman. Memory capacity in neural network. *Neural Networks*, **1**:223–38, 1988.

[185] C. M. Newman. *Topics in Disordered Systems*. Lectures in Mathematics ETH Zürich. Basel: Birkhäuser Verlag, 1997.

[186] C. M. Newman and D. L. Stein. Chaotic size dependence in spin glasses. In *Cellular Automata and Cooperative Systems* (Les Houches, 1992), vol. 396 of *NATO Adv. Sci. Inst. Ser. C Math. Phys. Sci.*, pp. 525–9. Dordrecht: Kluwer Academic Publishers, 1993.

[187] C. M. Newman and D. L. Stein. Thermodynamic chaos and the structure of short-range spin glasses. In *Mathematical Aspects of Spin Glasses and Neural Networks*, vol. 41 of *Progr. Probab.*, pp. 243–287. Boston, MA: Birkhäuser Boston, 1998.

[188] C. M. Newman and D. L. Stein. Equilibrium pure states and nonequilibrium chaos. *J. Statist. Phys.*, **94**(3–4):709–22, 1999.

[189] C. M. Newman and D. L. Stein. Metastable states in spin glasses and disordered ferromagnets. *Phys. Rev. E (3)*, 60(5, part A):5244–60, 1999.

[190] C. M. Newman and D. L. Stein. Short-range spin glasses: a status report. In *XIIth International Congress of Mathematical Physics (ICMP '97)* (Brisbane), pp. 167–172. Cambridge, MA: International Press, 1999.

[191] C. M. Newman and D. L. Stein. Local vs. global variables for spin glasses. In *Equilibrium and Dynamics of Spin-Glasses*. Berlin: Springer-Verlag, to appear.

[192] C. M. Newman and D. L. Stein. Short-range spin glasses: results and speculations. In *Equilibrium and Dynamics of Spin-Glasses*. Berlin: Springer-Verlag, to appear.

[193] H. Nishimori. *Statistical Physics of Spin Glasses and Information Processing*. International Series of Monographs on Physics 111. Oxford: Oxford University Press, 2001.

[194] E. Olivieri and P. Picco. On the existence of thermodynamics for the random energy model. *Comm. Math. Phys.*, **96**(1):125–44, 1984.

[195] L. Onsager. Crystal statistics, I. A two-dimensional model with an order-disorder transition. *Phys. Rev.*, **65**:117–49, 1944.

[196] G. Parisi and N. Sourlas. Random magnetic fields, supersymmetry, and negative dimensions. *Phys. Rev. Lett.*, **43**(11):744–5, 1979.

[197] L. A. Pastur and A. L. Figotin. Exactly soluble model of a spin glass. *Sov. J. Low Temp. Phys.*, **3**(6):378–83, 1977.

[198] L. A. Pastur and A. L. Figotin. On the theory of disordered spin systems. *Teoret. Mat. Fiz.*, **35**(2):193–210, 1978.

[199] L. A. Pastur and A. L. Figotin. Limit of the infinite range of interaction for a class of disordered systems. *Teoret. Mat. Fiz.*, **51**(3):380–8, 1982.

[200] L. A. Pastur, M. Shcherbina, and B. Tirozzi. The replica-symmetric solution without replica trick for the Hopfield model. *J. Statist. Phys.*, **74**(5–6):1161–83, 1994.

[201] R. Peierls. On Ising's model of ferromagnetism. *Proc. Cambridge Philos. Soc.*, **32**:477–81, 1936.

[202] D. Petritis. Thermodynamic formalism of neural computing. In *Dynamics of Complex Interacting Systems* (Santiago, 1994), vol. 2 of *Nonlinear Phenom. Complex Systems*, pp. 81–146. Dordrecht: Kluwer Academic Publishers, 1996.

[203] C.-E. Pfister. Thermodynamical aspects of classical lattice systems. In *In and Out of Equilibrium* (Mambucaba, 2000), vol. 51 of *Progr. Probab.*, pp. 393–472. Boston, MA: Birkhäuser Boston, 2002.

[204] S. A. Pirogov and Ja. G. Sinaĭ. Phase diagrams of classical lattice systems. *Teoret. Mat. Fiz.*, **25**(3):358–69, 1975.

[205] S. A. Pirogov and Ja. G. Sinaĭ. Phase diagrams of classical lattice systems. (Continuation). *Teoret. Mat. Fiz.*, **26**(1):61–76, 1976.

[206] J. Pitman. Combinatorial stochastic processes. In *Proceedings of the 2002 Ecole de'été de St. Flour*, Lecture Notes in Mathematics. Berlin: Springer-Verlag. To appear.

[207] C. J. Preston. *Gibbs States on Countable Sets*. Cambridge Tracts in Mathematics, No. 68. London: Cambridge University Press, 1974.

[208] C. J. Preston. *Random Fields*. Berlin: Springer-Verlag, 1976.

[209] E. Presutti. *Scaling Limits in Statistical Mechanics and Microstructures in Continuum Mechanics*. In preparation, 2005.

[210] S. I. Resnick. *Extreme Values, Regular Variation, and Point Processes*, vol. 4 of *Applied Probability. A Series of the Applied Probability Trust*. New York: Springer-Verlag, 1987.

[211] R. T. Rockafellar. *Convex Analysis*. Princeton Mathematical Series, No. 28. Princeton, NJ: Princeton University Press, 1970.

[212] G.-C. Rota. On the foundations of combinatorial theory. I. Theory of Möbius functions. *Z. Wahrscheinlichkeitstheorie und Verw. Gebiete*, **2** 340–68, 1964.

[213] D. Ruelle. Classical statistical mechanics of a system of particles. *Helv. Phys. Acta*, **36**:183–97, 1963.

[214] D. Ruelle. *Statistical Mechanics: Rigorous Results*. New York and Amsterdam: W. A. Benjamin, Inc., 1969.

[215] D. Ruelle. *Thermodynamic Formalism*, vol. 5 of *Encyclopedia of Mathematics and its Applications*. Reading, MA: Addison-Wesley Publishing Co., 1978.

[216] D. Ruelle. A mathematical reformulation of Derrida's REM and GREM. *Comm. Math. Phys.*, **108**(2):225–39, 1987.

[217] D. Ruelle. Some ill-formulated problems on regular and messy behavior in statistical mechanics and smooth dynamics for which I would like the advice of Yasha Sinai. *J. Statist. Phys.*, **108**(5–6):723–8, 2002.

[218] A. Ruzmaikina and M. Aizenman. Characterization of invariant measures at the leading edge for competing particle systems. *Ann. Probab.*, **33**(1):82–113, 2005.

[219] A. Scott and A. D. Sokal. The repulsive lattice gas, the independent set polynomials, and the Lovàsz local lemma. *J. Statist. Phys.*, **118**:1151–261, 2005.

[220] M. Shcherbina and B. Tirozzi. The free energy of a class of Hopfield models. *J. Statist. Phys.*, **72**(1–2):113–25, 1993.

[221] D. Sherrington and S. Kirkpatrick. Solvable model of a spin glass. *Phys. Rev. Letts.*, **35**:1792–96, 1972.

[222] A. N. Shiryayev. *Probability*, vol. 95 of *Graduate Texts in Mathematics*. New York: Springer-Verlag, 1984.

[223] J. W. Silverstein. Eigenvalues and eigenvectors of large-dimensional sample covariance matrices. In *Random Matrices and their Applications* (Brunswick, ME, 1984), vol. 50 of *Contemp. Math.*, pp. 153–9. Providence, RI: American Mathematical Society, 1986.

[224] B. Simon. *The statistical mechanics of lattice gases. Vol. I*. Princeton Series in Physics. Princeton, NJ: Princeton University Press, 1993.

[225] J. G. Sinaĭ. On the foundations of the ergodic hypothesis for a dynamical system of statistical mechanics. *Soviet Math. Dokl.*, **4**:1818–22, 1963.

[226] Y. G. Sinaĭ. *Theory of Phase transitions: Rigorous Results*, vol. 108 of *International Series in Natural Philosophy*. Oxford: Pergamon Press, 1982.

[227] D. Slepian. The one-sided barrier problem for Gaussian noise. *Bell System Tech. J.*, **41**:463–501, 1962.

[228] A. D. Sokal. Chromatic polynomials, potts models, and all that. *Physica A*, **279**:324–32, 1999.

[229] H. E. Stanley. *Introduction to Phase Transitions and Critical Phenomena*. Oxford: Oxford University Press, 1987.

[230] R. L. Stratonovič. A method for the computation of quantum distribution functions. *Dokl. Akad. Nauk SSSR (N.S.)*, **115**:1097–100, 1957.

[231] M. Talagrand. Concentration of measure and isoperimetric inequalities in product spaces. *Inst. Hautes Études Sci. Publ. Math.*, **81**:73–205, 1995.

[232] M. Talagrand. A new look at independence. *Ann. Probab.*, **24**(1):1–34, 1996.

[233] M. Talagrand. Rigorous results for the Hopfield model with many patterns. *Probab. Theory Related Fields*, **110**(2):177–276, 1998.

[234] M. Talagrand. The Sherrington–Kirkpatrick model: a challenge for mathematicians. *Probab. Theory Related Fields*, **110**(2):109–76, 1998.

[235] M. Talagrand. Exponential inequalities and convergence of moments in the replica-symmetric regime of the Hopfield model. *Ann. Probab.*, **28**(4):1393–469, 2000.

[236] M. Talagrand. Rigorous low-temperature results for the mean field p-spins interaction model. *Probab. Theory Related Fields*, **117**(3):303–60, 2000.

[237] M. Talagrand. Gaussian averages, Bernoulli averages, and Gibbs' measures. *Random Structures Algorithms*, **21**(3–4):197–204, 2002.

[238] M. Talagrand. Self organization in the low temperature phase of a spin glass model. *Rev. Math. Phys.*, **14**:1–78, 2003.

[239] M. Talagrand. *Spin Glasses: a Challenge for Mathematicians*, vol. 46 of *Ergebnisse der Mathematik und ihrer Grenzgebiete (3) [Results in Mathematics and Related Areas (3)]*. Berlin: Springer-Verlag, 2003.

[240] M. Talagrand. The Parisi solution. *Ann. Math.*, to appear, 2005.

[241] D. Ueltschi. Discontinuous phase transitions in quantum lattice systems. Ph.D. thesis, EPFL Lausanne, 1998.

[242] D. Ueltschi. Cluster expansions and correlation functions. *Mosc. Math. J.*, **4**(2):511–522, 576, 2004.

[243] H. van Beijeren. Interface sharpness in Ising systems. *Comm. Math. Phys.*, **40**:1–6, 1975.

[244] J. van den Berg and C. Maes. Disagreement percolation in the study of Markov fields. *Ann. Probab.*, **22**(2):749–63, 1994.

[245] A. van Enter, C. Maes, R. H. Schonmann, and S. Shlosman. The Griffiths singularity random field. In *On Dobrushin's Way. From Probability Theory to Statistical Physics*, vol. 198 of *Amer. Math. Soc. Transl. Ser. 2*, pp. 51–58. Providence, RI: American Mathematical Society, 2000.

[246] A. C. D. van Enter, I. Medved', and K. Netočný. Chaotic size dependence in the Ising model with random boundary conditions. *Markov Process. Related Fields*, **8**(3):479–508, 2002.

[247] A. C. D. van Enter, K. Netočný, and H. G. Schaap. On the Ising model with random boundary conditions. *Jour. Stat. Phys.*, **118**:997–1056, 2005.

[248] A. C. D. van Enter and H. G. Schaap. Infinitely many states and stochastic symmetry in a Gaussian Potts–Hopfield model. *J. Phys. A*, **35**(11):2581–92, 2002.

[249] A. C. D. van Enter and J. L. van Hemmen. The thermodynamic limit for long-range random systems. *J. Statist. Phys.*, **32**(1):141–52, 1983.

[250] A. C. D. van Enter, R. Fernández, and A. D. Sokal. Regularity properties and pathologies of position-space renormalization-group transformations: scope and limitations of Gibbsian theory. *J. Statist. Phys.*, **72**(5–6):879–1167, 1993.

[251] A. C. D. van Enter and R. B. Griffiths. The order parameter in a spin glass. *Comm. Math. Phys.*, **90**(3):319–27, 1983.

[252] A. C. D. van Enter and J. L. van Hemmen. Chopper model for pattern recognition. *Phys. Rev.*, **34**(A):2509–12, 1986.

[253] J. L. van Hemmen. Equilibrium theory of spin glasses: mean-field theory and beyond. In *Heidelberg Colloquium on Spin Glasses* (Heidelberg, 1983), vol. 192 of *Lecture Notes in Phys.*, pp. 203–33. Berlin: Springer-Verlag, 1983.

[254] J. L. van Hemmen. Spin-glass models of a neural network. *Phys. Rev. A (3)*, **34**(4):3435–45, 1986.

[255] J. L. van Hemmen, D. Grensing, A. Huber, and R. Kühn. Elementary solution of classical spin-glass models. *Z. Phys. B*, **65**(1):53–63, 1986.

[256] J. L. van Hemmen, A. C. D. van Enter, and J. Canisius. On a classical spin glass model. *Z. Phys. B*, **50**(4):311–36, 1983.

[257] L. van Hove. Sur l'intégrale de configuration pour les systèmes de particules à une dimension. *Physica*, **16**:137–43, 1950.

[258] J. Wehr and M. Aizenman. Fluctuations of extensive functions of quenched random couplings. *J. Statist. Phys.*, **60**(3–4):287–306, 1990.

[259] P. Weiss. L'hypothèse des champ moleculaire et la propriété ferromagnétique. *J. de Physique*, 6:661, 1907.

[260] Y. Q. Yin, Z. D. Bai, and P. R. Krishnaiah. On the limit of the largest eigenvalue of the large-dimensional sample covariance matrix. *Probab. Theory Related Fields*, **78**(4):509–21, 1988.

[261] V. V. Yurinskiĭ. Exponential inequalities for sums of random vectors. *J. Multivariate Anal.*, **6**(4):473–99, 1976.

[262] M. Zahradník. An alternate version of Pirogov–Sinaĭ theory. *Comm. Math. Phys.*, **93**(4):559–81, 1984.

[263] M. Zahradník. A short course on the Pirogov–Sinai theory. *Rend. Mat. Appl. (7)*, **18**(3):411–486, 1998.

[264] B. Zegarlinski. Random spin systems with long-range interactions. In *Mathematical Aspects of Spin Glasses and Neural Networks*, vol. 41 of *Progr. Probab.*, pp. 289–320. Boston, MA: Birkhäuser Boston, 1998.

Index

Aizenman, M., 72, 100, 111, 118, 167, 230
Amit, D. J., 271
average
   annealed, 98
   quenched, 98

Bassalygo, L. A., 106
Bauke, H., 287
Baxter, R. J., 48
Ben Arous, G., 170
Berretti, A., 107
binomial coefficient
   asymptotics, 41
Bodineau,Th., 72
Bogachev, L., 170
Borgs, Ch., 286
branching process
   Neveu's, 231
Brascamp–Lieb inequalities, 274
Bricmont, J., 111, 125
Brydges, D., 81, 112

canonical
   ensemble, 19
   partition function, 19
cavity method, 271, 280
central limit theorem
   functional, 253
chaining, 115, 258
Chayes, J. T., 286
Chebychev inequality, 259
cluster, 77
cluster expansion, 77
Comets, F., 167, 250
comparison lemma
   Gaussian, 197, 218, 228
concave hull, 190
concentration of measure, 113, 219, 264
conditional distribution, 56, 101
   regular, 56
conditional expectation, 56
contour models, 127
Contucci, P., 182

convexity, 7, 45
   local, 266
Cramèr entropy, 41
CREM, 195
Curie, P., 39
Curie–Weiss model, 39, 161
cylinder function, 53

Derrida, B., 165, 186, 196
dimensional reduction, 112
disordered system, 98
distance
   Hamming, 163
   lexicographic, 164
   ultrametric, 164
distribution
   micro-canonical, 10
DLR equations, 57
Dobrushin, R. L., 51, 62, 64, 72, 81, 106
Dobrushin states, 72

Edwards–Anderson model, 109
empirical distance distribution, 196, 202
ensemble
   canonical, 19
   equivalence of, 21
   micro-canonical, 13
enthalpy, 7
entropy, 5, 15
   Cramèr, 41
   of a measure, 18
   relative, 18
equation of state
   van der Waals, 29
ergodic theorem, 121
expansion
   cluster, 77
   high-temperature, 73
   low-temperature, 88
   Mayer, 77
extremal process, 187, 287
extreme value statistics, 166
extreme value theory, 220

ferromagnetism, 35
Figotin, A. L., 249
Fisher, D., 113
FKG inequalities, 68, 118
flow
　of probability measures, 207
fluctuations
　of REM partition function, 169
Fröhlich, J., 107, 113
free energy
　annealed, 98
　Gibbs, 7
　GREM, 192
　Helmholtz, 7
　in SK model, 218, 237
　of spin system , 35
　quenched, 98
frustration, 110

Gallavotti, G., 72, 81
Gardner, E., 196
Gaussian
　comparison lemma, 197, 228
　process, 163
Gayrard, V., 257
genealogy, 231
generalized random energy model, 186
generating functions, 121
Gentz, B., 258
Georgii, H.-O., 109
Ghirlanda, S., 182
Ghirlanda–Guerra relations, 182
　extended, 239
　SK model, 238
Gibbs measure, 19
　infinite-volume, 52, 57
　random, 99, 100
　　covariant, 119
　weak, 57
Gibbs specification
　random, 100
Gibbs state
　extremal, 108
Gielis, G., 106
GREM, 186, 232
　free energy, 192
　partition function, 192
Griffiths, R. B., 64, 100
ground-state, 89, 110, 126
Guerra, F., 163, 182, 218, 235
Guerra's bound, 235
Gutfreund, H., 271

Hebb, D., 248
Hebb's rule, 248
Higuchi, Y., 72
Hölder's inequality, 124
Hopfield, J. J., 247

Hopfield Hamiltonian, 249
Hopfield model, 162, 247

ideal gas, 9
Imbrie, J. Z., 107, 112
Imry–Ma argument, 111
inclusion-exclusion principle, 288
inequality
　Bonferroni, 288
　Brascamp–Lieb, 274
　Chebychev, 259
　concentration, 264
　FKG, 68
　Marcinkiewicz–Zygmund, 259
　Paley–Zygmund, 223
infinite volume, 51
integration by parts
　Gaussian, 182
　　multivariate, 197
interaction, 54
　absolutely summable, 54
　continuous, 54
　nearest-neighbor, 34
　random, 99
　regular, 54
internal energy, 6
interpolation
　Gaussian, 218
Ising, E., 34
Ising model, 35
　dilute, 107
　one dimensional, 37
　random field, 111
isotherm, 45

joint measure, 101

Kahane, J.-P., 197
Kirkpatrick, S., 163, 218
Klein, A., 107
Koch, H., 257
Kotecký, R., 81
Kotecký–Preiss criterion, 77
Külske, Ch., 104, 106
Kupiainen, A., 111, 125
Kurkova, I., 169, 186

Lanford, O. E., 51
large deviation, 250
　principle, 42, 250
lattice gas, 33
　ideal, 17
　Ising, 34
Lebowitz, J., 167
Legendre transform, 6, 20
Lieb, E. H., 8, 48
Lindeberg condition, 170
Lipschitz continuous, 113

local specification, 55
  Gibbs, 56
  random, 99, 100
Löwe, M., 169, 258
lumping, 243
lumps, 244, 253, 264

Maes, Ch., 106
magnetic field, 35
magnetisation, 35
Marcinkiewicz–Zygmund
  inequality, 259
martingale-difference sequence, 123
Mason, D. M., 258
Masooman, S., 107
Mayer expansion, 77
mean-field models, 40, 161
  disordered, 161
memory
  autoassociative, 248
  capacity, 249
Mertens, St., 285, 287
metastate, 99, 178, 253
  Aizenman–Wehr, 103, 256, 279
  empirical, 103
micro-canonical
  ensemble, 13
  partition function, 14
Miracle-Solé, S., 72, 81
Molchanov, S., 170
multi-index, 77

neural network, 247
Neveu, J., 167, 178
Newman, Ch., 100
non-convexity, 45
number partitioning, 285

Onsager, L., 47

Paley–Zygmund inequality, 223
Parisi, G., 112, 163, 235
Parisi solution, 231, 235
partition function, 14
  canonical, 19
  GREM, 192
  micro-canonical, 14
Pastur, L. A., 249, 272
Peierls, R., 64
Peierls argument, 107
phase separation, 45
phase transition, 7
  first order, 7
  higher order, 7
  liquid–vapour, 29
Piasko, J., 257
Picco, P., 257
Pirogov, S. A., 89

Pirogov–Sinai theory, 89
Pittel, J., 286
point process
  Poisson, 169
Poisson cascade, 187
Poisson–Dirichlet process, 206
Poisson process
  of extremes, 171, 287
polymer, 74
  connected, 74
  high-temperature, 74
  model
    abstract, 76
  representation, 75
potential
  chemical, 5
Preiss, D., 81
pressure, 5
Presutti, E., 31
probability measure
  flow of, 207
pure state, 178

quasi-local function, 53
quenched, 98

random energy model (REM), 165, 287
random-field Ising model, 111
random matrix, 260
  Marchenko–Pastur, 260
  sample covariance, 260
random model, 98
random overlap structure, 230
random process
  Gaussian, 163
random subsequences, 100
regular conditional distribution, 56
regular interaction
  random, 99
renormalization group, 125
  transformation, 127
replica
  method, 227
  symmetric solution, 271
  symmetry breaking, 228
Ruelle, D., 51, 167

second moment method, 167, 220
  truncated, 167
self-averaging, 113
Shcherbina, M., 257
Sherrington, D., 163, 218
Sherrington–Kirkpatrick (SK) model, 162
Sims, R., 230
Sinai, Ya. G., 89
Slepian, D., 221
Slepian's lemma, 220

Sokal, A., 82
Sompolinski, H., 271
Sourlas, N., 112
Spencer, T., 113
spin
   configuration, 34
   system, 35
   variable, 34
spin-glass, 109, 163
Starr, S. L., 230
Stein, D., 100
storage capacity, 258
subadditivity, 36, 219

tail sigma-algebra, 53
Talagrand, M., 113, 163, 167, 237, 264, 272
Talagrand's theorem, 237
Taylor expansion, 265
temperature, 5, 16
thermodynamic
   limit, 21
   potentials, 6
   variables, 5
      extensive, 5
      intensive, 5
thermodynamics, 5
   first law of, 5

Tirozzi, B., 257
Toninelli, F.-L., 163, 218, 237
topology
   product, 50
transfer matrix, 38
transformation
   Hubbard–Stratonovich, 46, 250
translation-covariant state, 118

Ueltschi, D., 82

van Beijeren, H., 72
van den Berg, J., 106
van der Waals
   equation of state, 29
   gas, 28
van der Waals, J. D., 28
van Enter, A. C. D., 100, 257
van Hemmen, J. L., 250, 257

weak convergence, 54
Wehr, J., 100, 111, 118
Weiss, P., 39

Yngvason, J., 8

Zahradník, M., 82
Zegarlinski, B., 107